Passive Design Strategies to Adapt to Heat Waves

Passive Design Strategies to Adapt to Heat Waves: Thermal Comfort, Energy Efficiency, and Building Resilience offers a comprehensive guide to passive design techniques tailored to various climate zones. Combining historical insights with modern advancements, this book explores practical strategies for enhancing thermal comfort, energy efficiency, and building resilience. Aimed at architects, engineers, researchers, and students, it bridges the gap between theory and application, providing real-world case studies and simulation methodologies. With a focus on reducing energy consumption and supporting global sustainability goals, it delivers actionable solutions for creating adaptive, energy-efficient buildings worldwide.

- Explores passive design techniques tailored to diverse climates, including hot and dry, hot and humid, Mediterranean, temperate, and continental regions.
- Integrates historical architecture with modern energy-efficient design, supported by real-world case studies and simulation methodologies.
- Emphasizes reducing energy consumption, enhancing thermal comfort, and building resilience to improve sustainability.

"*Passive Design Strategies to Adapt to Heat Waves* offers a rigorous and timely roadmap for making buildings both comfortable in everyday use and safe during extreme heat events. By uniting clear comfort-and-resilience frameworks with rare full-scale monitoring and a broad comparative literature, it convincingly bridges theory and practice. The book draws on cross-climate lessons and vernacular case studies to show how shading, thermal mass, natural ventilation, and allied passive measures can extend safe indoor conditions while reducing reliance on vulnerable mechanical cooling systems. It is immediately useful to researchers, practitioners, and policymakers alike. Above all, this is the kind of book that becomes a go-to reference on every resilient design shelf."
Dr. Hao Sun, *PhD, Research Fellow, Global Centre for Clean Air Research (GCARE), University of Surrey, UK, Part-time Research Associate, Building Energy & Environment Research Group, University of Nottingham*

"*Passive Design Strategies to Adapt to Heat Waves: Thermal Comfort, Energy Efficiency, and Building Resilience* is a timely and essential contribution to the growing discourse on climate-responsive architecture and sustainable urban development. As global temperatures rise and extreme heat events become more frequent and intense, the need for resilient building strategies has never been more urgent. This book addresses that challenge, offering a rich and multifaceted exploration of passive design techniques that enhance thermal comfort while reducing energy dependence. For professionals and researchers committed to advancing sustainable and climate-responsive design, *Passive Design Strategies to Adapt to Heat Waves* is more than a reference, it's a roadmap. Highly recommended for anyone seeking to make meaningful contributions to the future of resilient architecture."
Professor Eng. José Carlos Almeida, *Polytechnic University of Guarda, Portugal*

"This book is an extremely valuable source of knowledge for architects, engineers, urban planners, and anyone interested in the adaptation of buildings to extreme climatic conditions. The authors discuss in detail passive design strategies that increase thermal comfort, improve energy efficiency, and strengthen the resilience of buildings to heat waves in a changing climate, and the need to protect the environment. With practical tips and numerous case studies from around the world, this publication is an invaluable tool in the pursuit of sustainable development and the creation of more climate-resilient buildings. I recommend it to all professionals and students who want to deepen their knowledge of green and energy-efficient solutions in modern architecture."
Professor arch. Piotr Herbut, *University of Agriculture, Cracow, Poland*

"This book presents a holistic approach to the problem of heat waves, with useful practical adaptive strategies to deal with the difficult situations created by it. With an original blend of theory and practice, it provides an interesting perspective that can serve as a basis for significantly improving building construction solutions. The book creates a framework of knowledge with the more relevant and updated theories related to measuring heat wave impact on thermal comfort and improving thermal building resilience as well as a carefully selected international case studies applying

them, that can be very useful for all those involved in looking for the best design. Sustainable adaptation to the specific conditions requires local expertise that this book will help to develop."
M.A. Flórez de la Colina, *Architect, PhD, P.T.U. Professor, Building Technology Department (Departamento: Tecnología de la Edificación), Universidad Politécnica de Madrid, Spain*

"This book, *Passive Design Strategies to Adapt to Heat Waves: Thermal Comfort, Energy Efficiency, and Building Resilience*, offers a comprehensive exploration of passive cooling techniques, integrating empirical research, vernacular wisdom, and forward-looking life-cycle analyses. Its examination of thermal resilience metrics, including those from our work at Wuhan University, provides a framework for addressing the escalating challenges of overheating in a warming world. The authors' interdisciplinary approach makes this an important resource for architects, engineers, and policymakers striving for sustainable and resilient building design."
Professor Zhonghua Gou, *Wuhan University, China*

"This book offers an excellent synthesis of strategies to mitigate urban overheating and improve building resilience during heat waves. It bridges scientific knowledge with practical design solutions, making it a highly relevant resource for both researchers and practitioners."
Dr. Haochen Tan, *Assistant Atmospheric Scientist, Argonne National Laboratory, USA*

"This book is an essential guide for building design in a climate-changing world. It expertly combines a review of decades of research, with an emphasis on field experiments, and an in-depth analysis of passive design strategies from both vernacular and contemporary architecture. This book clearly states that climate resilience is directly linked to passive cooling strategies, which are a crucial foundation for having comfortable and safe buildings during extreme heat events. It is an invaluable resource for practitioners, researchers, and policymakers alike who are seeking to build a more resilient future."
Rafael Monge Palma, *Assistant Researcher, Department of Energy Engineering, School of Engineering, University of Seville, Spain*

"Overheating in buildings is one of the most pressing challenges of our time. This book offers not only a comprehensive resource but also an invitation to rethink how we design for a resilient future. Grounded in the principles of passive resilience, this work provides a bridge between theoretical analysis and practical experience, making a significant contribution to the ongoing dialogue on climate-adaptive architecture."
Silvana Flores Larsen, *Universidad Nacional de Salta—INENCO (UNSa-CONICET), Argentina*

"Kuczyński, Staszczuk, and Gortych's book provides a well-balanced combination of principles and practical ideas to design buildings that are better adapted to heat waves. Through its different chapters, it demonstrates that passive strategies are not just supplementary measures that can be implemented to enhance the performance of mechanical systems but can become foundational for adaptive building design in the twenty-first century. *Passive Design Strategies to Adapt to Heat Waves: Thermal Comfort, Energy Efficiency, and Building Resilience* is a welcome reference book for students, designers, and policymakers."

Pablo La Roche, *PhD LEED BD+C AIA Assoc Professor Architecture, Cal Poly Pomona, USA*

"This book highlights passive design strategies as a solution to the recent increase in energy consumption and carbon emissions. The extensive study on passive design strategies used in many areas around the world is important for the economy, health, and environmental sustainability. In this context, it is believed that this book will be an important reference source for architects, engineers, and other decision-makers in the field of sustainability."

Ruşen Ergün, *Assistant Professor, Faculty of Architecture, Dicle University, Diyarbakır, Türkiye*

Passive Design Strategies to Adapt to Heat Waves

Thermal Comfort, Energy Efficiency, and Building Resilience

Tadeusz Kuczyński, Anna Staszczuk,
and Marta Gortych

CRC Press
Taylor & Francis Group
Boca Raton London New York

CRC Press is an imprint of the
Taylor & Francis Group, an **informa** business

Designed cover image: Tadeusz Kuczyński

First edition published 2026
by CRC Press
2385 NW Executive Center Drive, Suite 320, Boca Raton FL 33431

and by CRC Press
4 Park Square, Milton Park, Abingdon, Oxon, OX14 4RN

CRC Press is an imprint of Taylor & Francis Group, LLC

ISBN: 978-1-041-08477-8 (hbk)
ISBN: 978-1-041-08478-5 (pbk)
ISBN: 978-1-003-64557-3 (ebk)

DOI: 10.1201/9781003645573

Typeset in Times
by KnowledgeWorks Global Ltd.

Contents

Preface

This book has been prepared as a response to the growing challenges facing contemporary construction in the context of overheating caused by global warming. Its origins are rooted both in international scientific research and in full-scale experiments carried out in western Poland. The combination of a global and a regional perspective became the key to formulating the questions and objectives that guided us during the work on this volume.

Our intention was to create a study that integrates theoretical analyses with practical experience. The literature provides extensive evidence for the effectiveness of passive strategies, yet a coherent framework that would connect results from different climates, case studies of vernacular architecture, and empirical data from the monitoring of real buildings has been lacking. We considered it necessary to bring together these three sources of knowledge in one narrative, in order to present not only the potential of individual methods but also their limitations and the conditions required for effective application.

The preparation of this book was an interdisciplinary process. It combined knowledge from architecture, building physics, environmental engineering, and the social sciences. Only such an approach makes it possible to understand the full context of the phenomena in which thermal comfort and building resilience become not only technical matters but also health-related, cultural, and social issues.

A special place in this book is given to the empirical research carried out at the scientific research center of the University of Zielona Góra in Nowy Kisielin, western Poland. These studies enabled the evaluation of passive strategies at full scale using high-resolution monitoring. The data provide a unique contribution to the debate on building design in a temperate climate. Their value also lies in the fact that they connect field observations with prospective analyses that include scenarios of climate change and energy transition.

This book is intended as an invitation to dialogue. We are aware that passive strategies are context-specific and that their effectiveness depends on local climatic, cultural, and economic conditions. Our aim has not been to propose a universal prescription, but rather to show that passive resilience should form the foundation of design in the face of climate change.

We express our gratitude to all institutions and individuals who supported our research and the preparation of this book. We thank our colleagues, students, and project partners whose engagement and critical remarks helped us refine the analyses presented. We are also grateful to the reviewers and to the publisher for their guidance and patience throughout the editorial process.

We hope that this book will serve as both inspiration and a tool for practitioners, researchers, and decision-makers. We believe that the analyses and case studies presented here will provide an impetus for further exploration and for the implementation of solutions that will allow buildings to respond more effectively to the challenges of the twenty-first century.

About the Authors

Prof. Tadeusz Kuczyński is Professor of Building Physics at the University of Zielona Góra, where he has served as Dean, Deputy-Rector, and Rector. His research focuses on indoor climate, building energy performance, passive design, and renewable energy systems. He has co-authored numerous articles in leading journals and coordinated high-impact projects. Prof. Kuczyński is recognized internationally as an expert in sustainable architecture, energy efficiency, and passive strategies for climate adaptation.

Prof. Anna Staszczuk is Associate Professor at the University of Zielona Góra, Institute of Building Engineering, and Head of the Department of General Construction and Building Physics. Her research interests include indoor climate, energy use in buildings, and heat storage. She has authored numerous research papers and a book, and has been involved in national and international projects. Her current work emphasizes sustainable construction, energy-efficient design, and passive technologies for protecting buildings against overheating. Prof. Staszczuk also teaches courses in building physics and sustainable architecture and construction.

Dr Marta Gortych is Assistant Professor at the University of Zielona Góra, Faculty of Engineering and Technical Sciences, where she serves as Vice-Dean for Student Affairs. She holds MSc degrees in Chemical and Process Engineering and Environmental Engineering, and a PhD in Technical Sciences. Her research focuses on building energy efficiency, heat transfer, and the application of phase change materials in construction and energy systems. Her current work particularly focuses on passive strategies for improving thermal comfort in buildings. Dr Gortych has published in leading international journals and actively participates in international research projects.

1 Introduction

1.1 CLIMATE CHANGE AND THE CHALLENGE OF OVERHEATING IN BUILDINGS

In recent decades, accelerating climate change has fundamentally transformed the conditions under which buildings are designed and operated. What was once perceived as a seasonal inconvenience, namely occasional overheating during the hottest months, has become one of the central challenges of contemporary architecture and engineering. Heat waves are growing longer, more frequent, and more intense. Increasingly, they coincide with peak electricity demand, overloading power systems. At the same time, the expansion of urban heat islands further elevates temperatures in city centers, intensifying health risks and increasing dependence on mechanical cooling.

This book responds to these challenges. It addresses building design not only in terms of annual energy efficiency but also through the combined lenses of thermal comfort and thermal resilience. Comfort refers to the baseline quality of the indoor environment in everyday use. Resilience refers to the capacity of buildings to maintain safe and habitable conditions during extreme events such as heat waves, cold spells, or power outages, when mechanical systems may be unavailable. These two objectives are inseparable and must be addressed together if buildings are to remain safe, sustainable, and prepared for the future.

Mechanical cooling is often regarded as the primary safeguard against building overheating; however, its widespread application carries significant environmental, social, and systemic costs. It increases energy demand during periods of peak load, raises greenhouse gas emissions, and releases waste heat that intensifies the urban heat island effect. Air conditioning also deepens social inequalities, protecting only those who can afford its installation and operation. Importantly, buildings heavily reliant on air conditioning are especially vulnerable to the risk of outages and power supply interruptions, the frequency of which is expected to rise under extreme heat conditions. These buildings were not designed to maintain thermal safety during crises without the support of mechanical systems.

In contrast, passive strategies such as shading, high-albedo surfaces, thermal mass, natural ventilation, and evaporative cooling address overheating at its source. They extend the period during which safe indoor conditions can be maintained without energy consumption. Their effectiveness, however, is highly contextual. Shading geometry must correspond to solar angles, ventilation must respond to prevailing winds and the distribution of ventilation corridors in the urban fabric, and thermal mass must align with diurnal and seasonal rhythms. When carefully selected and integrated, these strategies can deliver meaningful benefits across climate zones, enhancing both comfort and resilience.

This issue is particularly significant in cold and temperate climates. For decades, design priorities and regulations in these regions have focused primarily on reducing

DOI: 10.1201/9781003645573-1

heating demand, often treating overheating as a marginal concern. Yet as heat waves intensify, this approach is proving insufficient. Overheating now results in serious health risks, increased mortality, and economic losses. The ability of buildings to maintain habitable conditions during climatic extremes should therefore be recognized as a fundamental pillar of sustainable design, alongside energy efficiency, emissions reduction, and environmental responsibility.

The book rests on four pillars. First, it reviews 25 years of research on passive design and overheating mitigation, identifying strategies effective in diverse climates. Second, it presents full-scale empirical studies conducted on two experimental houses in western Poland, which provide rare field data showing how thermal mass and controlled ground coupling influence comfort and resilience. Third, it incorporates prospective life-cycle analyses, extending the perspective to include climate and energy scenarios that show how passive strategies perform over decades. Fourth, it engages in a dialogue between vernacular and contemporary architecture, highlighting continuity and reinterpretation of passive strategies across cultural and technological contexts.

1.2 OBJECTIVES OF THE BOOK

The book has six main objectives:

1. To define the framework for assessing thermal comfort and thermal resilience through a review of standards, indicators, and methods used in the literature.
2. To present and critically evaluate passive strategies such as shading, high-albedo surfaces, thermal mass, natural ventilation, and evaporative cooling, in both literature-based and empirical perspectives.
3. To develop the authors' proposal of a passive strategy based on increasing building thermal mass and allowing controlled heat exchange with the ground by removing floor insulation. This solution, once common in residential construction in temperate climates until the 1970s, is re-evaluated in the context of contemporary climate challenges.
4. To verify this strategy through full-scale empirical studies, assessing its impact on thermal comfort, resilience under extreme events, annual heating demand, and life-cycle environmental performance. These analyses allow evaluation of the strategy both in the short term and across a long time horizon.
5. To provide a critical comparison of vernacular and contemporary architecture, demonstrating the complexity of passive design from the scale of individual buildings to the urban fabric, and the continuity of solutions from tradition to modernity.
6. To link thermal resilience with environmental performance by means of prospective life-cycle analysis, indicating the long-term consequences of different strategies.

The scope of the book combines the global context of climate change with detailed regional analyses, integrating literature, field research, and contemporary case studies.

Its overarching aim is to provide both theoretical frameworks and practical tools that enable the design of buildings that are comfortable, resilient, and environmentally responsible.

The central message is clear: passive resilience is not an optional supplement to sustainable design, but one of its most important foundations.

1.3 STRUCTURE OF THE BOOK

The structure of the book reflects a logical progression from general to specific and from theory to practice.

- Chapters 1–5 define the analytical and methodological framework, including objectives, scope, guiding questions, the risks posed by heat waves, and the systems and metrics for assessing comfort and resilience, concluding with frameworks for the comprehensive evaluation of passive strategies.
- Chapters 6–12 focus on adaptive strategies, reviewing the influence of outdoor temperatures, the risks of increasing dependence on mechanical cooling, and the theoretical and practical foundations of passive methods in different climatic contexts.
- Chapters 13–16 present the results of scientific research, including the literature review, the authors' empirical studies, and analyses of seasonal energy performance and life-cycle impacts.
- Chapters 17–19 look to the future, presenting examples of contemporary projects based on the reinterpretation of vernacular strategies, and synthesizing key conclusions for design practice, regulation, and further research.

1.4 AUDIENCE AND RELEVANCE

The book is intended for a wide audience. Architects and engineers will find detailed evidence of the effectiveness of passive strategies. Students and researchers will benefit from the integration of monitoring data, modeling, and life-cycle scenarios. Policymakers and standardization bodies will gain arguments highlighting the need to extend current regulations so that they address thermal resilience alongside energy efficiency.

2 The Escalating Threat of Heat Waves in a Warming World

2.1 INTRODUCTION

In 2024, global temperature anomalies varied significantly across different regions, as illustrated in Figure 2.1 (Copernicus Climate Change Service [C3S], 2025).

The Arctic experienced the most pronounced deviations, with anomalies exceeding 4°C above pre-industrial baseline levels, highlighting the disproportionate

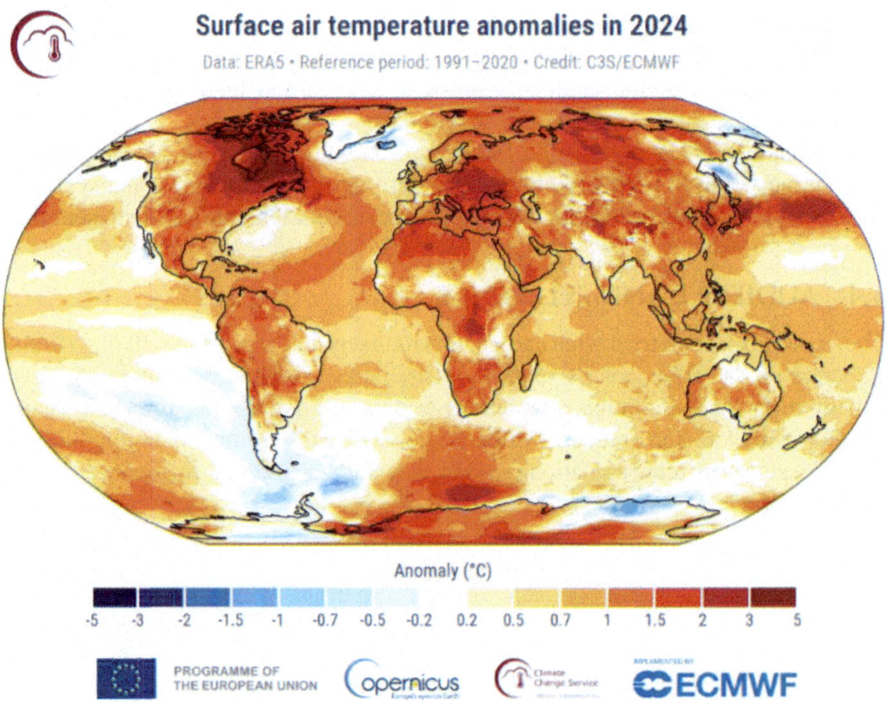

FIGURE 2.1 Global surface air temperature anomalies in 2024, highlighting substantial warming in densely populated regions such as Europe and South Asia.

(Source: Copernicus Climate Change Service (C3S), Global Climate Highlights 2024. Source: Copernicus Climate Change Service (C3S)/ECMWF. Data: ERA5. Reference period: 1991–2020.)

DOI: 10.1201/9781003645573-2

warming occurring in polar regions. However, densely populated areas—such as Europe and South Asia—faced the most severe challenges. In Europe, anomalies reached approximately 2°C, intensifying heat waves and putting strain on urban infrastructure. South Asia also recorded extreme temperature deviations, exacerbating heat stress in regions where millions of people lack access to cooling technologies. These variations underscore the urgent need to address the unequal impacts of global warming.

Despite the 2015 COP21 agreement—which aimed to limit global temperature increases to well below 2°C, with an aspirational target of 1.5°C above pre-industrial levels—recent analyses indicate that these thresholds are either being reached or exceeded. The Copernicus Climate Change Service (C3S) reported that 2024 was the warmest year in recorded history. For the first time, the global annual mean temperature surpassed 1.5°C above pre-industrial levels, reaching 1.6°C (Younger, 2025) (Figure 2.2).

Younger (2025), acknowledging potential uncertainties, estimated the 2024 global temperature anomaly at 1.47°C above the 1850–1900 average, with more than half of the monthly temperatures that year exceeding the critical 1.5°C threshold. These findings reflect a concerning trend: the past decade has been the warmest in recorded history, with the rate of global warming now surpassing 0.2°C per decade. The likelihood of permanently surpassing the 1.5°C limit before 2030 is considered high unless immediate measures are taken to reduce greenhouse gas emissions.

One of the most alarming consequences of global warming is the intensification of heat waves, defined as prolonged periods of extreme heat that are becoming more frequent and severe. Urban areas are particularly vulnerable due to the urban heat island effect, where materials such as concrete and asphalt absorb and retain heat,

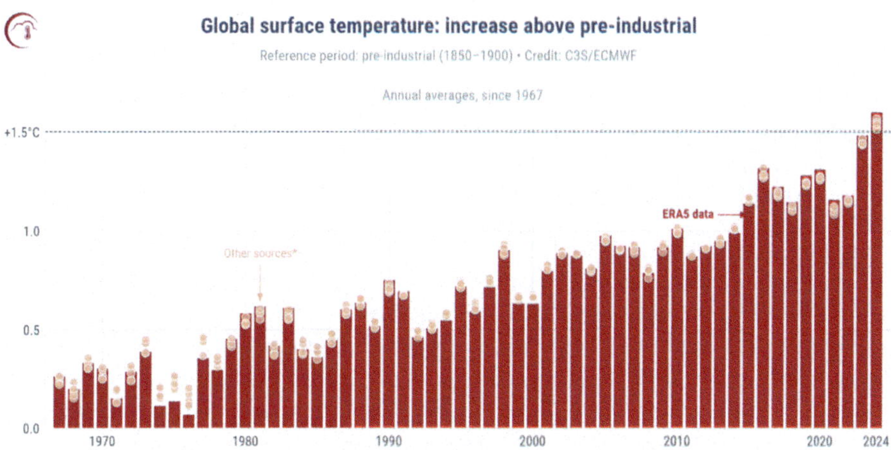

FIGURE 2.2 Global annual mean temperature anomaly in 2024, illustrating the first full-year exceedance of the 1.5°C pre-industrial threshold.

(Sources: Copernicus Climate Change Service (C3S), *Global Climate Highlights 2024.* Copernicus Climate Change Service (C3S)/ECMWF)

releasing it slowly at night. This exacerbates health risks by impeding the body's ability to cool down, increasing reliance on energy-intensive cooling systems, and further straining urban infrastructure and healthcare services.

Recent studies emphasize the growing risks of extreme heat. Lüthi et al. (2023) found that heat waves that occurred once every 100 years in 2000 now happen every 10 to 20 years under 2020 climate conditions. Their projections indicate that, with 1.5°C of warming, such extreme events may occur every 2 to 5 years; at 2°C of warming, they could become the norm. Rising temperatures not only increase the frequency of extreme heat but also amplify its health impacts. For example, extreme heat may account for over 10% of annual global mortality in some regions—a significant increase from current levels. Without substantial adaptation measures, these events will pose an escalating threat to public health and urban resilience.

Thompson et al. (2023) analyzed global trends to identify regions with a statistically heightened likelihood of experiencing record-breaking extreme heat events. Their findings suggest that historical climate data may underestimate the actual risks, as current extreme events exceed projected thresholds in 18 to 26% of global regions. The study highlights that these estimates may still be conservative, given the limited historical sampling of extreme events. Additionally, Thompson et al. (2023) stress the need to update heat action plans, especially in developing regions where high statistical risks, limited resources, and growing populations make communities highly vulnerable.

Regions such as Afghanistan and Central America are identified as particularly at risk. Without adaptation measures, these areas may face severe public health crises. The lack of resilient infrastructure, inefficient urban planning, and inadequate cooling access will exacerbate the consequences of intensifying heat waves. Without urgent intervention, including improved urban planning, enhanced cooling infrastructure, and targeted public health measures, heat waves will continue to grow in severity and impact, endangering both human lives and urban resilience worldwide.

2.2 DEFINITION AND QUANTIFICATION OF HISTORICAL AND FUTURE HEAT WAVE WEATHER SCENARIOS (BASED ON AL-ASSAAD et al., 2025)

A crucial step in assessing the thermal resilience of buildings is the selection of weather scenarios that reflect not only historical climate conditions but also projected future extremes. Typical meteorological years (TMYs) provide a useful baseline for typical conditions; however, they do not capture the extreme events that often determine whether buildings can maintain safe and habitable indoor environments. For resilience analysis, datasets must therefore include the most intense and prolonged heat waves expected in each climate zone across relevant time horizons.

The IEA EBC Annex 80 has developed a methodology that defines, detects, and characterizes heat wave events for use in building performance simulations (Machard et al., 2020, 2024). This approach combines high-resolution climate projections with a consistent event-detection method, enabling resilience assessments that are both regionally specific and globally comparable.

2.2.1 Climate Data Sources and Downscaling

Climate projections can be obtained from multiple sources, but their suitability for building-scale analysis varies. Global Climate Models (GCMs) are valuable for capturing large-scale trends but have coarse spatial resolution, making them unsuitable for modeling local microclimatic effects that significantly influence indoor thermal performance (Stone & Risbey, 1990).

Some approaches apply statistical downscaling techniques, such as morphing or stochastic weather generators (Al-Assaad et al., 2025), which modify historical datasets to reflect projected changes. While computationally efficient, these methods tend to assume that the temporal structure of future weather will resemble present-day patterns and can underestimate the frequency or severity of extreme events (Krelling et al., 2023, Stone & Risbey, 1990).

Annex 80 adopts dynamic downscaling using Regional Climate Models (RCMs), which preserve large-scale climate signals while adding higher spatial resolution. This enables the capture of local effects such as coastal influences, topography, and urban heat island patterns. In Europe, the EURO-CORDEX initiative provides RCM outputs at approximately 12-km resolution, with comparable datasets for other world regions (25 km for the Middle East and North Africa, 50 km elsewhere).

The Coordinated Regional Downscaling Experiment (CORDEX) supplies the climate projections used in this method. These datasets include variables such as dry-bulb temperature, humidity, solar radiation, and wind speed, and are based on the high-emission Representative Concentration Pathway RCP 8.5. Original data are provided in three-hourly time steps and are interpolated to hourly resolution for building simulations, with bias adjustment applied to align model outputs with observed conditions. The datasets are prepared for three distinct 20-year periods:

- Historical (2001–2020)
- Mid-term future (2041–2060)
- Long-term future (2081–2100)

2.2.2 Detection and Characterization of Heat Waves

With bias-adjusted, hourly datasets in place, heat wave events are identified using a percentile-based method adapted from Ouzeau et al. (2016):

- T_{HW} (detection threshold): 99.5th percentile of daily mean temperatures; exceeding this threshold indicates a heat wave is occurring.
- $T_{start/end}$ (start/end threshold): 97.5th percentile; defines the start and end of an event.
- T_{int} (interruption threshold): 95th percentile; merges consecutive hot periods without a significant cooling break.

Each detected event is characterized by:

1. Intensity—the maximum daily mean temperature recorded.
2. Duration—the total number of days.

3. Severity—cumulative temperature above the start/end threshold, expressed in °C·days.

For each climate zone and time period, the most intense heat wave is selected for simulation purposes, ensuring that resilience assessments test buildings against the most challenging conditions.

2.2.3 Trends across Climate Zones

The Annex 80 dataset shows a consistent pattern of increasing heat wave intensity and duration across most climate zones (Al-Assaad et al., 2025). For example, in warm-humid climate 3A, historical heat waves lasted around 15 days, whereas in the long-term future, they are projected to persist for up to 95 days. In hot-humid climate 2A, peak daily mean temperatures rise from 36.2°C to 43.4°C over the same time frame. Such changes imply not only higher thermal stress but also longer exposure periods, both of which challenge passive cooling strategies and building safety.

A synthetic index, the *normalized degree of shock* (doS), has been developed within Annex 80 to combine heat wave magnitude and duration into a single measure. While this report applies doS values in scenario selection, its calculation method and role in resilience assessment are presented in detail in Chapter 4, alongside other key thermal resilience indicators.

2.3 HEALTH, SOCIAL, AND ECONOMIC RISKS OF EXTREME HEAT (BASED ON SIU et al., 2025)

Extreme heat is among the most consequential climate-related hazards for human health, infrastructure, and economic stability. Beyond the immediate discomfort it causes, prolonged high temperatures can trigger a cascade of impacts that strain healthcare systems, disrupt essential services, and deepen social inequalities. Recent research confirms that these risks are already substantial and are projected to intensify markedly under future climate scenarios (Siu et al., 2025).

2.3.1 Mortality and Morbidity

Historically, single extreme heat events have been responsible for hundreds to thousands of deaths within individual cities. In larger-scale, non-localized events, fatalities can rise into the tens of thousands. The 2003 European heat wave remains one of the most severe examples, with an estimated 70,000 excess deaths across the continent. Contemporary studies indicate that even with improvements in forecasting, early warning systems, and medical care, heat-related mortality remains a significant concern.

The health burden of extreme heat is not evenly distributed. Older adults, particularly those aged 65 and above, are disproportionately affected due to reduced thermoregulatory capacity, a higher prevalence of chronic diseases, and limited mobility. Other vulnerable groups include individuals living alone, people with pre-existing mental illness, the uneducated, and those from low-income households.

These populations often have limited access to cooling technologies, live in poorly insulated or poorly ventilated dwellings, and may lack the resources to relocate during heat emergencies.

During heat waves, healthcare systems experience a measurable surge in demand. Analyses show increases in ambulance dispatches, emergency hotline calls, and emergency department visits, often accompanied by longer waiting times and delays in hospital admissions (Siu et al., 2025). Even in cases where daily patient volumes do not increase substantially, the age profile of patients shifts toward older demographics, and the incidence of heat-related illnesses such as heat stroke, dehydration, and exacerbation of cardiovascular and respiratory conditions rises sharply.

The economic consequences of these healthcare impacts can be significant. For example, between 2015 and 2019, heat wave–related morbidity in France was estimated to cost the healthcare system over €30 million (Siu et al., 2025). Similarly, the city of Adelaide, Australia, incurred more than AUD 1 million in heat wave-attributed emergency department costs over a three-year period. These figures do not capture indirect costs, such as reduced workforce productivity and long-term health deterioration.

2.3.2 Projected Future Risks

According to Siu et al. (2025), projections under high-emission scenarios indicate that the risks associated with extreme heat are expected to escalate sharply throughout the twenty-first century. Their analysis shows that, even under a relatively optimistic warming trajectory of just under 2°C by 2100, annual heat-related mortality among those aged 65 and older could rise substantially by mid-century, with the largest increases occurring in regions already experiencing high thermal stress. Under the high-emission Representative Concentration Pathway (RCP) 8.5 scenario, corresponding to a global warming range of approximately 2.4°C to 4.8°C, the projected mortality burden is even greater. These increases are driven not only by higher peak temperatures but also by the combined effects of longer heat wave durations and greater event frequency. As emphasized by Siu et al. (2025), prolonged exposure amplifies cumulative heat stress, undermines the body's capacity for recovery, and increases the likelihood of life-threatening health outcomes.

2.3.3 Impacts on Infrastructure and Services

The risks of extreme heat extend well beyond direct health effects. High temperatures place critical infrastructure under significant strain. Energy systems are particularly vulnerable, as demand for cooling peaks during heat waves, often coinciding with reduced generation efficiency in power plants and transmission systems. This mismatch between supply and demand can lead to blackouts, which in turn remove access to air conditioning and mechanical ventilation when they are needed most.

Transport systems also suffer. Asphalt roads can soften and deform under prolonged heat, and railway tracks may expand, causing buckling and delays. Water infrastructure is similarly stressed, as heat waves often coincide with drought conditions, increasing demand for potable water while reducing supply.

These cascading failures can exacerbate health risks by limiting access to healthcare facilities, cooling centers, and other essential services. They also intensify economic losses by interrupting commercial activity and disrupting supply chains. As noted by Siu et al. (2025), heat-induced reductions in labor productivity, particularly in sectors such as construction, manufacturing, and agriculture, can generate substantial economic impacts on a regional and national scale.

2.3.4 The Equity Dimension

Extreme heat is also an equity issue. The most severe impacts are borne by those least able to adapt. Low-income households often inhabit buildings with inadequate thermal performance and lack the financial means to purchase or operate cooling systems. Urban heat islands disproportionately affect densely populated neighborhoods with limited green space, compounding exposure for residents who are already socioeconomically disadvantaged.

These inequities extend to healthcare access. Vulnerable groups may be less able to seek timely medical care due to mobility limitations, lack of transportation, or mistrust in public health systems. In rural areas, sparse healthcare infrastructure can delay treatment for heat-related illnesses, increasing mortality risks.

Addressing these disparities requires integrating social vulnerability mapping into urban planning and public health strategies. Cooling centers, subsidized utility costs during heat events, and targeted outreach to at-risk groups are examples of interventions that can reduce the disproportionate burden.

2.3.5 Implications for Building Resilience

From the perspective of building science and design, the health risks associated with extreme heat underscore the importance of passive survivability and thermal resilience. As highlighted in later chapters, well-designed building envelopes, effective shading, and natural ventilation can delay the onset of unsafe indoor conditions, reducing dependence on active cooling systems that may fail during peak demand or power outages.

However, the intensification of heat waves projected under climate change scenarios means that building designs optimized for historical climate conditions may no longer provide adequate protection. Incorporating future climate projections into building codes, as well as considering long-duration heat wave scenarios, will be essential to ensure that thermal resilience is maintained throughout a building's lifespan.

2.3.6 Conclusion

The evidence is clear: extreme heat presents a multifaceted and escalating risk to human health, infrastructure, and economic stability. The greatest impacts are felt by vulnerable populations, and without targeted adaptation, both mortality rates and inequalities are likely to worsen significantly. While improvements in early warning systems, public health interventions, and infrastructure resilience have reduced fatalities in some regions, they remain insufficient in the face of rapidly intensifying climate extremes.

For the building sector, this means that adaptation cannot be an afterthought. Climate-informed design, performance-based codes, and equitable access to cooling must become central elements of resilience planning. In doing so, societies can mitigate the worst impacts of extreme heat, protect public health, and enhance the capacity of communities to withstand the challenges ahead.

2.4 HEAT WAVES: GLOBAL PATTERNS, REGIONAL IMPACTS, AND EVOLVING ADAPTATION

2.4.1 Introduction

Extreme heat events have become one of the most serious climate-related hazards, with both their frequency and intensity increasing in virtually all inhabited regions. Recent case studies from Southeast Asia, South Asia, Europe, and North America illustrate not only the diversity of impacts across climate zones but also the uneven progress in adaptation. Some regions have made substantial advances in early warning systems, infrastructure resilience, and public health preparedness, yet even the most proactive examples reveal persistent vulnerabilities.

The following synthesis brings together evidence from a series of major heat waves between 2003 and 2024, consolidating lessons from distinct climate contexts into a comparative framework. Four broad climate groupings are considered:

1. Tropical and humid climates, where high wet-bulb temperatures and persistent humidity combine to produce life-threatening conditions even at moderate air temperatures.
2. Hot and arid climates, where extreme dry-bulb temperatures and large diurnal ranges dominate, but with occasional intrusions of high humidity that exacerbate physiological stress.
3. Temperate and Mediterranean climates, where populations and infrastructure are less accustomed to sustained high temperatures, making such events disproportionately disruptive.
4. Hot continental and desert-edge climates of North America, where record-breaking daytime and nighttime heat, coupled with prolonged drought, create complex socio-economic and health challenges.

Within each block, historical and recent events are reviewed together, allowing for direct comparisons of meteorological characteristics, health impacts, infrastructure stresses, and adaptation responses. This integrated approach avoids repetition while retaining the specificity of each case study, ensuring that all published data and references remain intact.

2.4.2 Tropical and Humid Climates: Southeast Asia and Parts of India

Tropical and humid regions present a unique hazard profile during heat waves due to the combined effects of high air temperatures and elevated relative humidity. The 2016 Southeast Asia heat wave provides an early benchmark for the modern era, coinciding with a strong El Niño event that amplified baseline warming (Thirumalai et al., 2017). From March to June, temperatures reached 44.6°C in Mae Hong Son, Thailand, and 45°C in Myanmar. Urban centers such as Bangkok, Phnom Penh, Hanoi, and Yangon experienced daily maxima of 39–42°C with night-time minima as high as 28–30°C. Relative humidity frequently ranged from 70% to 90%, raising the heat index to levels equivalent to 50°C. These conditions severely limited nocturnal cooling and increased physiological strain, particularly in dense urban areas where the urban heat island effect compounded heat retention.

Public health consequences were severe. Thailand recorded more than 300 fatalities, while Myanmar experienced several hundred deaths, predominantly in rural communities. Medical facilities reported surges in heat stroke, dehydration, and cardiovascular complications, with rural areas facing particular difficulties due to limited access to cooling and potable water. The agricultural sector, especially rice cultivation, suffered widespread losses, while low river levels in the Mekong and Chao Phraya disrupted irrigation and drinking water supply.

The April 2023 Southeast Asia heat wave further demonstrated the intensifying nature of these events (Zachariah et al., 2023a). Temperatures surpassed previous records, with Tak province in Thailand reaching 45.4°C and the heat index in Bangkok peaking at 54°C. In Laos, Sainyabuli recorded 42.9°C, the highest in national history, while Vientiane reached 41.4°C. Similar extremes were observed in parts of Cambodia and Vietnam. As in 2016, health impacts were concentrated among the elderly and those with chronic illnesses, but the 2023 event also revealed gaps in medical infrastructure: in parts of Laos, delayed treatment due to scarce facilities significantly worsened outcomes.

Infrastructure failures exacerbated the crisis. In Thailand, power grid overloads caused extended outages, cutting off access to fans and air conditioning during the most critical hours. In urban environments, indoor temperatures exceeded safety thresholds for prolonged periods, contributing to increased incidence of heat stress. Agricultural productivity declined, with early harvesting in northern Laos and Thailand undertaken to avoid total crop failure, although quality losses were unavoidable (Japan Times, 2023).

Comparative analysis of 2016 and 2023 highlights both the persistence of core vulnerabilities and the limited effectiveness of adaptation measures implemented in the interim (Zachariah et al., 2023). Early warning systems were established in Thailand and Laos, public awareness campaigns expanded, and urban heat mitigation strategies such as increased green space and reflective construction materials were introduced (Sun et al., 2022). Nevertheless, these measures were not universally integrated into emergency response protocols, and they often failed to reach rural communities and outdoor workers, who faced the highest exposure risks (Mazdiyasni et al., 2017).

In the humid coastal regions of India and Pakistan, the challenges are similar but with added complexity due to population density and socio-economic conditions. The 2015 heat wave in Karachi and Andhra Pradesh occurred simultaneously with extreme heat in arid inland locations, yet the health burden was amplified in coastal areas where relative humidity reached 50–70%. In Karachi, daily highs of 44–46°C combined with night-time lows above 34°C, producing heat indices of 50–52°C. These conditions severely reduced the efficiency of evaporative cooling through sweating, increasing the incidence of heat exhaustion and heat stroke. Infrastructure limitations, including prolonged blackouts, intensified indoor heat stress. More than 1,000 deaths were recorded in Karachi alone, with an additional 852 in Andhra Pradesh, the majority among low-income outdoor laborers.

In 2024, parts of southern India, including Andhra Pradesh and Tamil Nadu, again experienced extreme humid heat during a record-breaking nationwide event. Humidity levels between 50% and 70% pushed the felt temperature above 50°C. While improved early warning systems, access to cooling centers, and water distribution reduced mortality compared to 2015, these advances did not eliminate health impacts. Hospitals continued to experience surges in heat-related admissions, and outdoor labor productivity was sharply reduced.

Across tropical and humid climates, the synthesis of these events confirms three critical points:

1. Humidity magnifies health risk by limiting evaporative cooling, making wet-bulb temperature a more relevant safety threshold than dry-bulb temperature alone.
2. Night-time conditions are decisive; lack of nocturnal relief exacerbates cumulative heat stress, especially in urban areas with strong heat island effects.
3. Adaptation progress is uneven; while urban early warning systems and public messaging have improved, rural populations and informal urban settlements remain underserved, and energy infrastructure often fails at peak demand.

2.4.3 Hot and Arid Climates: Inland India and Pakistan

Hot and arid climates present a distinctive set of challenges during extreme heat events. In these regions, high daytime temperatures combine with large diurnal ranges and very low humidity, producing conditions that are physiologically different from those in humid environments. While lower relative humidity can facilitate evaporative cooling, it also accelerates dehydration and electrolyte loss, especially when water access is limited. This section synthesizes findings from the 2015, 2023, and 2024 Indian and Pakistani heat waves, drawing comparisons between inland dry zones such as Rajasthan and Turbat and the more humid coastal regions of Karachi and Andhra Pradesh.

The 2015 heat wave in the cities of Turbat (Pakistan) and Rajasthan (India) was among the most severe on record for inland arid regions. Lasting approximately two weeks from mid-May to early June, with a peak between 20 and 27 May, it brought daytime temperatures up to 49°C in Turbat and 48°C in Rajasthan, while nighttime lows stayed between 30°C and 33°C (Mazdiyasni et al., 2017). The minimal nighttime cooling meant that the human body had limited recovery time, raising cumulative heat stress. Humidity levels remained extremely low, between 10% and 20%, which initially moderated the heat index but ultimately intensified water loss and dehydration. Vulnerable populations, especially outdoor workers in agriculture and construction, faced prolonged exposure without adequate shade or hydration. Hospitals were overwhelmed with cases of heat stroke and dehydration, and more than 650 fatalities were recorded, mostly among manual laborers, the elderly, and children (Mazdiyasni et al., 2017; Sun et al., 2022).

In contrast, hot and humid coastal zones during the same 2015 event, such as Karachi and Andhra Pradesh, experienced somewhat lower absolute temperatures (44–48°C) but with humidity levels reaching 50–70%. The combination of high heat and high humidity severely restricted sweat evaporation, causing heat index values of 50–52°C. Karachi alone recorded more than 1,000 deaths, with the majority occurring in low-income neighborhoods lacking air conditioning and access to safe drinking water (Mazdiyasni et al., 2017). These figures underscore the fact that while arid regions present dehydration risks, humid regions often lead to more rapid onset of heat illness due to impaired evaporative cooling.

By April 2023, India again faced extreme conditions, with maximum temperatures exceeding 44°C in large parts of Uttar Pradesh, Bihar, Odisha, and West Bengal, and heat index values surpassing 50°C in some areas (Zachariah et al., 2023a). Although not as prolonged as the 2015 or 2024 events, the 2023 heat wave coincided with significant failures in the electrical grid, especially in Bihar and West Bengal, which deprived large populations of cooling during critical hours. In Maharashtra, 13 deaths and over 650 hospitalizations from heat stroke were reported. The agricultural sector also suffered: rice crops faced heat-induced stress, prompting urgent irrigation interventions (Al-Masum Molla, 2023). Despite these impacts, the 2023 event also demonstrated some improvements in preparedness: targeted school closures, limited public health advisories, and the activation of coal-fired plants in Karnataka to stabilize power supply (*The Guardian*, 2023).

The 2024 heat wave was the longest and most widespread in India's history, lasting from March to June and peaking between 15 and 25 May. Rajasthan endured daytime highs exceeding 50°C and night-time temperatures above 35°C, conditions that rank among the hottest recorded globally. Central states such as Uttar Pradesh and Madhya Pradesh reported highs of 45–48°C with minimal night-time relief, while southern states like Andhra Pradesh and Tamil Nadu recorded 50–70% humidity, driving felt temperatures to 50–53°C. These extremes placed immense strain on agriculture, construction, and transport sectors, rendering much outdoor labor impossible for extended periods.

Notably, the 2024 event resulted in a dramatically lower death toll than in 2015, estimated at around 100 fatalities. This improvement was attributed to the implementation of integrated early warning systems, expanded access to cooling centers and drinking water, and significant investments in renewable energy infrastructure, particularly solar photovoltaic capacity, which stabilized the power grid during peak demand (Zachariah et al., 2023a). Air conditioning availability also rose from 8–10% of households in 2015 to 15–20% by 2024, though experts noted that mobile cooling stations and community water distribution played an even greater role in reducing fatalities among the poorest residents.

From a resilience perspective, the comparison across these events highlights both progress and persistent vulnerabilities. The combination of early warning systems, public awareness campaigns, and energy infrastructure upgrades has markedly reduced mortality in inland arid regions. Yet, the reliance on mechanical cooling in contexts of limited access means that any grid failure during prolonged events could still be catastrophic. Furthermore, while India's adaptation in 2024 was more successful than in 2015, the expansion of preparedness measures to rural and peri-urban areas remains uneven, leaving significant portions of the population at risk.

2.4.4 Temperate and Mediterranean Climates: Europe

European heat waves present a unique challenge because they affect both traditionally hot Mediterranean regions and temperate zones that historically experienced only moderate summer extremes. This dual exposure means that some areas have long-standing adaptation practices, while others have been caught unprepared. The heat waves of 2003, 2019, and 2023 provide a clear illustration of the trajectory from unpreparedness to partial adaptation, while also revealing that vulnerabilities persist.

The 2003 European heat wave remains one of the most catastrophic climate-related events in modern European history. Spanning late July to mid-August and peaking in the first ten days of August, it brought daytime temperatures above 40°C in Mediterranean cities such as Seville, Marseille, and Athens, and above 36–38°C in temperate cities such as Paris, Brussels, and London (Robine et al., 2008). Night-time temperatures often stayed between 25°C and 28°C in southern Europe and between 22°C and 26°C in temperate regions, preventing physiological recovery during rest periods. This sustained heat stress was exacerbated by high night-time humidity in some areas, which reduced the effectiveness of sweat evaporation. In densely built urban areas, the urban heat island effect amplifies indoor heat retention.

The human toll was extraordinary. An estimated 35,000 deaths occurred across Mediterranean Europe and a similar number in temperate Europe, with the most affected countries including France, Italy, Spain, Portugal, Greece, Belgium, the Netherlands, Germany, Switzerland, and the United Kingdom. In France alone, more than 15,000 deaths were recorded, many among elderly residents without air conditioning and in poorly ventilated housing (Robine et al., 2008). The healthcare systems in major cities faced unprecedented strain, with emergency departments overwhelmed by heat stroke, dehydration, and exacerbations of chronic cardiovascular and respiratory diseases. Infrastructure also suffered. Power outages, road surface deformation, rail buckling, and forest fires occurred across multiple countries, disrupting daily life and causing additional economic losses.

By 2019, Europe had made notable progress in heat wave preparedness. Early warning systems had been implemented in most countries, with alerts triggering public health advisories, adjustments to working hours, and the opening of cooling centers. Urban adaptation measures included increased green space, installation of reflective surfaces, and shading in public areas. These interventions coincided with one of the most intense heat waves in European history, which occurred in two main episodes in June and July 2019. Temperature records were broken in several countries: 46°C in Vérargues, France; 42.6°C in Lingen, Germany; 41.8°C in Begijnendijk, Belgium; and 40.7°C in Gilze-Rijen, the Netherlands. The United Kingdom recorded 38.7°C in Cambridge (Zachariah et al., 2023b).

Despite these extreme conditions, mortality in 2019 was significantly lower than in 2003. France reported around 1,500 deaths, the United Kingdom 900, Germany

500, and Belgium just over 700. This reduction was attributed to targeted adaptation measures, better coordination between meteorological services and health agencies, and improved public awareness. Nevertheless, the event revealed continuing weaknesses. Many vulnerable individuals, particularly the elderly living alone, did not receive adequate assistance. Night-time temperatures remained high, above 25°C in major cities such as Paris, Brussels, and Amsterdam, limiting the benefits of cooler daytime interventions.

The July 2023 heat wave further tested European adaptation capacity. It affected much of southern Europe and the Mediterranean basin, with temperatures exceeding 45°C in Spain, Italy, Greece, and the Balkans. Catalonia recorded its hottest day in history, while Greece and Italy endured multi-day periods above 40°C with night-time temperatures over 30°C (Zachariah et al., 2023b). These conditions significantly increase health risks, particularly for outdoor workers in agriculture and construction. Fatalities were reported among construction workers in Greece and Italy, and heat-related deaths occurred among elderly populations in Spain and Cyprus (Pinto et al., 2023).

Infrastructure impacts in 2023 were similar to previous events but occurred alongside newer challenges. Power grid overloads led to blackouts in parts of Greece and Italy (Otto et al., 2023). Road and rail deformation disrupted transport in Spain and Italy, and water restrictions were imposed in many Mediterranean cities due to prolonged drought conditions (Singh et al., 2023). Adaptation measures such as expanding air-conditioned public spaces, adjusting working hours in high-risk occupations, and enforcing outdoor labor restrictions during peak heat hours helped reduce mortality compared to past events. However, these interventions were insufficient to address the scale of the hazard, particularly in regions where poverty or housing quality limited access to cooling.

A comparison of these three events reveals both progress and persistent vulnerability. The mortality reduction between 2003 and later events demonstrates that early warning systems, public health campaigns, and urban cooling measures have been effective in saving lives. In Mediterranean countries, air conditioning penetration has increased to 40–50% of households, supported by improvements in renewable energy integration that enhance grid stability. Nevertheless, the 2023 heat wave highlighted remaining gaps, including water scarcity, uneven access to cooling technologies, and the need for further investment in healthcare capacity during extreme events.

From a resilience perspective, Europe's trajectory shows that targeted adaptation measures can significantly reduce the human toll of extreme heat, but complete protection is not yet achievable under current systems. As climate projections indicate that events similar to those in 2023 could occur every 2–5 years if global temperatures rise to 2°C above pre-industrial levels, maintaining and expanding adaptation will be essential. Future strategies should focus on integrating heat resilience into urban planning, enhancing building thermal performance, and ensuring equitable access to cooling resources.

2.4.5 North American Hot and Dry Events: The 2023 Heat Wave

North America's hot and dry climate zones, particularly the southwestern United States and northern Mexico, have long been subject to extreme summer heat.

However, the July 2023 heat wave stands out as one of the most intense and prolonged in recorded history, illustrating both the advances in adaptation and the substantial challenges that remain.

The event affected a vast area, from California and Nevada through Arizona, New Mexico, and Texas, extending south into the Mexican states of Baja California, Sonora, and Chihuahua. On 16 July, Death Valley, California, recorded 52°C, approaching the highest reliably measured temperatures on Earth. Across the southwestern United States, daily maximums frequently exceeded 45°C, while in northern Mexico, peaks reached 46°C (Zachariah et al., 2023b). The persistence of high nighttime temperatures was especially alarming. Phoenix, Arizona, recorded minimum temperatures above 32.2°C for 19 consecutive nights, breaking historical records and severely limiting nocturnal recovery from heat stress.

The health consequences were immediate and widespread. In the United States, more than 100 million people were exposed to extreme heat alerts, and thousands required hospital treatment for heat stroke, dehydration, and exacerbations of chronic illnesses (Zachariah et al., 2023b). Vulnerable populations, including the elderly, people with chronic diseases, outdoor laborers, and individuals experiencing homelessness, bore the brunt of the crisis. In Mexico, 211 heat-related deaths were reported, nearly half of them in the state of Nuevo León. Migrant populations attempting to cross desert areas along the U.S.–Mexico border were among the most at-risk, with confirmed fatalities in Arizona and Texas highlighting the lethal combination of exposure, dehydration, and lack of shelter.

Infrastructure faced significant stress during the event. Electricity demand surged as residents relied heavily on air conditioning, creating peak loads that strained and, in some areas, overwhelmed power grids. Blackouts occurred in parts of California and Texas, temporarily depriving residents of cooling at the most dangerous times. High pavement temperatures caused asphalt roads to soften, resulting in deformation and road closures, while steel railway tracks expanded and buckled, forcing speed restrictions and service interruptions. Wildfires, driven by the combination of prolonged drought and extreme heat, ignited across multiple states, degrading air quality and forcing thousands to evacuate (Zachariah et al., 2023b).

Adaptation measures implemented over the past decade likely prevented an even higher death toll. Early warning systems in the U.S. and Mexico issued heat advisories and public health alerts days in advance. In cities such as Phoenix and Las Vegas, municipal authorities opened air-conditioned cooling centers and extended their operating hours to accommodate larger numbers of residents. Targeted outreach programs distributed bottled water and information to homeless populations. In the Mexican states most affected, emergency water distribution points were established, and some municipalities introduced water conservation restrictions to prioritize human consumption and agricultural needs.

Despite these efforts, the 2023 heat wave exposed several persistent vulnerabilities. Cooling center access was uneven, with rural areas and smaller communities often lacking such facilities. Power grid resilience remained a concern, as even short-duration blackouts during peak heat can quickly become life-threatening, especially for medically vulnerable residents. Water scarcity, already an acute issue in many parts of the region, was exacerbated by the high evaporative demand of the heat wave, creating tension between agricultural, industrial, and residential water needs.

The event also underscored the social equity dimension of heat resilience. Lower-income households frequently occupy older, poorly insulated buildings with limited or no air conditioning, leaving residents exposed to prolonged indoor overheating. Energy cost burdens can deter households from running cooling systems even when available. In rural agricultural areas, outdoor workers faced extended exposure to dangerous conditions, and while some U.S. states and Mexican regions implemented work-hour restrictions or mandated cooling breaks, enforcement was inconsistent.

From a resilience perspective, the 2023 North American heat wave highlights both progress and gaps. The combination of advanced meteorological forecasting, public health preparedness, and targeted urban interventions reduced mortality compared to historic events of similar magnitude. However, the persistence of tropical night conditions, infrastructure vulnerabilities, and social inequities demonstrates that adaptation remains incomplete.

Future strategies for hot and dry North American climates will need to prioritize:

1. Grid resilience through diversification of energy sources, integration of renewable generation with storage, and demand-side management to reduce peak loads.
2. Expansion of passive cooling measures in building design, including reflective roofing, shading, and improved insulation, to reduce dependence on mechanical cooling.
3. Water resource management that balances human, agricultural, and ecological needs under worsening drought–heat wave combinations.
4. Labor protection policies that set enforceable limits on outdoor work during extreme heat and ensure compliance.
5. Equitable access to cooling through targeted subsidies, public cooling infrastructure, and retrofits in low-income housing.

As climate projections indicate that events like the 2023 heat wave could become significantly more frequent and intense under a 2°C warming scenario, integrating these measures into urban planning and policy frameworks will be essential. Without such action, even regions with advanced warning systems and high air conditioning penetration may continue to experience avoidable health impacts and economic losses during future heat waves.

Although the July 2023 event ranks among the most extreme in modern records, it represents only the tip of the iceberg in the history of U.S. heat extremes. In their 2025 study, Narayanan et al. (2025) compiled a comprehensive record of the most severe U.S. heat waves since the late 1980s, evaluated using the Heat Severity and Coverage Index (HSCI) and its humidity-adjusted counterpart Humidity-Adjusted Heat Severity and Coverage Index (HSCIH). Their analysis demonstrates that incorporating humidity can significantly change severity rankings, particularly in the eastern and southern regions where high atmospheric moisture markedly amplifies thermal stress. In some cases, the inclusion of humidity increases a heat wave's severity score several-fold, moving it from mid-ranking to among the most extreme on record. Several striking shifts emerge when humidity is included in severity assessments. Events in humid eastern and southern regions, such as the 2023 Southeastern U.S. Heat Wave, the 1999 Midwest and Eastern U.S. Heat Wave, and the 1995 Chicago Heat Wave, move dramatically upward in the HSCIH ranking. In contrast, dry-region events such as the 2003 Arizona Heat Wave and the 2020 Southwestern U.S. Heat Wave fall in relative position, reflecting humidity's limited role in those climates (Narayanan et al., 2025).

These results indicate that in humid climates, high atmospheric moisture can substantially increase physiological strain by reducing evaporative cooling, leading to an underestimation of hazard when only temperature-based indices are used. For public health and infrastructure planning, this means that early warning systems, heat action plans, and building design strategies in humid regions should be informed by humidity-adjusted metrics to better capture the true extent of risk.

This evidence also points to a broader global trend: when humidity is considered alongside temperature, severity rankings often shift substantially, particularly in climates where atmospheric moisture is consistently high. This pattern, evident in the United States, is also increasingly visible in other parts of the world.

2.4.6 Global Comparative Synthesis

The influence of humidity on heat wave severity, highlighted by the U.S. findings of Narayanan et al. (2025), is part of a wider global pattern. Across climate zones, high humidity during extreme heat events intensifies thermal stress and can reorder historical rankings, with the largest shifts occurring in tropical, monsoonal, and other moisture-rich regions.

The heat waves of 2003, 2019, and 2023 in Europe, along with recent extremes in India, North America, and Southeast Asia, show a clear trajectory: while adaptation measures such as early warning systems, public health interventions, and urban cooling have reduced mortality in some regions, they have not kept pace with the accelerating frequency, intensity, and duration of extreme heat events. The impacts are no longer confined to traditionally hot regions; areas such as Scandinavia and Canada, once considered low risk, now experience dangerous heat waves.

A consistent pattern emerges across climates. In humid regions, high wet-bulb temperatures limit the body's ability to cool, creating life-threatening conditions even at moderate air temperatures. In arid regions, extreme daytime heat and water scarcity combine with occasional humidity spikes to create equally dangerous conditions. In temperate and Mediterranean zones, populations and infrastructure often lack thermal resilience, amplifying the effects of prolonged events. Across all zones, persistent nighttime heat undermines recovery, and urban heat islands exacerbate risks.

Demographic change compounds the threat. By 2050, over 23% of people aged 69 and older are projected to live in regions where daily maximum temperatures exceed 37.5°C, up from 14% today, exposing an additional 177–246 million elderly individuals worldwide to dangerous heat levels (Falchetta et al., 2024). Asia and Africa are expected to bear the largest burden, with the elderly population at risk in Asia projected to triple. These figures underline the urgency of targeted adaptation strategies for vulnerable groups.

Recent projections and observed events confirm that traditional temperature-based metrics underestimate the health impacts of extreme heat, particularly where high humidity amplifies thermal stress. Integrating advanced indices such as wet-bulb temperature, Cooling Degree Days (CDD), and Hot Days (#HD) into public health systems and urban planning can improve preparedness (Cvijanovic et al., 2023).

Common adaptation priorities emerge across regions: expanding shaded areas and urban greenery, enhancing power grid resilience, improving water management, integrating passive and active cooling into building design, and ensuring equitable access to cooling resources. However, the 2023 events reveal persistent weaknesses in limiting economic losses, preventing infrastructure failures, and ensuring that protective measures reach rural communities and informal urban settlements.

The growing convergence of heat wave risks across climate zones reinforces the need for globally coordinated adaptation. Public communication should evolve beyond simple temperature forecasts to convey humidity-adjusted health risks, and infrastructure planning should prioritize passive survivability and long-term thermal resilience.

For the built environment, these lessons are critical. Heat waves are now a primary stress test for buildings, exposing design flaws, material vulnerabilities, and gaps in passive cooling capacity. Addressing these issues requires a systematic framework that links climate science with building performance evaluation.

This provides the foundation for the next chapter, which introduces the rating systems, standards, and performance metrics used to assess thermal comfort in buildings. Chapter 3 establishes the analytical tools required to evaluate overheating risk and indoor environmental quality, which are then applied in later chapters to examine thermal resilience, passive design strategies, and the interaction between outdoor heat extremes and indoor environments.

REFERENCES

Al-Assaad, D., Sengupta, A., An, P., Breesch, H., Afshari, A., Amaripadath, D., Attia, S., Baba, F., Corrado, V., Eli, L., Krelling, A.F., Lee, S.H., Levinson, R., Olinger, M., Tootkaboni, M.P., Wang, L., Zhang, C., & Zinzi, M. (2025). Resilient passive cooling strategies during heat waves: A quantitative assessment in different climates. *Building and Environment*, *274*, 112698. https://doi.org/10.1016/j.buildenv.2025.112698

Copernicus Climate Change Service. (2025). The 2024 Annual Climate Summary: Global Climate Highlights 2024. January 10, 2025. Updated January 17, 2025. https://climate.copernicus.eu

Cvijanovic, I., Mistry, M.N., Begg, J.D. et al. (2023). Importance of humidity for characterization and communication of dangerous heatwave conditions. *NPJ Climate and Atmospheric Science*, *6*, 33 (2023). https://doi.org/10.1038/s41612-023-00346-x

Falchetta, G., De Cian, E., Sue Wing, I. et al. (2024). Global projections of heat exposure of older adults. *Nature Communications*, *15*, 3678. https://doi.org/10.1038/s41467-024-47197-5

Lüthi, S., Fairless, C., Fischer, E.M. et al. (2023). Rapid increase in the risk of heat-related mortality. *Nature Communications*, 14, 4894. https://doi.org/10.1038/s41467-023-40599-x

Machard, A., Inard, C., Alessandrini, J.-M., Pelé, C., & Ribéron, J. (2020). A methodology for assembling future weather files including heatwaves for building thermal simulations from the EURO-CORDEX climate data. *Energies*, *13*(13), 3424. https://doi.org/10.3390/en13133424

Machard, A., Salvati, A., Tootkaboni, M.P., et al. (2024). Typical and extreme weather datasets for studying the resilience of buildings to climate change and heatwaves. *Scientific Data, 11*, 531. https://doi.org/10.1038/s41597-024-03319-8

Narayanan, A., Rezaali, M., Bunting, E.L., & Keellings, D. 2025. It's getting hot in here: Spatial impact of humidity on heat wave severity in the U.S. *Science of The Total Environment, 963*, 178397. https://doi.org/10.1016/j.scitotenv.2025.178397

Ouzeau, G., Soubeyroux, J.M., Schneider, M., Vautard, R., & Planton, S. (2016). Heat waves analysis over France in present and future climate. *Climate Services, 4*, 1–12. https://doi.org/10.1016/j.cliser.2016.09.002

Siu, C.Y., Brown, S., Brideau, S., Ji, L., O'Brien, W., Ferguson, A., Greco, E., Touchie, M., McLellan, C., Lepage, R., & Dmytrenko, N. (2025). Exploring the role of building codes in protecting occupants from overheating: A Canadian perspective. *Energy and Buildings, 338*, 115673. https://doi.org/10.1016/j.enbuild.2025.115673

Stone, P.H., & Risbey, J.S. (1990). On the limitations of general circulation climate models. *Geophysical Research Letters, 17*(12), 2173–2176. https://doi.org/10.1029/GL017i012p02173

Sun, X., Ge, F., Fan, Y., Zhu, S., & Chen, Q. (2022). Will population exposure to heat extremes intensify over Southeast Asia in a warmer world? *Environmental Research Letters, 17*(4), 044006. https://doi.org/10.1088/1748-9326/ac48b6

Thirumalai, K., DiNezio, P.N., Okumura, Y., & Deser, C. (2017). Extreme temperatures in Southeast Asia caused by El Niño and worsened by global warming. *Nature Communications, 8*, 15531. https://doi.org/10.1038/ncomms15531

Thompson, V., Mitchell, D., Hegerl, G.C., Collins, M., Leach, N.J., & Slingo, J.M. (2023). The most at-risk regions in the world for high-impact heatwaves. *Nature Communications*, 14, 2152. https://doi.org/10.1038/s41467-023-37554-1

Younger, S. (2025). 2024 Was the Warmest Year on Record. NASA Earth Observatory, January 21, 2025. https://earthobservatory.nasa.gov/images/153806/2024-was-the-warmest-year-on-record

Zachariah, M., Vautard, R., Chandrasekaran, R., Chaithra, S.T., & Kimutai, J. (2023a). Extreme Humid Heat in South and Southeast Asia in April 2023, Largely Driven by Climate Change, Detrimental to Vulnerable and Disadvantaged Communities. Report. https://doi.org/10.25561/104092

Zachariah, M., Philip, S., Pinto, I., Vahlberg, M.. Singh, R., Otto, F., Barnes, C., & Kimutai, J. 2023b. Extreme Heat in North America, Europe and China in July 2023 Made Much More Likely by Climate Change. https://doi.org/10.25561/105549

3 Frameworks for Assessing Building Thermal Comfort

Rating Systems, Standards, and Metrics

3.1 INTRODUCTION

3.1.1 Thermal Comfort in the Context of Rising Heat Exposure

The preceding chapter examined the growing prevalence and intensification of heat waves worldwide, highlighting their implications for energy infrastructure, public health, and the thermal performance of the built environment. It is now widely acknowledged that heat waves are no longer isolated meteorological anomalies, but a recurrent feature of a warming climate. As peak temperatures rise and extreme heat events become longer and more frequent, the capacity of buildings to maintain acceptable indoor conditions is being tested at an unprecedented scale.

In this evolving context, thermal comfort has emerged as a critical axis of climate resilience. While earlier assessments of thermal performance focused primarily on energy efficiency or passive survivability, contemporary research increasingly emphasizes the lived experience of building occupants. Indoor temperatures that remain within material or structural safety limits may still impose significant discomfort—or even health risks—especially for vulnerable individuals such as children, the elderly, or those with pre-existing conditions. Consequently, the evaluation of thermal comfort must move beyond technical metrics of energy demand or envelope performance and engage directly with how indoor environments are perceived, tolerated, or rejected by their users.

This chapter presents the core methodological frameworks used to assess thermal comfort in buildings. These include both normative standards—such as ASHRAE 55, ISO 7730, EN 16798-1, and CIBSE TM52/TM59—and a wide array of quantitative indicators that capture physiological strain, subjective satisfaction, and statistical deviation from comfort thresholds. Together, they form a comprehensive basis for evaluating the suitability of indoor conditions under both standard operation and stress scenarios such as heat waves or cooling system outages.

Importantly, the indicators introduced in this chapter serve a dual purpose. On the one hand, they allow researchers and practitioners to diagnose whether a given thermal environment supports occupant well-being under current climatic conditions.

DOI: 10.1201/9781003645573-3

On the other hand, they provide a basis for projecting how thermal comfort may deteriorate under future climate scenarios—a topic that will be explored in detail in Chapter 5. In this sense, thermal comfort assessment is not only a tool for understanding present-day indoor conditions but also a key input for climate adaptation planning, policy development, and performance-based design.

The sections that follow examine three core components of the thermal comfort framework. First, we review the leading international certification systems—Leadership in Energy and Environmental Design (LEED), Building Research Establishment Environmental Assessment Method (BREEAM), and WELL—that incorporate comfort-related criteria into building evaluation. Second, we introduce the scientific standards that define acceptable thermal conditions in different building types and usage contexts. Third, we present the most widely used thermal comfort indicators, including both steady-state models such as Predicted Mean Vote (PMV) and Predicted Percentage of Dissatisfied (PPD), and time-based or heat-stress-oriented metrics such as Standard Effective Temperature (SET), Discomfort Hours (DH), Heat Index (HI), and Wet-Bulb Globe Temperature (WBGT). By grounding these tools in the broader context of global heat trends and building resilience, the chapter aims to provide a structured foundation for evaluating the impact of climate change on indoor thermal environments.

3.2 BUILDING RATING SYSTEMS FOR THERMAL COMFORT

In response to the growing demand for buildings that safeguard occupant well-being under both normal and extreme thermal conditions, several international certification frameworks have integrated thermal comfort as a core evaluation criterion. These systems offer structured methodologies for verifying how effectively a building maintains habitable indoor environments across seasons, usage types, and climate zones. Among the most influential are the LEED, the BREEAM, and the WELL Building Standard. Each of these systems employs quantitative and procedural tools, ranging from thermal simulations and comfort models to post-occupancy surveys and health-based thresholds, to ensure that comfort is not an incidental outcome but a deliberate objective of climate-responsive design.

3.2.1 Leadership in Energy and Environmental Design

Originating in the United States and administered by the U.S. Green Building Council (USGBC), LEED is one of the most widely recognized and applied sustainability certification systems globally. It promotes environmentally responsible building practices by evaluating structures across multiple performance domains, including energy efficiency, water conservation, materials use, and indoor environmental quality (IEQ). Thermal comfort is addressed as part of the EQ credit category and is governed by compliance with ASHRAE Standard 55: *Thermal Environmental Conditions for Human Occupancy.*

Under LEED, buildings are required to demonstrate that at least 80% of regularly occupied spaces maintain thermal conditions acceptable to occupants. This is typically achieved through modelling based on the PMV and PPD indices. PMV values must remain within the range of −0.5 to +0.5, while PPD must not exceed 10%. For naturally ventilated buildings, LEED allows the use of adaptive comfort models that relate indoor thermal expectations to recent outdoor temperatures.

LEED encourages project teams to use whole-building simulation tools, such as EnergyPlus and IES-VE, to assess comfort across seasonal extremes, including scenarios involving solar gains, internal loads, and varying occupancy profiles. Beyond thermal modelling, LEED also recognizes the role of real-time monitoring and adaptive control strategies, such as occupant-adjustable thermostats, automated shading, and feedback-integrated Building Management Systems. Passive comfort solutions, including solar-responsive façades, thermal mass integration, and cross-ventilation, are explicitly promoted as strategies that enhance both comfort and energy performance.

LEED certification is awarded at four levels—Certified, Silver, Gold, and Platinum—each reflecting progressively stricter performance thresholds. Higher tiers typically require integrated approaches that simultaneously optimize indoor comfort, reduce carbon emissions, and anticipate future climatic stresses.

3.2.2 Building Research Establishment Environmental Assessment Method

Europe's leading contribution to sustainable building certification is embodied in BREEAM, a performance-based framework developed in the United Kingdom by the Building Research Establishment (BRE). BREEAM predates LEED and remains one of the most widely used assessment methods across Europe and parts of Asia. Its structure emphasizes environmental impacts across energy, health, materials, pollution, and resilience categories, of which thermal comfort forms a critical component within the Health and Wellbeing domain.

BREEAM approaches thermal comfort with a strong focus on seasonal adaptability, requiring that buildings demonstrate effective protection against both winter heat loss and summer overheating. This is achieved through the integration of high-performance building envelopes, shading and solar control, efficient glazing, and natural ventilation strategies. In contrast to LEED's strict reliance on ASHRAE standards, BREEAM allows—and often encourages—the use of EN 16798-1 and CIBSE TM52/TM59 for defining comfort thresholds, especially in residential and naturally ventilated contexts.

The framework places considerable emphasis on future climate resilience. Buildings are encouraged to undergo dynamic simulation analyses that incorporate climate projection data (e.g., UKCP18), enabling design teams to assess overheating risk under mid- and long-term scenarios. Adaptive comfort models, occupant-adjustable systems, and monitoring-based tuning are all considered valid approaches, provided that they demonstrably reduce discomfort over time.

Unlike LEED, which emphasizes conformity to fixed thermal thresholds, BREEAM supports occupant-centered flexibility. Greater credit is given to solutions that allow users to influence their thermal environment, whether through operable windows, zoning, or control interfaces, recognizing that comfort perception is inherently variable. Certification levels—ranging from Pass to Outstanding—reward projects that integrate thermal comfort within broader sustainability and usability goals, including post-occupancy feedback and long-term performance evaluation.

3.2.3 WELL Building Standard

Shifting the focus from environmental performance to human experience, the WELL Building Standard introduces thermal comfort as a determinant of physiological well-being and cognitive performance. Developed by the International WELL Building Institute, WELL evaluates buildings across ten core concepts, including Air, Water, Nourishment, Light, Movement, and Comfort. Within this structure, thermal comfort is explicitly linked to productivity, stress reduction, sleep quality, and health equity.

WELL-certified projects are required to comply with either ASHRAE 55 or ISO 7730, ensuring that environmental parameters such as temperature, humidity, and air speed fall within established PMV–PPD boundaries. However, WELL expands upon these requirements by promoting personalized climate control, thermal zoning, and occupant feedback loops. Unlike LEED or BREEAM, which allow for averaged thermal conditions across zones or populations, WELL advocates for localized interventions that respond to individual needs and preferences.

A defining feature of the WELL approach is its insistence on quantitative and experiential validation. Certification requires not only predictive modelling but also occupant surveys and, in some cases, continuous environmental monitoring using sensors installed at the breathing zone. Acceptability thresholds must be met not only statistically but in terms of perceived comfort, with particular emphasis on eliminating thermal drift, stratification, or stagnant conditions in high-occupancy zones.

WELL also recognizes the cognitive and behavioral implications of thermal discomfort. Variations in indoor temperature are understood to influence alertness, decision-making, and emotional regulation, particularly in workspaces, educational facilities, and healthcare settings. Design strategies that support stable, low-stress environments—including radiant systems, underfloor ventilation, and thermally inert materials—are therefore strongly encouraged.

Taken together, the LEED, BREEAM, and WELL systems reflect a spectrum of approaches to thermal comfort assessment. LEED emphasizes integrated simulation and passive design. BREEAM focuses on adaptive performance and climate preparedness. WELL centers on health metrics and occupant empowerment. Each framework embeds thermal comfort within a broader value system—be it sustainability, resilience, or well-being—and provides structured pathways for verification and improvement. The standards and indicators underlying these certifications are detailed in the following section, which outlines the thermal comfort thresholds and modeling principles that support compliance across building types and operational scenarios.

3.3 TECHNICAL STANDARDS FOR THERMAL COMFORT ASSESSMENT

While rating systems such as LEED, BREEAM, and WELL translate thermal comfort objectives into certification criteria, their underlying performance benchmarks are grounded in internationally recognized technical standards. These standards codify the environmental parameters, calculation models, and evaluation procedures that define acceptable indoor thermal conditions for different building types, occupancy patterns, and climatic contexts.

Among the most widely applied are ASHRAE Standard 55, ISO 7730, EN 16798-1, and the Chartered Institution of Building Services Engineers (CIBSE) guidelines TM52 and TM59. Together, they provide a scientific foundation for comfort assessment, drawing on decades of research in human thermoregulation, heat balance theory, and adaptive comfort modelling. They also serve as reference points for regulatory compliance, HVAC system design, and simulation-based performance verification.

Unlike voluntary rating systems, technical standards focus on measurable physical and physiological variables such as air temperature, mean radiant temperature, humidity, air velocity, metabolic rate, and clothing insulation. These inputs feed into predictive models that quantify comfort levels, identify overheating risk, or assess seasonal adaptability. Some standards, such as ASHRAE 55 and ISO 7730, adopt steady-state approaches, while others, such as CIBSE TM52/TM59, incorporate adaptive methods better suited to naturally ventilated buildings and variable climates.

The following subsections present the main features, modeling principles, and application contexts of these leading thermal comfort standards, beginning with ASHRAE Standard 55.

3.3.1 ASHRAE 55—Thermal Comfort Standard

ASHRAE Standard 55 is one of the foundational references for defining and quantifying thermal comfort in the built environment. It is widely adopted by both regulatory bodies and voluntary certification schemes, and it serves as a scientific baseline for HVAC design, simulation modelling, and compliance assessment. The standard formalizes the physical and physiological variables that influence human thermal perception and translates them into measurable parameters and predictive models.

Among the most influential technical frameworks for evaluating indoor thermal conditions is ASHRAE Standard 55, developed by the American Society of Heating, Refrigerating and Air-Conditioning Engineers. This standard defines the environmental and personal factors that govern thermal comfort in buildings, offering a quantitative methodology grounded in physiological heat balance theory. While many building certification systems refer to ASHRAE 55 for compliance purposes, the standard also underpins regulatory codes, HVAC system design, and comfort performance modeling across diverse climate zones.

ASHRAE 55 identifies six core parameters as determinants of thermal perception: air temperature, mean radiant temperature, air velocity, relative humidity, metabolic rate, and clothing insulation. Together, these variables shape the thermal exchange between the human body and its surroundings. The standard does not impose a single comfort threshold; instead, it provides models for calculating predicted occupant responses under varying indoor conditions.

At the core of ASHRAE 55 lies the steady-state heat balance model, originally developed by Fanger (1970), which assumes that thermal comfort is achieved when the body's metabolic heat production is balanced by its heat loss to the environment. This principle is mathematically expressed as:

$$M - W = E + C + R + S$$

where:

M—the metabolic rate [W/m^2],
W—the mechanical work performed [W/m^2],
E—the evaporative heat loss (sweat evaporation and respiration) [W/m^2],
C—the convective heat exchange with ambient air [W/m^2],
R—the radiative heat exchange with surrounding surfaces [W/m^2],
S—the rate of heat storage in the body [W/m^2].

For comfort conditions, the heat storage term is assumed to be zero ($S = 0$), meaning that internal body temperature remains stable over time.

3.3.1.1 Convective and Radiative Heat Exchange

The convective heat transfer component (C) is a function of air movement and the temperature difference between the skin and ambient air:

$$C = h_c\left(T_{skin} - T_{air}\right)$$

where:

C—convective heat loss from the skin [W/m^2],
h_c—convective heat transfer coefficient [W/(m^2·K)],
T_{skin}—skin temperature [°C],
T_{air}—air temperature in contact with the skin surface [°C].

Increased air velocity raises h_c, enhancing convective cooling under warm conditions but potentially causing discomfort in cooler environments. For this reason, ASHRAE 55 places limits on acceptable air speed depending on activity level, clothing insulation, and indoor temperature.

Radiative heat exchange is modeled as:

$$R = h_r\left(T_{skin} - T_{mrt}\right)$$

where:

T_{skin}—surface temperature of the skin [°C],
T_{mrt}—mean radiant temperature [°C],
h_r—radiative heat transfer coefficient [W/(m^2·K)].

In buildings with large glazed areas or high internal thermal mass, radiant asymmetry can significantly influence occupant comfort even when air temperature is well regulated.

3.3.1.2 Clothing Insulation and Humidity Effects

Clothing acts as a thermal barrier, modifying the rate at which the body exchanges heat with the environment. ASHRAE 55 quantifies this using the clo unit, where 1 clo corresponds to the insulation provided by typical business attire. Clothing insulation is expressed as:

$$I_{clo} = \frac{R_{cl}}{0.155}$$

where:

I_{clo}—clothing insulation expressed in clo (–),
R_{cl}—thermal resistance of the clothing layer [$m^2 \cdot K/W$].

Humidity influences the evaporative heat loss component of the model, particularly through sweat evaporation. The relevant equation is:

$$E = h_e \left(P_{skin} - P_{air} \right)$$

where:

E—evaporative heat loss from the skin [W/m^2],
h_e—evaporative heat transfer coefficient [$W/(m^2 \cdot kPa)$],
P_{skin}—water vapor pressure at the skin surface [kPa],
P_{air}—water vapor pressure in the surrounding air [kPa].

High humidity limits evaporation, reducing the body's ability to self-regulate temperature. For this reason, ASHRAE 55 recommends maintaining relative humidity between 30% and 60%, unless thermal neutrality is achieved through air motion or radiant cooling.

3.3.1.3 Thermal Comfort Models and Application

ASHRAE Standard 55 operationalizes the heat balance model through two quantitative comfort indices. The first is the PMV, which provides a statistical estimate of the average thermal sensation of occupants on a seven-point scale ranging from –3, indicating cold discomfort, to +3, indicating hot discomfort. The second is the PPD, a complementary metric that represents the proportion of occupants likely to experience thermal dissatisfaction under the prevailing conditions.

According to the standard, an indoor environment is considered thermally acceptable when the PMV falls between –0.5 and +0.5, a range that corresponds to a PPD of no more than 10%. These criteria are commonly applied in mechanically ventilated or HVAC-controlled buildings and serve as foundational benchmarks in certification frameworks such as LEED and WELL.

3.3.1.4 Adaptive Thermal Comfort in Naturally Ventilated Buildings

Recognizing that occupants of naturally ventilated buildings adapt their expectations based on recent weather, ASHRAE 55 incorporates an adaptive comfort model for free-running buildings. This model relates the acceptable indoor comfort temperature to the mean outdoor temperature over the preceding 7–30 days:

$$T_{comfort} = a\, T_{out} + b$$

where:

$T_{comfort}$—indoor comfort temperature [°C],
T_{out}—mean outdoor air temperature used in the adaptive model [°C],
a—empirically derived slope coefficient of the adaptive comfort relationship [–],
b—empirically derived intercept representing the base comfort temperature [°C].

where the coefficients a and b are empirically derived based on geographic location and building typology. The adaptive model allows a broader comfort band, acknowledging that users in naturally ventilated environments tolerate higher summer temperatures and lower winter temperatures than those in fully conditioned spaces.

This adaptive formulation applies only under conditions in which the building operates without mechanical heating or cooling and occupants are free to adjust their environment through operable windows or vents. It requires that the indoor climate be governed by natural forces rather than by HVAC controls, since active conditioning would suppress the behavioral and physiological adaptations on which the model is based. When these requirements are met, the adaptive approach provides a realistic representation of comfort in free-running buildings and captures how occupants respond to gradual changes in outdoor temperature.

3.3.1.5 Conclusion

ASHRAE 55 provides a comprehensive and scientifically grounded method for assessing thermal comfort across building types, usage patterns, and climatic conditions. By integrating the principles of human heat balance, clothing physiology, humidity regulation, and adaptive behavior, the standard supports both precision modeling and contextual interpretation. Whether applied to mechanical system design, simulation-based certification, or real-time monitoring, ASHRAE 55 remains a cornerstone of performance-based comfort evaluation in contemporary architecture and building science.

3.3.2 ISO 7730—Thermal Comfort and Assessment Methodology

Complementing the ASHRAE 55 framework, ISO 7730 provides an internationally harmonized methodology for evaluating thermal comfort in buildings. Developed by the International Organization for Standardization, it is grounded in the steady-state thermal balance model introduced by Fanger and applies the same six core environmental and personal variables: air temperature, mean radiant temperature, air velocity, relative humidity, metabolic rate, and clothing insulation. These parameters describe the physical conditions of the indoor environment and the physiological characteristics of the occupants, forming the basis for calculating comfort indices.

The defining feature of ISO 7730 is its structured classification system, which introduces three thermal comfort categories. Each category specifies acceptable ranges for the PMV and associated limits for the PPD, allowing designers to align comfort expectations with building function and user sensitivity.

3.3.2.1 Thermal Comfort Categories

Category	PMV Range	PPD Limit	Typical Applications
A	$-0.2 \le PMV \le +0.2$	<6%	High-precision or sensitive environments (e.g., hospitals, laboratories)
B	$-0.5 \le PMV \le +0.5$	<10%	Standard comfort spaces (e.g., offices, dwellings)
C	$-0.7 \le PMV \le +0.7$	<15%	Industrial, warehouse, or transitional spaces

Categories A–C enable clear specification of comfort targets at the design stage. Category A is typically reserved for spaces requiring strict control of environmental variables, while Category C accommodates more variable conditions where comfort is less critical.

3.3.2.2 Local Discomfort Factors

ISO 7730 gives explicit attention to localized discomfort, recognizing that even when global PMV and PPD values meet target thresholds, factors such as air drafts, vertical temperature gradients, and radiant asymmetry can undermine perceived comfort.

Air drafts are quantified using the Draft Rating (DR) equation:

$$DR = (34 - T_a)(V - 0.05)^{0.62}\left(0.37V\,I + 3.14\right)$$

where:

DR—percentage of occupants affected by draft [%],
T_a—air temperature [°C],
V—local air velocity [m/s],
I—turbulence intensity [%].

A DR value below 15% is recommended for sedentary activity levels. A DR below 15% is recommended for sedentary zones.

Vertical temperature gradient is defined as the temperature difference between head level (1.1 m) and foot level (0.1 m):

$$\Delta T = T_{head} - T_{feet}$$

where:

T_{head}—air temperature measured at 1.1 m above the floor [°C],
T_{feet}—air temperature measured at 0.1 m above the floor [°C].

ISO 7730 recommends that ΔT should not exceed 3°C to avoid discomfort caused by excessive vertical temperature gradients.

Radiant asymmetry represents another key contributor to local discomfort. It may arise from cold window surfaces, overheated façades, or unshielded heat sources, creating uneven radiant temperature fields that are not captured by PMV-based assessments. ISO 7730 addresses radiant asymmetry through qualitative evaluation and commissioning procedures, ensuring that design and operational measures limit excessive directional radiative differences.

3.3.2.3 Comparison with ASHRAE 55

Although ISO 7730 and ASHRAE Standard 55 are grounded in the same theoretical basis—primarily Fanger's heat balance model—their practical implementation differs in several important respects. ISO 7730 adopts a categorical framework, defining three comfort categories (A, B, and C) with progressively relaxed limits. Category A imposes particularly stringent PMV and PPD requirements, which are more restrictive than the general comfort allowances typically permitted under ASHRAE 55. This tiered structure enables regulators and designers to articulate comfort expectations with greater precision and to differentiate clearly between space types or occupancy conditions.

A further distinction lies in the treatment of naturally ventilated buildings. ASHRAE 55 explicitly incorporates an adaptive comfort model for free-running spaces, allowing acceptable temperature ranges to vary in response to outdoor conditions and occupant behavioral adaptation. ISO 7730, in contrast, retains fixed comfort limits regardless of the ventilation strategy, resulting in a more prescriptive and less adaptive approach for buildings without mechanical conditioning.

The standards also diverge in their treatment of local discomfort. ISO 7730 formalizes the assessment of draft risk, vertical temperature gradients, radiant asymmetry, and floor surface temperature, providing quantitative criteria intended to be applied systematically. ASHRAE 55 includes similar concepts but treats them as optional, context-dependent considerations to be evaluated only when relevant to the project's scope or the specific characteristics of the occupied space.

Taken together, these differences reflect distinct philosophies: ISO 7730 emphasizes standardization, regulatory clarity, and explicit performance categorization, whereas ASHRAE 55 prioritizes flexibility, adaptability, and alignment with observed occupant behavior. As a result, the choice between the two standards often depends on the regulatory environment, the design intent, and the degree of adaptability required in the thermal comfort assessment.

3.3.2.4 Applications in Design and Compliance

ISO 7730 is widely referenced in European building regulations, certification frameworks such as BREEAM and HQE, and public procurement processes. Category A is often mandated for healthcare and laboratory environments, Category B for offices and education buildings, and Category C for industrial facilities.

Dynamic simulation tools such as EnergyPlus, IES-VE, and TRNSYS incorporate ISO 7730 calculations, allowing comfort compliance to be assessed at early design stages. The standard's prescriptive structure makes it well suited for projects where clear, documented guarantees of comfort performance are required.

3.3.2.5 Conclusion

ISO 7730 offers a rigorous, hierarchically structured approach to thermal comfort assessment. By combining global comfort indices with limits on local discomfort, it enables the precise specification and verification of indoor environmental conditions. Its fixed-threshold methodology supports compliance and quality assurance, particularly in settings with sensitive or high-demand occupants.

3.3.3 EN 16798-1—Adaptive Thermal Comfort Model

In climates where natural ventilation is a viable design strategy, the assumption that indoor thermal comfort must always be maintained within narrow, static temperature limits becomes increasingly difficult to justify. The EN 16798-1 standard, developed by the European Committee for Standardization (CEN), addresses this challenge by introducing an adaptive model of thermal comfort that links acceptable indoor conditions to prevailing outdoor temperatures and occupant expectations. This approach marks a conceptual departure from the steady-state thermal balance models used in ASHRAE 55 and ISO 7730; it offers greater flexibility in buildings that are not mechanically conditioned.

EN 16798-1 is part of a broader framework, formerly EN 15251, for evaluating IEQ, including thermal, acoustic, and air quality dimensions. It is particularly relevant in the context of low-energy and passive building design, where occupants interact with their environment and adapt their behavior, for example by opening windows or adjusting clothing, in response to seasonal and daily variations.

3.3.3.1 Adaptive Comfort Model: Core Assumptions and Equation

The adaptive model in EN 16798-1 is based on extensive field studies conducted across Europe, which demonstrated that thermal comfort in naturally ventilated spaces is influenced not only by physical conditions, but also by thermal history and psychological adaptation. Instead of assuming a universal optimal temperature, the model defines the neutral or comfort temperature T_{comfort} as a linear function of the mean outdoor air temperature $T_{out,mean}$:

$$T_{\text{comfort}} = 0.33\ T_{out,mean} + 18.8$$

where:

T_{comfort}—the indoor comfort temperature [°C],

$T_{out,mean}$—the exponentially weighted running mean of the daily outdoor air temperature over the previous seven days [°C].

This formulation implies that if the mean outdoor temperature is 25°C, the acceptable indoor comfort temperature is approximately 27.1°C. Conversely, when the outdoor mean is 10°C, the comfort expectation drops to 22.1°C. Such variability aligns with empirical evidence that users acclimate to seasonal norms and do not require constant indoor temperatures year-round.

3.3.3.2 Comfort Categories and Allowable Temperature Bands

EN 16798-1 defines four comfort categories (I–IV) that establish acceptable deviations from the calculated neutral temperature. These categories are analogous to ISO 7730's thermal classifications but are structured specifically for the adaptive model:

Category	Acceptable Range	Application Context
I	±2°C	Sensitive occupants (e.g., hospitals, museums)
II	±3°C	Normal expectations (e.g., offices, schools, dwellings)
III	±4°C	Limited expectations (e.g., workshops, storage)
IV	>±4°C	Temporary or non-compliant environments

For example, if the mean outdoor temperature is 26°C, the comfort temperature is 27.4°C. Under Category II, acceptable indoor temperatures would therefore range between 24.4°C and 30.4°C. These broader bands contrast with the narrow PMV-based thresholds in ASHRAE 55 and ISO 7730, making the EN model more forgiving in buildings without full HVAC control.

3.3.3.3 Applicability and Boundary Conditions

The adaptive comfort model in EN 16798-1 is intended for use only when specific boundary conditions are met. The building must operate in a naturally ventilated mode during the evaluation period, without active heating or cooling that would override the influence of the outdoor climate. Occupants must also have meaningful control over operable windows or ventilation openings, enabling behavioral adaptation. Under these conditions, indoor temperatures are driven primarily by external climatic variations and occupant interaction with the building envelope, rather than by mechanical conditioning systems.

When these requirements are not satisfied—for example in sealed, mechanically cooled offices—the adaptive model is no longer valid. In such cases, comfort must instead be evaluated using fixed PMV/PPD-based criteria as defined in ISO 7730 or ASHRAE Standard 55. The adaptive approach is therefore most applicable in temperate and transitional climates, where seasonal variability in outdoor temperature allows passive strategies to meaningfully influence indoor conditions. Its usefulness diminishes in extreme climates, such as hot-humid or very cold regions, where the building may be unable to maintain indoor conditions within even the expanded adaptive limits.

Beyond its boundary conditions, the adaptive model has important implications for both design and simulation practice. It encourages the use of low-energy strategies by allowing wider indoor temperature ranges, especially during shoulder seasons. This flexibility reduces reliance on mechanical cooling and makes it easier to apply passive measures such as night-time ventilation, solar shading, and thermal mass. Because the comfort limits are tied to occupants' behavioral responses, the model also accommodates window-opening actions without treating them as comfort violations, aligning performance targets more closely with how buildings are actually used.

From a modeling perspective, simulation tools such as EnergyPlus, TRNSYS, and IES-VE incorporate algorithms for calculating the running mean outdoor temperature and deriving the corresponding adaptive comfort thresholds. These features support compliance assessments under European performance standards and are commonly used to evaluate thermal resilience in low-energy and passive house projects. EN 16798-1 is referenced in a range of certification frameworks and national regulations—including BREEAM, Passivhaus methodologies, and UK guidance such as Part L, TM52, and TM59. Within these contexts, the adaptive limits provide the benchmark against which exceedance metrics, including DH or Hours of Exceedance (HE), are calculated during overheating assessments.

3.3.4 CIBSE TM52—Overheating Assessment in Non-Residential Buildings

While adaptive comfort models such as those defined in EN 16798-1 provide a framework for estimating acceptable indoor temperatures, they do not, by themselves, define when a building is at risk of overheating. In response to this limitation, the CIBSE introduced Technical Memorandum 52 (TM52), a performance-based standard designed to assess the frequency, intensity, and duration of overheating events in non-residential buildings. TM52 builds on the adaptive thermal comfort methodology; however, it extends this approach by offering a set of operational criteria that allow overheating to be explicitly diagnosed from simulation or monitored data.

The guidance provided in TM52 has become especially relevant in the context of climate adaptation and net-zero building design, where increased reliance on passive or mixed-mode strategies raises the risk of internal heat accumulation. TM52 addresses this concern by establishing clear, testable thresholds that distinguish between acceptable thermal variation and conditions that may endanger occupants, particularly in schools, offices, and public-use buildings where cooling may be intermittent or deliberately omitted for energy efficiency.

3.3.4.1 Overheating Criteria: A Multi-Dimensional Framework

TM52 establishes a three-criterion framework for evaluating overheating risk in naturally ventilated and free-running buildings. The approach is based on adaptive comfort temperatures derived from the EN 15251 / EN 16798-1 model. A building is deemed at risk of overheating if it fails two or more criteria during the occupied season, typically defined as May to September.

Criterion 1: HE

The first criterion limits the frequency of overheating. The operative temperature T_{oper} must not exceed the adaptive comfort temperature T_{comf} by more than 1°C for more than 3% of the annual occupied hours:

$$\frac{\sum\left(T_{\text{oper}} > T_{\text{comf}} + 1\right)}{T_{\text{occupied}}} \leq 0.03$$

where:

T_{oper}—operative indoor temperature [°C],
T_{comf}—adaptive comfort temperature [°C],
$T_{occupied}$—total number of annual occupied hours.

The fraction $\Sigma(T_{oper} > T_{comf} + 1)$ represents the number of hours during which the operative temperature exceeds the comfort limit by more than 1°C. According to EN 16798-1, this fraction must not exceed 3%.

This criterion reflects the overall frequency of overheating events and is conceptually similar to duration-based metrics such as DH or HE. It captures the cumulative burden of elevated indoor temperatures across the season.

Criterion 2: Daily Severity Limit

The second criterion addresses the intensity of overheating. At no point may the difference between the operative temperature and the adaptive comfort limit exceed 4°C:

$$\Delta T_{\text{max}} = T_{\text{oper}} - T_{\text{comf}} \leq 4°\text{C}$$

where:

ΔT_{max}—maximum allowable indoor temperature exceedance relative to adaptive comfort [°C],
T_{oper}—the indoor operative temperature [°C],

T_{comf}—adaptive comfort temperature defined by the selected comfort model [°C],4°C is the upper exceedance limit permitted by the adaptive comfort criteria,

$\Delta T_{max} \leq 4°C$—operative temperature may exceed the adaptive comfort temperature by no more than four degrees Celsius.

This captures peak intensity, identifying buildings that may occasionally become acutely uncomfortable even if most days remain within acceptable ranges.

Criterion 3: Weighted Exceedance (We)The cumulative daily discomfort over occupied hours must not exceed 6 degree-hours on more than 40 days during the warm season:

$$W_e = \sum_i \left(T_{oper,i} - T_{comfort,i}\right)\Delta t$$

$$W_e \leq 6°\mathrm{C} \cdot h/day$$

where:

W_e—cumulative temperature exceedance expressed in degree-hours [°C·h],
$T_{oper,i}$—operative temperature at hour i [°C],
$T_{comfort,i}$—adaptive comfort temperature at hour i [°C],
Δt—time interval used in the summation, typically 1 hour [h].

According to EN 16798-1, the weighted exceedance must not exceed 6°C·h per day. This metric captures moderate but persistent overheating that may not trigger the daily peak criterion yet still results in significant cumulative discomfort.

Together, these three criteria provide a comprehensive assessment of overheating risk by considering both short-term intensity and long-term persistence of elevated indoor temperatures. Falling short on two out of the three criteria indicates the need for corrective measures in building design, façade, and shading strategy, ventilation approach, or operational controls.

3.3.4.2 Design, Simulation, and Practical Implications

CIBSE TM52 is principally intended as a design-stage methodology for evaluating overheating risk in naturally ventilated and mixed-mode non-residential buildings. Its application relies on dynamic thermal simulation, typically undertaken with tools such as EnergyPlus, DesignBuilder, or IES-VE. These simulations must incorporate realistic boundary conditions, including credible internal gains, occupancy patterns, ventilation schedules, and window-opening behavior. TM52 further emphasizes the use of future weather files, such as CIBSE's DSY1, DSY2, and DSY3 datasets, to ensure that designs are resilient under representative heat-wave conditions rather than only during historically typical summers.

From a design perspective, TM52 highlights several strategies that can significantly mitigate overheating. Night-time ventilation is fundamental, as it enables the purging of heat stored in thermal mass and improves the building's capacity to buffer daytime temperature swings. External solar control—such as brise-soleil, deep reveals, fixed overhangs, or operable blinds—reduces short-wave gains and prevents façade overheating. Additional measures include the use of high-albedo materials to limit solar absorption, zonal control systems that allow targeted interventions in vulnerable areas, and mixed-mode ventilation strategies capable of switching between natural and mechanical operation as conditions change. The aim of these measures is not to eliminate indoor temperature variation, but to maintain it within physiologically tolerable and contextually acceptable limits.

In contrast to comfort standards such as ISO 7730 or ASHRAE Standard 55, TM52 does not depend on the PMV–PPD framework and therefore does not require inputs relating to metabolic activity or clothing insulation. Instead, it is rooted in an adaptive understanding of thermal comfort, reflecting the reality that occupants in naturally ventilated buildings respond dynamically to environmental conditions through behavioral and psychological adaptation. As a result, TM52 is not appropriate for sealed or continuously conditioned buildings, laboratories, or environments requiring tight temperature control; its strength lies specifically in the assessment of free-running or hybrid spaces where air movement, outdoor variability, and occupant interaction define thermal experience.

One of the principal advantages of TM52 is its alignment with real-world occupant perceptions, particularly during periods of extreme heat. The exceedance-based framework provides a transparent account of how often and by how much operative temperatures surpass adaptive comfort levels, offering a measure that is easily understood by clients, policymakers, and design teams. When paired with visual tools—such as exceedance charts, overheating maps, or cumulative discomfort curves—the methodology becomes a powerful means of communicating risk and evaluating the impact of design decisions.

Taken as a whole, TM52 offers a rigorous and empirically informed approach to diagnosing overheating risk during the warm season. By providing simulation-ready criteria tied directly to adaptive comfort models, it supports both regulatory compliance and optimization of passive and mixed-mode design strategies. In the context of rising summer temperatures and increasing concern over the energy and carbon implications of mechanical cooling, TM52 remains a vital reference point for designers seeking to balance occupant comfort, thermal resilience, and environmental performance.

3.3.5 CIBSE TM59—Overheating Assessment in Residential Buildings

Residential buildings pose a distinct set of thermal challenges, particularly during summer heat waves when occupants may spend extended periods indoors and often lack access to mechanical cooling. In recognition of these vulnerabilities, the CIBSE introduced Technical Memorandum 59 (TM59) as a complementary methodology to TM52, with a specific focus on overheating risk in dwellings. While TM52 is oriented toward non-residential buildings with variable occupancy patterns, TM59 addresses the domestic context, where occupants such as children, older adults, and individuals with pre-existing health conditions may be exposed to elevated indoor temperatures for prolonged periods.

The development of TM59 was driven by growing evidence that conventional thermal performance indicators, such as annual energy use or compliance with Part L of the Building Regulations, do not adequately reflect the acute, time-sensitive nature of overheating risk. As a result, the memorandum adopts a performance-based framework that draws on adaptive thermal comfort theory but is tailored to residential occupancy profiles and room typologies.

3.3.5.1 Dual-Criterion Framework for Overheating Assessment

TM59 defines two principal criteria for assessing overheating risk in residential buildings, and a dwelling is considered non-compliant if it breaches either one. The first criterion applies to daytime living spaces, such as living rooms or open-plan living, dining, and kitchen areas. In these rooms, the operative temperature must remain within the adaptive comfort band during the majority of the warm season. Compliance is evaluated by calculating the proportion of occupied hours between May and September during which the operative temperature exceeds the adaptive comfort temperature by more than 1°C. This requirement is formally expressed as:

$$\frac{\sum\left(T_{\text{oper}} > T_{\text{comf}} + 1\right)}{T_{\text{occupied}}} \leq 0.03$$

where:

T_{oper}—operative indoor temperature [°C],
T_{comf}—adaptive comfort temperature [°C],
$T_{occupied}$—total number of occupied hours during the analysis period [h].

This ensures that such exceedances represent no more than 3% of the total occupied hours, aligning directly with the adaptive comfort methodology in EN 16798-1.

This metric is analogous to the TM52 Criterion 1 but uses a slightly narrower temperature band, reflecting the lower tolerance for temperature excursions in spaces used continuously for sedentary activity.

A second formulation is used to evaluate nighttime conditions, particularly in bedrooms, where thermal tolerance is more limited and overheating poses increased risks to sleep quality and occupant health. TM59 therefore introduces a stricter limit on nighttime exceedances, defined as the proportion of nighttime hours defined as 22:00 to 07:00 in which the operative temperature rises above 26°C. The corresponding condition is given by:

$$\frac{\sum\left(T_{\text{oper}} > 26^{\circ\text{C}}\right)}{T_{\text{night}}} \leq 0.01$$

where:

T_{oper}—operative indoor temperature [°C],
T_{night}—total number of night-time occupied hours [h].

According to EN 16798-1, no more than 1% of night hours may exceed an operative temperature of 26°C.

This criterion restricts exceedances above 26°C to no more than 1% of the total nighttime hours within the assessment period. Together, these two equations capture both daytime adaptation and nighttime vulnerability, providing a robust basis for determining whether a dwelling maintains acceptable thermal conditions under summer operation.

By targeting both daytime and nighttime occupancy, TM59 provides a comprehensive thermal risk profile that extends beyond average temperatures or annual metrics. In contrast to TM52, which requires failure of two out of three criteria, TM59 is more conservative: failure of either one implies non-compliance, underscoring the higher health stakes in domestic environments.

3.3.5.2 Simulation Protocols and Assumptions

Overheating risk must be assessed using dynamic thermal simulation, typically with hourly or sub-hourly time steps. TM59 prescribes the use of Design Summer Year (DSY1) weather files, such as those based on 2020s or 2050s climate scenarios for London, Birmingham, or Manchester, to represent severe but plausible summer conditions.

Key inputs include:

- Internal heat gains from occupants, lighting, and appliances, based on SAP or BREDEM assumptions,
- Ventilation rates determined by window opening schedules, trickle vents, or MVHR systems,
- Solar gain control measures such as shading devices, glazing g-values, and orientation,
- Thermal zoning to reflect independent room use and solar exposure.

Design teams are encouraged to simulate multiple design options, including natural ventilation only, limited mechanical assistance, and user behavior scenarios (e.g., windows closed at night for noise or security).

3.3.5.3 Design Implications, Mitigation Strategies, and Policy Relevance of TM59

TM59 provides a dedicated framework for evaluating and mitigating overheating risk in residential buildings, responding to the unique thermal characteristics and occupancy patterns of domestic environments. Unlike non-residential guidance, TM59 places equal emphasis on daytime comfort and nighttime thermal conditions, acknowledging that prolonged exposure to elevated temperatures, particularly during sleep, has direct consequences for health, well-being, and productivity. Its criteria therefore serve not only as technical performance indicators but also as safeguards against the unacceptable thermal burdens that may arise in modern, airtight housing.

A central component of TM59 is its emphasis on passive and low-energy design measures. External shading is highlighted as a primary means of limiting solar gains, particularly on south- and west-facing façades that receive the most intense afternoon radiation. By reducing incident solar load, shading devices—such as overhangs, deep reveals, fixed louvres, or operable blinds—substantially lower the thermal stress placed on indoor spaces. Natural ventilation strategies form an equally important aspect of TM59-compliant design. Cross-ventilation, the use of aligned openings, and night-time purge ventilation all support the removal of accumulated heat, enabling the building to recover overnight and enter the following day in a cooler, more resilient state. The strategic use of thermal mass further enhances this effect by absorbing heat during the day and releasing it when outdoor temperatures drop, thereby moderating indoor temperature fluctuations.

High-performance glazing systems contribute additional benefits by limiting solar heat gains without compromising daylight quality, while reducing internal gains from lighting, appliances, and equipment helps to lower the overall thermal load. When implemented collectively, these measures form an integrated passive cooling strategy that aligns directly with the TM59 performance thresholds. They support the broader objective of achieving acceptable thermal conditions through architectural design rather than mechanical intervention.

In high-density urban developments, however, the feasibility of natural ventilation may be limited by constraints such as noise pollution, poor outdoor air quality, or security concerns. TM59 acknowledges that in such contexts, passive strategies alone may not suffice. Hybrid approaches—combining passive cooling with demand-controlled ventilation, mechanical assistance, or automated window control systems—may therefore be necessary to meet both comfort and regulatory performance goals.

Beyond providing design guidance, TM59 is closely tied to a wider network of standards and regulatory instruments. It draws on the same adaptive comfort framework as EN 16798-1 but adapts its application through room-specific thresholds and temporal distinctions between daytime and nighttime occupancy. TM59 also plays a significant role within the UK regulatory environment, informing compliance with the London Plan, Building Regulations (particularly Part O in England), and BREEAM residential schemes. As such, it serves as a central benchmark in contemporary housing policy, shaping the design and evaluation of dwellings across England and the wider UK.

The relevance of TM59 is likely to intensify under projected climate change scenarios. Many new homes are being constructed with high insulation levels and increased airtightness to meet decarbonization and energy-efficiency targets. While these measures reduce heating demand in winter, they also diminish the building's ability to expel unwanted heat in summer, raising the risk of overheating in periods of warm or extreme weather. TM59 therefore, functions as a critical counterbalance to low-carbon design, ensuring that the pursuit of operational efficiency does not inadvertently undermine thermal resilience or occupant well-being.

3.3.6 Computational Modeling Methods for Thermal Comfort Standards

Thermal comfort standards such as ASHRAE Standard 55, ISO 7730, EN 16798-1, and the CIBSE TM52/TM59 guidelines rely increasingly on computational modeling to assess IEQ under present and future climatic conditions. Dynamic simulation is now the primary tool for evaluating thermal comfort because it can represent the transient and spatially variable behavior of buildings, including thermal mass, solar gains, ventilation patterns, internal loads, and occupant-driven variability. As Attia (2018) notes, comfort assessments are credible only when they integrate building physics with realistic user behavior and climate-sensitive boundary conditions.

Across different regulatory contexts, modeling approaches vary considerably. Some national methodologies require multi-zone, hourly dynamic simulation, while others rely on simplified, quasi-static models that cannot resolve inter-zonal differences, rapid temperature fluctuations, or adaptive occupant responses. Dynamic thermal simulation has therefore become central to performance-based design, offering the temporal resolution and modeling flexibility required to satisfy comfort standards in both free-running and mechanically conditioned buildings (Clarke, 2001; Hensen & Lamberts, 2019).

Modern simulation engines—including EnergyPlus, TRNSYS, IES-VE, and Modelica-based platforms—support multiple thermal comfort models, such as PMV–PPD, adaptive comfort limits from EN 16798-1, and overheating criteria used in TM52 and TM59. The growing interest in health-informed metrics has also introduced cumulative exposure indicators that may require minute-level time steps, extending modeling requirements beyond traditional hourly simulation.

The selection of weather data is critical to accurate comfort predictions. Many regulatory frameworks continue to rely on typical meteorological years, which smooth climate variability and underrepresent extreme heat events. To address future climatic risks, designers increasingly adopt design summer years, extreme weather files, or climate-projection datasets derived from downscaling, morphing, or stochastic generation methods (Belcher et al., 2005). Such scenario-based modeling aligns with broader approaches to climate-sensitive analysis and supports resilience assessment under anticipated warming trends (Morales et al., 2014).

Urban microclimate conditions—such as the urban heat island effect, local shading, radiative trapping, and street-canyon geometry—further influence thermal comfort outcomes. Established approaches include adjusting weather files with urban corrections, incorporating on-site measurements, or coupling building-scale models with urban microclimate tools. Models such as ENVI-met, SOLENE-Microclimat, UWG, and UrbClim provide detailed representations of local radiation balance, vegetation effects, and airflow, although they require significantly greater computational effort and expertise (Mirzaei & Haghighat, 2010).

In naturally ventilated and mixed-mode buildings, the representation of air movement is essential. Standard building performance simulation tools implement airflow models based on empirical or pressure-driven equations, while more advanced workflows employ Computational Fluid Dynamics (CFD) to capture drafts, airflow distribution, vertical gradients, and radiant asymmetries. Although CFD–BPS coupling is typically reserved for complex or high-performance projects, it provides the level of spatial detail often required to fully evaluate local discomfort factors.

Occupant behavior remains one of the largest sources of uncertainty in thermal comfort modeling. Most compliance tools rely on deterministic schedules for occupancy, internal gains, clothing insulation, and metabolic rate. However, field studies show that occupants in free-running and hybrid buildings often adjust windows, blinds, clothing, and ventilation in ways that cannot be represented with static profiles. Recent developments include probabilistic routines, Markov-based models, and agent-based approaches that simulate adaptive behaviors (Hong et al., 2017), providing a more realistic representation of user–environment interactions.

Despite technological advances, several methodological gaps persist. Internal gains are often represented with oversimplified schedules. Model validation and calibration procedures are inconsistently applied, reducing the reliability of simulation-based decisions (de Wilde, 2018). Metadata and documentation practices vary widely, limiting transparency and reproducibility. Finally, interoperability between building-scale simulation engines and urban-scale models remains limited, restricting the integration of district-level microclimate effects.

In summary, the computational implementation of thermal comfort standards requires flexible and multi-layered workflows that integrate appropriate comfort models, realistic occupant behavior, high-quality weather data, and—in urban settings—microclimate effects. Dynamic simulation provides the technical foundation for such analysis, enabling designers to transition from compliance-focused evaluations toward predictive and resilience-oriented modeling. As climate change increases the frequency and severity of extreme heat events, the convergence of regulatory practice and state-of-the-art simulation methods becomes essential for ensuring thermal comfort, energy efficiency, and long-term building resilience.

3.3.7 Thermal Comfort Indicators: Methods and Applications

As discussed in the preceding section, the limitations of traditional thermal comfort standards cannot be resolved without examining the indicators used to quantify thermal conditions. Across the broader body of research, a consistent observation emerges: commonly applied metrics such as PMV, PPD, or simple exceedance hours provide only a partial picture of thermal risk, particularly in heat-vulnerable spaces or during prolonged periods of high temperatures. These limitations have led to the development of a new generation of comfort indicators that account for duration, severity, physiology, and spatial variability—criteria that are increasingly important in the context of climate change and low-carbon, highly insulated buildings.

A major line of innovation involves time-integrated indicators that capture not only whether a comfort threshold has been exceeded, but also for how long and by how much. Weighted exceedance metrics—such as Daily Weighted Exceedance or cumulative discomfort degree-hours—assign greater influence to severe or persistent overheating, making them particularly suitable for multi-day heat events. These indicators address the central weakness of binary metrics by linking overheating to both exposure duration and intensity.

Concerns about nighttime thermal stress, especially in bedrooms, have prompted studies to propose stricter temperature limits for sleeping conditions. These approaches often use absolute exceedance thresholds (e.g., 26°C–28°C) and apply persistence criteria, so that overheating is registered only when elevated temperatures continue beyond a minimum duration. Such methods help distinguish short, inconsequential fluctuations from events that carry genuine physiological or behavioral consequences.

To reinforce physiological validity, researchers increasingly incorporate health-based indices, originally developed for occupational or outdoor heat exposure. The HI integrates temperature and humidity and is particularly useful in humid climates or poorly ventilated interiors. The WBGT, while more complex due to its sensitivity to radiant heat and air movement, offers additional insight for spaces with high solar exposure, such as sunrooms or highly glazed living areas. These indices provide thresholds linked to thermoregulatory strain, offering a pathway toward comfort criteria that reflect heat-health risks more directly.

Other work focuses on composite perceptual indices such as the SET, which integrates air temperature, humidity, radiant temperature, and air speed into a single equivalent temperature. SET is particularly valuable in free-running or mixed-mode buildings, where rapid environmental changes and significant radiant asymmetries diminish the reliability of PMV-based assessments. While SET requires detailed input data and numerical calculation, it provides a far more realistic basis for evaluating comfort in dynamic indoor environments.

Beyond environmental and physiological metrics, a third group of indicators seeks to capture thermal dissatisfaction over time. Perceptual-weighted measures—such as cumulative PPD-based indicators—avoid the binary nature of exceedance counts by weighting each timestep according to predicted dissatisfaction. These metrics align more closely with comfort experience and allow designers to distinguish between marginal discomfort and extreme or persistent conditions.

Finally, research on thermal resilience underscores the value of zone-specific and scenario-based indicators. For example, cumulative hours above 26°C in bedrooms, metrics describing the rate at which temperatures rise during heat-wave or blackout conditions, and measures of passive survivability provide direct insight into the protective capacity of the building envelope. These indicators are increasingly relevant in low-energy housing, where airtightness and insulation can elevate indoor temperatures during extreme weather events.

In summary, the evolution of thermal comfort assessment reflects several converging trends:

- a shift from binary metrics toward time-integrated severity measures,
- growing use of health-based indices that incorporate humidity, radiation, and metabolic strain,
- increasing attention to perceptual and behavioral dimensions of comfort, and
- heightened emphasis on space-specific and scenario-driven indicators that address thermal vulnerability.

Together, these innovations offer a more realistic and operationally useful framework for evaluating thermal comfort than conventional PMV/PPD-based approaches. They also align with the broader objective of developing buildings that remain safe and comfortable under dynamic, uncertain, and increasingly extreme climatic conditions. The following subsections provide detailed definitions and applications of the principal indicators introduced here.

3.4 THERMAL COMFORT INDICATORS: METHODS AND APPLICATIONS

The limitations of current thermal comfort standards, as discussed in the preceding section, cannot be fully addressed without reconsidering the indicators used to operationalize them. Conventional metrics such as PMV, PPD, or simple counts of hours above a temperature threshold provide only a partial view of thermal performance.

They are often insufficient for assessing overheating risk in heat-vulnerable spaces or under dynamic climate conditions.

Recent research responds to these limitations by proposing indicators that are more temporally sensitive, cumulative, or health-oriented. Time-integrated metrics such as Daily Weighted Exceedance (WEx) and Discomfort Degree-Hours (Dh_EN) account not only for whether a threshold is exceeded, but also for the duration and severity of the exceedance, making them better suited to multi-day heat events. In parallel, health-based indices such as the HI and WBGT incorporate humidity and radiant load, offering more physiologically relevant risk markers. Other developments include perceptual-weighted metrics, such as Global Thermal Overload, and zone-specific measures for sensitive spaces like bedrooms, where both frequency and persistence of discomfort are critical.

In operational terms, thermal comfort indicators provide a quantitative basis for translating complex interactions between environmental parameters, building characteristics, and human thermoregulation into interpretable values. They enable objective comparison across building types, climates, and ventilation strategies, as well as longitudinal tracking of discomfort or risk. Indicators differ in their theoretical foundation and scope: some are derived from rational heat balance models (e.g., PMV), others from empirical survey data, and still others integrate exposure, adaptation, and health effects. Their selection should reflect the building typology, the evaluation context, whether it is steady-state design analysis or dynamic risk assessment, and the time horizon, from present-day operation to future climate extremes.

The following subsections present the principal thermal comfort indicators used in both research and practice, outlining their theoretical basis, variables, and interpretation. Each is discussed in terms of its formulation, measurement requirements, and applicability within existing standards or resilience frameworks. Together, they form the analytical toolkit for quantifying design performance, indoor safety, and climate adaptability.

3.4.1 Predicted Mean Vote

The PMV is one of the most widely used thermal comfort indices and serves as the foundation of thermal comfort assessment in several international standards, including ASHRAE 55 and ISO 7730. Developed by Fanger in the 1970s, the PMV model estimates the average thermal sensation of a large group of people exposed to a given indoor environment. It is based on the principles of human heat balance and assumes that thermal comfort is achieved when the internal heat production of the body equals the heat loss to the environment. The PMV index ranges from –3 to +3 and is interpreted as a seven-point scale of subjective thermal sensation, where –3 represents "cold," 0 denotes "neutral," and +3 indicates "hot."

The PMV is calculated using a steady-state energy balance model that accounts for the primary physical and physiological parameters affecting heat exchange between the human body and its environment. These include metabolic heat production, mechanical work, heat transfer through convection and radiation, evaporative heat loss, clothing insulation, and environmental factors such as air temperature, mean radiant temperature, relative humidity, and air velocity.

The general formulation of PMV is expressed as:

$$PMV = \left(0.303e^{-0.036M} + 0.028\right)\left[(M - W) - E_d - E_r - C - R\right]$$

where:

PMV—Predicted Mean Vote (–),
M—metabolic rate [W/m²],
W—external work performed by the body [W/m²],
E_d—heat loss due to skin diffusion of water vapor [W/m²],
E_r—heat loss due to sweating in regulatory conditions [W/m²],
C—convective heat loss [W/m²],
R—radiative heat loss [W/m²].

The equation combines convective and radiative heat loss, evaporation from skin and respiratory tract, latent heat loss due to sweating, and heat loss by water vapor diffusion. The exponential coefficient reflects the diminishing sensitivity of the body's thermal regulation response at higher metabolic rates. The PMV model assumes steady-state conditions, fixed metabolic activity, and a homogeneous indoor environment. These assumptions make the model most suitable for mechanically ventilated or HVAC-controlled spaces.

The PMV value is often interpreted alongside the PPD, which estimates the percentage of occupants likely to be thermally uncomfortable at a given PMV value. Standards such as ISO 7730 define thermal comfort categories based on PMV and PPD thresholds. For instance, Category A corresponds to PMV values between −0.2 and +0.2, with PPD below 6%; Category B allows PMV values between −0.5 and +0.5, with PPD below 10%.

Although the PMV model has been foundational in thermal comfort assessment, it also has well-known limitations. It assumes fixed clothing and activity levels, does not account for occupant adaptation or psychological factors, and is less accurate in naturally ventilated or free-running buildings. Nonetheless, it remains a key metric in building energy codes, simulation software, and green building certifications such as LEED and BREEAM, where thermal comfort compliance is assessed through PMV-based thresholds under standardized conditions.

As reviewed by Enescu (2017), several adaptations of the PMV model have been proposed to address its limitations, particularly in non-conditioned or hybrid buildings. A commonly discussed extension is the Adaptive PMV (aPMV), which modifies the PMV output by applying a correction factor derived from the difference between the Actual Mean Vote (AMV) and PMV. The AMV represents the average thermal sensation reported by building occupants in field studies. Empirical comparisons have shown that PMV often overestimates discomfort in free-running buildings, where adaptation and expectation play significant roles. The aPMV model introduces a scaling coefficient (β) that reflects the degree of contextual divergence between modeled and perceived comfort.

Another variation is the Expectancy-Corrected PMV, which incorporates a psychological adjustment factor to account for cultural or seasonal expectations. For example, people in hot-humid regions may perceive a given indoor condition as acceptable even if the PMV indicates slight warmth. This correction improves the model's realism in naturally ventilated settings, where behavioral adaptation is frequent.

Some studies also propose the use of PMV as a reference for deriving regression-based comfort indicators or machine-learning-driven classification systems. In such cases, PMV acts as a benchmark for developing surrogate models that integrate dynamic behavior, occupant feedback, or stochastic thermal conditions. Enescu emphasizes that PMV remains a useful foundational tool, but its interpretive value increases when complemented by field-calibrated adjustments or hybrid models that incorporate adaptive variables.

For designers and energy modelers, these refinements suggest that PMV should not be treated as a universally accurate predictor but rather as a controlled reference point. In mechanically ventilated zones with stable internal loads, it provides robust guidance. However, in naturally ventilated spaces, hybrid ventilation schemes, or buildings designed for occupant adaptability, PMV-based analysis should be supplemented by direct occupant surveys, adaptive algorithms, or time-integrated discomfort metrics.

As a result, standards such as EN 16798-1 or CIBSE TM52 increasingly encourage the use of adaptive comfort models in parallel with, or in place of, PMV, particularly when evaluating indoor comfort in non-mechanically cooled buildings. Nonetheless, understanding the assumptions, calculation method, and correction possibilities for PMV remains essential for all professionals engaged in comfort-based building design or certification.

3.4.2 Predicted Percentage of Dissatisfied

The PPD is a thermal comfort index used to estimate the percentage of occupants likely to feel thermally uncomfortable under given indoor environmental conditions. Originally developed by Fanger as an extension to the PMV model, the PPD index provides a statistical prediction of discomfort based on the calculated deviation from thermal neutrality. While PMV represents the average thermal sensation of a population on a seven-point scale, ranging from cold to hot, PPD quantifies the proportion of people expected to be dissatisfied with the thermal environment, even if the average vote appears neutral.

The mathematical formulation of PPD is a nonlinear, parabolic function of PMV. It is defined as:

$$PPD = 100 - 95e^{-0.03353\,PMV^4 - 0.2179\,PMV^2}$$

where:

PPD—Predicted Percentage of Dissatisfied [%],
PMV—Predicted Mean Vote (–).

The PPD value is expressed as a percentage and assumes a minimum of 5% dissatisfaction, even under ideal thermal conditions (PMV = 0). This reflects the empirical observation that complete satisfaction is statistically unattainable in any real-world environment, given the variability of human thermal perception.

The PPD function is symmetric and increases rapidly as PMV deviates from zero in either direction. A PMV of +0.5 or −0.5 corresponds to a PPD of roughly 10%, while a PMV of ±1.0 yields dissatisfaction levels of about 25%. At PMV values around ±2.0, more than 75% of occupants are predicted to be dissatisfied. This steep relationship makes PPD a convenient performance indicator in certification frameworks and comfort classifications. According to ISO 7730, buildings that maintain PPD levels below 10% satisfy Category B requirements, while values below 6% qualify as Category A.

Although PPD is most commonly used alongside PMV, it provides a more intuitive decision-making basis, especially in regulatory and compliance contexts where thresholds expressed as percentages are easier to communicate to non-technical stakeholders. However, PPD shares the underlying assumptions of the PMV model. It is based on steady-state heat balance, fixed metabolic and clothing conditions, and uniform environmental parameters. As a result, it performs best in tightly controlled HVAC environments where air temperature, humidity, mean radiant temperature, and air movement remain relatively stable.

Because of its standardized formulation, PPD has been incorporated into building simulation tools such as EnergyPlus, IES-VE, and TRNSYS, and is widely used in performance-based design assessments, post-occupancy evaluations, and thermal comfort audits. It remains one of the most recognized metrics for quantifying expected dissatisfaction due to thermal conditions in conditioned spaces.

Decades of field research, however, have shown that the PMV–PPD model tends to underestimate actual dissatisfaction in free-running or naturally ventilated buildings. Studies by Humphreys, Nicol, de Dear, Brager, and others consistently demonstrate that occupants in adaptive environments exhibit higher variability in thermal preference and often report comfort at temperature levels that the PPD curve would classify as unacceptable—or, conversely, report discomfort at higher rates than predicted when indoor conditions fluctuate significantly throughout the day.

Further limitations arise from the idealized nature of the PPD formulation. The curve was originally derived under laboratory conditions with homogeneous indoor environments. Real buildings introduce non-uniformities such as drafts, vertical temperature gradients, radiant asymmetry, local heat sources, and spatial variability in airflow patterns. These factors can produce dissatisfaction even when PMV is near zero, yet they are not captured in the PPD equation. Research in thermal comfort literature repeatedly highlights that local discomfort can dominate thermal sensation in practice.

PPD also does not distinguish between opposite discomfort sensations (feeling too hot versus too cold), nor does it incorporate psychological, cultural, or contextual factors that influence comfort expectations. Numerous studies show that thermal acceptability depends heavily on previous exposure, personal control, socioeconomic context, seasonal adaptation, and expectations regarding indoor climate—variables that lie outside the PMV–PPD framework.

For these reasons, contemporary comfort research recommends using PPD cautiously and in combination with other indicators, especially in adaptive or hybrid buildings. In free-running environments, adaptive comfort models, occupant surveys, subjective feedback (AMV), and exceedance-based metrics provide a more realistic representation of thermal experience. PPD retains value in controlled, mechanically conditioned spaces but is insufficient as a stand-alone measure in buildings where behavioral adaptation, variable environmental conditions, or non-uniform microclimates play a significant role.

In summary, the PPD index remains a widely adopted, standardized measure of expected thermal dissatisfaction and is central to many certification frameworks. However, its limitations—particularly in naturally ventilated or adaptive contexts—necessitate a complementary approach that integrates both physical and behavioral aspects of thermal comfort. For designers, engineers, and building evaluators, PPD should therefore be applied as one component within a broader suite of comfort indicators rather than as the sole measure of thermal acceptability.

3.4.3 Standard Effective Temperature

The SET is a comprehensive thermal comfort index that represents the combined influence of air temperature, humidity, air movement, radiant conditions, clothing insulation, and metabolic activity on human thermal sensation. Developed by Gagge et al. (1986), the index is grounded in thermophysiological modeling and is designed to capture transient heat exchanges that occur under real, time-varying indoor conditions.

3.4.3.1 Foundational Heat-Balance Relationships

SET is based on a dynamic heat-balance model that divides the human body into a core and a skin compartment. The overall heat balance is expressed through the following relationship:

$$H_{\text{res}} + H_{\text{sk}} = M - W$$

where:

H_{res}—respiratory heat loss [W/m²],
H_{sk}—skin heat loss [W/m²],
M—metabolic heat production [W/m²],
W—external mechanical work [W/m²].

This equation states that the total heat dissipated via respiration and the skin must match the metabolic heat that is not converted into mechanical work.

Skin heat loss comprises three components—convective, radiative, and evaporative—summarized as:

$$H_{\text{sk}} = C_{\text{sk}} + R_{\text{sk}} + E_{\text{sk}}$$

where:

C_{sk} represents convective heat loss [W/m²] and is typically expressed as

$$C_{\text{sk}} = h_c\left(T_{\text{sk}} - T_a\right)$$

R_{sk} represents radiative heat loss [W/m²] and is typically expressed as

$$R_{\text{sk}} = h_r\left(T_{\text{sk}} - T_r\right)$$

E_{sk} represents evaporative heat loss [W/m²] and is commonly expressed as

$$E_{\text{sk}} = w\,E_{max}$$

T_{sk}—skin temperature [°C],
T_a—air temperature [°C],
T_r—mean radiant temperature [°C],
h_c and h_r —convective and radiative heat transfer coefficients [W/(m²·K)],
w—skin wettedness (0–1),
E_{max}—maximum evaporative capacity [W/m²].

These terms capture the primary pathways through which the human body exchanges heat with its surroundings, and they demonstrate why SET is highly sensitive to humidity, radiant asymmetry, and air velocity.

The formal definition of SET is expressed through an implicit equivalence condition:

$$H_{\text{sk}}^{\text{ref}}(SET) + H_{\text{res}}^{\text{ref}}(SET) = H_{\text{sk}}^{\text{actual}} + H_{\text{res}}^{\text{actual}}$$

where the left-hand side denotes heat loss in a standardized reference environment (50% relative humidity, air velocity 0.1 m/s, and mean radiant temperature equal to air temperature), and the right-hand side represents heat loss in the actual environment.

Because this relationship must be satisfied iteratively, SET cannot be derived from a closed-form equation. Instead, it is obtained by numerically solving the coupled differential equations that govern heat storage and exchange in the core and skin nodes.

3.4.3.2 Physiological Modeling and Computational Procedure

The two-node model calculates transient heat storage in the core and skin, accounting for sweating, vasodilation, skin wettedness, and evaporative cooling. These processes dominate thermal regulation under warm or humid conditions, making SET more physiologically meaningful than steady-state indices such as PMV. The algorithm adjusts the skin and core temperatures over time and identifies the reference-environment temperature that produces an equivalent heat loss, which is the reported SET value.

Due to its computational complexity, SET is implemented primarily through simulation tools rather than manual calculation. It is included in the ASHRAE Thermal Comfort Tool, EnergyPlus (via Thermal Comfort Report outputs), CONTAM, and the CBE Comfort Tool. These platforms allow designers to use SET in evaluating environments with fluctuating ventilation rates, high radiant loads, evaporative cooling systems, ceiling fans, and other dynamic thermal phenomena.

3.4.3.3 Interpretation and Practical Application

SET is expressed in degrees Celsius and is comparable in scale to dry-bulb temperature, although the values may diverge considerably in conditions of high humidity or significant air movement. Typical neutral comfort thresholds range from 22°C to 26°C, but the acceptable range widens in naturally ventilated or mixed-mode buildings, where occupants adapt to environmental variations.

Compared with steady-state indicators, SET offers superior realism in transient or asymmetrical environments. It captures physiological processes that strongly influence thermal sensation—particularly skin wettedness and evaporative regulation—and therefore provides more accurate predictions in hot-humid climates, in hybrid ventilation scenarios, or in spaces where mean radiant temperature diverges from air temperature.

However, SET also has limitations. It requires detailed input data, consistent assumptions regarding clothing and metabolic rate, and computational resources to solve the underlying dynamic equations. This complexity has limited its adoption in regulatory frameworks and certification schemes, which tend to favour simpler indices such as PMV, PPD, or adaptive comfort criteria. Nevertheless, SET remains a valuable tool for advanced comfort studies, for evaluating passive cooling strategies, and for assessing dynamic indoor environments that cannot be adequately characterized by steady-state approaches.

When used alongside adaptive comfort models or occupant-reported satisfaction data, SET helps bridge the gap between simulated performance and real-world comfort perception. For designers, it provides a nuanced and context-sensitive metric that supports the creation of low-energy, climate-responsive buildings capable of maintaining thermal comfort across a wide range of environmental conditions.

3.4.4 Cooling Degree Days

Cooling Degree Days (CDD) is a cumulative thermal index used to estimate the demand for cooling energy in buildings. Unlike PMV or SET, which focus on short-term thermal sensation, CDD is a climate-based indicator that quantifies how much and for how long the outdoor temperature exceeds a predefined base threshold during a specified period (typically a day or a season). It is widely used in energy modeling, regional climate comparisons, and HVAC sizing procedures to predict seasonal cooling loads and assess overheating risks.

The general formula for calculating CDD is:

$$CDD = \sum_{days} \max\left(0, T_{mean} - T_{base}\right)$$

where:

T_{mean}—daily mean outdoor air temperature [°C],

T_{base}—base temperature used for cooling degree calculations, typically 18°C.

Only days with T_{mean} greater than T_{base} contribute to the CDD value.

If the daily mean temperature is below the base value, the contribution for that day is zero. The base temperature represents the outdoor temperature above which cooling is assumed to be required to maintain indoor comfort. Typical base temperatures vary by region and application:

In practice, the selection of the base temperature varies according to climate and building context. A value of 18°C is commonly adopted in ASHRAE-based methods and in many applications for temperate climates. In hot or humid regions, or when passive cooling strategies are under consideration, higher base temperatures in the range of 22°C–26°C are often more appropriate. In some cases, the base temperature is defined in a building-specific manner, for example, by aligning it with the actual cooling setpoints or with comfort thresholds relevant to the intended use of the space. This approach ensures that temperature-based indices more accurately reflect local climatic conditions and the thermal behavior of the building under study.

CDD can also be expressed in hourly resolution when using simulation data, in which case the metric is referred to as Cooling Degree Hours (CDH):

$$CDH = \sum_{hours} \max\left(0,\ T_{out,i} - T_{base}\right)$$

where:

$T_{out,i}$—outdoor air temperature at hour i [°C],
T_{base}—base temperature used for cooling degree-hour calculations, typically 18°C.

CDD and CDH are two complementary indicators widely used in the assessment of climate-driven cooling demand and the exposure of buildings to overheating. Both metrics rely on outdoor air temperature, yet they differ fundamentally in temporal resolution and interpretive scope. CDD aggregates the exceedance of a selected base temperature over longer periods, typically at daily or monthly scales, while CDH expresses the cumulative number of hours above a comfort threshold. As a result, CDD is more suitable for energy modeling and climate zoning, whereas CDH is more sensitive to heatwave dynamics and short-term overheating peaks.

CDD values vary substantially across regions. In Northern Europe, typical annual totals remain between 0 and 300°C·days·yr^{-1}, whereas in Southern Europe, the Middle East or the southern United States, they often exceed 1500–2000°C·days·yr^{-1}. In humid tropical climates, annual CDD can surpass 3000°C·days·yr^{-1} depending on the base temperature and analytical method. CDH exhibits even stronger climatic gradients, since it reacts to both absolute temperatures and the persistence of nighttime heat; in Mediterranean and subtropical regions, long sequences of nocturnal CDH accumulation are common during heatwaves, indicating elevated stress for naturally ventilated dwellings and vulnerable occupants.

CDD remains a simple yet informative descriptor of external thermal forcing. As noted in the literature, it correlates well with cooling energy consumption and is widely used in regional to continental assessments of cooling demand (e.g., NOAA, Eurostat, EEA). However, because it reflects only outdoor temperature, CDD does not incorporate solar gains, humidity, wind exposure or building-specific moderating factors such as thermal mass, shading systems, or ventilation strategies. Two buildings situated in the same climatic context may therefore share identical CDD values yet experience profoundly different indoor thermal outcomes.

CDH partially mitigates this limitation. By counting the hours when outdoor temperature exceeds a comfort or threshold value, CDH provides a more granular picture of heat stress intensity, diurnal asymmetry and night-time recovery potential.

This makes CDH particularly relevant for studies of overheating, sleep disruption and urban heat island amplification.

Both indicators depend critically on the choice of the base or threshold temperature. While traditional CDD often uses Tb = 18.3°C or 22°C for energy calculations (ASHRAE, Eurostat), research increasingly recommends adaptive thresholds linked to comfort temperature ranges (such as those defined in EN 16798-1). Using adaptive thresholds makes both CDD and CDH more relevant for naturally ventilated or mixed-mode buildings, where indoor comfort expectations shift with outdoor conditions. Nonetheless, the selection of thresholds remains a methodological challenge, and poorly chosen values may lead to over- or underestimation of cooling needs, especially in low-energy and passively optimized buildings.

As climate indicators, CDD and CDH quantify the intensity of external heat stress rather than the indoor thermal response. They should therefore be used in conjunction with dynamic modeling tools and indoor comfort metrics such as Indoor Overheating Degree (IOD), HE, PMV/PPD, or more recent resilience-based indices. This combination provides a balanced interpretation by linking climatic forcing with the ability of buildings to buffer, delay, or attenuate heat.

In summary, CDD and CDH offer complementary insights into climatic drivers of overheating. CDD provides an accessible, planning-oriented perspective for estimating cooling energy demand and comparing climate zones. CDH adds temporal resolution and enables the characterization of heatwave severity and night-time stress, which are critical for assessing passive cooling potential and thermal resilience. When applied together and coupled with dynamic indoor metrics, these indicators support robust evaluations of overheating risk and help guide the design of climate-adaptive buildings.

3.4.5 Discomfort Hours

DH is a time-based thermal comfort indicator that quantifies the number of hours during which indoor environmental conditions exceed predefined thermal comfort thresholds. Unlike indices such as PMV or SET, which represent continuous-scale comfort sensations, DH is a binary metric that registers whether the indoor temperature is within or outside the comfort range at each time step. It is widely used in overheating risk assessments, post-occupancy evaluations, and building performance simulations to identify periods of thermal discomfort over a specific timeframe, typically a day, season, or full year.

The general form of the DH metric is given by:

$$DH = \sum_{hours} H\left(T_{oper,i} > T_{max}\right) + H\left(T_{oper,i} < T_{min}\right)$$

where:

DH—total number of Discomfort Hours [h],

$T_{oper,i}$—operative indoor temperature at hour i,

T_{max} and T_{min}—upper and lower comfort thresholds [°C],

$H(x)$—an indicator function equal to 1 if the condition is true, 0 otherwise.

In many applications, the lower bound is not considered if the focus is on summer overheating, resulting in a simplified form:

$$DH = \sum_{hours} H\left(T_{oper,i} > T_{max}\right)$$

where:

DH—number of hours in which operative temperature exceeds the upper comfort threshold [h],

$T_{oper,i}$—operative indoor temperature at hour *i* [°C],

T_{max}—upper comfort threshold [°C],

H(x) is an indicator function equal to 1 when the condition is true and 0 otherwise.

The operative temperature used in DH calculations may be derived from simulations or field measurements and includes the combined effect of air temperature and mean radiant temperature. The choice of threshold values T_{max} and T_{min} depends on the selected comfort framework. In PMV-based standards such as ASHRAE 55, thresholds are tied to fixed comfort bands (e.g., 23–26°C). EN 16798-1 uses an adaptive comfort model, in which acceptable temperature bands shift in response to the running mean outdoor temperature. Overheating assessments in CIBSE TM52 and TM59 rely on dynamic exceedance limits derived from adaptive comfort thresholds.

DH can be evaluated for an entire building or for specific zones and may be adjusted using occupancy schedules to reflect actual exposure. Some studies extend the metric by applying severity weighting, as in Degree-Hours or Weighted Discomfort Hours, but in its fundamental form, DH remains a binary time-counting approach.

Because DH is straightforward to compute and interpret, it is widely integrated into building performance simulation workflows. Tools such as EnergyPlus, IES-VE, and DesignBuilder generate hourly or sub-hourly indoor temperatures that can be directly counted as DH. Designers frequently use DH to evaluate the performance of passive cooling strategies, shading devices, insulation systems, and natural ventilation configurations.

3.4.5.1 Strengths and Limitations

DH is one of the most widely adopted comfort indicators in both research and regulation because of its clarity and communicative power. It effectively highlights the frequency of thermally uncomfortable periods and provides an accessible measure for clients, planners, and policymakers. Nonetheless, important limitations must be acknowledged.

First, the metric is highly sensitive to the selection of threshold temperatures. Thresholds that are too lenient may underreport discomfort, whereas overly restrictive ones can exaggerate overheating. Fixed thresholds (such as 26°C) do not account for behavioral adaptation or climate variation, while adaptive thresholds improve realism but increase methodological complexity.

Second, DH measures only the duration of discomfort but not its intensity. A one-degree exceedance and a five-degree exceedance contribute equally, despite having very different implications for comfort, performance, and health. Severity-weighted variants address this issue but are less commonly used.

Third, the calculated DH value depends on the temporal resolution of the data. Hourly data may overlook short but frequent discomfort episodes, while fine-resolution data (e.g., 5–10-minute intervals) produce more accurate estimates but require additional processing. This is particularly relevant in passive or naturally ventilated buildings, where thermal conditions may fluctuate throughout the day.

Despite these limitations, DH remains a practical, transparent, and simulation-friendly indicator that plays a central role in design-stage overheating assessments. Although it does not capture physiological complexity in the way that PMV or SET do, it provides actionable insight into the distribution and duration of thermal discomfort. When used alongside adaptive thresholds or severity-weighted metrics, DH forms an essential component of comprehensive comfort evaluation in low-energy and climate-responsive building design.

3.4.6 Indoor Overheating Degree

IOD is a cumulative thermal discomfort indicator that quantifies the total magnitude of indoor temperature exceedance above a given thermal comfort threshold. Unlike binary indicators such as DH, which only register the duration of discomfort, IOD incorporates both the duration and the intensity of overheating. It is therefore classified as a time-integrated overheating metric and is particularly useful for performance-based building design, passive cooling assessment, and overheating resilience analysis.

The general formulation of IOD is expressed as:

$$IOD = \sum_{i=1}^{N} \max\left(0, T_{in,i} - T_{lim}\right) \cdot \Delta t$$

where:

IOD—indoor overheating degree-hours,

$T_{in,i}$—indoor air temperature at time step i [°C],

T_{lim}—threshold temperature used to define overheating [°C],

Δt—time step of the analysis, typically 1 hour [h]

Only hours for which $T_{in,i}$ exceeds T_{lim} contribute to the IOD value.

The max(0, …) function ensures that only values exceeding the threshold contribute to the sum, effectively ignoring periods of thermal neutrality or comfort. The resulting IOD value represents the accumulated degree-hours of overheating during the evaluation period.

The use of the *max(0, …)* operator ensures that only those hours in which the indoor temperature exceeds the selected threshold contribute to the cumulative total, while periods of thermal neutrality or comfort are excluded from the calculation.

The choice of the threshold temperature T_{lim} depends on the analytical context and the normative framework adopted. Fixed thresholds, typically 26°C or 28°C, are often applied in simplified assessments or in studies conducted in warm climate regions. In free-running buildings, adaptive thresholds derived from the running mean outdoor temperature are commonly used in accordance with EN 16798-1 or

CIBSE TM52/TM59. In resilience assessments or analyses of heat-wave conditions, health-based thresholds—such as 30°C or values derived from biometeorological heat-stress indices—may be more appropriate.

IOD may be calculated over various temporal scales, including hourly, daily, seasonal or annual periods. Daily values are useful for detecting acute overheating episodes, whereas cumulative seasonal or annual values provide insight into long-term exposure and building performance under current or future climatic conditions.

Hamdy et al. (2017) introduced IOD as a baseline indicator for evaluating overheating in free-running buildings and demonstrated its value in assessing thermal resilience. As Enescu (2017) notes, IOD offers several advantages over duration-only indicators such as Degree Hours (DH). It captures both the duration and severity of temperature exceedances, allows more refined differentiation between design alternatives that may yield similar DH values, and integrates effectively with outputs from dynamic building performance simulation tools.

However, the interpretability of IOD is strongly dependent on the selected threshold T_{lim}. An inappropriate threshold may misrepresent overheating severity, particularly when comparing different climates, occupancy patterns, or building types. For this reason, alignment with established standards such as EN 16798-1 or ASHRAE 55, or the use of occupancy-specific thresholds for sensitive spaces such as bedrooms, is recommended.

IOD does not account for physiological parameters such as relative humidity, acclimatization, or vulnerability within specific population groups. Consequently, it is increasingly complemented by biometeorological indicators such as HIHH or WBGT, especially in assessments focused on health outcomes. In comparative studies, IOD is sometimes normalized by floor area ($°C \cdot h \cdot m^{-2}$) to facilitate benchmarking across buildings of different sizes.

In summary, OID is a flexible and severity-sensitive indicator that complements duration-based metrics and supports detailed analysis of overheating dynamics. Its simplicity, adaptability, and compatibility with simulation-based workflows make it a core metric in passive design evaluation, regulatory assessments, and climate-resilience research, particularly in contexts where heat stress varies significantly over time or across building zones.

3.4.7 Hours of Exceedance

The HE is a time-based thermal indicator used to quantify the total number of hours during which the operative indoor temperature exceeds a defined comfort threshold. Unlike cumulative indices such as IOD, HE reflects only the duration of thermal discomfort, without accounting for the magnitude of the temperature exceedance. It is commonly used in building performance simulation, overheating assessments, and design compliance verification, particularly in relation to adaptive comfort models.

The general formulation of HE is given by:

$$HE = \sum_{i=1}^{n} H\left(T_{in,i} - T_{lim}\right)$$

where:

HE—Hours of Exceedance [h],

$T_{in,i}$—operative indoor temperature at time step *i* [°C],

T_{lim}—time-varying or fixed comfort threshold [°C],

$H(x)$—indicator function equal to 1 if the condition is true, and 0 otherwise,

n—number of time steps in the assessment period.

Operationally, HE corresponds to the number of discrete hours during which indoor temperatures rise above the selected threshold. This threshold may take the form of a fixed limit, such as 26°C or 28°C; a dynamic comfort temperature defined by adaptive models, including those in EN 16798-1 or ASHRAE Standard 55; or a zone-specific requirement, such as lower nighttime thresholds for bedrooms recommended in CIBSE TM59.

HE is typically computed from building performance simulation outputs—for example, from EnergyPlus or IES-VE—or from measured indoor temperatures in post-occupancy studies. Its simplicity and intuitive interpretation have made it a widely adopted indicator in both design practice and regulatory compliance.

The metric is embedded in several established comfort assessment methodologies. CIBSE TM52 specifies that indoor temperatures should remain within the adaptive comfort band for at least 97% of occupied hours. CIBSE TM59 extends this approach to residential buildings, imposing criteria for both daytime living spaces and nighttime bedroom conditions, such as limiting nighttime exceedance of 26°C. Sun et al. (2021) apply HE in health-oriented simulations, using thresholds such as 30°C or 32°C to examine overheating risks during extreme heat events or power outages.

Although HE does not quantify the severity of the exceedance—how far above the threshold temperatures rise—it provides a clear indication of exposure duration. It is therefore often used in combination with cumulative measures such as IOD or HIHH, which capture intensity in addition to duration.

As noted by Enescu (2017), HE is closely related to the DH metric. The principal distinction lies in the threshold definition: HE commonly employs dynamic or adaptive limits, whereas DH relies on fixed temperature values. This characteristic makes HE especially suitable for naturally ventilated or hybrid buildings in which comfort expectations vary with outdoor conditions.

HE is also sensitive to occupancy schedules and zone function. An exceedance of 28°C may be inconsequential in an unoccupied utility room but critical in a bedroom during nighttime hours. For this reason, HE is frequently reported separately for different zone types, time periods, or occupancy categories.

Time resolution is another important consideration. Most HE assessments are based on hourly data. When sub-hourly data are used—such as 10-minute measurements—the HE value must be scaled appropriately, for instance by dividing the number of exceedance intervals by six. This approach allows more precise identification of short but recurrent overheating episodes.

HE additionally serves as a useful comparative metric for evaluating performance under different design strategies or climate scenarios. For example, future-weather simulations based on SSP pathways may show substantial increases in HE for passive buildings under high-emission scenarios, highlighting the potential vulnerability of the building stock. Conversely, HE reductions following interventions such as external shading, improved insulation, or night ventilation quantify the effectiveness of passive strategies.

In summary, HE is a robust, duration-based indicator of thermal discomfort that complements cumulative indices such as IOD. Its adaptability to different comfort models, occupancy patterns, and climate scenarios, as well as its integration into standards such as TM52 and TM59, make it a central metric in performance-based design and overheating resilience assessment.

3.4.8 Wet-Bulb Globe Temperature

The WBGT is a composite thermal index that quantifies the combined effect of air temperature, humidity, solar radiation, and wind speed on human heat stress. Originally developed for occupational and military settings, WBGT has become a widely adopted metric for assessing outdoor and non-air-conditioned indoor environments in terms of thermal discomfort and health risk. It is referenced in several standards, including ISO 7243, ASHRAE, WHO guidelines, and U.S. OSHA regulations.

WBGT integrates three different measurements:

- T_{wb}—Natural wet bulb temperature (reflecting humidity and evaporative cooling),
- T_g—Globe temperature (representing radiant heat from the environment),
- T_{db}—Dry bulb (air) temperature.

Depending on whether the environment is exposed to direct solar radiation or not, WBGT is calculated using different empirical formulas:

For outdoor conditions with solar exposure:

$$WBGT = 0.7T_{wb} + 0.2T_g + 0.1T_{db}$$

For indoor or shaded conditions:

$$WBGT = 0.7T_{wb} + 0.3T_g$$

where:

T_{wb}—natural wet-bulb temperature [°C],
T_g—globe temperature [°C],
T_{db}—dry-bulb air temperature [°C].

These weighted averages reflect the relative physiological contributions of evaporative, radiant, and convective heat exchange to human thermoregulation. The dominance of the wet-bulb component (70%) arises from the central role of evaporative cooling in maintaining thermal balance under heat stress. However, as noted by Lemke and Kjellstrom (2012), the natural wet-bulb temperature (T_{nwb}) can be substantially higher than the psychrometric wet-bulb temperature (T_{pwb}) in conditions of strong solar radiation or low wind speed. In extreme cases, this difference may reach up to 10°C, which significantly elevates the resulting WBGT value and demonstrates that simplified models relying solely on Tpwb can seriously underestimate actual heat stress, particularly in environments with high radiative load or poor air movement. Such findings underscore the importance of using valid wet-bulb estimations when interpreting the physiological burden represented by WBGT.

Unlike indices such as PMV or SET, the WBGT does not explicitly incorporate clothing insulation or metabolic rate in its core formulation. These variables are addressed indirectly through exposure limits defined for different clothing ensembles and workload categories, such as light, moderate, or heavy activity. ISO 7243 specifies upper permissible WBGT values for safe exposure, above which protective measures—including work–rest cycles, hydration, shading, and targeted ventilation—must be implemented.

From a building-performance perspective, WBGT is particularly useful for assessing thermal risk in non-conditioned indoor environments such as warehouses and workshops, for evaluating outdoor comfort in courtyards, balconies, and urban microclimates, for supporting occupant-safety protocols during heat waves or blackout scenarios, and for informing mitigation strategies that involve shading, natural ventilation, or vegetation.

Although not originally developed for building simulation, WBGT can be estimated using standard simulation outputs when air temperature, relative humidity (for calculating wet-bulb temperature), radiant or mean radiant temperature (for approximating globe temperature), and wind speed are available. Several modeling platforms, including ENVI-met, OpenFOAM, and advanced building-performance simulation tools, allow users to approximate WBGT in localized indoor or outdoor settings, particularly when microclimate modeling is integrated.

In her comprehensive review, Enescu (2017) notes that WBGT is well suited for assessing acute heat exposure, especially in contexts where health and occupational safety are primary concerns. She highlights its practicality in hot–humid climates, where limited evaporative cooling substantially increases thermal stress. At the same time, several limitations are identified: WBGT is empirical and context dependent and may require calibration for specific populations or environmental conditions. It may underestimate thermal risk in high-mass indoor environments where radiant heat accumulates, and it does not explicitly account for thermal adaptation or subjective comfort, which limits its application in long-term comfort optimization.

Despite these constraints, WBGT remains a valuable supplementary indicator in heat-risk assessments. Although it does not directly evaluate the thermal resilience of buildings, it provides critical insight into occupant vulnerability—particularly in passive buildings, dense urban environments, and during periods of power loss—making it an indispensable tool in both occupational safety and climate-resilience research.

3.4.9 Summary of Thermal Comfort Indicators and Transition to Resilience Assessment

Thermal comfort indicators provide a structured foundation for evaluating IEQ under standard operating conditions. They encompass steady-state indices such as

PMV and PPD, adaptive or dynamic indices such as the SET, and time-based or cumulative measures including DH, HE, CDD, and IOD. In addition to these comfort-oriented indicators, the WBGT has gained relevance in indoor heat-risk assessments, especially in buildings without mechanical cooling.

Together, these metrics offer a versatile toolkit for characterizing thermal performance. PMV and PPD remain effective in mechanically conditioned buildings, where environmental conditions are relatively stable. Adaptive and exceedance-based measures such as HE, DH, and IOD provide stronger diagnostic insight in free-running or mixed-mode buildings, where indoor temperatures track outdoor variability. SET delivers a more physiologically meaningful representation of comfort under transient or asymmetric conditions, while WBGT bridges the gap between indoor comfort assessment and heat-stress evaluation.

While highly effective in conventional design contexts, these indicators have inherent limitations when applied to conditions involving extreme heat or mechanical system failures. Traditional comfort metrics were developed for occupied, operational buildings and assume a stable supply of cooling, ventilation, and electricity. In a warming climate, these assumptions are increasingly challenged. Longer and more intense heat waves, coupled with the rising likelihood of grid stress and power outages, demand tools that address occupant safety as well as comfort.

Under such conditions, the central question shifts from "Is the environment comfortable?" to "How long can the environment remain safe?" Conventional comfort indicators—whether steady-state, adaptive, or cumulative—do not fully capture the thermal dynamics that matter most during extreme or disruptive events. Evaluating passive survivability, indoor heat-retention characteristics, and the temporal progression of hazardous temperature thresholds requires a different class of metrics.

The following chapter introduces this second group of indicators. These include time-integrated overheating measures, threshold-timing metrics, composite exposure indices, and health-based safety thresholds that reflect both acute physiological strain and longer-term adaptive capacity. This progression from comfort assessment to resilience assessment reflects the broader shift toward performance frameworks capable of addressing thermal safety in a changing climate. It establishes the conceptual bridge between everyday comfort evaluation and the emerging imperative to design buildings that maintain habitable, safe indoor conditions during extreme weather or system disruption.

REFERENCES

Attia, S., Benzidane, C., Rahif, R., Amaripadath, D., Hamdy, M., Holzer, P., Koch, A., Maas, A., Moosberger, S., Petersen, S., Mavrogianni, A., Hidalgo-Betanzos, J. M., Almeida, M., Akander, J., Khosravi Bakhtiari, H., Kinnane, O., Kosonen, R., & Carlucci, S. (2023). Overheating calculation methods, criteria, and indicators in European regulation for residential buildings. *Energy and Buildings*, *292*, 113170. https://doi.org/10.1016/j.enbuild.2023.113170

Belcher, S. E., Hacker, J. N., & Powell, D. S. (2005). Constructing design weather data for future climates. *Building Services Engineering Research and Technology*, *26*(1), 49–61. https://doi.org/10.1191/0143624405bt112oa

Clarke, J. A. (2001). *Energy simulation in building design* (2nd ed.). Routledge.

de Wilde, P. (2018). *Building performance analysis*. John Wiley & Sons. https://doi.org/10.1002/9781119341901

Enescu, D. (2017). A review of thermal comfort models and indicators for indoor environments. *Renewable and Sustainable Energy Reviews*, *79*, 1353–1379. https://doi.org/10.1016/j.rser.2017.05.175

Fanger, P. O. (1970) *Thermal comfort*. Danish Technical Press.

Gagge, A., Fobelets, A., & Berglund, L. (1986). A standard predictive index of human response to the thermal environment. *ASHRAE Transactions*.

Hensen, J. L. M., & Lamberts, R. (Eds.). (2019). *Building performance simulation for design and operation* (2nd ed.). Routledge. https://doi.org/10.1201/9780429402296

Hong, T., Yan, D., D'Oca, S., & Chen, C. (2017). Ten questions concerning occupant behavior in buildings: The big picture. *Building and Environment*, *114*, 518–530. https://doi.org/10.1016/j.buildenv.2016.12.006

Lemke, B., & Kjellstrom, T. (2012). Calculating workplace WBGT from meteorological data: A tool for climate change assessment. *Industrial Health*, *50*(4), 267–278. https://doi.org/10.2486/indhealth.MS1352

Mirzaei, P. A., & Haghighat, F. (2010). Approaches to study Urban Heat Island: Abilities and limitations. *Building and Environment*, *45*(10), 2192–2201. https://doi.org/10.1016/j.buildenv.2010.04.001

Morales, R. D., Audenaert, A., & Verbeke, S. (2025). Thermal comfort and indoor overheating risks of urban building stock: A review of modelling methods and future climate challenges. *Building and Environment*, *269*, 112363. https://doi.org/10.1016/j.buildenv.2024.112363

4 Thermal Resilience Systems and Indicators

4.1 INTRODUCTION

While thermal comfort assessment focuses on maintaining acceptable indoor environmental conditions under normal operation, thermal resilience extends this scope to evaluate performance during abnormal or disruptive events. A thermally resilient building is one that can sustain safe and habitable conditions when conventional energy supplies or mechanical conditioning systems are unavailable, whether due to extreme weather, grid outages, or other emergencies.

This shift in emphasis reflects an important evolution in building performance objectives. Comfort metrics such as Predicted Mean Vote (PMV), Predicted Percentage of Dissatisfied (PPD), or adaptive models describe conditions in controlled environments, but they do not capture a building's capacity to protect occupants during prolonged heat waves, cold spells, or multi-day power failures. Thermal resilience metrics address this gap by combining passive design strategies, envelope performance, and, where appropriate, backup energy provisions to ensure life-supporting conditions are maintained without reliance on active systems.

The following sections present leading frameworks and methodologies for assessing thermal resilience. They include both standardized approaches, such as the LEED Passive Survivability credit, and research-based indicators designed to quantify a building's ability to buffer indoor environments against external extremes.

4.2 THERMAL RESILIENCE SYSTEMS

4.2.1 LEED Passive Survivability—Thermal Resilience Criteria

The LEED Passive Survivability credit (LEED BD+C v4.1, Credit: Passive Survivability and Backup Power During Disruptions) is one of the first standardized frameworks to formally incorporate thermal resilience into building certification. Its primary objective is to ensure that buildings can maintain thermally survivable indoor conditions and preserve operational functionality during disruptions to external energy supply, especially under extreme weather scenarios such as prolonged heat waves or cold spells.

The credit combines passive and active design strategies, requiring that occupied spaces remain within a specified thermal safety envelope in the event of power failure. This is operationalized through the use of Standard Effective Temperature (SET), where indoor conditions must be maintained within the 12.2°C–30°C range. A building is deemed non-compliant if cumulative exceedances of these thresholds surpass 120 degree-hours (°C·h) during a continuous seven-day outage. This criterion

DOI: 10.1201/9781003645573-4

reflects not only comfort boundaries but also minimum thermal safety requirements essential for occupant health.

In addition to passive performance, LEED requires provisions for backup energy systems capable of supporting critical loads. These include ventilation, limited space conditioning, lighting, water supply systems, medical equipment, and communication infrastructure. Acceptable solutions may involve conventional generators powered by diesel or natural gas, hybrid systems, or renewable sources such as photovoltaic panels coupled with battery storage. The standard emphasizes energy autonomy as a resilience strategy, particularly in settings vulnerable to infrastructure disruptions.

To verify compliance, project teams must conduct detailed dynamic simulations representing a seven-day power outage under extreme seasonal conditions. These simulations must evaluate the building's capacity to retain acceptable temperatures using only passive strategies. Model scenarios typically incorporate both summer heat waves and winter cold events and require an analysis of insulation effectiveness, thermal mass behavior, and envelope thermal transmittance. The influence of passive cooling methods such as natural cross-ventilation, night flushing, and solar control features (including shading elements and reflective surfaces) must also be addressed. In cold conditions, particular attention is given to envelope integrity and the ability to reduce thermal losses.

The LEED credit strongly promotes the use of passive measures to enhance thermal resilience. Effective insulation in walls and roofs is considered essential to reduce undesired heat exchange. Solar-reflective materials and cool roofs limit heat gain in warm climates, while energy-efficient glazing systems, including low-emissivity coatings and integrated shading, improve the building's overall thermal envelope. Natural ventilation strategies are encouraged where feasible, with emphasis on stack effect cooling and optimized airflow paths. High thermal mass elements, such as concrete or phase change materials, are also recommended for their capacity to stabilize internal temperatures by absorbing and releasing heat over time.

Beyond its technical requirements, LEED Passive Survivability reflects a broader shift in building standards toward climate-adaptive design. By encouraging buildings to maintain life-supporting indoor environments without mechanical conditioning, the standard addresses both acute risk reduction and long-term sustainability goals. As climate change increases the likelihood of extreme weather events and grid instability, passive survivability becomes not only a safeguard for occupant health but also a critical tool for reducing reliance on carbon-intensive systems. Through the integration of thermal autonomy, emergency preparedness, and low-energy strategies, this credit contributes to wider decarbonization efforts and enhances the future readiness of the built environment.

4.2.2 RELi Resilience Action List: A Comprehensive Approach to Thermal Resilience

The RELi Resilience Action List is a sustainability and resilience rating system originally developed by the Institute for Market Transformation to Sustainability and currently managed by the U.S. Green Building Council. Unlike conventional green certification schemes, RELi places equal emphasis on sustainable design principles and strategies for climate adaptation, disaster preparedness, and post-crisis recovery. Its objective is to promote buildings and communities that are not only environmentally responsible but also capable of withstanding and adapting to extreme weather events, infrastructure disruptions, and longer-term climate stressors.

Thermal resilience plays a central role in the RELi framework. The system encourages the design of buildings that can maintain internal safety and functionality during scenarios such as heat waves, cold spells, power outages, floods, and droughts. To meet RELi requirements, buildings must incorporate both passive and active design elements that contribute to energy independence and occupant protection under duress. Key strategies include the use of durable, climate-resistant materials and the integration of backup power systems, ranging from diesel or gas generators to photovoltaic arrays with battery storage and hybrid energy technologies. These systems are expected to support critical infrastructure functions, including ventilation, thermal conditioning, emergency lighting, and water supply.

In contrast to systems such as LEED Passive Survivability, RELi does not define a universal comfort threshold in terms of a fixed temperature range. Instead, it allows for climate-responsive adaptation, encouraging design targets based on local conditions, seasonal extremes, and the specific needs of building users. Thermal safety is interpreted broadly, encompassing both heat stress and cold exposure. The absence of rigid temperature limits enhances the flexibility of the system while reinforcing the importance of contextual design.

A critical component of RELi is its promotion of passive design strategies that enhance thermal survivability. These include the use of high-performance insulation to reduce unwanted heat exchange; reflective coatings on roofs and façades to limit solar absorption; energy-efficient glazing systems to mitigate internal heat gain or loss; and natural ventilation configurations that leverage cross-ventilation and the stack effect for passive cooling. The use of climate-adaptive materials, particularly those capable of responding dynamically to thermal gradients, is also encouraged. Through these elements, RELi enables buildings to maintain habitable conditions even in the absence of mechanical systems.

In addition to design strategies, RELi incorporates scenario-based risk assessment as a core requirement. This involves identifying environmental hazards relevant to the building site, modeling the impacts of climate change projections, and assessing the potential consequences for thermal performance. Designers are encouraged to account for long-term phenomena such as urban heat island effects and to plan for temperature and humidity shifts over the building's lifecycle. The system prioritizes proactive risk mitigation and the ability to recover from failures, rather than merely reacting to past patterns.

What distinguishes RELi from other certification systems is its explicit integration of social and community-based resilience. The framework requires that buildings contribute to broader crisis management strategies and provide safe environments not only for regular occupants, but also for at-risk groups during emergencies. Special attention is given to vulnerable populations, including seniors, individuals with disabilities, children, and residents of low-income households who may lack reliable access to heating or cooling resources. RELi also promotes the long-term planning of essential services, such as decentralized energy systems, resilient food storage, and potable water supply, thereby supporting autonomous operation in post-disaster conditions.

At its core, RELi is structured around a multi-criteria approach to resilience. It combines thermal comfort and energy autonomy with structural robustness, health and safety considerations, and adaptive capacity. Rather than relying on singular metrics, the system encourages integrated solutions that support both environmental performance and social well-being. In this way, RELi responds not only to the technical demands of climate adaptation but also to the ethical imperative of protecting people and communities under escalating climate risk.

As the built environment faces intensifying challenges from global warming, extreme weather, and aging infrastructure, RELi provides a flexible yet comprehensive pathway for designing buildings that are future-ready. By aligning passive thermal strategies with broader goals of sustainability, autonomy, and community protection, it offers a valuable tool for developers, architects, and policymakers committed to long-term resilience.

4.2.3 Annex 80: Resilient Cooling of Buildings

Annex 80, titled *Resilient Cooling of Buildings*, is a collaborative initiative under the International Energy Agency's Energy in Buildings and Communities (IEA EBC) Program. Developed in response to the increasing frequency and intensity of extreme weather events, Annex 80 addresses the critical need for buildings that can maintain thermal comfort while reducing energy consumption and greenhouse gas emissions. Its core mission is to establish effective strategies for resilient cooling, defined as the capacity to ensure indoor thermal comfort during heat waves, power outages, and climate-induced disruptions without relying solely on conventional mechanical systems.

The framework introduced in Annex 80 promotes a spectrum of passive and low-energy cooling approaches. These strategies aim not only to reduce indoor heat stress but also to alleviate peak loads on electricity grids, particularly during hot spells. Emphasis is placed on cooling solutions that remain functional under conditions of grid instability, which is especially relevant in the context of compound climate risks. Although the primary focus of the Annex lies in cooling performance, it also considers winter thermal resilience by including envelope insulation, air tightness, and solar management strategies that minimize heat loss. A core principle in the Annex is to balance insulation and reflectivity, recognizing that over-insulated envelopes may contribute to overheating in summer, while overly reflective surfaces may reduce beneficial solar gains in winter.

To quantify thermal resilience in operational terms, Annex 80 introduces three key performance indicators (KPIs). The first is the Thermal Resilience Index (TRI), which expresses the proportion of time a building remains within a defined thermal comfort zone during a disruption event. A TRI value approaching 1.0 indicates that the building maintained comfort throughout most of the disruption period, while lower values signify a higher share of time spent in thermally unsafe conditions.

The second indicator is Hours of Safety (HST), which calculates the number of hours indoor conditions remain within acceptable limits before reaching critical thermal thresholds. This metric accounts for different thresholds depending on occupant vulnerability and local climatic expectations. Typical safe ranges are defined as 17°C–27°C, although narrower bands may be used when assessing elderly, ill, or otherwise vulnerable occupants.

The third indicator, SET Degree-Hours, reflects cumulative thermal stress over time by incorporating multiple environmental variables, including air temperature, humidity, mean radiant temperature, and air velocity. Thresholds above 30°C are used to assess overheating risk, while values below 12.2°C capture exposure to cold-related stress.

All three indicators are intended to be evaluated through dynamic building simulations that replicate extreme scenarios, such as prolonged blackouts during summer or abrupt cold snaps in winter. Simulation outputs should be benchmarked against international comfort standards, including ASHRAE 55, EN 16798-1, and ISO 7730, to ensure methodological consistency and global applicability.

In terms of practical application, Annex 80 encourages the use of combined passive and hybrid cooling strategies. Key elements include solar-reflective materials on roofs and façades, which reduce external heat gains, and energy-efficient glazing systems that limit conductive and radiative transfer through fenestration. Thermal mass is also emphasized for its ability to delay indoor temperature peaks by storing excess heat, a feature particularly advantageous during daily load fluctuations. Hybrid systems that combine passive design with efficient mechanical solutions, such as photovoltaic-powered fans or evaporative cooling, are promoted as viable pathways to achieve both resilience and energy efficiency. Where off-grid operation is necessary, photovoltaic systems with battery storage provide autonomous cooling capacity that can extend survivability during power disruptions.

Annex 80 strongly advocates for user-centered thermal design, especially in contexts where building occupants may have limited access to active cooling technologies. Particular consideration is given to sensitive groups such as elderly individuals, children, persons with chronic health conditions, and low-income populations. The inclusion of such criteria underscores the social dimension of resilience and positions Annex 80 as more than a purely technical framework.

Ultimately, Annex 80 delivers a coherent model for integrating thermal resilience into building design and operation. It combines simulation-based assessment, performance indicators, passive strategies, and adaptation planning to ensure that buildings can withstand evolving climate risks. By aligning its metrics with established standards and promoting design solutions that serve both environmental and social goals, the initiative offers a replicable and scientifically grounded approach to future-proofing the built environment.

4.2.4 National Resilience Rating System: Thermal Resilience Assessment

The National Resilience Rating System (NRRS) is a U.S.-based evaluation framework designed to assess and enhance the capacity of buildings to withstand climate-related stressors, including extreme weather events and infrastructure disruptions. Developed as a proactive response to the escalating challenges of climate change, NRRS focuses on ensuring occupant safety, maintaining functional integrity during crises, and promoting long-term adaptation through the integration of passive and durable design strategies. It extends beyond traditional definitions of energy efficiency by placing resilience and recovery at the core of building performance.

A central element of NRRS is the assessment of thermal resilience. Within this framework, thermal resilience refers to a building's ability to maintain healthy indoor temperatures during disruptive events, such as heat waves or cold spells, particularly in the absence of fully functioning mechanical systems. This capacity is evaluated based on both passive thermal performance and the redundancy of active systems. The framework encourages the use of thermal insulation, reflective coatings, energy-efficient glazing, and natural ventilation to improve thermal autonomy and reduce dependence on active conditioning.

To quantify thermal performance under stress, NRRS employs several key indicators. The TRI measures the proportion of time that indoor conditions remain within a defined comfort range during an extreme event. A TRI value close to 1.0 indicates that the building sustained comfort for most of the disruption period, while lower values reveal prolonged exposure to thermal discomfort. The HST indicator complements TRI by calculating the number of hours during which indoor temperatures remain within safe thermal thresholds before reaching critical limits. These limits are typically defined within a 17°C–27°C range, although narrower bands, such as 18°C–25°C, are recommended for vulnerable populations, including elderly occupants or young children.

Another important metric used in NRRS is Overheating Hours (OH), which captures the cumulative number of hours during which indoor temperatures exceed comfort thresholds, particularly during heat waves. This indicator helps identify buildings with insufficient passive protection and supports interventions such as improved insulation, exterior shading, enhanced cross-ventilation, or better-performing fenestration. To assess the physiological impact of thermal exposure, NRRS incorporates SET Degree-Hours, a cumulative measure that accounts for air temperature, humidity, radiant conditions, and air movement. Thresholds above 30°C are used to indicate heat stress, while those below 12.2°C identify exposure to cold-related health risks.

Crucially, NRRS introduces a performance benchmark for cumulative thermal stress. A maximum of 120°C·h of SET exceedance over a continuous seven-day period is considered tolerable. Surpassing this value triggers the need for design improvements, such as enhanced insulation, improved ventilation systems, or the selective introduction of active cooling. Compliance with these requirements must be demonstrated through dynamic thermal simulations, incorporating inputs from local climate data, urban heat island models, and future weather projections.

To ensure operational continuity during power disruptions, NRRS mandates the inclusion of reliable backup energy systems. These may consist of diesel or gas generators, photovoltaic installations with battery storage, or other renewable hybrid solutions. Backup systems are expected to maintain essential functions, including thermal conditioning, ventilation, emergency lighting, and potable water supply. The level of energy resilience achieved through such systems directly affects the overall NRRS score, reinforcing the framework's focus on continuity during crisis events.

NRRS is aligned with multiple international standards, including ASHRAE Standard 55 for thermal comfort, ISO 7730 for indoor environmental quality, and

World Health Organization guidelines for heat wave preparedness. By integrating key thermal indicators such as TRI, HST, OH, and SET Degree-Hours with dynamic simulation and climate-responsive planning, NRRS offers a comprehensive method for climate-resilient building design.

Beyond technical metrics, NRRS contributes to broader climate adaptation by promoting the transformation of the built environment. It supports the creation of buildings that remain habitable during systemic failures, protect occupant health under extreme thermal conditions, and reduce dependency on vulnerable energy infrastructure. Through its emphasis on passive cooling, thermal buffering, and decentralized energy systems, NRRS helps to advance long-term sustainability and enhance the resilience of communities in a changing climate.

4.2.5 Multi-Criteria Thermal Resilience Certification System

The multi-criteria thermal resilience certification system proposed by Amaripadath et al. (2024) offers a comprehensive framework for evaluating a building's ability to maintain thermal comfort and safeguard occupant health during extreme heat events. Departing from single-metric overheating assessments that rely primarily on indoor temperature thresholds, this system introduces a broader, integrative methodology. It draws upon multiple performance indicators to generate a resilience classification that reflects both physical performance and occupant-centered outcomes. The approach is grounded in internationally recognized standards, including ISO 17772-1, CIBSE TM52, LEED, and RELi, ensuring compatibility with established practices in thermal comfort assessment.

The framework considers four main dimensions of thermal performance: overheating severity, thermal comfort frequency, physiological heat stress, and perceived hygrothermal discomfort. By integrating these aspects, the system captures both short-term exceedances and long-term exposure patterns across a wide range of climatic conditions. This makes it particularly useful for differentiating buildings not only by technical attributes but also by occupant vulnerability and regional thermal risk.

Assessment is based on four core KPIs, each evaluated on a normalized scale from 0 to 1. These are Indoor Overheating Degree (IOD), Hours of Exceedance (HE), Wet-Bulb Globe Temperature (WBGT), and the Heat Index (HI). IOD, as introduced by Hamdy et al. (2017), captures cumulative overheating intensity in °C·h and is especially relevant for free-running buildings or structures with substantial thermal inertia. HE, following the formulation of Tao et al. (2025), quantifies the duration during which operative temperatures exceed established comfort thresholds. This is particularly applicable to sleeping areas or buildings with continuous occupancy, where even small deviations can have significant health implications.

WBGT serves as a physiological index, combining air temperature, humidity, radiant heat, and air velocity to assess total body heat stress. Commonly used in occupational safety contexts, WBGT is adapted here for residential and public buildings to evaluate vulnerability during high-humidity heat waves. The HI, derived from public health research, complements WBGT by highlighting the perceived experience of thermal discomfort based on air temperature and relative humidity.

While WBGT quantifies physiological strain, HI captures subjective discomfort and is particularly relevant in urban areas where evaporative cooling is limited.

Each of the four indicators contributes one point to the building's thermal resilience score. A score of 1 indicates full compliance with the defined threshold throughout the evaluation period, while a score of 0 indicates failure. The cumulative score, ranging from 0 to 4, determines the building's resilience classification. A total of 4 points corresponds to the highest resilience rating (five stars), whereas a score of 0 indicates minimal thermal protection. This classification is designed to be both analytically robust and intuitively understandable.

To ensure flexibility, the framework allows for variable weighting of the indicators. Although equal weighting is the default, modifications are encouraged to reflect project-specific conditions or local climate risks. For instance, WBGT may carry greater weight in humid regions due to the increased risk of heat stress, while in arid environments, IOD and HI may offer a more accurate reflection of thermal discomfort. This flexibility enhances the system's relevance across geographic and socio-economic contexts.

Amaripadath et al. (2024) validated the certification system through case studies conducted on both new and existing buildings across a range of climate zones and urban-rural settings. These included passive houses and thermally inefficient structures with limited passive design features. The results revealed clear performance differentiation. High-mass buildings with inadequate ventilation performed poorly on IOD and HE, while lightweight buildings with night ventilation and solar shading achieved higher scores, especially in WBGT and HI. Moreover, tailoring KPI weights to local climatic conditions improved the consistency of outcomes, confirming the value of the system's adaptability.

By combining empirical thermal indicators with international benchmarks and simulation-based modeling, this multi-criteria framework provides a rigorous and context-sensitive tool for thermal resilience assessment. It supports both passive and active design strategies and offers a means to evaluate how well buildings can protect occupants as heat-related climate risks intensify. Its structured yet adaptable approach positions it for integration into broader sustainability certification systems and climate policy tools. In a future where extreme heat poses a growing threat to human health and well-being, this framework contributes a critical tool for thermal safety, equitable design, and climate-responsive innovation.

4.3 THERMAL RESILIENCE INDICATORS

4.3.1 INTRODUCTION

As energy infrastructure becomes increasingly vulnerable to climate-induced disturbances, the capacity of buildings to passively maintain safe indoor conditions during extreme heat or cold has emerged as a critical feature of climate-responsive design. In recent years, a growing body of research has aimed to quantify this capability through the development of thermal resilience indicators. These metrics extend beyond conventional energy or comfort benchmarks to evaluate how buildings perform under thermal stress, system failure, and prolonged exposure.

These indicators signal a shift in focus from optimization under normal operating conditions to performance under constraint and disruption. Unlike traditional metrics such as heating degree days or PMV, resilience-oriented indicators capture the duration, intensity, recovery trajectory, and health impact of deviations from safe or comfortable indoor temperatures. They are particularly relevant in scenarios involving power outages, heat waves, and cold spells, where mechanical systems may be limited or inactive and passive survivability becomes essential.

This chapter presents a structured framework for thermal resilience indicators, organized into distinct functional categories. The first group includes cumulative intensity metrics such as the IOD, which quantifies the total thermal exposure above defined comfort thresholds. Complementary indicators in this category include the Ambient Warmness Degree (AWD) and the Overheating Escalation Factor (OEF), which assess the magnitude of external thermal forcing and the building's relative sensitivity or amplification response.

The second group comprises regression-based and memory-informed metrics. These include the Accumulation Factor (γ_{IOD}) and the Solar Radiation Factor (β_{IOD}), which capture lagging effects of heat retention and the influence of solar gains, particularly in buildings with high thermal inertia. Also included in this category are intuitive time-resolved indicators such as Time-to-threshold (T_X) and Temperature-after-X-hours ($T(t_X)$), which are designed to monitor short-term thermal decay under blackout conditions.

A third category addresses physiological and health-related impacts. This includes metrics such as Heat Index Hazard Hours (HIHH) and HE, both of which align thermal exposure with established health risk thresholds. These indicators are particularly applicable in vulnerable residential contexts and have been employed in studies of nearly zero-energy buildings and thermally retrofitted dwellings.

Additional indicators support assessment under cold weather conditions. These include the Weighted Unmet Thermal Performance (WUMTP) and the Degree of Disruption (DoD), both of which integrate discomfort magnitude and exposure duration across different disruption scenarios. Operational tools such as the HST provide actionable information for planning and managing blackout survivability.

Together, these indicators enable a multidimensional approach to thermal resilience assessment. They are suitable for both simulation-based and empirical studies, compatible with building certification systems, and adaptable to design, policy, and operational planning. The sections that follow present formal definitions, mathematical formulations, and usage contexts for each metric. These serve as the analytical foundation for comparative assessments presented in Chapters 6 and 14.

4.3.2 Methodological Framework for Assessing Thermal Sensitivity: The IOD–AWD–α_{IOD} Model (Hamdy et al., 2017)

The methodology introduced by Hamdy et al. (2017) provides a compact yet powerful analytical framework for quantifying overheating risk in residential buildings under current and future climate conditions. The approach is structured around three core equations that capture, respectively, the indoor thermal response of the building, the climatic severity of outdoor warm periods, and the building's sensitivity to changes in external forcing. Together, these indicators allow engineers to evaluate not only the magnitude of overheating but also the degree to which a building amplifies or dampens external thermal stress. The following subsections introduce these equations and describe how they jointly form a coherent and practice-oriented thermal resilience assessment method.

4.3.2.1 IOD: Quantifying Indoor Thermal Stress

The IOD expresses the total magnitude and duration of overheating during occupied hours. Unlike simple exceedance-hour metrics, IOD integrates both intensity and persistence of temperature elevations above a comfort threshold, and does so independently for each thermal zone in the building. This formulation acknowledges the differences between day-use and night-use spaces and enables zone-weighted aggregation across the entire dwelling.

(Equation 1) Indoor Overheating Degree

$$IOD = \frac{\sum_{z=1}^{Z}\sum_{i=1}^{N_{occ}(z)}\left[\max\left(T_{fr,i,z} - T_{L,comf,i,z}, 0\right)t_{i,z}\right]}{\sum_{z=1}^{Z}\sum_{i=1}^{N_{occ}(z)} t_{i,z}}$$

where:

IOD—Indoor Overheating Degree [°C·h],

$T_{fr,i,z}$—free-running indoor operative temperature at time step i in zone z [°C],

$T_{L,comf,i,z,}0$—comfort temperature limit at time step i in zone z [°C],

$t_{i,z}$—duration of time step i in zone z [h],

$N_{occ(z)}$—number of occupied hours in zone z,

Z—total number of thermal zones in the dwelling.

Equation (1) yields a dimensionally consistent measure of accumulated overheating. IOD represents the accumulated overheating load experienced by occupants. Larger values signify both higher indoor temperatures and longer duration of exposure. Because the denominator normalizes by total occupied hours, IOD supports fair comparison among buildings with different schedules and zoning configurations. In simulation-based assessments, IOD is extracted from free-running temperature trajectories obtained through dynamic thermal modeling.

4.3.2.2 AWD: Quantifying External Forcing

Indoor overheating cannot be meaningfully interpreted without a corresponding measure of outdoor climatic severity. The AWD provides such a measure by aggregating the positive differences between outdoor air temperature and a fixed base temperature of 18°C. AWD therefore characterizes the magnitude and duration of warm outdoor conditions, serving as an indicator of the climate's thermal forcing on the building envelope.

(Equation 2) Ambient Warmness Degree

$$AWD_{18^\circ C} = \frac{\sum_{i=1}^{N} (T_{a,i} - T_b) t_i}{\sum_{i=1}^{N} t_i}$$

where:

$AWD_{18°C}$—Ambient Warmness Degree [°C·h],
$T_{a,i}$—outdoor dry-bulb temperature at time step i [°C],
T_b—base temperature (18°C),
t_i—duration of time step i [h],
N—number of time steps such that $T_{a,i} \geq 18°C$.

As a normalized degree-hour indicator, AWD provides a concise representation of the climatic warmness driving indoor overheating. Higher AWD values correspond to hotter summers, reduced diurnal cooling potential, and more persistent warm periods. This metric is crucial because it isolates the climatic driver of overheating, thereby enabling the subsequent sensitivity analysis.

4.3.2.3 OEF (α_{IOD}): Quantifying Thermal Sensitivity

The OEF (α_{IOD}) is the central innovation of the Hamdy methodology. It provides a normalized measure of how strongly a building amplifies or attenuates outdoor thermal stress. By dividing IOD by AWD, the metric expresses the building's indoor thermal response per unit of climatic forcing. This allows direct comparison across building types, vintages, orientations, shading configurations, and ventilation strategies.

(Equation 3) Overheating Escalation Factor

$$\alpha_{IOD} = \frac{IOD}{AWD_{18^\circ C}}$$

α_{IOD}—Overheating Escalation Factor [–],
IOD—Indoor Overheating Degree [°C·h],
$AWD_{18°C}$—Ambient Warmness Degree with base temperature 18°C [°C·h].

A value of $\alpha_{IOD} < 1$ indicates that the building moderates the external thermal forcing (thermal damping). Values near or above unity reveal that the indoor environment responds almost proportionally—or even more strongly—than the outdoor climate, indicating elevated sensitivity. In the work of Hamdy et al. (2017), 97% of analysed dwellings achieved $\alpha_{IOD} < 1$, although certain designs approached unity under warmer climate scenarios.

4.3.2.4 How the Method Is Applied in Practice

In practical engineering workflows, the three indicators are computed sequentially. Dynamic thermal simulations produce hourly operative temperatures for each zone, from which IOD is derived using Equation (1). Weather files representing current or future climates are processed to obtain AWD using Equation (2). Finally, Equation (3) relates indoor overheating to external forcing, providing the building's sensitivity factor α_{IOD}.

This process offers a clear diagnostic pathway. High IOD with low AWD suggests design-related vulnerabilities (e.g., insufficient shading, inadequate ventilation). Conversely, high AWD combined with modest IOD indicates a robust envelope capable of damping climatic load. The α_{IOD} metric makes these relationships explicit and quantifiable.

4.3.2.5 Key Findings from the Simulation Study

The large-scale simulation study conducted by Hamdy et al. (2017) demonstrates the diagnostic power of the IOD–AWD–α_{IOD} framework. The authors evaluated 36,864 combinations of dwelling archetypes, construction periods, ventilation regimes, shading strategies, internal gains, and climate scenarios. Their results show that natural ventilation and external shading are the dominant moderators of overheating. Dwellings with limited air-exchange capacity or unshaded glazing consistently reached the highest IOD values and displayed escalation factors approaching unity.

Interestingly, newer highly insulated dwellings (2005–2012) were often more sensitive to climate warming than older, leakier buildings. In the absence of effective solar control and ventilation, insulation contributes to heat retention rather than protection. By contrast, dwellings equipped with external shading, cross-ventilation potential, and balanced glazing achieved significantly lower IOD and α_{IOD} values. Under the "Worst Future" climate scenario, even the most resilient variants exhibited upward shifts in indoor temperatures, underscoring the need for integrated passive cooling strategies in future building codes.

4.3.2.6 Advantages and Limitations of the Method

The major strength of the Hamdy et al. (2017) methodology lies in its parsimony: three compact equations capture the essential interplay between building physics and climatic forcing. The method is transparent, scalable, and directly compatible with dynamic simulation environments. It permits cross-building benchmarking and supports decision-making at both design and policy levels.

The approach is not without limitations. Because it focuses exclusively on temperature-based metrics, it does not incorporate humidity or physiological stress indicators. It assumes passive free-running operation, omitting behavioral adaptation and blackout scenarios. Adaptive comfort limits are simplified and may differ from international standards. Nevertheless, for comparative overheating assessments under climate change, the IOD–AWD–α_{IOD} framework remains one of the most effective, interpretable, and computationally efficient tools currently available.

4.3.3 Regression-Based Indicators and Solar–Thermal Dynamics (Flores-Larsen et al., 2023)

The regression-based framework developed by Flores-Larsen et al. (2023) extends the original IOD–AWD–α_{IOD} methodology of Hamdy et al. (2017) by incorporating daily-scale processes that are critical in passive and free-running buildings exposed to heat waves. While Hamdy's model captures the relationship between indoor overheating and outdoor warmness, it assumes that daily IOD depends linearly on daily AWD alone. Flores-Larsen et al. demonstrated, through combined simulations and field measurements, that this assumption is insufficient in buildings with high thermal inertia, significant solar exposure, or prolonged heat retention. Their contribution introduces a regression-based model that accounts for solar radiation and thermal memory effects, enabling more robust predictions of overheating dynamics under extreme heat events.

4.3.3.1 Indoor Overheating Degree

The authors retained the zonal formulation of the IOD, calculated on an hourly basis across all occupied spaces. However, to align the overheating metric with physiological stress, they replaced comfort limits based on adaptive or PMV models with a health-based threshold of 26.7°C. This value corresponds to the "Caution" category of the U.S. National Weather Service Heat Index and reflects the dry and hot climate of Victorica, where relative humidity is low and dry-bulb temperature closely approximates perceived thermal stress. Under these boundary conditions, IOD provides a sensitive measure of indoor exposure during heat waves.

(Equation 1) Indoor Overheating Degree

$$IOD = \frac{\sum_{z=1}^{Z}\sum_{i=1}^{N_{occ}(z)}\left[\left(T_{fr,i,z} - T_{L,comf,i,z}\right)t_{i,z}\right]}{\sum_{z=1}^{Z}\sum_{i=1}^{N_{occ}(z)} t_{i,z}}$$

where:

IOD—the Indoor Overheating Degree in degree-hours [°C·h],
$T_{\mathrm{fr},i,z}$—the free-running operative temperature in zone z at time step i [°C],
$T_{\mathrm{L,comf},i,z}$—the comfort limit temperature in zone z at time step i [°C],
$t_{i,z}$—the duration of the time step [h],
$N_{\mathrm{occ}}(z)$—the number of occupied hours in zone z [h],
Z—the total number of thermal zones [–].

Equation (1) preserves the scalar structure of the original framework while adapting the comfort threshold to a heat-stress-oriented limit. The indicator quantifies the cumulative magnitude of overheating during occupied hours, supporting fair comparisons between buildings with different schedules and zoning patterns. As in the original formulation, only positive exceedances above the comfort limit contribute to the calculation.

4.3.3.2 Ambient Warmness Degree

The severity of external thermal stress was quantified using the AWD, calculated during occupied hours only. Flores-Larsen et al. confirmed that the conventional use of a low base temperature (18°C) may significantly distort AWD in hot, arid climates by inflating the magnitude of warmness. Because this threshold lies far below meaningful comfort conditions, $AWD_{18°C}$ becomes dominated by routine summer weather rather than extreme heat events. To resolve this inconsistency, the authors set the base temperature equal to the comfort threshold used for IOD, adopting $T_b = T_{comf} = 26.7°C$.

(Equation 2) Ambient Warmness Degree

$$AWD_{26.7°C} = \frac{\sum_{i=1}^{N} (T_{a,i} - T_b) t_i}{\sum_{i=1}^{N} t_i}$$

where:

AWD—Ambient Warmness Degree over occupied hours [°C·h],
$T_{a,i}$—outdoor air temperature at time step *i* [°C],
T_b—base temperature (26.7°C in this study) [°C],
t_i—time-step duration [h],
N—number of time steps with $T_{a,i} > T_b$ [h].

This adjustment ensures conceptual consistency between indoor and outdoor indicators: both IOD and AWD are referenced to the same comfort-based temperature. AWD then serves as a meaningful representation of the climatic forcing acting on the building during heat events.

4.3.3.3 OEF (α_{IOD})

The OEF α_{IOD} expresses the sensitivity of indoor overheating to outdoor thermal forcing. Following the original definition by Hamdy et al. (2017), α_{IOD} is computed from the slope relating daily IOD and daily AWD. This interpretation remains valid when both metrics share a consistent temperature limit.

(Equation 3) Overheating Escalation Factor

$$\alpha_{IOD} = \frac{IOD}{AWD_{26.7°C}}$$

where:

α_{IOD}—the Overheating Escalation Factor [–],
IOD—the daily Indoor Overheating Degree [°C·h],
AWD—the Ambient Warmness Degree referenced to the same comfort temperature [°C·h].

Values of α_{IOD} below unity indicate that the building attenuates outdoor thermal stress, while values above unity reflect amplification. Flores-Larsen et al. found that the studied bioclimatic building consistently exhibited $\alpha_{IOD} < 1$, confirming that passive measures—such as thermal mass, shading, ventilation pathways, and reflective surfaces—effectively moderated the outdoor heat load. However, the correlation between daily IOD and AWD was weak in free-running conditions ($R^2 = 0.30$), suggesting that additional variables contribute to overheating dynamics beyond dry-bulb temperature alone.

4.3.3.4 Extended Regression Model: Solar and Memory Effects

To address the limitations of single-variable regression, Flores-Larsen et al. introduced a multivariable model that captures the influence of daily solar radiation and the persistence of thermal conditions from one day to the next. The extended model reflects the fact that massive envelopes accumulate heat during peak events and release it gradually, creating day-to-day thermal memory that is not represented in AWD.

(Equation 4) Multiple Regression Model for Daily Overheating

$$IOD_i = a_0 + \alpha_{IOD}\, AWD_i + \beta_{IOD}\, H_i + \gamma_{IOD}\, IOD_{i-1}$$

where:

IOD_i—overheating degree on day i [°C·h],
AWD_i—ambient warmness degree on day i [°C·h],
H_i—daily global horizontal irradiance [W·m^{-2}·day^{-1}],
IOD_{i-1}—previous day's overheating degree [°C·h],
a_0—regression intercept [°C·h],
α_{IOD}—the sensitivity coefficient [–],
β_{IOD}—Solar Radiation Factor [°C·h per (W·m^{-2}·day^{-1})],
γ_{IOD}—Accumulation Factor describing thermal persistence [–].

Regression analysis revealed that $\gamma_{IOD} = 0.763$ was highly significant, indicating that the previous day's conditions exerted a dominant influence on present indoor temperatures. In contrast, β_{IOD} was statistically insignificant ($p > 0.8$), which the authors attributed to the effective external shading and reflective finishes of the building. The final model displayed high predictive accuracy ($R^2 = 0.984$), outperforming single-variable regressions by an order of magnitude.

4.3.3.5 Classification and Resilience Diagnostics

The inclusion of γ_{IOD} and β_{IOD} provides a richer diagnostic basis for thermal resilience assessment. The AF γ_{IOD} differentiates buildings with rapid thermal recovery from those with prolonged overheating following heat events—an essential insight when assessing passive survivability under blackout or free-running conditions. While β_{IOD} was negligible for the Victorica building, it may be critical in designs with extensive solar exposure or insufficient shading. Together, these parameters allow engineers to distinguish between climate-driven, solar-driven, and history-driven overheating behaviors.

4.3.3.6 Relevance, Applications, and Future Integration

The regression-enhanced IOD model broadens the applicability of the IOD–AWD–α_{IOD} framework for climates dominated by solar irradiance and high thermal inertia. Because it is fully compatible with widely used simulation tools such as EnergyPlus and OpenStudio, it can be integrated smoothly into resilience assessments based on current and future weather files, including SSP-conditioned climate scenarios.

When combined with health-oriented indicators (e.g., HI), exceedance-based metrics (HE), or WUMTP, the extended approach provides an enriched view of both thermal comfort and thermally induced health risk. Future research may expand this direction by exploring nonlinear accumulation behaviors, occupant-driven adaptive responses, or real-time monitoring to support operational resilience strategies.

4.3.4 Health- and Discomfort-Based Time-Integrated Indicators (Sun et al., 2021)

Sun et al. (2021) developed a set of four time-integrated thermal resilience indicators intended to quantify indoor heat exposure during extreme heat waves, including periods with and without grid power. Their methodology focuses on vulnerable residential buildings in disadvantaged communities and aims to evaluate how passive cooling measures modulate indoor heat stress under exceptionally challenging conditions. In contrast to traditional thermal comfort metrics, these four indicators incorporate both discomfort and health-relevant thresholds, making them suitable for assessing passive survivability during blackouts and prolonged heat waves. The indicators—Unmet Degree-Hours (UDH), HIHH, Standard Effective Temperature Unmet Degree-Hours (SETUDH), and Predicted Mean Vote Exceedance Hours (PMVEH)—capture complementary aspects of indoor thermal burden and are time-integrated to represent the cumulative physiological impact on occupants.

4.3.4.1 Unmet Degree-Hours

The first metric, UDH, quantifies the cumulative overheating based solely on indoor air temperature. It integrates the temperature exceedance above a selected limit—typically the cooling setpoint—over time, thus accounting for both intensity and duration of overheating. UDH is particularly useful for evaluating passive survivability in grid-off conditions, where indoor temperatures can rise far above typical comfort limits.

(Equation 1) Unmet Degree-Hours

$$UDH = \int_{t_1}^{t_2} \left[T(t) - T_{\text{limit}}\right]^{+} dt$$

where:

$T(t)$—indoor air temperature at time t [°C],

T_{limit}—temperature threshold [°C],

$[t]^{+}$—part of t (equal to t when $t > 0$, otherwise 0),

t—time [h].

By weighting exceedances in °C·h, UDH provides a more comprehensive representation of overheating than simple unmet-hours counts. In the simulations of Sun et al., UDH allows clear differentiation between passive measures that reduce temperature peaks and those that merely delay overheating. It also reflects scenarios in which active cooling systems are undersized or unavailable.

4.3.4.1.1 Heat Index Hazard Hours

While UDH focuses on dry-bulb temperature, HIHH incorporate both air temperature and humidity, offering a more physiologically meaningful indicator of heat stress. The HI, used operationally by NOAA, reflects perceived temperature by accounting for reduced evaporative cooling at high humidity levels. HIHH accumulates the number of hours during which the HI exceeds defined risk thresholds, which correspond to "Caution" and "Danger" levels for human health.

(Equation 2) Heat Index Hazard Hours

$$HIHH = \int_{t_1}^{t_2} f\left(HI(t) - HI_{\text{threshold}}\right) dt$$

where:

$HI(t)$—Heat Index at time t [°C],
$HI_{\text{threshold}}$—selected hazard threshold (typically 27°C or 39°C) [°C],
$f(x)$ equals 1 when $x > 0$ and 0 otherwise,
t—time [h].

HIHH explicitly captures latent-heat contributions and is particularly relevant in humid climates or in poorly ventilated interiors where moisture accumulation exacerbates heat stress. Sun et al. showed that HIHH is a sensitive indicator under grid-off conditions, providing insight into physiological risk levels that exceed what dry-bulb temperature alone would suggest.

4.3.4.2 Standard Effective Temperature Unmet Degree-Hours

The SET integrates air temperature, humidity, air velocity, mean radiant temperature, metabolic rate, and clothing insulation into a single composite measure of thermal stress. SET is widely used in resilience studies and is endorsed in the LEED Passive Survivability credit to define "livable temperatures." SETUDH extends this concept by integrating only those periods in which SET exceeds comfort or health-based limits.

(Equation 3) SET Unmet Degree-Hours

$$SETUDH = \int_{t_1}^{t_2} \left[SET(t) - SET_{\text{threshold}}\right]^{+} dt$$

where:

$SET(t)$—Standard Effective Temperature at time t [°C],
$SET_{\text{threshold}}$—upper acceptability or survivability threshold [°C],
t—time [h].

Sun et al. (2021) adopted $SET_{\text{threshold}}$ = 28°C for grid-on scenarios and 30°C for grid-off scenarios to reflect the "livable temperature" criterion used by the U.S.

Green Building Council. Because SET accounts for both radiant load and humidity, SETUDH is especially suitable for evaluating passive survivability in buildings with large solar gains or elevated radiant temperatures.

4.3.4.2.1 Predicted Mean Vote Exceedance Hours

PMVEH provide a time-integrated discomfort measure derived from the PMV model standardized in ASHRAE 55 and ISO 7730. PMVEH accumulates the time during which PMV exceeds comfort or safety thresholds, capturing both environmental and physiological contributors to heat stress. Because PMV reflects the combined effects of clothing, metabolism, radiant load, air speed, and humidity, PMVEH is sensitive to building design features and occupant characteristics.

(Equation 4) PMV Exceedance Hours

$$PMVEH = \int_{t_1}^{t_2} f\left(PMV(t) - PMV_{\text{threshold}}\right) dt$$

where:

$PMVEH$—cumulative PMV exceedance duration [h],
$PMV(t)$—Predicted Mean Vote at time t [–],
$PMV_{\text{threshold}}$—selected comfort or survivability limit [–],
$f(x)$ equals 1 when $x > 0$ and 0 otherwise,
t—time [h].

Thresholds adopted by Sun et al. include PMV = +0.7 for moderate discomfort in grid-on conditions and PMV = +3 for potentially life-threatening heat exposure in grid-off conditions. Because PMV is highly sensitive to clothing insulation and metabolic rate, PMVEH is particularly informative for buildings such as nursing homes, where occupants have limited capacity for adaptive behavior.

4.3.4.3 Synthesis and Application

The four indicators proposed by Sun et al. (2021) offer a multi-dimensional assessment of indoor heat resilience that bridges the gap between conventional comfort metrics and health-oriented risk modeling. UDH captures dry-bulb thermal load; HIHH identifies humidity-driven hazard escalation; SETUDH represents composite heat stress with strong radiant and convective relevance; and PMVEH translates environmental exposure into a physiologically interpretable scale. Used together, these metrics support a robust, time-resolved evaluation of passive thermal performance under both grid-on and grid-off scenarios.

The framework has influenced subsequent resilience studies, including Tao et al. (2025), who adopted HIHH as a core metric for heat-health classification systems. The indicators are especially suited for passive survivability evaluations in disadvantaged communities, where air-conditioning access may be limited and envelope performance is often inadequate. Although each indicator has limitations—whether in the treatment of humidity, radiant load, or human behavior—their combined application provides a comprehensive basis for assessing building-level and district-scale heat resilience.

4.3.5 Scenario-Based Health Risk Scoring and Application (Tao et al., 2025)

The framework proposed by Tao et al. (2025) builds directly upon the time-integrated heat-exposure metrics developed by Sun et al. (2021), extending them into a scenario-based scoring system suitable for evaluating indoor thermal safety under extreme heat waves and power outages. Their method integrates four established indicators—HIHH, HE, IOD, and WBGT—into a unified thermal-resilience index applicable across projected future climate scenarios. Rather than introducing new physiological or thermal metrics, the authors operationalize existing ones within a structured rating framework that permits cross-scenario and cross-building comparison.

4.3.5.1 Heat Index Hazard Hours

Originally formulated by Sun et al. (2021), the HIHH indicator captures cumulative exposure to elevated humidity-dependent heat stress. Tao et al. adopt the same NOAA-based thresholds—27°C for "caution" and 39°C for "danger"—and apply HIHH in parametric simulations of a near-zero-energy dwelling exposed to combined heatwave and blackout scenarios. According to the results, HIHH proved highly sensitive to humidity-driven risk escalation and particularly effective for identifying elevated hazard conditions for older adults, whose impaired thermoregulation makes them more vulnerable during periods of prolonged high humidity and limited air movement.

4.3.5.2 Hours of Exceedance

To complement HIHH, Tao et al. employ the HE metric, which measures the cumulative time during which indoor operative temperature surpasses static or adaptive comfort thresholds. This definition is consistent with ISO 7730 and CIBSE TM52 criteria. In contrast to HIHH, which highlights latent-heat-driven physiological risk, HE represents dry-bulb discomfort and is widely used in comfort certification and residential design standards. Tao et al. demonstrate that combining HE with HIHH provides a more complete representation of indoor heat exposure, capturing both sensible and latent components of thermal risk.

4.3.5.3 Composite Thermal Resilience Score

The principal contribution of Tao et al. (2025) lies in their development of a multi-indicator scoring system that consolidates IOD, HE, WBGT, and HIHH into an intuitive resilience scale. In their formulation, each indicator contributes one point if its value remains within predefined acceptable thresholds during a heatwave–blackout event, yielding a cumulative index ranging from 0 to 4. This index is then mapped to a five-star resilience classification. A score of 4 corresponds to a fully resilient dwelling, while lower scores reflect increasing degrees of thermal vulnerability.

Application of this framework to a lightweight, highly insulated near-zero-energy dwelling in Davis, California demonstrated a marked deterioration in thermal resilience under future climate conditions. Under the SSP5–8.5 high-emission scenario for 2080, the HIHH accumulated to more than 230°C·h and HE exceeded 140 h, resulting in resilience scores near the minimum, despite the presence of night ventilation and passive shading devices. The analysis shows that conventional high-efficiency design strategies may be insufficient to ensure safe indoor environments during compound heatwave-and-power-failure events.

4.3.5.4 Relevance and Implementation

The scenario-based scoring system developed by Tao et al. provides a rigorous and highly practical framework for evaluating the future readiness of residential buildings subjected to extreme heat and infrastructure disruptions. By integrating multiple health- and discomfort-oriented metrics within a unified classification scheme, the method enables benchmarking across different building types, climatic contexts, and emissions trajectories. It also offers a transparent basis for validating passive cooling strategies in dynamic simulation environments such as EnergyPlus, TRNSYS, or OpenStudio.

Although Tao et al. do not introduce new mathematical indicators, their contribution lies in translating physiological and comfort-based exposure data into an actionable rating system that bridges thermal simulation, climate adaptation planning, and public-health-oriented design guidance. This approach is particularly relevant for regions where combined heatwaves and power outages are anticipated to intensify, and where vulnerable populations—including the elderly—may face disproportionate risks.

4.3.6 Winter Discomfort and Cold-Climate Indicators (Homaei & Hamdy, 2021)

While most thermal resilience metrics focus on overheating during summer heat waves, buildings in cold climates face significant risks during winter power outages. In such scenarios, indoor temperatures may fall rapidly below safe habitability levels, particularly for elderly or mobility-limited occupants. To address these risks, Homaei and Hamdy (2021) developed a structured winter-resilience methodology combining a simulation-based test framework with a phase-weighted thermal performance indicator known as the WUMTP. This approach captures not only the magnitude of temperature decline during a disruption, but also the timing, duration, severity and recovery behavior of the building once heating is restored.

4.3.6.1 The Multiphase Resilience Curve

The basis of the method is the multiphase resilience curve—a dynamic trajectory of indoor operative temperature across a predefined winter disruption scenario. The curve comprises four stages: an initial state under normal heating setpoints, Phase I representing the disruptive period (typically a heating-system failure), Phase II capturing the passive thermal recovery once heating resumes, and a final state reflecting re-stabilization. Three temperature thresholds delineate performance zones: a setpoint temperature T_{SP}, a robustness threshold T_{RT}, and a habitability threshold T_{HT}, below which indoor conditions are considered unsafe. Together, these thresholds separate acceptable, habitable, and uninhabitable indoor environments.

4.3.6.2 The WUMTP Metric: Definition and Structure

The WUMTP indicator quantifies thermal resilience by integrating the temperature deficit below comfort thresholds, weighted by phase (disruption or recovery), hazard category (acceptable, habitable, uninhabitable) and exposure time. The full winter-resilience test divides the event timeline into 12 segments—6 belonging to the disruption period (Phase I) and 6 to the recovery period (Phase II). Each segment is assigned a triplet of penalties reflecting the psychological, physiological, and temporal burden of cooling-related discomfort.

(Equation 1) WUMTP for a single thermal zone

$$WUMTP = \sum_{i=1}^{12} S_i \, W_{P,i} \, W_{H,i} \, W_{E,i}$$

$S_{(i)}$—temperature–time area of segment i [°C·h],
$W_{P,i}$—phase penalty for segment i [–],
$W_{H,i}$—hazard penalty for segment i [–],
$W_{E,i}$—exposure-time penalty for segment i [–].

The combined weighting system enables WUMTP to distinguish between two buildings with identical minimum temperatures but differing temporal profiles. A dwelling experiencing brief shallow dips receives a much lower WUMTP value than one that spends prolonged periods below habitability thresholds.

4.3.6.2.1 Penalty Structure and Segment Weighting

The WUMTP metric aggregates the deviation of indoor operative temperature below setpoints, weighted by phase, severity, and exposure time. The framework divides the performance timeline into 12 segments (6 in Phase I and 6 in Phase II), each assigned a phase penalty, a hazard penalty, and an exposure time penalty. These penalty values reflect the psychological and physiological burden associated with the timing, magnitude, and duration of temperature drops. The adopted values follow Table 4.1, which distinguishes clearly between acceptable, habitable, and uninhabitable thermal conditions.

In this structure, the hazard penalty is defined as 0.1 for acceptable conditions, 0.2 for habitable conditions and 0.7 for uninhabitable conditions. The phase penalty is set to 0.6 for Phase I and 0.4 for Phase II, while the exposure time penalty ranges from 2 to 40 depending on the segment. These penalty definitions enable WUMTP to differentiate between short, moderate deviations and prolonged, hazardous thermal exposures.

4.3.6.2.2 Multizone Aggregation

For multizone dwellings, WUMTP values are aggregated using zone floor areas as weights. This ensures that large rooms with prolonged cold conditions contribute proportionally more to the overall thermal resilience evaluation.

TABLE 4.1
Penalty Structure for the 12-Segment Resilience Test Framework (Homaei and Hamdy, 2021); Creative Commons CC-BY license

Segment	Penalties		
	Phase penalty (W_P)	Hazard penalty (W_H)	Exposure time penalty (W_E)
S1	0.6	0.1	2
S2	0.6	0.1	8
S3	0.6	0.2	10
S4	0.6	0.2	20
S5	0.6	0.7	20
S6	0.6	0.7	40
S7	0.4	0.7	40
S8	0.4	0.7	20
S9	0.4	0.2	20
S10	0.4	0.2	10
S11	0.4	0.1	8
S12	0.4	0.1	2

(Equation 2) Overall WUMTP

$$WUMTP_{\text{overall}} = \frac{\sum_{z=1}^{Z} WUMTP_z}{\sum_{z=1}^{Z} A_z}$$

where

$WUMTP_z$—weighted unmet thermal performance for zone z [°C·h],
A_z—floor area of zone z [m²],
Z—number of thermal zones [–].

4.3.6.3 Simulation Test Framework

To operationalize the winter-resilience methodology, Homaei and Hamdy defined a "four-day test" simulating a typical cold-season blackout. It includes a one-day initial period under normal heating, four days without heating (Phase I), four days of passive thermal recovery (Phase II), and a final stabilization period. Thresholds for each zone—such as living rooms, bedrooms and bathrooms—are assigned to reflect Nordic comfort practices and WHO habitability guidelines (e.g., 18°C robustness threshold, 15°C habitability threshold). The resilience curve obtained from simulation captures the building's cooling trajectory as well as the speed and extent of its thermal recovery.

4.3.6.4 Resilience Labeling and the RCI Index

To support decision-making and scenario comparison, the authors introduced the Resilience Class Index (RCI), conceptually analogous to energy labeling. The RCI expresses how well a design resists cold-climate disruptions relative to a reference building.

(Equation 3) Resilience Class Index

$$RCI = \frac{WUMTP_{ref}}{WUMTP_{overall}}$$

where:

RCI—resilience class index [–],

$WUMTP_{ref}$—reference weighted unmet thermal performance [°C·h/m²],

$WUMTP_{overall}$—weighted unmet thermal performance of the evaluated design [°C·h/m²].

Higher RCI values indicate greater resilience (lower cold-stress exposure). Resilience classes range from F (least resilient) to A+, offering a straightforward means to compare envelope upgrades, battery storage systems, photovoltaic solutions, or whole-house retrofits.

4.3.6.5 Application and Implications

Applying the winter-resilience framework to a calibrated two-storey house in Oslo showed that upgrading from a TEK17-standard dwelling to a passive-house specification reduced WUMTP by 71%, substantially improving resilience during prolonged outages. Adding photovoltaic or battery systems further enhanced performance, though their relative value was highest in less robust designs. The method revealed that envelope improvements alone cannot guarantee winter safety unless accompanied by strategies for limiting exposure duration and supporting post-outage thermal recovery.

Compared with simpler metrics such as passive survivability or thermal autonomy, WUMTP provides a nuanced, time-weighted evaluation of winter discomfort. Its multiphase structure captures not only the coldest temperature reached but also the timing and persistence of hazardous exposure. This makes the framework particularly valuable in cold-climate regions where health risks from insufficient indoor temperatures escalate rapidly during outages. With local calibration, the approach can be applied broadly across climate zones and occupancy profiles.

4.3.7 Thermal Safety Metric for Blackouts (Wijesuriya et al., 2024)

Wijesuriya et al. (2024) introduced the HST as a thermal resilience indicator expressly intended for residential buildings subjected to extreme heat waves or winter storms under blackout conditions. In contrast to comfort-based metrics such as IOD or HE, HST captures the duration for which indoor conditions remain within medically acceptable limits after the loss of mechanical systems. The metric is grounded in evidence-based safety thresholds derived from WHO guidelines and national heat-stress classifications, allowing direct interpretation in terms of occupant survival and safe shelter-in-place duration.

4.3.7.1 Definition and Threshold Criteria

HST is defined as the period between the onset of a power outage and the moment when indoor air temperature or the indoor HI first exceeds a specified safety threshold. In winter blackout simulations the threshold is set at 15°C, corresponding to the boundary between "no thermal stress" and "mild cold stress" in public-health guidance. During heatwave scenarios, the threshold follows the NOAA HI scale, with 33°C marking the transition from "caution" to "extreme caution." These thresholds provide a physiologically meaningful basis for assessing how long a dwelling can protect its occupants during an abrupt failure of heating or cooling systems.

(Equation 1) Hours of Safety

$$HST = t_{\text{threshold}} - t_0$$

where

HST—hours of safety after the onset of the outage [h],

t_0—time at which the power outage begins [h],

$t_{\text{threshold}}$—time at which indoor temperature or HI first exceeds the safety threshold [h].

The HST formulation yields a clear and actionable indicator: higher values signify greater capacity to maintain safe indoor conditions without mechanical heating or cooling, while lower values indicate rapid decline toward unsafe zones. Because it is based on a single exceedance event rather than accumulation, the metric provides a conservative estimate of survivability in blackout scenarios.

4.3.7.2 Simulation Protocol and Parametric Basis

The authors evaluated HST using whole-building dynamic simulations in EnergyPlus 22.1 applied to a statistically representative single-family residence in Houston, Texas, derived from the NREL ResStock database. The living zone comprises approximately 184 m² of conditioned floor area with envelope properties typical for hot-humid U.S. regions. Extreme weather files, constructed by modifying TMY3 records, were combined with total HVAC shutdown to emulate compound events such as heat waves or winter storms coinciding with grid failure.

The parametric evaluation included envelope retrofits (to IECC 2021 standards), infiltration reduction (from ACH50 = 10 to 0.6), integration of phase-change materials (PCMs) with melting points tailored to expected event conditions, and preheating or precooling strategies to increase initial thermal stability. This allowed assessment of both component-level and whole-building interventions across a range of hazard intensities.

4.3.7.3 Winter Conditions and Results

Under winter storm conditions, characterized by outdoor temperatures dropping below –6°C, the baseline dwelling reached the 15°C safety threshold in only 2 h after the outage. Introducing PCM layers (melting point 17°C–21°C; latent heat 200 kJ/kg; thickness 1.9 cm) to exterior walls and ceilings increased HST to approximately 44

h. Extending PCMs to include internal partitions raised HST to 70–72 h, demonstrating strong benefit from added heat-storage capacity.

Air sealing substantially influenced resilience: reducing infiltration to ACH50 = 2.5 doubled HST compared to insulation-only strategies, while combining PCMs with passive-house-level sealing (ACH50 = 0.6) extended HST to around 100 h. Preheating the building by 5°C further improved conditions, with preheated PCM designs achieving HST values near 130 h, even during prolonged exposure to sub-freezing temperatures.

4.3.7.4 Heatwave Conditions and Results

During extreme heatwave conditions, with outdoor temperatures exceeding 42°C and high humidity, the indoor HI in the baseline building reached the 33°C safety threshold within 12 h. Applying PCMs with melting ranges of 25°C–29°C extended HST to 37 h, while increasing PCM coverage and tightening infiltration reduced the rate of indoor heat accumulation. Passive-house-level sealing combined with preconditioning enabled HST to exceed 130 h even in the absence of active cooling.

The results highlight that both heat-storage materials and envelope airtightness play complementary roles in heat-wave survivability. PCMs delay the onset of unsafe thermal conditions, while air sealing limits moisture infiltration and convective heat gains.

4.3.7.5 Significance and Application

The HST metric provides a direct and intuitive measure of passive survivability under blackout conditions. Its reliance on medical safety thresholds rather than comfort criteria enables clear communication to policymakers, emergency planners and non-technical stakeholders. The indicator is adaptable across climate zones and building typologies, provided that appropriate cold-stress and heat-stress thresholds are selected for each event type.

Furthermore, HST supports retrofit prioritization by revealing which passive strategies yield the greatest protective value under compound hazard conditions. The findings of Wijesuriya et al. (2024) confirm that while insulation and efficient envelopes improve overall thermal stability, PCMs, infiltration control and preconditioning provide the most substantial extensions of safety time, especially in climates with highly variable diurnal or seasonal extremes.

4.3.8 Disruption Severity Index: DoD by Rehman et al. (2024)

Rehman et al. (2024) introduced the DoD as a comparative, dimensionless indicator designed to quantify the severity of indoor thermal degradation during winter power outages. The metric was developed in response to the need for a simple yet informative scalar measure capable of describing both the magnitude of temperature deviation from comfort conditions and the time over which this deviation persists. While many existing resilience indices emphasize threshold exceedance or time-to-failure behavior, DoD captures the combined performance gap by comparing the real thermal trajectory during an outage with an idealized reference state. This makes the indicator particularly useful for evaluating heating resilience across building vintages, envelope strategies, renewable energy configurations and future climate-year scenarios.

4.3.8.1 Conceptual Basis and Motivation

The conceptual foundation of DoD lies in representing thermal disruption as a two-dimensional performance deficit: a deviation in indoor temperature relative to the intended setpoint, and the duration over which this deficit occurs. Rather than expressing discomfort strictly as unmet hours or minimum temperatures, the indicator scales the severity of indoor cooling by the length of the blackout period, yielding a unitless measure that can be compared across climates and scenario durations. The approach thus bridges the gap between purely temperature-driven indices and more complex time-weighted metrics, enabling a consistent evaluation framework for cold-climate heating failures.

4.3.8.2 Mathematical Formulation

The DoD is defined as the product of a normalized temperature deviation and a normalized time factor:

$$DoD = \frac{P_{disruption} - P_{reference}}{P_{reference}} \times \frac{T_{disruption}}{T_{reference}}$$

where:

$P_{disruption}$—indoor operative temperature during the outage (mean or minimum) [°C],
$P_{reference}$—target indoor setpoint in normal operation (21.5°C in the study) [°C],
$T_{disruption}$—duration of the outage [h],
$T_{reference}$—total evaluation period [h].

This expression can be interpreted as the normalized area between the measured indoor temperature curve during disruption and the idealized setpoint line. A DoD value of zero indicates that comfort conditions were fully maintained throughout the outage, while larger values represent increasingly severe and prolonged deviations. Because the metric is dimensionless, it allows direct comparison between houses, simulations and scenario configurations.

4.3.8.3 Application to Parametric Resilience Scenarios

Rehman et al. applied the DoD metric using high-resolution simulations in IDA ICE v4.8, modeling two representative Nordic single-family houses: a 1970s reference dwelling and a modern TEK17-standard low-energy building. Each was assessed under a suite of climate-year conditions, including a typical design year, an extreme cold year, and an extreme warm year. The disturbance scenario consisted of a 31-h winter blackout, corresponding to realistic outage durations experienced in Norway during severe storms. Additional variations included the presence or absence of photovoltaic generation, battery storage, or combined PV–battery systems.

Across all scenarios, DoD proved highly sensitive to building age and envelope performance. In the baseline configuration without renewable energy systems, the 1970s house yielded a DoD of 0.357, indicating pronounced deviation from target comfort conditions. The TEK17 dwelling under identical climate forcing achieved a lower DoD of 0.271, reflecting better envelope insulation and reduced heat loss. Integration of PV and battery systems significantly improved both buildings' resilience, with DoD values approaching zero in some cases. These results demonstrate that DoD effectively captures the benefits of envelope upgrades, renewable-energy autonomy and hybrid resilience strategies.

4.3.8.4 Classification and Decision Support

DoD also serves as the basis for a visual classification scheme, enabling rapid comparative assessment of multiple resilience scenarios. By mapping DoD values across combinations of climate year and energy-autonomy configuration, the authors created an interpretable disruption matrix, ranging from severe disruption (red) to minimal disruption (green). This matrix provides actionable insight for design teams, revealing which combinations of envelope performance, PV systems and battery storage offer the greatest return on resilience investment. Because the indicator is dimensionless, scenario matrices can be generated consistently across studies, portfolios, or building stocks.

While DoD does not quantify health-specific exposure in the manner of HIHH or WUMTP, it excels as an early-stage screening tool for envelope optimization, renewable-energy sizing, and winter-resilience planning. The metric's simplicity allows integration into multi-criteria evaluations that combine resilience, energy cost, life-cycle carbon, and operational performance.

4.3.8.5 Relevance and Future Integration

The DoD metric is particularly well suited to cold-climate regions where electrified heating systems dominate and where winter power outages represent a growing risk under climate change. By incorporating future climate-year datasets, such as ECY or SSP-based warming projections, the methodology enables long-term resilience evaluations that anticipate shifts in baseline outdoor temperatures and blackout severity. Its compatibility with simulation tools, including IDA ICE and EnergyPlus, makes DoD a promising candidate for adoption in resilience-focused labeling schemes or municipal energy-planning frameworks.

As thermal resilience becomes a core component of low-carbon design practice, the DoD offers a transparent, intuitive and analytically efficient method for comparing heating-resilience performance across diverse building typologies and retrofitting strategies. Its ability to condense the thermal impact of both magnitude and duration into a single interpretable value ensures its relevance in future research and policy development.

4.3.9 Heatwave Impact, Absorptivity, and Recovery Indicators (Al-Assaad et al., 2025)

Within the framework of IEA EBC Annex 80, Al-Assaad et al. (2025) proposed an extended methodology for characterizing thermal resilience during extreme heatwaves. Their approach builds on existing indicators such as the IOD, AWD and the OEF (α_{IOD}), but advances resilience assessment by introducing three dynamic performance dimensions: Heatwave Impact (SET DH), Absorptivity Time (t_{abs}), and Recovery Time (t_{rec}). The authors further incorporated the Degree of Shock (doS), a measure of heatwave severity that synthesizes intensity and duration. Together, these indicators allow a nuanced evaluation of how buildings respond to, absorb, and recover from thermal shocks.

4.3.9.1 Heatwave Impact: SET Degree-Hours (SET DH)

The first indicator, SET DH, quantifies the cumulative overheating above the SET alert threshold. Based on the two-node physiological model (Gagge et al. (1972), ASHRAE 55), SET captures the combined influence of air temperature, humidity, radiant load, air velocity, metabolism, and clothing. Al-Assaad et al. (2025) standardized SET DH as the primary shock-impact metric across climates.

(Equation 1) SET Degree-Hours

$$SET\ DH = \sum \left(SET(t) - SET_{alert}\right)^{+} \Delta t$$

where:

$SET\ DH$—accumulated SET degree-hours above the alert threshold [°C·h],
$SET(t)$—Standard Effective Temperature at time t [°C],
SET_{alert}—alert threshold (28°C in this study) [°C],
Δt—simulation timestep (typically 1 h) [h].

Al-Assaad et al. (2025) adopted $SET_{alert} = 28°C$, consistent with Annex 80 guidance distinguishing between comfortable, alert and emergency physiological zones. Overheating is declared when SET DH exceeds 230°C·h for young adults or 117°C·h for older adults. SET DH thus quantifies the magnitude and persistence of heat stress with physiological relevance.

4.3.9.2 Absorptivity Rate (t_{abs}): Quantifying Heat Uptake Dynamics

The absorptivity dimension evaluates how quickly indoor conditions deteriorate after crossing the heat-alert threshold. In Annex 80, absorptivity is operationalized as the time needed for SET to rise from the alert threshold to its peak value during the heatwave. It captures the building's ability to buffer heat gains before reaching the critical maximum temperature.

(Equation 2) Absorptivity Time

$$t_{abs} = t_{peak} - t_{start}$$

where:

t_{abs}—absorptivity time [h],
t_{peak}—time at which peak SET is reached [h],
t_{start}—first hour when SET exceeds $\mathrm{SET_{alert}}$ after at least 24 h below the threshold [h].

Longer t_{abs} values correspond to slower heat buildup and greater buffering capacity, whereas shorter values indicate rapid escalation. In the simulations, t_{abs} increased significantly with rizing doS (severity), showing that buildings tend to remain in stressed conditions longer as heatwaves intensify.

4.3.9.3 Recovery Rate (t_{rec}): Return to Safe Conditions

Recovery time measures how quickly indoor conditions stabilize after the heatwave peak. It describes the ability of ventilation, cooling systems and envelope properties to dissipate accumulated heat. As defined in the Annex 80 methodology, recovery ends once SET remains below the alert threshold for a continuous 24-h period.

(Equation 3) Recovery Time

$$t_{rec} = t_{below} - t_{peak}$$

where:

t_{rec}—recovery time [h],
t_{peak}—hour of maximum SET [h],
t_{below}—time when SET drops below SET_{alert} and remains there for ≥ 24 h [h].

Shorter recovery periods imply higher thermal resilience and strong cooling effectiveness. Longer recovery periods, especially under high doS conditions, show that some buildings accumulate heat faster than they can dissipate it—often due to high thermal mass, inadequate ventilation or insufficient shading.

4.3.9.4 doS: Severity of the Heatwave

The doS, originally defined by Sengupta et al. (2023–2024), quantifies heatwave severity by combining intensity (temperature anomaly) and duration. Al-Assaad et al. (2025) adopted doS as the external forcing variable against which the three resilience aspects (SET DH, t_{abs}, t_{rec}) are evaluated.

(Equation 4) Normalized Degree of Shock

$$doS = \left(\frac{T_{shock} - T_{ref}}{T_{ref}} \right) \times \left(\frac{t_{shock}}{t_{ref}} \right)$$

where:

T_{shock}—mean dry-bulb temperature during the heatwave [°C],
T_{ref}—corresponding mean temperature from the TMY baseline [°C],
t_{shock}—duration of the heatwave [h],
t_{ref}—duration of the longest heatwave in the reference period [h].

doS captures the combined intensity and duration of heatwaves, allowing consistent comparison across historical, mid-century, and late-century climate projections. Higher doS values correspond to more severe shocks with reduced night-time cooling and greater solar loading.

4.3.9.5 Synthesis and Relevance

Together, SET DH, t_{abs}, t_{rec}, and doS establish a dynamic resilience framework capable of distinguishing between shock impact, heat-uptake capacity, and thermal recovery. SET DH quantifies total heat stress, t_{abs} characterizes the speed of heat accumulation, t_{rec} describes the building's cooling and flushing effectiveness, and doS represents the underlying climatic forcing. This combination offers richer diagnostic insight than traditional metrics such as IOD or HE, which evaluate only static or threshold-based exceedance.

The study demonstrated that as doS increases, thermal resilience deteriorates sharply: absorptivity and recovery times lengthen, SET DH grows nonlinearly, and building design parameters shift in relative importance. Window-to-wall ratio, shading operation, cooling capacity, and the availability of night ventilation emerged as the most influential design parameters. These findings underline the need for integrated passive design—shading, low-WWR facades, ventilative cooling—and targeted mechanical support to sustain habitable indoor environments under future extreme heat.

4.3.10 Application of Dynamic Heatwave Indicators to Passive Cooling Strategies (Al-Assaad et al., 2025)

The dynamic heatwave resilience indicators formalized within IEA EBC Annex 80—namely the cumulative impact measure SET DH, the Absorptivity Time t_{abs}, the Recovery Time t_{rec}, and the doS—provide a rigorous analytical framework for evaluating building behavior under extreme heat events. While Section 4.3.10 introduced these indicators conceptually and mathematically, their full explanatory power becomes evident when they are applied to comparative analyses of passive cooling strategies. Al-Assaad et al. (2025) demonstrated this by evaluating five passive strategies across five representative ASHRAE climate zones and three heatwave periods (historical, mid-century, late-century). Their approach reveals how each indicator captures a distinct dimension of heatwave performance, and how passive design measures modify thermal trajectories under increasing climatic forcing.

4.3.10.1 Heatwave Impact (SET DH): Cumulative Exposure Across Climates

SET DH represents the total degree-hour severity of heatwave exposure above a physiologically meaningful SET alert threshold. When applied across a wide climatic dataset, the indicator revealed a strong monotonic relationship with the doS. As doS increased due to higher peak temperatures and longer event durations, SET DH escalated non-linearly, especially in humid climates where latent load exacerbates air–skin heat exchange. This behavior underscores the sensitivity of cumulative thermal stress to both solar and atmospheric moisture forcing.

When passive strategies were introduced, SET DH provided a clear ranking of their cooling effectiveness. Shading consistently produced the largest reductions in cumulative thermal stress, particularly in climates 2A and 3A, where solar gains and high humidity combine to limit evaporative cooling. Conversely, in dry, high-insolation climates such as 3B and 4A, high-albedo envelopes and reduced glazing ratios demonstrated substantial SET DH reductions due to lowered shortwave absorption. The indicator thus operationally quantifies how much cumulative thermal load a passive measure prevents across diverse climatic contexts.

4.3.10.2 Absorptivity Time (t_{abs}): Transient Heat Uptake and Escalation Dynamics

Absorptivity time characterizes the transient interval between the moment the indoor environment first exceeds the alert threshold and the moment of peak thermal stress. Physically, t_{abs} reflects the net heat-gain rate into the indoor air node and its coupling to solar radiation, internal gains and radiant asymmetry. Short values correspond to steep SET trajectories, indicating that the building absorbs heat rapidly once exposed to peak external forcing.

Across the climates examined, Al-Assaad et al. observed that t_{abs} was particularly sensitive to glazing ratio, envelope reflectivity and shading control. Reducing the window-to-wall ratio or enhancing reflectivity significantly prolonged absorptivity time, especially in dry regions where direct and diffuse solar radiation dominate daytime heat gains. In contrast, humid climates showed smaller gains in t_{abs}, as the high latent load reduces the indoor–outdoor temperature differential and limits the building's capacity for delayed thermal response. The indicator therefore captures how transient heat vectors interact with envelope properties under different heatwave forcing conditions.

4.3.10.3 Recovery Time (t_{rec}): Post-Peak Cooling and Thermal Decay

The recovery time measures how quickly the indoor environment returns to safe conditions after peak SET levels are reached. This indicator effectively quantifies the cooling time constant of the building, governed by ventilation availability, envelope thermal mass, outdoor diurnal amplitude and radiant heat loss. A small t_{rec} indicates rapid thermal decay, while larger values reveal buildings that retain heat and remain stressed for extended periods.

Ventilative cooling emerged as the most influential mechanism for reducing recovery time across all climates. In 3B and 4A, where dry night-time conditions produce high radiative and convective cooling potential, operable windows and stack-assisted ventilation achieved the fastest post-peak decay. Humid climates displayed the opposite trend: limited night-time cooling restricted heat removal, resulting in t_{rec} values that increased sharply with doS. Reflective building skins also accelerated recovery by reducing radiant heat storage during the peak period, indicating that post-event cooling is driven by the building's interaction with nocturnal climate conditions as much as by envelope thermal mass.

4.3.10.4 doS: Normalizing External Forcing for Comparative Assessment

The doS provides a non-dimensional representation of heatwave intensity and duration. Al-Assaad et al. used doS as the independent forcing variable for plotting SET DH, t_{abs}, and t_{rec}, revealing how heatwave severity shapes indoor thermal behavior irrespective of building configuration. As doS increases, SET DH rises, absorptivity times shorten and recovery times lengthen—a pattern consistently observed across climates and design strategies.

Using doS as the organizing parameter creates a unified comparison space, allowing results from different climates and future warming scenarios to be evaluated within the same analytical framework. This represents a key methodological advancement, effectively transforming heterogeneous climate-building interactions into a common resilience scale.

4.3.10.5 Synthesis and Implications for Passive Strategy Design

The combined application of SET DH, absorptivity and recovery metrics demonstrates that different passive strategies influence different segments of the heatwave trajectory. Shading primarily reduces cumulative heat stress (SET DH), reflective envelopes flatten the transient escalation phase (increasing t_{abs}), and ventilative cooling accelerates post-peak recovery (reducing t_{rec}). No single strategy dominates across all heatwave phases or climates, indicating that effective resilience design requires integrated combinations of envelope, shading and ventilation interventions.

Furthermore, the clear dependence of all three indicators on the doS reveals that the resilience of even optimized passive designs is strongly conditioned by climate-change-driven intensification of heatwave events. Buildings that perform adequately under historical conditions may become insufficiently resilient under mid- or late-century forcing. Consequently, dynamic resilience metrics, when analysed jointly with doS, provide a more robust basis for future-ready passive design than static comfort or cumulative degree-hour metrics alone.

4.3.11 Thermal Resilience Indicators: Development, Scope, and Application

With the increasing frequency, duration, and intensity of heat waves, the ability of buildings to maintain acceptable indoor conditions during power outages has emerged as a central dimension of climate adaptation. In response, research over the past decade has produced a diverse suite of thermal resilience indicators that extend far beyond traditional comfort metrics. These indicators quantify not only overheating severity but also dynamic thermal response, energy autonomy, health relevance, and event-level forcing conditions. This section synthesizes the development and application of these metrics, tracing their evolution from cumulative degree-hour indices to multidimensional, scenario-based resilience tools.

The earliest contributions focused on cumulative overheating. Hamdy et al. (2017) introduced the Indoor Overheating Degree (IOD) to quantify the magnitude and duration of temperature exceedances above comfort thresholds. Two companion metrics, the Ambient Warmness Degree (AWD) and the Overheating Escalation Factor (α_{IOD}), extended this framework by relating indoor overheating to outdoor thermal forcing and indoor sensitivity to external heat stress. Flores-Larsen et al. (2023) further generalized this approach using regression-based formulations, showing that daily overheating depends not only on AWD but also on the thermal history of the building. Within this framework, the Accumulation Factor (γ_{IOD}) captures heat carry-over effects, while the Solar Radiation Factor (β_{IOD}) accounts for solar-driven contributions to overheating. Together, these indicators enable a time-aware analysis of daily overheating dynamics under extreme heat conditions.

A parallel stream of research addressed physiological risk directly. Sun et al. (2021) introduced two health-relevant indicators: the HIHH, a humidity-sensitive measure based on the NOAA Heat Index, and the SETUDH, incorporating radiant load and evaporative resistance. Tao et al. (2025) subsequently operationalized these metrics into a multi-indicator Thermal Resilience Score (0–4), integrating HIHH, HE, IOD, and WBGT. This scoring framework enabled comparative assessment across emission scenarios, passive-design strategies, and vulnerability groups.

Cold-climate research introduced complementary methods focused on heating resilience. Homaei and Hamdy (2021) proposed the WUMTP, a multiphase index combining severity, duration, and sequence of thermal deficit. Wijesuriya et al. (2024) introduced the HST, a time-to-threshold metric indicating how long a dwelling remains within medically safe temperature limits during blackouts. Gortych and Kuczyński (2025) expanded this logic with two physically intuitive indicators—Time-to-Threshold (T_x) and Temperature-after-x-hours ($T(t_x)$)—usable in both simulation and empirical monitoring.

Rehman et al. (2024) introduced the DoD, a normalized scalar index comparing comfort degradation across buildings or retrofit strategies. Because it aggregates both intensity and duration into a single dimensionless value, DoD enables rapid scenario screening and supports multi-criteria evaluation when combined with cost, carbon, or energy-autonomy metrics. Its standard deviation (σDoD) further enables spatial equity analysis across zones or units.

The most recent developments originate from IEA EBC Annex 80, where Sengupta et al. (2024) proposed a dynamic suite of heatwave-resilience indicators. The Heat Wave Impact metric (SET DH) standardizes cumulative exposure above the SET alert threshold. Two newly defined transient indicators—the Absorptivity Rate (Δabs) and the Recovery Rate (Δrec)—quantify, respectively, the rate of deterioration once conditions become unsafe and the rate at which the building returns to safer temperatures. In parallel, the doS, introduced earlier by Sengupta et al. (2024), provides a normalized measure of the heatwave's inherent intensity and duration. Together, these indicators capture the full dynamic trajectory of a heatwave event: onset, escalation, peak stress, and recovery.

Table 4.2 summarizes these indicators, their definitions, units, primary applications, and sources.

Together, these indicators illustrate the increasing methodological sophistication of thermal resilience assessment. No single metric captures the full complexity of indoor thermal dynamics; however, their combined use provides a powerful diagnostic and design framework. As shown throughout this chapter, the evolution from single-threshold metrics to dynamic, event-based tools parallels the broader development of resilience-driven performance standards. Their integration into simulation workflows, experimental studies and regulatory evaluation systems will be essential for climate-responsive design in both warm and cold environments.

Finally, the inclusion of dynamic heatwave indicators such as SET DH, Δabs, Δrec, and doS expands resilience assessment beyond cumulative exposure, enabling the evaluation of not only how severely a building overheats, but also how rapidly conditions deteriorate and how effectively they recover. This multidimensional approach will be crucial for advancing next-generation resilience labeling and performance-based design frameworks.

TABLE 4.2
Thermal Resilience Indicators Used in the Literature

Indicator	Definition	Unit	Application	Sources
IOD	Cumulative indoor overheating above a comfort threshold	°C·h	Free-running buildings, passive cooling	Hamdy et al. (2017)
AWD	External warmness severity	°C·h	Climate forcing assessment	Hamdy et al. (2017)
OEF / α_{IOD}	Sensitivity of indoor overheating to ambient conditions	–	Climate–envelope amplification	Hamdy et al. (2017); Flores-Larsen et al. (2023)
γ_{IOD}	Accumulated overheating from previous days	–	High-mass buildings	Flores-Larsen et al. (2023)
β_{IOD}	Contribution of solar gains	–	Glazed façades, solar exposure	Flores-Larsen et al. (2023)
HE	Hours above adaptive limit	h	Warm climates, comfort evaluation	Sun et al. (2021)
HIHH	Hours above Heat Index thresholds	°C·h	Health-based stress classification	Sun et al. (2021)
SETUDH	SET-based degree-hours above safety limit	°C·h	Humid climates, radiant sensitivity	Sun et al. (2021)
Thermal Resilience Score	Composite HIHH–HE–IOD–WBGT score (0–4)	–	Scenario comparison, SSP trajectories	Tao et al. (2025)
WUMTP	Dynamic winter-resilience metric	°C·h/m²	Cold-climate blackouts	Homaei & Hamdy (2021)
HST	Time until unsafe temperature is reached	h	Blackouts, PCM retrofits	Wijesuriya et al. (2024)
T_x / $T(t_x)$	Threshold time / temperature after x hours	h / °C	Passive buffering assessment	Gortych & Kuczyński (2025)
DoD	Normalized disruption severity	–	Retrofit screening, cross-climate comparison	Rehman et al. (2024)
SET DH	SET degree-hours above alert threshold	°C·h	Heatwave impact	Sengupta et al. (2024)
Δabs	Rate of overheating onset	°C/h	Dynamic heat uptake	Sengupta et al. (2024)
Δrec	Rate of thermal recovery	°C/h	Post-peak cooling effectiveness	Sengupta et al. (2024)
doS	Severity of heatwave forcing	–	Event-level risk classification	Sengupta et al. (2024)
TA	Passive-comfort autonomy	%	Free-running performance	Ubbelohde (2013)

REFERENCES

Al-Assaad, D., Sengupta, A., An, P., Breesch, H., Afshari, A., Amaripadath, D., Attia, S., Baba, F., Corrado, V., Eli, L., Krelling, A. F., Lee, S. H., Levinson, R., Olinger, M., Tootkaboni, M. P., Wang, L., Zhang, C., & Zinzi, M. (2025). Resilient passive cooling strategies during heat waves: A quantitative assessment in different climates. *Building and Environment, 274*, 112698. https://doi.org/10.1016/j.buildenv.2025.112698

Amaripadath, D., Azar, E., Singh, M. K., & Attia, S. (2024). Heat exposure mitigation in renovated nearly zero-energy dwellings during concurrent heat waves and power outages. *Journal of Building Engineering, 91*, 109655. https://doi.org/10.1016/j.jobe.2024.109655

Flores-Larsen, S., Filippín, C., & Bre, F. (2023). New metrics for thermal resilience of passive buildings during heat events. *Building and Environment, 230*, 109990. https://doi.org/10.1016/j.buildenv.2023.109990

Gagge, A. P., Stolwijk, J. A. J, & Nishi, Y. (1972). An effective temperature scale based on a simple model of human physiological regulatory response. *Memoirs of the Faculty of Engineering, Hokkaido University, 13*, 21–36.

Gortych, M., & Kuczyński, T. (2025). Winter thermal resilience of lightweight and ground-coupled mediumweight buildings: An experimental study during heating outages. *Energies, 18*(15), 4022. https://doi.org/10.3390/en18154022

Hamdy, M., Carlucci, S., Hoes, P.-J., & Hensen, J. L. M. (2017). The impact of climate change on the overheating risk in dwellings—a Dutch case study. *Building and Environment, 122*, 307–323. https://doi.org/10.1016/j.buildenv.2017.06.031

Homaei, S., & Hamdy, M. (2021). Thermal resilient buildings: How to be quantified? A novel benchmarking framework and labelling metric. *Building and Environment, 201*, 108022. https://doi.org/10.1016/j.buildenv.2021.108022

Rehman, H. U., Nik, V. M., Ramesh, R., & Ala-Juusela, M. (2024). Quantifying and rating the energy resilience performance of buildings integrated with renewables in the Nordics under typical and extreme climatic conditions. *Buildings, 14*(9), 2821. https://doi.org/10.3390/buildings14092821

Sengupta, A., Al Assaad, D., Kazanci, O. B., Shinoda, J., Breesch, H., & Steeman, M. (2024). Building and system design's impact on thermal resilience to overheating during heatwaves: an uncertainty and sensitivity analysis. *Building and Environment, 265*, 112031. https://doi.org/10.1016/j.buildenv.2024.112031

Sun, K., Zhang, W., Zeng, Z., Hong, T., & Jain, R. K. (2021). Passive cooling designs to improve heat resilience of homes in underserved and vulnerable communities. *Energy and Buildings, 252*, 111383. https://doi.org/10.1016/j.enbuild.2021.111383

Tao, M., Gou, Z., & Ma, N. (2025). Assessment of thermal comfort and thermal resilience in dwellings during heat waves: A case study of a near-zero energy house. *Journal of Building Engineering, 109*, 113052. https://doi.org/10.1016/j.jobe.2025.113052

Ubbelohde, S. (2013). Thermal autonomy as metric and design process. Retrieved August 1, 2025, from https://www.academia.edu/74045062/Thermal_Autonomy_as_Metric_and_Design_Process

Wijesuriya, S., Kishore, R. A., Bianchi, M. V. A., & Booten, C. (2024). Enhancing thermal resilience of US residential homes in hot humid climates during extreme temperature events. *Cell Reports Physical Science*, 101986. https://doi.org/10.1016/j.xcrp.2024.101986

5 Comprehensive Assessment of Passive Strategies

Environmental, Energy, and Economic Perspectives

5.1 INTRODUCTION

Passive design strategies are widely recognized for their potential to reduce energy demand and improve indoor comfort. Yet their evaluation must go beyond energy savings alone. True sustainability requires consideration of environmental impacts across the life cycle, energy resilience in the face of climate change, and long-term financial viability.

This chapter develops a comprehensive framework for assessing passive strategies through three interconnected dimensions: environmental, energy, and economic. The environmental analysis focuses on life cycle performance, embodied and operational carbon, and circular economy integration. The energy analysis examines demand reduction, efficiency gains, and resilience against climatic extremes, supported by advanced simulation methods. The economic analysis considers investment costs, operational savings, material recovery, and long-term financial performance.

By integrating these three perspectives, the chapter provides a structured basis for evaluating the full potential of passive strategies. This holistic approach allows designers, engineers, and policymakers to identify solutions that deliver environmental responsibility, operational efficiency, and financial resilience, thereby ensuring that passive design plays a central role in the transition to sustainable and climate-adaptive architecture.

To establish whether passive strategies are truly sustainable, the analysis must begin with their environmental dimension. This perspective sets the foundation by examining embodied and operational impacts across the life cycle and by situating passive measures within broader ecological and circular economy frameworks.

5.2 ENVIRONMENTAL ANALYSIS IN THE ASSESSMENT OF PASSIVE STRATEGIES

5.2.1 INTRODUCTION

Passive design strategies are often highlighted for their potential to reduce operational energy demand, yet a comprehensive evaluation must extend beyond this

DOI: 10.1201/9781003645573-5

immediate benefit. Environmental performance depends not only on energy savings but also on the materials applied, their production processes, durability, and potential for reuse or recycling. To ensure genuine sustainability, passive strategies should be examined across the entire building life cycle, with particular attention to embodied impacts, long-term durability, and compatibility with circular economy principles.

This section presents a structured approach to environmental analysis, positioning Life Cycle Assessment (LCA) as the central methodology. It considers the balance between embodied and operational carbon, material sustainability, ecosystem impacts, and climate adaptation. By integrating these elements, environmental evaluation helps identify strategies that are both energy-efficient and ecologically responsible, supporting design decisions that contribute to resilience and long-term sustainability.

5.2.2 Life Cycle Assessment as a Core Methodology

LCA provides a robust, standardized framework for quantifying environmental impacts at every stage of a building's existence: material extraction, processing, construction, operation, and end-of-life. It captures a wide spectrum of indicators, including global warming potential, resource depletion, water use, acidification, and pollution.

In the context of passive strategies, LCA enables comparison of the long-term environmental implications of different materials and design solutions. For instance, a wall assembly with high thermal mass may deliver superior energy performance but also carry substantial embodied carbon due to cement or brick production. Conversely, bio-based materials such as hempcrete or cellulose insulation may minimize embodied emissions but require careful detailing to ensure protection from moisture or biological degradation.

A critical concept within LCA is the Carbon Payback Time (CPT), which expresses the period required for operational carbon savings to offset the initial embodied emissions of a strategy. CPT provides a time-sensitive measure of the environmental Return on Investment (ROI) and highlights the importance of choosing systems that deliver net benefits within the expected service life of a building.

5.2.3 Balancing Embodied and Operational Carbon

One of the central challenges in environmental assessment is balancing embodied and operational carbon. Passive components such as advanced insulation, thermal mass, or ventilated façades may require more material inputs and specialized products, resulting in higher embodied emissions. However, these elements often deliver long-term reductions in heating and cooling demand, making their overall contribution positive over time.

Achieving this balance requires careful optimization. Materials such as cross-laminated timber (CLT) can sequester biogenic carbon while providing both structural and thermal benefits. Mineral wool or aerogel insulation, on the other hand, offers excellent thermal resistance but at the cost of higher embodied energy. Regional context is also decisive: in areas where the electricity grid remains carbon-intensive, operational savings have greater relative weight, while in highly decarbonized grids, embodied impacts dominate the carbon profile of buildings.

5.2.4 Material Sustainability and Circular Economy Integration

Material selection has a profound influence on the environmental footprint of passive strategies. Sustainable materials are not defined only by low embodied carbon but also by attributes such as durability, ease of maintenance, recyclability, and local availability.

Circular economy principles amplify the benefits of passive design by closing material loops and enabling future adaptability. Design for Deconstruction (DfD) is especially relevant for components integrated into building envelopes and structures. Prefabricated insulation panels, modular shading devices, or timber frame assemblies can be designed for reversibility, allowing easier upgrading or dismantling at the end of service life. Such strategies reduce demolition waste and support the reuse of materials, contributing to a regenerative building culture.

5.2.5 Ecosystem Impact and Biodiversity Considerations

The environmental performance of passive design must also be evaluated in relation to broader ecological systems. Many strategies provide co-benefits for biodiversity, microclimate regulation, and water management. For example, green roofs and vegetated façades offer both thermal buffering and enhanced habitat quality, while also reducing stormwater runoff. Similarly, permeable pavements and bioswales promote groundwater recharge and reduce flooding risks in urban areas.

Passive shading, daylight-optimized façades, and natural ventilation schemes not only reduce operational energy demand but also mitigate urban heat islands and lower artificial lighting loads. These indirect effects reduce outdoor air pollution and light disturbance to nocturnal species. Comprehensive assessment should therefore account for ecological co-benefits as well as carbon-related impacts.

5.2.6 Climate Adaptation and Future-Proofing

With accelerating climate change, passive strategies must demonstrate adaptive as well as static performance. Rising temperatures, shifting precipitation patterns, and more frequent extreme events demand buildings that can sustain environmental goals while protecting occupant comfort.

Future-oriented passive solutions include dynamic shading systems, phase change materials (PCMs), operable windows, and thermally responsive façades. These strategies allow buildings to adjust their behavior to seasonal and daily variations, enhancing comfort while reducing dependence on mechanical systems.

Incorporating climate modelling into environmental assessments is essential. Future weather files and Representative Concentration Pathways (RCPs) provide realistic scenarios for testing resilience. This ensures that passive systems remain effective and environmentally sound throughout the building's service life.

5.2.7 Conclusion

Environmental analysis forms a vital dimension of assessing passive building strategies. It expands the evaluation beyond short-term operational savings to encompass embodied impacts, material sustainability, ecological integration, and climate adaptability. By employing LCA as the methodological foundation and embracing circular economy principles, designers can identify strategies that are environmentally regenerative, materially efficient, and resilient over time.

When analyzed holistically, passive strategies emerge as not only efficient tools for energy reduction but also key enablers of long-term ecological sustainability. Their careful integration reinforces the transition to low-carbon, circular, and adaptive building practices, aligned with the environmental challenges of the twenty-first century.

While the environmental perspective highlights life cycle impacts and ecological integration, it must be complemented by an assessment of direct energy performance. The next section therefore focuses on how passive strategies reduce energy demand, moderate indoor conditions, and enhance resilience against climatic extremes.

5.3 ENERGY ANALYSIS IN THE ASSESSMENT OF PASSIVE STRATEGIES

5.3.1 Introduction

Evaluating the energy performance of passive strategies is essential for understanding their effectiveness in reducing energy demand, improving thermal comfort, and enhancing resilience. Unlike conventional assessments that focus primarily on operational consumption, a comprehensive approach must account for both operational and embodied energy. This ensures that passive measures provide sustained efficiency across the building's entire life cycle while reducing reliance on mechanical systems.

By integrating advanced simulation methods, real-world monitoring, and key energy metrics, energy analysis establishes a quantitative framework for optimizing passive design. This section explores the role of passive measures in reducing heating and cooling loads, stabilizing indoor environments, and supporting long-term resilience against climate variability and energy price fluctuations. Closely linked to environmental and economic evaluations, energy analysis ensures that passive strategies are technically effective, financially viable, and aligned with sustainability goals.

5.3.2 Reducing Energy Demand through Passive Design

The primary objective of passive strategies is to limit dependence on mechanical heating, cooling, and ventilation. Substantial reductions in total energy use are achieved through climate-responsive solutions that harness natural regulation of heat, airflow, and solar radiation.

Natural ventilation is central to passive cooling. Cross-ventilation, stack effect, and pressure differentials provide airflow that improves indoor comfort while lowering cooling demand. In warm climates, night ventilation removes stored heat and pre-cools the building fabric for the following day.

Thermal mass materials such as stone, rammed earth, and PCMs moderate indoor fluctuations by absorbing and slowly releasing heat. This stabilies internal temperatures and reduces peaks in heating, ventilation, and air conditioning (HVAC) demand, particularly in regions with large diurnal temperature ranges.

Solar control elements further support energy reduction. Dynamic shading, ventilated façades, and advanced glazing minimize unwanted heat gain while maintaining daylight. Optimized building orientation and façade design enhance solar heating in cold climates and reduce overheating in hot climates.

The effectiveness of these measures depends on detailed climate analysis and site-specific adaptation, ensuring that energy savings do not compromise comfort.

5.3.3 Quantifying Energy Performance and Efficiency

To evaluate passive strategies objectively, energy analysis relies on performance indicators that capture efficiency gains and operational improvements.

- Energy Savings (ΔE): Reduction in annual energy demand compared to a baseline design (kWh or GJ per year).
- Energy Use Intensity (EUI): Annual energy consumption per unit of floor area (kWh/m^2/year), a standard benchmark for cross-comparison.
- Peak Load Reduction (ΔP): Change in maximum hourly demand, indicating how passive strategies alleviate pressure on HVAC systems and grids.
- Heating and Cooling Degree Days (HDD/CDD): Climatic indicators that guide optimization of insulation, thermal mass, and solar control.
- CPT: Period required for operational savings to offset embodied emissions of passive components.

These metrics allow stakeholders to compare strategies, assess long-term benefits, and select solutions that balance efficiency, comfort, and cost.

5.3.4 Simulation-Based Energy Modeling

Advanced modeling tools are indispensable for predicting performance and testing design alternatives.

- EnergyPlus and TRNSYS provide dynamic thermal simulations, capturing heat transfer, solar gains, and HVAC interactions.
- OpenStudio and IES VE integrate thermal comfort, daylighting, and whole-building energy analysis for comprehensive evaluation.
- Computational Fluid Dynamics (CFD) models airflow, pressure fields, and natural ventilation dynamics, refining passive cooling strategies.

Climate data from typical meteorological year (TMY) files or future weather projections allow realistic scenario testing. Parametric modelling identifies optimal configurations for different climates, occupancy patterns, and building typologies, supporting robust decision-making.

5.3.5 Enhancing Energy Resilience through Passive Design

Passive strategies contribute not only to energy savings but also to resilience. By reducing reliance on mechanical systems and fossil fuels, they maintain functionality under extreme weather or grid failures.

Buildings with high thermal mass and adaptive envelopes retain stable indoor conditions even during outages. Passive solar heating and shading systems balance seasonal demands without significant operational energy input. Combined with renewable technologies such as photovoltaics and thermal storage, passive measures support building self-sufficiency and limit vulnerability to energy price volatility.

This resilience dimension underscores the importance of integrating passive design into long-term adaptation strategies.

5.3.6 Integrating Energy Analysis with Environmental and Economic Assessments

Energy analysis gains full relevance when linked to environmental and economic dimensions. Reduced demand lowers carbon emissions and operating costs simultaneously, reinforcing ecological and financial benefits.

- Environmental perspective: Lower energy consumption directly contributes to emission reduction, aligning with climate targets.
- Economic perspective: Energy savings translate into cost reductions, improved asset value, and longer system lifespans.

Integration with LCA and Life Cycle Cost Analysis (LCCA) ensures that passive strategies deliver balanced benefits across performance, sustainability, and affordability.

5.3.7 Conclusion

Energy analysis provides a quantitative foundation for evaluating passive strategies, demonstrating their role in reducing demand, moderating peak loads, and enhancing resilience. With advanced simulation tools and appropriate performance metrics, designers can optimize passive solutions for diverse climates and future scenarios.

When linked with environmental and economic evaluations, energy analysis reinforces the status of passive design as a cost-effective, sustainable, and adaptive pathway for twenty-first-century architecture.

Energy analysis reveals the operational benefits of passive measures, but these advantages only translate into real-world adoption when they are financially feasible. The following section evaluates the economic dimension, showing how capital investment, operational savings, and material recovery determine the long-term viability of passive strategies.

5.4 ECONOMIC ANALYSIS IN THE ASSESSMENT OF PASSIVE STRATEGIES

5.4.1 Introduction

The financial feasibility of passive building strategies is a critical factor for their large-scale adoption. While passive design is strongly associated with energy efficiency and environmental benefits, its economic viability determines whether such solutions can realistically be implemented. A comprehensive economic analysis extends beyond initial costs to include life cycle considerations such as operational expenses, long-term ROI, and material recovery potential.

This section outlines how passive strategies influence cost savings, investment returns, and financial resilience. It introduces key methods such as LCCA and Total Cost of Ownership (TCO), alongside financial indicators that support decision-making in both design and policy contexts.

5.4.2 Capital Expenditures and Initial Investment Costs

One of the main barriers to the uptake of passive strategies is the perception of high upfront costs. Advanced insulation, adaptive shading, or thermal mass systems typically require larger initial investments than conventional solutions. However, these costs must be assessed in relation to long-term operational savings and maintenance reductions.

For instance, ventilated double-skin façades and solar shading devices increase construction costs but significantly reduce cooling demand, leading to lower energy bills throughout the building's life. Similarly, high-performance insulation materials such as vacuum insulation panels or bio-based alternatives may require higher initial spending but provide durable energy savings.

A balanced cost-benefit assessment should therefore compare initial capital expenditures with projected savings, maintenance profiles, and increased property value, ensuring that investment decisions are grounded in life cycle economics.

5.4.3 Operational Cost Reduction through Passive Design

Passive strategies substantially reduce operating costs by lowering reliance on active HVAC systems. This results in:

- Reduced annual energy costs, due to lower heating, cooling, and ventilation demand.
- Lower maintenance expenses, since passive elements such as thermal mass or fixed shading systems require minimal upkeep compared to mechanical systems.
- Extended service life of HVAC systems, as reduced loads decrease wear and replacement frequency.

These benefits can be quantified through LCCA and TCO frameworks, which capture both direct and indirect financial savings. Predictive models further allow planners to account for future variability in energy prices, carbon taxation, and financial incentives, thereby strengthening the economic case for passive design.

5.4.4 Return on Investment and Long-Term Economic Benefits

Beyond operational savings, passive strategies contribute to improved asset value, reduced vacancy rates, and increased market competitiveness. Energy-efficient buildings often achieve higher rental income and resale prices, offering tangible financial advantages.

Certification schemes such as Leadership in Energy and Environmental Design (LEED), Building Research Establishment Environmental Assessment Method (BREEAM), and WELL Building Standard (WELL) amplify these benefits by providing recognied quality labels that attract investors and occupants. Moreover, improved indoor environmental quality, daylighting, and thermal stability enhance occupant productivity and well-being, which translates into indirect economic value in workplaces, schools, and healthcare facilities.

From a broader perspective, passive buildings also contribute to energy security by reducing dependence on fossil fuels, thus stabilizing long-term operating costs and lowering exposure to market volatility.

5.4.5 End-of-Life Cost Considerations and Material Recovery

The economic assessment of passive strategies must extend to the end-of-life stage. DfD and material recovery can reduce demolition costs and preserve financial value through reuse and recycling.

- Modular systems allow components to be dismantled and reused.
- Recyclable insulation and bio-based materials lower disposal costs.
- Adaptive façades and glazing systems may be upgraded rather than discarded, extending their economic life.

Integrating these considerations ensures that passive buildings maintain financial value even at the end of their operational life, aligning with circular economy principles and reducing overall life cycle costs.

5.4.6 Key Economic Performance Indicators for Passive Buildings

To evaluate financial viability, economic analysis applies several key indicators:

- LCCA: compares passive and conventional strategies over the full building life cycle.
- TCO: quantifies cumulative costs, including operation and end-of-life phases.
- Net Present Value (NPV): measures the discounted long-term ROI, accounting for inflation and energy price trends.
- Payback Period (PP): indicates the time required for energy savings to offset initial costs.

- ROI: evaluates the overall financial gain from passive strategies, including operational savings and property value increases.

These indicators provide stakeholders with clear metrics for comparing design options and prioritizing interventions.

5.4.7 Integrating Economic, Energy, and Environmental Assessments

Economic analysis achieves its full relevance when integrated with environmental and energy perspectives. Lower energy use translates directly into reduced operating costs and fewer emissions, reinforcing both financial and ecological outcomes.

- From an environmental perspective, energy savings reduce carbon footprints and resource depletion.
- From an economic perspective, these savings increase financial resilience and building value.

By combining LCCA, TCO, and LCA, passive strategies can be assessed holistically, ensuring that design decisions deliver robust financial, ecological, and operational benefits.

5.4.8 Conclusion

Economic analysis provides essential evidence of the financial viability of passive strategies. By considering initial costs, operational savings, and end-of-life recovery, it supports the case for long-term resilience and cost-effectiveness.

Taken together, the environmental, energy, and economic perspectives form a comprehensive framework for assessing passive strategies. By integrating these dimensions, decision-makers can identify solutions that balance ecological responsibility, technical performance, and financial resilience, ensuring that passive design contributes meaningfully to sustainable and climate-adaptive architecture.

This comprehensive three-dimensional framework also provides a foundation for the subsequent chapters, where policy recommendations and design guidance are developed to translate these analytical insights into practical implementation strategies.

6 Impacts of Outdoor Temperatures on Indoor Environments

6.1 OVERHEATING RISKS IN LONDON APARTMENTS: A STUDY ON INDOOR TEMPERATURES AND PASSIVE COOLING LIMITATIONS

Pathan et al. (2017) conducted a study to assess the risk of overheating in 122 apartments located in London during the summer periods of 2009 and 2010. Indoor temperatures were monitored in living rooms and bedrooms at ten-minute intervals, and the collected data were analyzed using both static and adaptive thermal comfort criteria, such as American Society of Heating, Refrigerating and Air-Conditioning Engineers (ASHRAE) and Chartered Institution of Building Services Engineers (CIBSE) standards. Additionally, the study included surveys on the characteristics of the apartments, resident behaviors, and the use of household appliances.

The findings revealed that overheating was particularly pronounced in bedrooms, with 75% exceeding the permissible temperature thresholds. The most significant temperature exceedances occurred during two heatwaves: 22 June–23 July and 9–16 July 2010. The results were confirmed using adaptive thermal comfort criteria according to ASHRAE Standard 55. Similar to the findings of Beizaee et al. (2013), apartments built after 1996 exhibited higher indoor temperatures compared to older buildings. Detached and terraced houses were found to be less vulnerable to overheating than apartments in multi-story blocks or semi-detached houses.

The study concluded that overheating is a widespread issue in London apartments, affecting both older and newer buildings. The latter are particularly susceptible due to higher energy efficiency standards, which inadvertently reduce passive cooling capabilities. The authors emphasize the importance of incorporating passive design strategies in both new construction and renovation projects to mitigate the risk of overheating, particularly given the projected effects of climate change.

6.2 OVERHEATING IN ENGLISH HOMES: LESSONS FROM THE HOTTEST SUMMER

Lomas et al. (2021) monitored indoor temperatures during the hottest summer in recorded meteorological history in England in 2018. The study covered 750 households, measuring temperatures in living rooms and bedrooms. The summer of 2018 is considered representative of future climatic conditions projected for the 2050s.

DOI: 10.1201/9781003645573-6

Climate models from UKCP18 suggest that such hot seasons, with maximum temperatures reaching 35.3°C, will become more frequent due to global warming.

The study found that permissible temperature thresholds were exceeded in 15% of living rooms and 19% of bedrooms. Particularly high indoor temperatures were recorded in London, where, due to the urban heat island (UHI) effect and dense building development, excessive temperatures were observed in 28% of living rooms and 32% of bedrooms. Similar to the findings of Beizaee et al. (2013) and Pathan et al. (2017), buildings with high energy efficiency standards were more likely to overheat compared to less energy-efficient buildings. This was primarily because these buildings were often apartments in multi-story blocks, which are more prone to overheating. The most vulnerable groups were low-income households and elderly individuals aged 75 and above.

The analysis also highlighted the limitations of static overheating criteria, such as the 26°C threshold for bedrooms, which led to an overestimation of the problem. The use of adaptive thermal comfort criteria (CIBSE TM52) provided a more realistic assessment of overheating, considering residents' ability to adapt to higher temperatures. The results indicate the need for more reliable adaptive criteria in evaluating overheating risks, especially in the context of climate change.

The authors suggest that adaptation measures should focus on apartments in multi-story buildings and social housing, particularly in London. They also emphasize the necessity of education and support for elderly individuals in managing indoor temperatures, as well as the implementation of passive cooling strategies in highly energy-efficient buildings.

6.3 NIGHTTIME OVERHEATING IN GERMAN BEDROOMS DURING HEATWAVES

Beckmann et al. (2021) conducted a large-scale study monitoring indoor air temperatures during a heatwave in Augsburg, Germany. The analysis included data collected from 427 households, where bedroom temperatures were recorded at 15-minute intervals during the heatwave in July 2019. Additionally, surveys were conducted to assess heat perception.

The study found that the average nighttime temperature in bedrooms during the heatwave was 27.38°C, exceeding recommended values in most overheating standards. In 81.5% of bedrooms, the temperature exceeded 26°C, and in 93%, it surpassed 25°C. In contrast, only 1.6% of bedrooms recorded temperatures below 24°C, highlighting the scale of the overheating issue. The findings suggest that the 26°C threshold, commonly used in WHO guidelines, is regularly exceeded during heatwaves, emphasizing the need for adapting residential spaces to rising temperatures (Beckmann et al., 2021).

An analysis of residents' perceptions revealed that subjective thermal stress was significantly higher in bedrooms where average temperatures exceeded 26.4°C. High temperatures had the most severe impact on individuals with chronic illnesses. Older adults, despite experiencing health problems related to heat, reported lower levels of thermal stress than younger age groups, possibly indicating a lack of awareness of heat-related risks in this demographic (Beckmann et al., 2021).

Building factors such as floor level, building age, and construction type significantly influenced thermal stress levels. Residents living on higher floors and in densely built urban areas reported greater heat stress. Unlike the findings of Beizaee et al. (2013), Pathan et al. (2017), and Lomas et al. (2021), apartments in newer buildings (constructed after 2001) exhibited lower levels of thermal stress compared to older buildings.

The authors emphasize that their findings are crucial for developing thermal adaptation plans and managing heat stress risks, particularly in the context of climate change and rising temperatures. They recommend expanding educational efforts and risk communication targeted at vulnerable groups, such as elderly individuals and those with chronic illnesses, and promoting passive strategies such as improved air circulation, thermal insulation, and shading (Beckmann et al., 2021).

6.4 OVERHEATING TRENDS IN AMSTERDAM: CROWDSOURCED AND MONITORED DATA

Peerlings et al. (2024) conducted a similarly large-scale study in Amsterdam during a September 2023 heatwave, utilizing two complementary data collection methods: a professional monitoring network covering 92 households and a short-term crowdsourcing campaign involving 571 city residents.

The study revealed significant indoor temperature variations, with average bedroom temperatures rising from 22.1°C on 4 September to 27.9°C on the hottest day, 10 September. The lowest recorded bedroom temperature was 22.7°C, while the highest reached 34.4°C, highlighting substantial differences in heat accumulation across monitored apartments.

Additionally, the crowdsourcing campaign recorded a higher median indoor temperature of 28.0°C, confirming widespread overheating in most residential units.

An analysis of residents' perceptions from the crowdsourcing campaign revealed that 59% of participants found coping with the heat very difficult, reporting an average temperature of 28.6°C. Another 29.8% considered the conditions acceptable at an average of 26.3°C, while only 11.2% described the heat as easy to endure at 24.9°C.

The authors emphasize that the findings indicate a significant overheating issue in Amsterdam, particularly in bedrooms, which heated up more rapidly than living rooms. This aligns with previous studies on overheating in London but contradicts findings from Arnhem and Groningen, where living rooms tended to reach higher temperatures than bedrooms. The elevated bedroom temperatures could be attributed to smaller room sizes and faster heat accumulation (Peerlings et al., 2024).

The study also highlights differences between professional monitoring data and crowdsourced reports, suggesting that professional indoor temperature monitoring systems remain essential. However, crowdsourcing campaigns provide valuable insights into residents' perceptions and enable rapid data collection during extreme weather events such as heatwaves (Peerlings et al., 2024).

In conclusion, the authors stress the urgent need to implement adaptation strategies to mitigate indoor overheating, particularly given the projected increase in the frequency of European heatwaves.

6.5 RESIDENTIAL OVERHEATING IN FINLAND: LARGE-SCALE CLIMATE ADAPTATION STUDY

Velashjerdi Farahani et al. (2023) conducted an unprecedented large-scale study analyzing the risk of residential overheating in the Helsinki region of Finland. The research included real air temperature measurements in apartments during the typical summer of 2020 and the exceptionally hot summer of 2021. The study encompassed a total of 10,648 apartments in 2020 and 10,597 apartments in 2021, located in 6 cities: Helsinki, Vantaa, Espoo, Järvenpää, Kerava, and Tuusula, and examined buildings constructed between 1902 and 2016, with apartment sizes ranging from 20 to 232 m^2, allowing for a broad assessment of various housing typologies and building characteristics.

After an initial data processing phase, the cleaned dataset included 578 buildings in the summer of 2020 and 560 buildings in the summer of 2021, resulting in temperature records for 6,974 apartments in 2020 and 6,057 apartments in 2021, monitored from 15 May to 31 August. The large sample size, combined with the broad range of building ages and geographical coverage, makes this study highly representative of the entire Helsinki region (Velashjerdi Farahani et al., 2023).

To assess the extent of overheating, hourly indoor temperature distributions were analyzed under two occupancy scenarios: continuous presence in the apartment and nighttime conditions between 9:00 PM and 8:00 AM. The results indicate significant differences in overheating risk between the summers of 2020 and 2021. In 2020, with an average outdoor temperature of 16.3°C, indoor temperatures exceeded 27°C in 76% of apartments, 30°C in 8% and 32°C in 1% of cases. In 2021, when the average outdoor temperature was 1.1°C higher, these figures increased to 96%, 32%, and 4%, respectively, highlighting a substantial deterioration in indoor thermal conditions during the hotter summer.

During the summer of 2021, indoor temperatures exceeded 27°C at least occasionally in nearly all apartments, while one-third of cases experienced maximum hourly temperatures above 30°C. Notably, nearly half of the apartments maintained temperatures above 27°C for more than seven consecutive days, significantly increasing health risks, particularly for elderly residents.

The analysis revealed a strong correlation between outdoor and indoor temperatures, with the highest correlation coefficient observed for the five-day average outdoor temperature. This suggests a delayed effect in how buildings respond to extreme temperatures. This delay means that peak indoor temperatures occur several days into a heatwave, allowing residents to implement adaptive strategies such as nighttime ventilation or the use of cooling devices. The authors suggest that these findings could serve as a basis for better adaptation policy planning, particularly in light of climate change and the growing risk of heatwaves (Velashjerdi Farahani et al., 2023).

6.6 FUTURE OVERHEATING IN FINNISH CLT APARTMENTS: THERMAL SIMULATIONS

Sukanen et al. (2023) conducted a study on the risk of residential overheating in Finland under both current and future climate conditions using dynamic thermal simulations in the IDA-ICE software. The analysis focused on three apartments in a cross-laminated timber (CLT) building in Jyväskylä.

The study aimed to evaluate the intensity of overheating, expressed in degree-hours (°C·h) above the 27°C threshold, as well as maximum indoor temperatures under different climate scenarios, including the present climate and projections for 2030, 2050, and 2100. The results indicate that all apartments significantly exceeded the permissible overheating threshold of 150°C·h, both under current and future climate conditions. Studios and living rooms in one-bedroom apartments exhibited the highest overheating risk, reaching 15,129°C·h and 16,127°C·h, respectively, by 2100, with maximum temperatures exceeding 33°C and 34°C. In contrast, two-bedroom apartments, particularly bedrooms, were the least affected by overheating.

The reasons for this variation include smaller glazed areas and favorable window orientations, such as north- and west-facing windows, which reduce solar heat gain. Additionally, bedrooms, which are less intensively used during the day, accumulated less internal heat from occupants, appliances, and lighting. The layout of two-bedroom apartments also allowed for better heat dissipation compared to the more concentrated spaces in studios and one-bedroom apartments (Sukanen et al., 2023).

The authors emphasize the necessity of updating Finnish building regulations to account for future climate changes. They propose introducing policies that promote passive adaptation measures such as natural ventilation and adaptive thermal comfort, which could mitigate overheating risk and reduce reliance on active cooling systems such as air conditioning.

The analysis demonstrated that future climate change will substantially increase the risk of overheating in all studied apartments. The highest overheating values were observed in rooms with southern and western exposure, underscoring the importance of considering building orientation and window sizing in architectural design. The results confirm the need for implementing effective adaptation strategies to mitigate future overheating risks (Sukanen et al., 2023).

6.7 WINDOW-TO-WALL RATIOS AND OVERHEATING IN NORWEGIAN HOMES

Tian et al. (2020) explored the overheating risks in typical Norwegian residential buildings by examining both current climate conditions and future projections for 2050 and 2080. Their study focused on buildings constructed in accordance with Norwegian building regulations (Pbl/TEK17), incorporating variations in window-to-wall ratios (WWR) and airtightness levels. The simulated residential structure reflected contemporary construction and energy efficiency standards in Norway.

The research encompassed 72 distinct scenarios, analyzing four WWR configurations (0.35, 0.5, 0.75, and 0.9), three levels of airtightness (high, medium, and standard), and two primary building orientations (north and south). Simulations were carried out for current climatic conditions as well as projected climates for 2050 and 2080. This approach allowed for identifying which architectural and climatic factors contribute most significantly to overheating risks.

To accurately assess thermal performance, the researchers created a detailed simulation of the building's interior layout, focusing on individual rooms. The kitchen and dining areas were positioned on the ground floor, while three bedrooms were located upstairs. A representative bedroom was selected for in-depth analysis, with its thermal conditions simulated for both north- and south-facing orientations.

Findings indicated that buildings with higher WWR values (0.75 and 0.9) are particularly prone to overheating, especially in south-facing bedrooms, where increased solar heat gains significantly affect indoor temperatures. Projections for 2080 suggest that this issue will become even more pronounced, particularly in highly airtight buildings where limited natural ventilation leads to heat build-up (Tian et al., 2020).

The study also highlighted variations in the effectiveness of different overheating assessment methods. Adaptive criteria such as CIBSE TM59 revealed that north-facing buildings are more susceptible to nighttime overheating due to delayed heat build-up caused by prolonged daylight exposure in Norway's climate. Conversely, fixed temperature criteria Passive House Planning Package (PHPP) were effective in estimating overall overheating levels but did not account for variations in room usage patterns, limiting their applicability for nighttime comfort assessment.

To mitigate overheating risks, the authors recommend minimizing excessive glazing and implementing adaptive strategies, including natural nighttime ventilation and shading solutions. However, as climate conditions continue to evolve, passive measures alone may not be sufficient. Future-proofing Norwegian residential buildings may require the adoption of active cooling solutions, such as air conditioning or cooling panels. Strengthening overheating regulations for new constructions and introducing large-scale retrofitting programs for older buildings could significantly enhance the climate resilience of Norway's housing stock.

6.8 OVERHEATING IN NEARLY ZERO-ENERGY BUILDINGS (nZEBS) IN IRELAND

In another study, Colclough and Salaris (2024) investigated overheating in nearly zero-energy buildings (nZEB) in Ireland, focusing on real indoor temperatures in both newly constructed and retrofitted dwellings. Their research assessed compliance with overheating standards set by CIBSE, the Passive House Institute (PHI), and the World Health Organization (WHO). A total of 57 residential buildings were monitored, representing diverse housing types and heating systems, including social housing, owner-occupied homes, and student residences.

To account for different building conditions, the researchers categorized properties into seven groups based on location, layout, ventilation approach, and resident demographics. Among these, 44 dwellings met passive house standards: 33 were newly built, while 11 had undergone deep energy retrofitting (DER). The remaining 13 buildings adhered to the minimum nZEB requirements, allowing for a comprehensive assessment of how different construction strategies impact thermal comfort.

Temperature data were collected between 2016 and 2021, with most buildings equipped with individual sensors recording temperatures every five minutes. In student residences, readings were taken every minute, while in EnerPHit DER retrofitted homes, temperature data were aggregated at the building level. Alongside quantitative measurements, qualitative data on thermal comfort were gathered through occupant satisfaction surveys (OSS) and the Building User Survey (BUS) methodology.

The overheating assessment combined static and adaptive thermal comfort criteria. Static criteria relied on fixed temperature thresholds, whereas adaptive standards considered occupants' ability to acclimate to varying temperatures. The analysis incorporated evaluations based on CIBSE TM59, which uses both static (26°C) and adaptive thresholds for living spaces, as well as WHO guidelines, which classify indoor temperatures above 24°C as potentially harmful to health.

Results showed that 18% of nZEB buildings failed to meet CIBSE TM59 overheating criteria, with all cases linked to elevated temperatures in bedrooms. In 38% of buildings, temperatures exceeded 24°C for at least 10% of the year, a threshold that WHO associates with adverse health effects. Among passive house-certified buildings, 5% failed to meet overheating criteria, while 11% were classified as "acceptable," suggesting a need for further design optimization.

Newly constructed buildings were more prone to overheating than retrofitted ones. Among new constructions, 26% failed to meet CIBSE TM59 criteria, compared to just 9% among retrofitted buildings. This finding implies that energy retrofitting strategies may be more effective in mitigating overheating risks, which has important implications for Ireland's future housing policies.

At the typology level, apartments and newly built terraced houses were most susceptible to overheating. In the case of new residential buildings, 46% of apartments and 58% of terraced homes exceeded WHO overheating thresholds, while 31% of apartments failed to meet CIBSE TM59 standards. These results emphasize the need for more detailed overheating risk assessments during the design phase, particularly for smaller dwellings with limited cross-ventilation.

The study also shed light on the role of occupant behavior. In some cases, overheating was not solely due to design limitations but was exacerbated by improper heating system use and individual temperature preferences. Some occupants intentionally maintained higher indoor temperatures, perceiving them as more comfortable—highlighting the need for a better alignment between overheating criteria and real-world user behaviors.

The authors conclude that Ireland's building regulations require further refinement regarding overheating standards for nZEB buildings. Currently, overheating assessments rely on the DEAP methodology, which has not been fully integrated with CIBSE TM59—leading to potential underestimation of overheating risks at the design stage. Delays in adopting improved assessment frameworks could result in an increased number of homes vulnerable to overheating, particularly as climate change drives higher temperatures in the coming decades.

6.9 CLIMATE CHANGE AND OVERHEATING IN FINNISH RESIDENTIAL BUILDINGS

Velashjerdi Farahani et al. (2024) examined the resilience of residential buildings in Finland, particularly in Helsinki, to overheating during both current and future climate scenarios. Their study combined real data from the exceptionally hot summer of 2018 with climate simulations for 2050 and 2080 under the high-emission RCP8.5 scenario. The results revealed significant differences in how various building types responded to external temperatures and how indoor conditions evolved under extreme heat events.

The study followed a structured process that began with the collection of real meteorological data, followed by thermal modelling of buildings, and concluded with an assessment of their reactions to extreme temperatures. Three primary residential building types were analyzed: senior housing facilities, multi-family apartment buildings, and single-family homes—providing a comprehensive evaluation of how well Finland's current building standards can withstand rising temperatures.

The study included both older buildings constructed between the 1960s and 1980s and newer structures built according to Finnish building codes introduced after 2012. A similar assessment was carried out for multi-family housing.

Simulations demonstrated that lightweight structures, such as single-family homes, responded to external temperature changes twice as fast as newer buildings designed to meet the latest construction standards—such as modern apartment complexes and senior housing facilities.

While these newer buildings generally maintained indoor temperatures below Finland's critical health thresholds (30°C for elderly residents and 32°C for healthy adults), older structures frequently exceeded these limits, even under current climatic conditions (Velashjerdi Farahani et al., 2024).

The greatest overheating risks were observed in older buildings that had not undergone energy retrofits. During heatwaves, these structures posed a direct health threat to residents—especially seniors, who are particularly vulnerable to extreme temperatures. The findings also indicated that older buildings had significantly higher cooling energy demands than modern energy-efficient constructions. Future projections suggest that cooling demand in older buildings will increase by 40% to 160%, depending on the building type and climate scenario.

Current Finnish building regulations cap the allowable number of hours exceeding critical temperatures at 150 hours above 27°C for new multi-family buildings and 150 hours above 25°C for senior housing during the summer months. However, the study suggests that these limits may become inadequate in the face of future climate change. There is a pressing need to revise them and introduce stricter design requirements that incorporate climate simulations for 2050 and 2080 during the building design phase.

The authors argue that mandatory renovations for older buildings should be considered to improve their resilience to overheating—particularly in Helsinki, where buildings from the 1960s–1980s dominate the housing stock. Their findings highlight the urgent necessity of integrating thermal comfort requirements into building design to mitigate health risks and enhance energy efficiency in a warming climate (Velashjerdi Farahani et al., 2024).

6.10 OVERHEATING IN DUTCH HOMES: MULTI-SCENARIO CLIMATE ASSESSMENT

Hamdy et al. (2017) investigated the impact of climate change on residential overheating risks in the Netherlands. Their study covered 9,216 different building configurations, accounting for a wide range of structural and operational parameters, such as building orientation, insulation levels, shading options, and ventilation strategies. The research was conducted under four climate scenarios, including historical weather conditions and future projections developed by the Royal Netherlands Meteorological Institute. Numerical modeling was performed using IDA-ICE 4.6, with additional calculations supported by MATLAB scripts—enabling a precise assessment of building vulnerability to overheating and the effectiveness of various adaptation strategies.

The study encompassed eight different housing typologies, including detached homes, semi-detached houses, and apartments in multi-story residential buildings. Each typology featured distinct geometries and thermal properties, allowing for a broad evaluation of how architectural factors influence overheating risks.

A key part of the study focused on the overheating risks associated with various building types. Detached houses—which have greater exposure to solar radiation—were particularly prone to overheating, especially under future climate projections. In contrast, semi-detached houses and apartments—while less exposed to direct sunlight—often suffered from inadequate cross-ventilation, which could negatively impact indoor comfort.

The analysis revealed that older buildings, particularly those constructed before 1964, exhibited greater thermal resilience due to their heavier construction, which allowed for natural heat storage and slower temperature fluctuations. In contrast, modern buildings designed to meet more recent insulation standards were highly energy-efficient in winter but struggled to dissipate heat during summer—significantly increasing the risk of overheating (Hamdy et al., 2017).

The study also examined the role of ventilation in mitigating overheating. While increased ventilation rates could effectively reduce indoor temperatures, their effectiveness diminished as external temperatures continued to rise in future climate scenarios. Under extreme future conditions, passive ventilation alone was insufficient to ensure thermal comfort—highlighting the necessity for active cooling solutions in modern residential buildings.

The findings suggest that building design in the Netherlands must account for climate change projections, incorporating adaptive architectural solutions to mitigate future overheating risks. In particular, external shading systems were identified as highly effective in reducing solar heat gains—making them a recommended strategy for both new and existing buildings.

6.11 OVERHEATING RISKS IN IRISH HOMES: NEED FOR CLIMATE-RESPONSIVE RENOVATION

Mulville et al. (2025) conducted a comprehensive assessment of overheating risks in residential buildings in Ireland, focusing on the impacts of climate change and large-scale renovation projects. Their research paid particular attention to different housing typologies, especially top-floor apartments, which are often more susceptible to overheating. The study was based on an extensive literature review and dynamic simulation modeling (DSM), analyzing key housing characteristics alongside internal and external factors influencing thermal conditions. In addition, existing regulatory frameworks were evaluated using climate projections from Irish meteorological services, allowing the researchers to simulate future overheating scenarios.

Their findings indicated that by 2050, approximately 10% of Ireland's residential buildings could be at risk of excessive overheating, affecting an estimated 51,920 residents. This number is expected to rise significantly by 2100, raising concerns about indoor comfort and overall living conditions. The study underscores the necessity of stricter regulations that account for evolving climatic conditions. The authors emphasize that modernization strategies must integrate overheating risk mitigation, requiring a holistic approach that combines passive cooling measures, urban planning considerations, and improved ventilation strategies. From a regulatory perspective, the study suggests that new standards for solar shading, ventilation, and energy efficiency will be crucial in minimizing overheating risks in future housing stock (Mulville et al., 2025).

6.12 OVERHEATING IN CENTRAL AND EASTERN EUROPE: POLICY GAPS AND STRATEGIES

Taylor et al. (2023) examined the broader context of overheating in Central and Eastern Europe, highlighting the growing frequency and intensity of heatwaves and the corresponding rise in residential overheating risks. Despite these trends, the study found that knowledge of effective mitigation strategies remains limited, and policy measures addressing overheating are still in their early stages. The authors note that in many countries within the region, traditional housing designs have not prioritized passive cooling strategies, and there is a general lack of public awareness regarding appropriate adaptive behaviors.

The study also documented the alarming increase in heat-related mortality across Central and Eastern Europe. While the six-year moving average of heat-related deaths in Europe rose by 33% between 2005 and 2019, this increase was significantly higher in countries such as Germany, Poland, and Ukraine. Heat-related health risks are not evenly distributed across the population, with older adults being particularly vulnerable due to their reduced physiological ability to cope with high temperatures. Additionally, individuals with pre-existing cardiovascular and respiratory conditions, as well as those experiencing social isolation, face greater health risks during extreme heat events.

Taylor et al. (2023) argue that existing research disproportionately focuses on outdoor temperatures, while studies incorporating measured indoor thermal conditions remain scarce. They advocate for a stronger emphasis on the effects of indoor overheating on occupant health, particularly in residential buildings where people spend the majority of their time during heatwaves.

According to the authors, the rising threat of overheating in Northern and Central European countries calls for the urgent implementation of effective heat management strategies in housing design. Historically, residential construction in these regions has been primarily centered on reducing energy consumption for heating, an approach that is increasingly inadequate in the face of escalating summer temperatures. Future housing strategies should incorporate lessons from countries with warm and hot climates, alongside intensified scientific research on overheating prevention.

While technical and engineering solutions are critical to mitigating overheating, the study emphasizes that behavioral and lifestyle factors also play a crucial role. Thus, efforts to address overheating risks must be accompanied by modifications to urban planning and building codes. The study further highlights the importance of public education campaigns aimed at raising awareness of appropriate heatwave responses. Given the growing reliance on air conditioning, the authors stress the importance of minimizing future cooling demand through effective building design, thereby reducing dependence on energy-intensive cooling solutions (Taylor et al., 2023).

6.13 HEATWAVES AND THERMAL DISCOMFORT IN ATHENS' LOW-STANDARD HOUSING

Sakka et al. (2012) conducted an empirical study on thermal conditions in 50 low-standard residential buildings in Athens, Greece, during the anomalously hot summer of 2007, which was the warmest on record in the region. The research focused on dwellings situated within urban heat island (UHI) zones, characterized by poor thermal performance due to inadequate insulation, single-glazed windows, and high air infiltration rates. The majority of these buildings were constructed prior to 1980, preceding the implementation of mandatory insulation standards in Greece.

The results indicated that indoor temperatures in these dwellings frequently reached critical levels, with recorded maxima of 40°C and minimum nighttime temperatures consistently above 28°C. During periods of thermal anomalies, the mean indoor temperature increased by 4.2 K relative to standard climatic conditions, with a daily increment ranging between 0.2 K/day and 0.8 K/day at the peak of the heatwaves.

Prolonged exposure to elevated temperatures was a defining feature of the study. Indoor temperatures remained above 30°C for more than 85% of the monitored period, with 27% of the time exceeding 33°C. Notably, sustained overheating events lasted up to 216 consecutive hours above 30°C, with some dwellings experiencing temperatures above 33°C for six successive days. Such persistent heat exposure poses significant health risks, particularly for vulnerable populations, including elderly individuals, children, and persons with pre-existing medical conditions (Sakka et al., 2012).

The authors emphasized the necessity of implementing integrated heat mitigation strategies to improve indoor thermal conditions in low-standard housing. They proposed that a combination of solar shading devices, reflective roof coatings, and materials with high thermal mass could substantially reduce heat gains and enhance indoor environmental quality.

Furthermore, the study underscored the importance of comprehensive adaptation measures aimed at protecting low-income residents, who are disproportionately affected by the increasing frequency and intensity of heatwaves.

6.14 REASSESSING OVERHEATING CRITERIA FOR SUBTROPICAL AUSTRALIAN HOMES

Kim et al. (2023) investigated the applicability of the CIBSE TM59 overheating criteria in Australian residential buildings located in a subtropical climate. The study involved a longitudinal assessment of thermal comfort and indoor temperatures in 162 dwellings across the Sydney–Wollongong and Brisbane–Gold Coast regions over a two-year period.

Temperature measurements were recorded at 15-minute intervals in living rooms and bedrooms, complemented by occupant surveys evaluating perceived thermal comfort and sleep quality.

The analysis revealed that 68.9% of surveyed bedrooms exceeded the static overheating threshold of 26°C for more than 1% of nighttime hours. Furthermore, 17.5% of living rooms and 19.2% of bedrooms during the day failed to meet the adaptive overheating criterion established by CIBSE TM59, which defines overheating as exceeding the adaptive thermal comfort limit by more than 3.5 K for over 3% of occupied hours. On average, bedrooms exhibited a greater frequency of exceedance (1.9%) compared to living rooms (1.4%), with the most extreme case registering an exceedance rate of 22.8% in a single bedroom.

Despite these measured exceedances, survey data indicated a discrepancy between objective thermal criteria and subjective occupant perceptions. Approximately 80% of respondents reported acceptable indoor thermal conditions during both daytime and nighttime periods.

Regression analysis further demonstrated that thermal discomfort and sleep quality were not statistically correlated with temperatures above 26°C in bedrooms. A statistical model projected that the probability of experiencing thermal discomfort remained below 20% even at temperatures approaching 29°C, suggesting an adaptive behavioral response among residents in subtropical climatic conditions (Kim et al., 2023).

The authors concluded that the static 26°C threshold prescribed in CIBSE TM59 for bedrooms may be overly conservative in subtropical environments and does not align with empirically observed thermal acceptability. They argued for a revision of overheating criteria to better reflect regional climatic conditions and occupant adaptation mechanisms.

Furthermore, the study highlighted the potential unintended consequences of applying excessively stringent overheating standards, as such policies could inadvertently promote increased reliance on mechanical cooling rather than fostering the adoption of passive design strategies such as natural ventilation and climate-responsive architectural solutions.

6.15 OVERHEATING IN ANDALUSIAN SOCIAL HOUSING: PASSIVE COOLING STRATEGIES

Caro and Sendra (2024) examined the overheating risks in social housing units located in historic buildings in Andalusia, Spain, during the extreme summer heat of 2022. Their study analyzed three apartments within two heritage-listed buildings in Córdoba and Granada, which are subject to the highest level of conservation protection under Spanish law. The research spanned a full year and incorporated data on indoor thermal conditions, including air temperature, relative humidity, and CO_2 concentration.

The study revealed that none of the monitored rooms met the CIBSE TM52 and TM59 overheating criteria. In the living rooms of the Córdoba apartments (C1 and C2), the overheating threshold (He), set at 3% of occupied hours, was significantly exceeded, reaching 18% and 15%, respectively. In the C2 bedroom, which lacked air conditioning, temperatures ranged from 32°C to 34°C and, at times, peaked at an extreme 40°C. Additionally, in apartments C1 and C2, the daily exceedance threshold of 6°C·h, as defined by the second criterion of CIBSE TM52, was surpassed for 24 and 25 consecutive days, respectively (Caro & Sendra, 2024).

Even extensive air conditioning use in the living rooms (44% and 52% of occupied hours in C1 and C2, respectively) failed to prevent overheating. The C2 dwelling, characterized by high thermal mass and an east-facing orientation, exhibited greater susceptibility to overheating compared to C1, which had a lighter construction and a south-facing orientation.

The authors concluded that historic social housing in Mediterranean climates, such as Andalusia, requires adaptive strategies to mitigate overheating risks. They emphasized the importance of improving natural ventilation and implementing passive cooling measures, which could reduce heat retention and improve indoor comfort and safety during extreme heat events.

6.16 HEAT VULNERABILITY IN AUSTRALIAN SOCIAL HOUSING: LARGE-SCALE STUDY

Haddad et al. (2022) conducted a large-scale field study assessing the impact of extreme climatic conditions on energy efficiency and occupant well-being in low-income housing in the subtropical climate of Australia. The research analyzed 106 dwellings between 2018 and 2019, with an extended summer study in 2019 covering 109 Australian households in New South Wales. The study collected demographic, health, and housing condition data, in addition to evaluating thermal comfort using the seven-point ASHRAE scale. Residents were surveyed about environmental health concerns, and a total of 98 subjective responses were gathered.

Additionally, indoor temperature, humidity, and air quality were continuously monitored.

Findings indicated that thermal conditions in all surveyed dwellings failed to meet even minimal comfort standards. During summer, indoor temperatures peaked at 39.8°C, while the number of hours exceeding 26°C ranged from 416 to 2,150 in western Sydney. In the most overheated apartments, temperatures surpassed 30°C for 141 hours and 26°C for 997 hours. Income disparities were found to have a direct impact on indoor heat exposure, with lower-income households facing higher overheating risks (Haddad et al., 2022).

The authors stressed the urgent need for stricter building standards and urban adaptation measures to enhance living conditions and public health in Australian social housing. Their research underscored the importance of integrating passive cooling strategies and improving construction quality to reduce reliance on air conditioning. The study called for immediate legislative action to ensure that future social housing developments provide thermally safe and livable environments.

6.17 OVERHEATING IN TEGUCIGALPA: THE ROLE OF BUILDING TYPOLOGY

Gamero-Salinas et al. (2020) investigated residential overheating risks in Tegucigalpa, Honduras, during the hottest period of the tropical climate season. Their study analyzed 12 buildings divided into 2 primary categories: single-family dwellings (SD) and multi-family apartment buildings (AT). The study examined how different building typologies responded to high temperatures and humidity levels, assessing variations in indoor thermal performance based on architectural design and ventilation strategies.

The study spanned 41 days of indoor temperature monitoring and applied the CIBSE TM52 methodology alongside adaptive approaches from EN15251 and ASHRAE 55 to assess overheating risks.

The results revealed that overheating hours (He) reached up to 20.3% (133 hours) in SD buildings and 12.5% (61 hours) in AT buildings. The significantly higher overheating risk in SD structures was attributed to their direct exposure to solar radiation and low thermal insulation in walls and roofs. Meanwhile, in AT buildings, limited natural ventilation—especially in cases where cross-ventilation was not possible due to single-sided façades—exacerbated nighttime overheating risks. As a result, only 56% of AT residents reported opening windows at night, compared to 100% of SD residents (Gamero-Salinas et al., 2020).

The authors recommended that in tropical regions such as Tegucigalpa, adaptive strategies should include roof insulation, high-albedo surfaces, ventilated roofing systems, and improved shading techniques. Such measures could mitigate overheating risks and enhance thermal comfort, particularly in areas where air conditioning access is limited.

6.18 FUTURE OVERHEATING IN SPANISH SOCIAL HOUSING: NEED FOR RETROFITTING

Escandón et al. (2022) conducted a simulation-based study assessing the risk of overheating in social housing in southern Spain under both current climatic conditions and projected climate change scenarios for 2050. Their analysis focused on residential buildings in Seville, representing approximately 42,000 dwellings constructed between 1940 and 1980, before global energy demand regulations were introduced. These buildings, primarily built during the post-war urban expansion, now constitute a particularly vulnerable segment of Spain's housing stock, prone to prolonged exposure to extreme temperatures.

The study highlighted that these structures generally lack thermal insulation, have inefficient cooling systems, and present limited opportunities for retrofitting due to the low-income status of their residents. Many of these dwellings are inhabited by elderly individuals, who are especially susceptible to the adverse health effects of prolonged overheating. To develop a representative simulation model, the research focused on the most prevalent social housing typology in Seville—linear multi-family buildings, which account for over 40% of the city's housing stock.

Findings revealed that the majority of the analyzed buildings already fail to meet the overheating criteria defined by CIBSE (including TM52 and TM59). Under present climatic conditions, nearly 38% of the buildings exceed two out of three overheating criteria, classifying them as at risk of overheating. Projections for 2050 indicate that this percentage will rise to nearly 100%, meaning that all analyzed buildings will be classified as overheating-prone (Escandón et al., 2022).

Nocturnal overheating emerged as a particularly critical issue, given the stricter nature of adaptive thermal comfort criteria during nighttime hours. In the current climate, the median percentage of nighttime overheating hours (OHn) is approximately 36%, whereas by 2050, this figure is expected to rise to 78%. These projections highlight a substantial increase in thermal discomfort, especially among the most vulnerable groups, including older adults and low-income residents.

While adaptive comfort models offer a more dynamic representation of indoor thermal conditions in naturally ventilated buildings, the study found that even with a nighttime ventilation rate of 6 h^{-1}, overheating risk was reduced by only 10%, leaving 99.7% of buildings categorized as high-risk.

The authors argue that failing to implement large-scale renovation measures for social housing in southern Spain will result in a significant deterioration of living conditions and an increased likelihood of heat-related health issues. They emphasize the urgent need for adaptation strategies that not only reduce energy consumption and emissions but also ensure adequate thermal comfort for residents.

6.19 POWER OUTAGES AND OVERHEATING IN GUANGZHOU DURING HEATWAVES

Similar concerns regarding overheating risks in warm climates have been documented in other regions, as demonstrated by the work of Chen et al. (2024), who investigated the impact of heatwaves on residential buildings in Guangzhou, China. Their study involved 3,720 simulation cases across three distinct building types: urban village houses (UVH), old houses (OH), and contemporary high-rise apartments (CH). The research also incorporated the analysis of heat blackout incidents—scenarios in which power outages eliminate access to air conditioning.

The severity of overheating under power failure conditions was exemplified by an extreme event in July 2019, when a two-day heatwave with peak temperatures of 38°C caused widespread electricity outages, rendering air conditioning systems inoperative. The study found that urban village houses (UVH) were the most vulnerable to overheating, with a 100% overheating probability after three days without electricity. Old houses (OH) also exhibited high susceptibility, with 78% classified as overheating-prone after five days without cooling. Even modern apartments (CH), which generally feature better insulation, faced significant overheating risks, with 58% reaching critical temperature levels after five days of power loss.

In extreme cases, residents' core body temperatures rose to 37.81°C for older adults and 37.28°C for young adults, while cumulative water loss exceeded 30% of body weight after seven days without cooling—posing severe health risks.

The authors stress the urgent need to establish standardized overheating assessment criteria for the most at-risk regions of China. Implementing such guidelines would improve risk management strategies and support the development of more climate-resilient buildings (Chen et al., 2024).

A comparison of both studies underscores that overheating in social housing is a growing concern in the context of climate change, particularly for older, thermally uninsulated buildings inhabited by low-income populations. Both studies highlight the necessity of implementing adaptation measures, including building retrofits, improved insulation, enhanced ventilation strategies, and stricter thermal protection standards for residents.

6.20 HEATWAVE CLASSIFICATION AND OVERHEATING RISKS IN CHINESE HOMES

Liu et al. (2024) conducted a quantitative study aimed at classifying heatwaves and assessing their impact on indoor overheating in residential buildings. Their analysis was based on two decades of meteorological data from eight cities in China, enabling a comprehensive characterization of heatwave patterns and their influence on indoor thermal conditions. The classification methodology incorporated both heatwave duration and intensity, allowing the identification of four distinct categories: short and moderate, short and intense, long and moderate, and long and intense. The results indicated that short and moderate heatwaves were the most frequent events, while long and intense heatwaves had the most pronounced impact on indoor overheating, underscoring their significance in assessing building resilience.

The methodological framework outlines the sequential steps of the study, from data acquisition to indoor overheating assessment. The initial phase involved collecting historical meteorological data from 2003 to 2022 to identify real-world heatwaves. These events were subsequently classified using a statistical analysis of duration and intensity distributions. The final step consisted of simulating indoor thermal conditions in buildings to evaluate their susceptibility to overheating under different extreme temperature scenarios.

The study examined three dominant residential building typologies in China: row houses, low-rise buildings, and high-rise buildings. To simulate realistic overheating conditions, the buildings were modelled in a free-running state, without active cooling systems, allowing for the assessment of heat stress during power outages or in residences lacking air conditioning. The simulations, conducted using the DeST-h tool, calculated the Indoor Overheating Degree (IOD) index, which accounted for both the intensity and frequency of thermal comfort exceedances.

Simulation results indicated that high-rise buildings were the most susceptible to overheating, primarily due to their low surface-to-volume ratio and the increased presence of west-facing windows, which amplified solar heat gains. Furthermore, limited natural ventilation on upper floors hindered nighttime heat dissipation, exacerbating overheating risks. While row houses and low-rise buildings exhibited lower overall overheating risks, they remained vulnerable during prolonged and intense heatwaves.

The precise classification of heatwaves enabled a more accurate prediction of overheating conditions in buildings and facilitated the adaptation of resilience strategies to extreme temperature scenarios. The findings suggest that building design standards should incorporate differentiated responses to various heatwave types, enhancing passive cooling solutions, solar shading systems, and optimizing natural ventilation strategies. Liu et al. (2024) emphasize that assessments of building thermal resilience should account not only for average summer conditions but also for extreme weather events, which are expected to play an increasingly significant role in occupant comfort.

Their research aligns with previous studies on building overheating in Spain (Escandón et al., 2022) and China (Chen et al., 2024), offering a novel framework for quantitatively assessing thermal risks in residential architecture. Integrating heatwave classification into building design and retrofitting efforts could substantially improve climate resilience while minimizing adverse health effects for residents.

6.21 THERMAL COMFORT AND OVERHEATING RISKS IN PAMPLONA: A DATA-DRIVEN CLASSIFICATION APPROACH

Gamero-Salinas et al. (2024) conducted an empirical and simulation-based study assessing the risk of overheating and thermal discomfort in residential buildings during heatwaves in Pamplona, Spain. The study combined long-term indoor temperature monitoring with unsupervised machine learning techniques to classify dwellings according to their thermal performance and explore the underlying causes and consequences of indoor overheating.

The authors monitored 43 dwellings across Pamplona during the summer of 2022, a period marked by extreme heat events, including a prolonged heatwave in July with daily temperatures exceeding 37°C. The monitoring campaign recorded indoor air temperature and humidity, which were used to compute the Standard Effective Temperature (SET*) for each dwelling. Additionally, a post-occupancy survey was conducted to evaluate user perceptions, thermal discomfort, behavioral adaptation, and the presence or absence of mechanical cooling systems.

Using the collected data, the authors applied K-means clustering to identify five thermal performance categories (clusters) ranging from well-performing to highly overheated homes. The classification was based on cumulative SET*, daily overheating hours, and peak indoor temperatures. The results revealed significant disparities in indoor thermal resilience, largely associated with construction period, envelope quality, and dwelling orientation. Dwellings built before 1980, typically lacking thermal insulation, were overrepresented in the most overheated clusters.

Notably, more than half of the monitored dwellings experienced SET* values above 28°C for extended periods, exceeding recommended comfort thresholds. Residents in these homes reported higher thermal discomfort, increased use of adaptive behaviors (e.g., fans, shading, changes in clothing), and greater vulnerability during heatwaves. In contrast, newer dwellings or those with retrofitted envelopes maintained more stable indoor conditions, often staying within acceptable comfort ranges despite high outdoor temperatures.

The analysis also identified a mismatch between objective indoor conditions and occupant perceptions in certain cases. Some users underestimated their exposure to overheating, while others reported discomfort despite measured values remaining within standard comfort thresholds. This finding underscores the complexity of evaluating thermal resilience based solely on physical metrics.

Gamero-Salinas et al. (2024) concluded that building typology and construction period are key determinants of indoor thermal performance during extreme heat. They advocated for the inclusion of post-occupancy monitoring and behavioral surveys in building evaluations and called for retrofitting programs targeted at low-performing dwellings, particularly in areas lacking mechanical cooling. Significantly, the authors argued that, in the context of escalating climate extremes—even in the temperate climate of Pamplona—passive strategies alone may no longer suffice, and the adoption of mechanical cooling systems may become necessary to ensure thermal safety and comfort. The study demonstrated the utility of unsupervised learning techniques in identifying thermal performance patterns and recommended their integration into larger-scale resilience assessments across urban housing stocks.

6.22 SUMMARY OF OVERHEATING RISKS ACROSS CLIMATIC CONTEXTS

The growing threat of residential overheating has become evident across all climate zones, as demonstrated by empirical and simulation-based studies conducted in 21 distinct geographic, architectural, and socio-economic contexts. While the mechanisms and manifestations of overheating differ between hot, temperate, and cold climates, the overarching trend points to a convergence of vulnerabilities driven by climate change, urbanization, and outdated building stock. This section summarizes key findings by climate type, highlighting both shared and region-specific challenges.

6.22.1 Hot Climates: Persistent Thermal Stress and Infrastructure Limitations

In hot and subtropical regions such as Andalusia (Spain), Queensland (Australia), Tegucigalpa (Honduras), Guangzhou (China), and southern China, the risk of overheating is not episodic but structural. Studies in these regions consistently report indoor temperatures well above critical comfort and health thresholds, with limited capacity for nocturnal heat dissipation.

In Andalusian social housing (Caro & Sendra, 2024), even intensive air conditioning usage failed to prevent rooms from reaching 40°C. Similarly, Australian low-income households experienced temperatures exceeding 30°C for hundreds of hours per season (Haddad et al., 2022). In Tegucigalpa, overheating reached 20.3% of occupied hours, disproportionately affecting single-family homes with poor insulation (Gamero-Salinas et al., 2020). Chinese studies (Chen et al., 2024; Liu et al., 2024) further revealed extreme vulnerability during power outages, with core body temperatures and dehydration rates reaching dangerous levels after just three days without cooling.

A common theme across hot-climate studies is the combination of weak building envelopes, inadequate shading, and limited ventilation, which leads to chronic overheating even in the absence of heatwaves. The adaptive capacity of low-income populations is particularly constrained by infrastructure gaps and energy insecurity. Passive strategies such as roof insulation, high-albedo surfaces, and ventilated roofs are repeatedly cited as cost-effective and urgently needed solutions. However, their implementation is hindered by economic and regulatory barriers.

Additionally, studies from China and Honduras emphasize the role of building typology and orientation. High-rise and west-facing units are significantly more prone to overheating, with limited cross-ventilation exacerbating nighttime thermal stress. These insights underscore the importance of climate-sensitive design at both building and urban scales in hot environments.

6.22.2 Temperate Climates: Escalating Risks in Energy-Efficient and Heritage Buildings

In temperate climates such as the UK, Ireland, the Netherlands, Germany, and northern Spain, recent research documents a rapid emergence of overheating as a major challenge, especially in urban areas and energy-efficient dwellings.

In London (Pathan et al., 2017) and across England (Lomas et al., 2021), newer buildings constructed to high insulation standards often experienced greater overheating than older stock, particularly during nighttime hours. Similar patterns were observed in Amsterdam (Peerlings et al., 2024) and Dublin (Colclough & Salaris, 2024), where nearly zero-energy buildings (nZEBs) exhibited critical temperature exceedances despite meeting sustainability benchmarks. Notably, bedrooms—often smaller and poorly ventilated—were the most vulnerable zones in all cases.

Overheating in heritage and social housing presents a dual challenge: buildings in Andalusia (Caro & Sendra, 2024) and Seville (Escandón et al., 2022) lack mechanical cooling and modern insulation, yet are often protected from retrofit due to conservation regulations.

Similarly, low-income households in Ireland and Pamplona reported high discomfort levels even when measured temperatures appeared acceptable, underscoring the gap between standardized thresholds and occupant perception (Gamero-Salinas et al., 2024).

Several studies stress that passive strategies alone may no longer suffice, particularly in compact dwellings with limited cross-ventilation. Across cases, heatwaves are increasingly exceeding building resilience, even in regions previously considered climatically benign. The need for regulatory reform is repeatedly highlighted, with calls for integrating overheating risk into building codes, renovation schemes, and thermal performance standards.

Machine learning approaches (e.g., Pamplona study) have also shown promise in identifying thermal vulnerability at the housing-stock level, offering scalable tools for prioritizing retrofitting interventions.

6.22.3 Cold Climates: Emerging Overheating in Nordic Regions and Cold Continental Zones

Traditionally, cold regions such as Finland and Norway are now facing overheating risks, particularly in summer months and during extreme weather events. While the absolute temperature levels are lower than in hot climates, the thermal stress is often amplified by building design choices optimized for winter performance.

In Finland, studies conducted across over 20,000 apartments (Velashjerdi Farahani et al., 2023, 2024) found that lightweight single-family houses responded more rapidly to outdoor heat, while older buildings lacking retrofits exhibited sustained overheating. In newer constructions, overheating was delayed but eventually accumulated, particularly in the context of prolonged heatwaves.

Simulation studies (Sukanen et al., 2023) confirm that even highly insulated CLT apartments will exceed critical comfort thresholds by 2100, with living spaces reaching 16,000°C·h of overheating annually. Night ventilation and favorable orientations helped reduce exposure but were insufficient under future warming scenarios. Similarly, research in Norway (Tian et al., 2020) showed that buildings with high WWR, especially those facing south, were significantly more susceptible to heat gains. Adaptive assessment criteria, such as CIBSE TM59, captured these nuances more effectively than static ones.

Importantly, current building regulations in Nordic countries—focused on winter heating—do not yet reflect the emerging summer risks. Authors across all studies urge the inclusion of overheating simulations and climate projections (2050, 2080) in building design and permit processes. Recommendations include improved passive cooling, design for cross-ventilation, and, in some cases, selective use of mechanical cooling systems to future-proof the housing stock.

6.22.4 Conclusion: Overheating as a Cross-Climate Challenge

Although the drivers and dynamics of overheating differ across climate zones, the collective evidence from these 21 studies demonstrates a global shift in building vulnerability. Whether due to urbanization, energy retrofits, climate change, or socioeconomic disparities, overheating now represents a critical risk to occupant health, comfort, and energy security in all regions.

- In hot climates, overheating is persistent and exacerbated by infrastructure deficits and energy poverty.
- In temperate climates, it is often a by-product of efforts to improve winter energy performance, highlighting the need for balanced design strategies.
- In cold climates, it is an emerging but growing concern, requiring anticipatory policy responses and design adaptation.

To address these converging challenges, future adaptation must prioritize context-specific strategies, integrate behavioral and perceptual factors, and embed thermal resilience into every stage of building lifecycle planning.

REFERENCES

Beckmann, S. K., Hiete, M., & Beck, C. (2021). Threshold temperatures for subjective heat stress in urban apartments: Analysing nocturnal bedroom temperatures during a heat wave in Germany. *Climate Risk Management*, *32*, 100286. https://doi.org/10.1016/j.crm.2021.100286

Beizaee, A., Lomas, K. J., & Firth, S. K. (2013). National survey of summertime temperatures and overheating risk in English homes. *Building and Environment*, *65*, 1–17. https://doi.org/10.1016/j.buildenv.2013.03.011

Caro, R., & Sendra, J. J. (2024). Effects of an extreme heat event on indoor conditions of social heritage housing in Mediterranean climate. *Energy Reports*, *12*, 2328–2345. https://doi.org/10.1016/j.egyr.2024.08.037

Chen, S., Zhao, H., Luo, K., Li, Q., Nie, Z., & Zhao, L. (2024). Health risk assessment of residential overheating under the heat waves in Guangzhou. *Building and Environment*, *266*, 112089. https://doi.org/10.1016/j.buildenv.2024.112089

Colclough, S., & Salaris, C. (2024). Quantifying overheating in nZEB Irish residential buildings: An analysis of recorded interior temperatures of Irish newbuild and retrofit residential buildings against CIBSE, Passive House and WHO overheating criteria and recorded occupant satisfaction. *Energy and Buildings*, *303*, 113571. https://doi.org/10.1016/j.enbuild.2023.113571

Escandón, R., Suárez, R., Alonso, A., & Mauro, G. M. (2022). Is indoor overheating an upcoming risk in southern Spain social housing stocks? Predictive assessment under a climate change scenario. *Building and Environment*, *207*(Pt. B), 108482. https://doi.org/10.1016/j.buildenv.2021.108482

Gamero-Salinas, J., López-Hernández, D., González-Martínez, P., Arriazu-Ramos, A., Monge-Barrio, A., & Sánchez-Ostiz, A. (2024). Exploring indoor thermal comfort and its causes and consequences amid heatwaves in a southern European city: An unsupervised learning approach. *Building and Environment*, *265*, 111986. https://doi.org/10.1016/j.buildenv.2024.111986

Gamero-Salinas, J. C., Monge-Barrio, A., & Sánchez-Ostiz, A. (2020). Overheating risk assessment of different dwellings during the hottest season of a warm tropical climate. *Building and Environment*, *171*, 106664. https://doi.org/10.1016/j.buildenv.2020.106664

Haddad, S., Paolini, R., Synnefa, A., De Torres, L., Prasad, D., & Santamouris, M. (2022). Integrated assessment of the extreme climatic conditions, thermal performance, vulnerability, and well-being in low-income housing in the subtropical climate of Australia. *Energy and Buildings*, *272*, 112349. https://doi.org/10.1016/j.enbuild.2022.112349

Hamdy, M., Carlucci, S., Hoes, P.-J., & Hensen, J. L. M. (2017). The impact of climate change on the overheating risk in dwellings: A Dutch case study. *Building and Environment*, *122*, 307–323. https://doi.org/10.1016/j.buildenv.2017.06.031

Hamdy, M., Carlucci, S., Hoes, P.-J., & Hensen, J. L. M. (2017). The impact of climate change on the overheating risk in dwellings, a Dutch case study. *Building and Environment*, *122*, 307–323. https://doi.org/10.1016/j.buildenv.2017.06.031

Kim, J., Xiong, J., de Dear, R., Parkinson, T., Jeong, B., Wu, Z., Sadeghi, M., & Chen, D. (2023). Testing the applicability of CIBSE overheating criteria to Australian subtropical residential contexts. *Building and Environment*, *246*, 110987. https://doi.org/10.1016/j.buildenv.2023.110987

Liu, X., Liu, Z., Wu, Y., Hu, S., Bu, F., An, J., Zhou, X., & Yan, D. (2024). A novel quantitative method of heatwave classification for building resilience analysis. *Sustainable Cities and Society*, *112*, 105603. https://doi.org/10.1016/j.scs.2024.105603

Lomas, K. J., Watson, S., Allinson, D., Fateh, A., Beaumont, A., Allen, J., Foster, H., & Garrett, H. (2021). Dwelling and household characteristics' influence on reported and measured summertime overheating: A glimpse of a mild climate in the 2050s. *Building and Environment*, *201*, 107986. https://doi.org/10.1016/j.buildenv.2021.107986

Mulville, M., Harrington, S., Li, C., Raushan, K., Essien-Thompson, E., & Ahern, C. (2025). Dwelling overheating risk in cool climates: Assessing the risk in the context of retrofit and climate change in Ireland. *Indoor Environments*, *2*(1), 100072. https://doi.org/10.1016/j.indenv.2025.100072

Pathan, A., Mavrogianni, A., Summerfield, A., Oreszczyn, T., & Davies, M. (2017). Monitoring summer indoor overheating in the London housing stock. *Energy and Buildings*, *141*, 361–378. https://doi.org/10.1016/j.enbuild.2017.02.049

Peerlings, E., Vranić, S., Ommer, J., Kalas, M., & Steeneveld, G.-J. (2024). Indoor heat in Amsterdam: Comparing observed indoor air temperatures from a professional network and from a citizen science approach. *City and Environment Interactions*, *24*, 100173. https://doi.org/10.1016/j.cacint.2024.100173

Sakka, A., Santamouris, M., Livada, I., Nicol, F., & Wilson, M. (2012). On the thermal performance of low income housing during heat waves. *Energy and Buildings*, *49*, 69–77. https://doi.org/10.1016/j.enbuild.2012.01.023

Sukanen, H., Taylor, J., & Kaasalainen, T. (2023). Passive mitigation of overheating in Finnish apartments under current and future climates. *Indoor and Built Environment*, *32*(7). https://doi.org/10.1177/1420326X231160977

Sukanen, H., Taylor, J., & Kaasalainen, T. (2023). Passive mitigation of overheating in Finnish apartments under current and future climates. *Indoor and Built Environment*, *32*(7). https://doi.org/10.1177/1420326X231160977

Taylor, J., McLeod, R., Petrou, G., Hopfe, C., Mavrogianni, A., Castaño-Rosa, R., Pelsmakers, S., & Lomas, K. (2023). Ten questions concerning residential overheating in central and northern Europe. *Building and Environment*, *234*, 110154. https://doi.org/10.1016/j.buildenv.2023.110154

Taylor, J., McLeod, R., Petrou, G., Hopfe, C., Mavrogianni, A., Castaño-Rosa, R., Pelsmakers, S., & Lomas, K. (2023). Ten questions concerning residential overheating in central and northern Europe. *Building and Environment*, *234*, 110154. https://doi.org/10.1016/j.buildenv.2023.110154

Tian, Z., Zhang, S., Deng, J., & Hrynyszyn, B. D. (2020). Evaluation on overheating risk of a typical Norwegian residential building under future extreme weather conditions. *Energies*, *13*(3), 658. https://doi.org/10.3390/en13030658

Velashjerdi Farahani, A., Jokisalo, J., Korhonen, N., Jylhä, K., & Kosonen, R. (2024). Simulation analysis of Finnish residential buildings' resilience to hot summers under a changing climate. *Journal of Building Engineering*, *82*, 108348. https://doi.org/10.1016/j.jobe.2023.108348

Velashjerdi Farahani, A. V., Kravchenko, I., Jokisalo, J., Korhonen, N., Jylhä, K., & Kosonen, R. (2023). Overheating assessment for apartments during average and hot summers in the Nordic climate. *Building Research & Information*, *52*(3), 273–291. https://doi.org/10.1080/09613218.2023.2253338

7 Active Strategies for Mitigating Heat Stress in Buildings

7.1 PROJECTIONS FOR AIR CONDITIONING DEMAND AND ENERGY CONSUMPTION

Castaño-Rosa et al. (2021) provide a comprehensive analysis of the projected expansion of the air conditioning market in buildings and its associated energy consumption across a wide range of climates, from polar to hot Mediterranean. This approach offers valuable insight into how cooling demand may evolve under different climatic conditions and helps anticipate future trends in other regions with similar profiles.

The study examined 19 cities across seven countries—Cyprus, Finland, Greece, Israel, Portugal, Slovakia, and Spain—encompassing ten distinct climate zones based on the Köppen–Geiger classification. While air conditioning usage is already widespread in some regions, such as Cyprus, where 84% of households have air conditioning, a considerable portion of residents (28.2%) cannot afford to use it due to high energy costs. In Greece, 59.5% of households have air conditioning, compared to 92% in Israel. In contrast, Spain and Portugal show lower adoption rates, with only 35.5% and 13.8% of households, respectively, using air conditioning. Finland and other Nordic countries have historically relied solely on heating, with minimal cooling infrastructure in place.

Projections suggest that in southern European countries, the number of installed air conditioning units may double by 2050. In Portugal, cooling demand is expected to increase by 56–69%. In northern European countries, where air conditioning has not been a standard feature, installation rates are expected to rise significantly, driven by warming trends and the lack of passive cooling systems in many existing buildings. Finland and other Nordic nations, traditionally focused on heating requirements, will need to adapt their building stock to meet growing demand for cooling.

Recent findings by Zhang et al. (2024) reinforce these projections through empirical evidence from the Yangtze River Delta in China. In their two-decade study of 27 cities with hot summers and no central heating, they found that a 1% increase in cooling degree days (CDDs) led to a 0.6% increase in air conditioning ownership among urban households and a 2.2% increase among rural households. This suggests that rural populations, even with lower baseline access, are more responsive to temperature stress when income allows. Furthermore, they observed strong lagged effects: a hot summer caused not only immediate purchases but also sustained growth in ownership over the following three to four years. A 1 °C rise in summer temperature triggered a 21.8% increase in rural ownership in the first year and 58.4%

DOI: 10.1201/9781003645573-7

cumulatively over three years. These patterns indicate that Europe's future air conditioning market will likely be shaped not only by immediate discomfort but also by accumulated heat exposure and delayed behavioral adaptation.

Overall electricity consumption for air conditioning is expected to quadruple, rising from 35 TWh in 2015 to 137 TWh in 2050. In cities such as Madrid, Athens, and Thessaloniki, the number of hours with operative temperatures above 32.2 °C could increase sixfold, substantially increasing the load on cooling systems. Changing climatic conditions in northern Europe, where cooling was previously unnecessary, will necessitate building adaptation and a rise in energy demand for cooling.

In southern Europe, the projected growth in air conditioning demand is expected to increase summer peak electricity loads by 20–30%, requiring significant new generation capacity—for example, an estimated 13 GW in Italy and 10 GW in Spain. In India, peak demand could grow by more than 35% in states such as Punjab and Maharashtra. These projections highlight the increasing stress that cooling-related demand will place on electricity systems in both high- and middle-income countries by mid-century, particularly under high-warming scenarios (Colelli et al., 2023).

A recent global-scale assessment by De Cian et al. (2025) further confirms these projections. Using survey data from 25 countries representing 62% of the global population and over 70% of residential electricity use, the study finds that ownership of air conditioning increases household electricity consumption by an average of 36%, with impacts reaching up to 57% in hotter regions. This increase is not uniform: low-income countries exhibit both lower initial access and greater vulnerability to cost shocks, underscoring the need for differentiated policy responses.

Projections to 2050 under moderate and high-warming scenarios (SSP2–4.5 and SSP5–8.5) suggest that global residential cooling electricity demand will double or triple, rising to 976–1393 TWh annually. This is equivalent to the current total electricity use of countries such as India. India, Indonesia, and China will see the most significant increases—not only due to rising temperatures but also because of large populations and rapid economic growth. In contrast, Africa is expected to have the lowest adoption levels, with only 9–15% penetration by 2050, reflecting affordability barriers and infrastructure limitations.

These findings underscore the long-term infrastructural implications. In India, for instance, meeting additional cooling needs may require a 75–120 GW increase in peak generation capacity. Without significant investments in grid resilience, demand-side efficiency, and distributed generation, cooling-driven peaks could strain electricity systems in many emerging economies.

An earlier analysis by Peacock et al. (2010) projected that, in the near future, the penetration rate of air conditioning in European residential buildings would reach levels observed in the United States for locations with similar cooling degree days. Based on similar assumptions, Jakubcionis and Carlsson (2018) estimated that air conditioning adoption in Poland and Germany, where it is currently rare, would rise to 44.5% and 46.1%, respectively—a tenfold increase. These assumptions have been echoed in official studies commissioned by the UK government, particularly in the Research into Overheating in New Homes—Phases 1 and 2 (Ministry of Housing, Communities & Local Government 2019a, 2019b).

The rapid expansion of the residential air conditioning market, especially in moderate and cold European climates, aligns with the worst-case scenario described by Andreou et al. (2020). Their panel data analysis, based on historical trends across the EU-28, suggests that by 2050, Europe's air conditioning market could reach saturation rates similar to those observed in the United States, assuming rapid adoption in new and retrofitted buildings. Between 2000 and 2015, approximately 3% of all residential buildings in the EU-28 were either newly constructed or renovated annually. These buildings typically exhibit higher rates of air conditioning adoption than unrenovated structures. In 2015, 17.1% of new buildings were equipped with air conditioning, compared to just 8.9% of all existing buildings.

Market development trends observed in the United States suggest that increased adoption in new buildings can encourage further installations in the existing building stock, ultimately transforming the entire European residential cooling market. In moderate climates, where installation rates were previously low, projections indicate that market saturation could reach 87.5% by 2050. The share of moderate and cold climate countries in total European residential cooling energy demand would then rise to 60.1%. Under this scenario, total residential cooling energy consumption in the EU-28 would increase to approximately 3.7% of overall energy use and 11.7% of electricity consumption in buildings by 2050.

The potential rapid increase in air conditioning installations in moderate European climates must also consider its broader implications for total energy demand. Reducing adoption rates could yield greater energy savings than simply limiting operating hours through predefined thresholds. In addition to direct consumption, every new installation adds embedded energy demand related to manufacturing and installation. Another critical consideration is the urban heat island (UHI) effect, in which widespread use of air conditioning in dense urban environments raises nighttime temperatures and further exacerbates cooling demand.

7.2 BEHAVIORAL AND MARKET FACTORS INFLUENCING AIR CONDITIONING USE

Studies indicate that air conditioning adoption in moderate-climate European countries will largely depend on individual homeowner decisions, particularly during prolonged and intense heatwaves, which are expected to become more frequent. Historical data suggest that extreme heat events can accelerate adoption, as seen in Athens, where homes traditionally had only one air-conditioned room. Following the exceptionally hot summer of 2007, installing air conditioning throughout the home became a common practice (Puckett and Gethering, 2019).

Furthermore, once installed, air conditioning units are often used more frequently than necessary. Research shows that usage patterns are not always based on actual thermal comfort requirements but are shaped by user expectations and behavior. A study conducted in the United Kingdom by He et al. (2005) found that air conditioning was activated at indoor temperatures between 22.7 °C and 25 °C, maintaining interior conditions between 18.8 °C and 22.1 °C. The average nightly operation time was seven hours. Similarly, Pathan et al. (2008) reported that UK residents typically switched on air conditioning when indoor temperatures reached 24–25 °C, with an average daily operation time of five hours and seven hours at night.

Zhang et al. (2024) provide further insights into behavioral and economic drivers of cooling behavior. In their study of the Yangtze River Delta (YRD), rural households displayed an income elasticity for air conditioning ownership significantly greater than one, confirming that cooling remains a luxury good in these settings. By contrast, urban households showed an elasticity below one, suggesting that air conditioning has become a necessity in more developed areas. This pattern is highly relevant for European contexts, where affordability continues to limit access.

Additionally, Zhang et al. (2024) found that the frequency of extremely hot days, measured using temperature bins exceeding 30 °C, was a statistically significant driver of new installations, even after controlling for income and population factors. In contrast, intra-seasonal temperature variation—such as spring or autumn fluctuations—had no measurable effect on ownership rates. These results indicate that consumer responses are closely linked to acute thermal stress rather than seasonal averages or moderate discomfort, echoing findings from other global regions.

Even by 2050, approximately 640 million people in India and 60 million in Europe are projected to remain without access to air conditioning, despite being exposed to daily maximum temperatures above 24 °C. As shown in the study by Colelli et al. (2023), this persistent gap highlights enduring inequalities in adaptive capacity between and within regions, particularly where income constraints and infrastructural deficits limit access to cooling.

According to Falchetta et al. (2024), the number of people exposed to above-average annual cooling degree days (CDDs) without access to air conditioning could rise from 2.5 billion in 2020 to 3.2 billion by 2050 under the SSP2–4.5 scenario. Under the more pessimistic SSP3–7.0 pathway, this figure may exceed 4 billion, with the largest exposure gaps concentrated in sub-Saharan Africa and South Asia. These projections highlight the growing global scale of the cooling gap and the disproportionate burden faced by vulnerable populations. Falchetta and colleagues also document the regressive distribution of air conditioning electricity use in regions such as Latin America, the Middle East, and North Africa—and even Europe—where low-income households often consume similar or greater amounts of electricity for cooling than wealthier ones due to residence in high-exposure areas. As a result, they bear a disproportionately high energy burden relative to income, further entrenching inequality.

The marginal cost of reducing heat exposure also varies substantially between regions. Colelli et al. (2023) estimate that achieving a one-unit reduction in person-degree day (PDD) exposure requires approximately 0.53 kWh of electricity in Europe, compared to just 0.10 kWh in India. In emissions terms, this equates to 0.16 kg of CO_2 per PDD avoided in Europe and 0.09 kg in India. These differences reflect disparities in climatic baselines, income levels, and the structure and carbon intensity of electricity systems in the two regions.

A global analysis by De Cian et al. (2025), based on data from nearly 700,000 households across 25 countries, offers additional depth. The study finds that once air conditioning is adopted, it becomes the dominant driver of household electricity use. Ownership is associated with a 36% increase in annual consumption, with the effect scaling nonlinearly with temperature exposure—reaching up to 57% under high CDD conditions. Crucially, this intensive-margin effect varies by income group: wealthier households not only operate more or larger units, but also maintain usage at higher CDD thresholds. By contrast, poorer households tend to reach peak usage at lower heat exposures, indicating financial constraints on sustained operation.

These disparities in usage result in unequal adaptive capacity. For instance, air conditioning users in the lowest global income quintile consume on average 679 kWh annually for cooling, while users in the wealthiest quintile exceed 1,400 kWh. Despite owning cooling devices, lower-income households may refrain from using them due to electricity cost concerns—a phenomenon identified by the authors as "cooling poverty." This concept captures the financial burden of maintaining thermal comfort, particularly for households in emerging economies where cooling-related electricity costs may exceed 5–8% of total expenditure. The regressive nature of this burden is evident in multiple countries, including China, Pakistan, and India, where poor households allocate a disproportionately high share of their budget to cooling needs.

The study also demonstrates that both ownership and usage patterns are highly sensitive to economic and climatic factors. By 2050, under moderate and high-warming scenarios, air conditioning ownership is projected to rise significantly even in countries with historically low penetration. However, without mitigation through policy, the associated increase in electricity expenditure may worsen existing inequalities. The analysis also identifies a mitigating role for decentralized renewables. For example, households in regions with above-median solar PV generation consume 25% less electricity for cooling—likely due to self-generation and higher price responsiveness. This finding highlights the potential for distributed energy systems to alleviate cost-related barriers to thermal adaptation.

According to UNICEF (2023), approximately 2 billion air conditioning units are currently in operation worldwide, contributing to an estimated 1 gigaton of CO_2 emissions annually. China remains the largest market, accounting for 37% of global installed cooling capacity in 2020 (Zhao et al., 2025). Between 2000 and 2021, the number of air conditioning units in China increased sixfold, with further rapid expansion projected. In 2020, air conditioning-related emissions in China totaled 378.9 million tons of CO_2. By 2040, cooling energy consumption in the country is expected to double compared to 2020 levels. By 2100, projected CO_2 emissions from cooling systems may reach 1,095.4 million tons under the SSP2–4.5 scenario—assuming no improvements in energy efficiency or grid decarbonization.

In the United States, data from the Residential Energy Consumption Survey (RECS, U.S. Energy Information Administration 2020) show that air conditioning is used in 83% of homes built before 1950 and 93% of homes built between 2010 and 2020. Over the past decade, heat pumps have become a more popular alternative to traditional air conditioning, offering both heating and cooling capabilities. However, adoption remains relatively slow: annual sales of heat pumps total around 3 million units, compared to over 5 million conventional air conditioning units sold each year. Policymakers and industry experts suggest that widespread adoption of heat pumps could be accelerated through targeted incentives—such as direct subsidies or tax credits for each installed unit.

7.3 THE MITIGATION–ADAPTATION TRADE-OFF IN RESIDENTIAL COOLING

The widespread adoption of air conditioning in response to rising temperatures offers clear adaptation benefits, yet it also poses a substantial challenge for climate change mitigation. As cooling demand grows, so too does electricity consumption, much of which is still generated from carbon-intensive sources. This duality—between protecting public health through thermal comfort and accelerating emissions from increased energy use—forms the core of the mitigation–adaptation trade-off now confronting residential cooling systems.

Recent empirical analysis by Colelli et al. (2023) illustrates this tension through projections of air conditioning expansion in two contrasting regions: Europe and India. By 2050, under a high-warming SSP5–8.5 scenario, the share of households with air conditioning is expected to rise from 19% to 41% in Europe and from 10% to 40% in India. This growth would reduce daily heat exposure, measured in person-degree days, by 150 million in Europe and 3.8 billion in India. However, the adaptation benefit comes at a cost: annual electricity demand would increase by 34 TWh in Europe and 168 TWh in India, representing 2% and 15% of each region's current electricity use, respectively. If powered by today's energy mix, this would result in an additional 130 million metric tons of CO_2 emissions, of which 120 $MtCO_2$ would come from India.

The cost of reducing heat exposure through air conditioning varies substantially across contexts. On average, a one-unit reduction in person-degree day exposure in Europe requires five times more electricity and twice the carbon emissions than in India—0.53 kWh and 0.16 kg CO_2 in Europe versus 0.10 kWh and 0.09 kg CO_2 in India. These differences reflect both climatic and socioeconomic asymmetries: Europe benefits from a cleaner electricity mix and milder baseline temperatures, while India faces greater cooling needs and a more carbon-intensive grid. Moreover, within both regions, the burden of this trade-off scales with income. Wealthier jurisdictions face 50–75% greater increases in electricity demand and emissions per unit of adaptation, except in locations with large hydropower shares.

A key insight from the study is the distinction between extensive and intensive adaptation margins. In India, income-driven increases in air conditioning ownership—the extensive margin—are the main contributor to electricity demand growth, accounting for more than two-thirds of the total increase in most states. In Europe, by contrast, higher usage of existing units—the intensive margin—plays a larger role, particularly in southern countries where air conditioning is already common. Future warming will amplify both margins: rising temperatures will increase cooling needs, while rising incomes and urbanization will drive new installations. Even after accounting for market saturation and economic disparities, total electricity demand is projected to increase by 34 TWh in Europe and 168 TWh in India (see also Fig. 4 and Table 1 in Colelli et al., 2023).

Moderating the mitigation–adaptation trade-off will require a multi-pronged strategy. First, decarbonizing electricity generation is essential. Even modest reductions in the carbon intensity of Europe's power sector—from 270 to 265 g CO_2/kWh—and more substantial ones in India—from 775 to 700 g CO_2/kWh—could significantly offset emissions associated with rising demand. Second, improving the energy efficiency of cooling appliances can reduce electricity use without sacrificing comfort. Raising the seasonal energy efficiency ratio (SEER) to best available technology levels could cut electricity demand by 50% in Europe and 40% in India.

Third, behavioral changes such as increasing thermostat setpoints or combining air conditioning with fans can be highly effective. Colelli et al. (2023) estimate that using fans to enhance perceived cooling could reduce household electricity demand by 40–60% in Europe and 50–60% in India, depending on usage patterns. These strategies are especially effective during moderate heat events, which are projected to increase in frequency. Time- and space-shifting behaviors—such as using shared cooled spaces or avoiding outdoor activity during peak heat hours—provide additional opportunities for adaptation without increasing energy demand.

Structural interventions will also be crucial. Building energy codes, reflective roofing, passive ventilation, urban greening, and smart grid integration can reduce baseline cooling needs and peak electricity loads, easing the tension between adaptation and decarbonization. Despite these options, projections indicate that by mid-century, nearly 640 million people in India and 60 million in Europe will still be exposed to high temperatures without access to air conditioning. This underscores the urgency of equitable and sustainable cooling strategies.

Global projections from Falchetta et al. (2024) reinforce this urgency. They estimate that electricity demand for residential air conditioning will rise from 1,220 TWh in 2020 to between 1,590 and 2,377 TWh by 2050, depending on socioeconomic and climate scenarios. Associated CO_2 emissions are projected to reach 590 to 1,365 Mt CO_2e annually, with Asia and the Americas accounting for the majority. While these results broadly align with IEA forecasts, they underscore the cumulative impact of cooling on energy systems and the need for systemic responses.

As this chapter transitions toward passive cooling strategies, the findings outlined here serve as a caution. Mechanical air conditioning, while effective in the short term, is not a sustainable universal solution. The path to resilient and low-carbon adaptation will require integrated design, policy, and behavioral strategies that reduce emissions while maintaining thermal protection.

7.4 RECONSIDERING ADAPTATION: TOWARD A PASSIVE COOLING PARADIGM

While mechanical air conditioning is increasingly viewed as the default adaptation strategy in response to extreme heat, its growing dominance may not be inevitable, nor universally optimal. A comparative study by De Cian et al. (2019) across eight OECD countries highlights the importance of behavioral, demographic, and cultural factors in shaping household responses to rising temperatures. In countries such as France, the Netherlands, and Switzerland, adoption of air conditioning has historically remained low despite rising heat exposure. Instead, passive measures such as thermal insulation and natural ventilation are deeply embedded in building traditions and policy frameworks, leading to different trajectories of thermal adaptation.

The study found that in higher-income European contexts, variables such as household composition, age of occupants, environmental attitudes, and urban density significantly influenced adaptation choices. Notably, older individuals—despite being more vulnerable to heat—were less likely to use air conditioning, whereas families with children exhibited higher adoption rates, motivated by concerns over comfort and perceived health risks. These patterns underscore that thermal comfort is not determined solely by temperature or income, but by cultural norms, risk perceptions, and long-standing housing preferences.

De Cian et al. (2025) also emphasize that passive strategies, such as thermal insulation, offer dual benefits: they reduce exposure to heat and lower the operational energy demand for cooling. Unlike air conditioning, which is reactive and energy-intensive, insulation represents a structural solution that simultaneously supports mitigation. Yet its uptake appears less sensitive to short-term weather extremes and more dependent on policy instruments—such as building codes, subsidies, and renovation programs. This distinction reinforces the idea that adaptation is not only a technical decision but also an institutional and temporal one.

Findings by Colelli et al. (2023) further underscore the value of passive and behavioral approaches. Their mid-century projections for Europe and India demonstrate that the mitigation–adaptation trade-off becomes particularly severe under high air conditioning prevalence, unless it is counterbalanced by decarbonization and end-use efficiency. However, the study identifies several promising alternatives. Passive cooling strategies such as reflective roofs, vegetative infrastructure, and thermally efficient envelopes are among the most effective ways to reduce baseline cooling needs. Combined approaches—for example, pairing air conditioning with fans—can reduce electricity use by up to 60% without compromising comfort. These interventions, along with time-shifting and demand-side flexibility, form the foundation for a more sustainable cooling paradigm.

Falchetta et al. (2024) highlight another key insight: income growth, not climate warming, is the main driver of increased air conditioning adoption and electricity use. This finding suggests that adaptation strategies must address the underlying socioeconomic drivers of demand. In this context, passive and behavioral solutions may offer more equitable and climate-aligned pathways to long-term resilience.

As the discussion transitions to the next chapter, which presents the theoretical underpinnings of passive cooling, these findings prompt a re-evaluation of the current paradigm. They suggest that the future of climate adaptation in the built environment may not rest solely on expanding access to mechanical systems, but rather on rethinking how buildings are designed, insulated, and regulated to provide comfort with resilience. The challenge is not merely to cool buildings, but to do so affordably, equitably, and in a manner consistent with long-term sustainability goals.

7.5 CONCLUSION

The accelerating demand for air conditioning across climate zones signals a pivotal shift in how societies adapt to thermal stress. As this chapter has demonstrated, mechanical cooling is no longer a marginal or regionally confined phenomenon. Its diffusion is reshaping energy infrastructures, amplifying peak electricity demand, and exacerbating adaptation inequalities. Although access to air conditioning provides measurable benefits in reducing heat-related health risks, the patterns of ownership and use are uneven, and their long-term impacts complex.

A key finding emerging from this analysis is that air conditioning adoption is not driven solely by climatic need, but also by economic growth, behavioral norms, and infrastructural readiness. This creates a tension between mitigation and adaptation goals that cannot be resolved through technology alone. In wealthier contexts, increased usage of existing systems (the intensive margin) dominates energy growth, while in lower-income settings, ownership expansion (the extensive margin) is the primary driver. Both trends contribute to rising emissions, albeit through different mechanisms.

Crucially, the burden of this transformation is distributed unequally. Lower-income households often face disproportionate energy costs relative to their comfort levels, a condition increasingly described as "cooling poverty." These disparities are compounded by regional differences in electricity grid resilience, carbon intensity, and public policy frameworks.

The chapter also underscores that the cost-effectiveness and emissions intensity of cooling vary widely across geographies. In Europe, achieving the same reduction in heat exposure requires substantially more electricity and emissions than in India, highlighting the need for locally calibrated efficiency measures and energy mixes.

Looking ahead, the challenge is not only technological but institutional. Incentivizing low-carbon cooling solutions, including the adoption of high-efficiency equipment, building energy codes, and equitable access to renewables, will be critical. At the same time, behavioral and architectural strategies must be mainstreamed—from shifting cultural expectations about indoor temperatures to integrating passive cooling in building regulations and urban planning.

The growing reliance on mechanical cooling should not obscure the potential of alternative pathways. As shown in the concluding sections, passive strategies offer a resilient and culturally embedded foundation for managing heat in both new and existing buildings. In this light, cooling should not be framed merely as a matter of consumption, but as a design and policy opportunity. The success of future adaptation will depend on our ability to shift from reactive, energy-intensive systems to integrated, adaptive approaches that address comfort, equity, and climate together.

REFERENCES

Andreou, A., Barrett, J., Taylor, P. G., Brockway, P. E., & Wadud, Z. (2020). Decomposing the drivers of residential space cooling energy consumption in EU-28 countries using a panel data approach. *Energy and Built Environment*, *1*(4), 432–442. https://doi.org/10.1016/j.enbenv.2020.03.005

Castaño-Rosa, R., Barrella, R., Sánchez-Guevara, C., Barbosa, R., Kyprianou, I., Paschalidou, E., Thomaidis, N. S., Dokupilova, D., Gouveia, J. P., Kádár, J., Hamed, T. A., & Palma, P. (2021). Cooling degree models and future energy demand in the residential sector. A seven-country case study. *Sustainability*, *13*(5), 2987. https://doi.org/10.3390/su13052987

Colelli, F. P., Wing, I. S., & Cian, E. D. (2023). Air-conditioning adoption and electricity demand highlight climate change mitigation–adaptation tradeoffs. *Scientific Reports*, *13*, 4413. https://doi.org/10.1038/s41598-023-31469-z

De Cian, E., Falchetta, G., Pavanello, F., Romitti, Y., & Sue Wing, I. (2025). The impact of air conditioning on residential electricity consumption across world countries. *Journal of Environmental Economics and Management, 131*, 103122. https://doi.org/10.1016/j.jeem.2025.103122

De Cian, E., Pavanello, F., Randazzo, T., Mistry, M. N., & Davide, M. (2019). Households' adaptation in a warming climate. Air conditioning and thermal insulation choices. *Environmental Science & Policy, 100*, 136–157. https://doi.org/10.1016/j.envsci.2019.06.015

Falchetta, G., Cian, E. D., & Pavanello, F. et al. (2024). Inequalities in global residential cooling energy use to 2050. *Nature Communications, 15*, 7874. https://doi.org/10.1038/s41467-024-52028-8

He, J., Young, A. N., Pathan, A., & Oreszczyn, T. (2005). "Air Conditioning Energy Use in Houses in Southern England." In Proceedings of the DYNASTEE Conference: Dynamic Analysis, Simulation and Testing Applied to the Energy and Environmental Performance of Buildings, Athens, Greece,12-14 October 2025.

Jakubcionis, M., & Carlsson, J. (2018, February). Estimation of European Union service sector space cooling potential. *Energy Policy, 113*, 223–231. https://doi.org/10.1016/j.enpol.2017.11.012

Ministry of Housing, Communities & Local Government. 2019a. Research into Overheating in New Homes—Phase 1 Report. Accessed 3.01.2025 https://assets.publishing.service.gov.uk/government/uploads/system/uploads/attachment_data/file/835240/Research_into_overheating_in_new_homes_-_phase_1.pdf

Ministry of Housing, Communities & Local Government. 2019b. Research into Overheating in New Homes—Phase 2 Report. Accessed 3.01.2025 https://assets.publishing.service.gov.uk/government/uploads/system/uploads/attachment_data/file/845483/Research_into_overheating_in_new_homes_-_phase_2.pdf

Pathan, A., Young, A., & Oreszczyn, T. 2008. "UK Domestic Air Conditioning: A Study of Occupant Use and Energy Efficiency." In Proceedings of Conference: Air Conditioning and the Low Carbon Cooling Challenge, Network for Comfort and Energy Use in Buildings, London, UK.

Peacock, A. D., Jenkins, D. P., & Kane, D. (2010). Investigating the potential of overheating in UK dwellings as a consequence of extant climate change. *Energy Policy, 38*(7), 3277–3288. https://doi.org/10.1016/j.enpol.2010.01.021

Puckett, K., & Gethering, W. (2019). *Design for climate change* (1st ed.). RIBA Publishing. https://doi.org/10.4324/9780429347641

U.S. Energy Information Administration (EIA). 2020. Residential Energy Consumption Survey (RECS), 2020. https://www.eia.gov/consumption/residential/data/2020/

Zhang, M., Ma, X., Wang, W., Sheng, J., Cao, J., Cheng, Z., & Zhang, X. (2024). Climate adaptation investments: Short-term shocks and long-term effects of temperature variation on air conditioning adoption. *Sustainable Cities and Society, 108*, 105493. https://doi.org/10.1016/j.scs.2024.105493

Zhao, M., Yan, B., Cai, W., & Zhang, C. (2025). Projection of trade-offs of commercial air conditioning: Increasing carbon emission and reducing heat exposure. *Applied Energy, 382*(15), 125332. https://doi.org/10.1016/j.apenergy.2025.125332

8 Passive Cooling as an Alternative to Mechanical Air Conditioning Systems

Theoretical Foundations

8.1 THEORETICAL FOUNDATIONS OF SHADING SYSTEMS

8.1.1 Introduction

Shading systems play a crucial role in passive heat management in buildings, reducing excessive solar heat gains and improving indoor thermal comfort. By limiting the amount of solar radiation reaching building surfaces, these systems effectively mitigate overheating—particularly in hot and temperate climates.

The fundamental mechanisms of shading systems include:

- Reduction of direct solar radiation, preventing excessive heat accumulation on glazing and building envelopes.
- Minimization of surface heating, thereby reducing cooling energy demand.
- Optimization of building heat balance, adapted to seasonal and geographical conditions.

To fully understand the effectiveness of shading systems, it is essential to examine the fundamental physics of solar radiation propagation, its interaction with building surfaces, and its impact on the building's energy balance.

8.1.2 Solar Radiation Propagation and Its Impact on Buildings

Solar radiation reaching a building consists of three primary components:

- Direct radiation (I_{dir})— sunlight that directly strikes the building surface.
- Diffuse radiation (I_{diff})—solar radiation scattered by the atmosphere, clouds, and other objects.
- Reflected radiation (I_{refl})—energy reflected from surrounding surfaces such as adjacent buildings, terrain, or water.

The total incident radiation on a building surface is expressed as:

$$I_{\{tot\}} = I_{\{dir\}} + I_{\{diff\}} + I_{\{refl\}}$$

 DOI: 10.1201/9781003645573-8

The effectiveness of shading systems depends on their ability to control the solar radiation flux, particularly direct and reflected components.

This equation defines the total solar irradiance incident on a building surface as the sum of three components: direct radiation, diffuse sky radiation, and radiation reflected from surrounding surfaces. Because each of these components interacts with the building envelope in different ways, the equation provides a complete picture of the solar energy that contributes to the heating of exterior and interior surfaces. Direct radiation produces the highest and most concentrated heat load, diffuse radiation represents the scattered component that reaches the surface even under overcast skies, and reflected radiation captures the influence of local surroundings such as neighboring buildings, pavements, or water bodies. By expressing total irradiance as the combination of these three elements, the equation forms the basis for analysing solar heat gains, evaluating shading performance, and assessing the impact of orientation, glazing design, and external obstructions on the thermal behavior of buildings. It is therefore fundamental to any assessment of overheating risk and passive cooling potential in architectural design.

8.1.3 Calculation of Solar Heat Gains through Glazing

Without shading, the amount of heat transmitted through a window is determined using the Solar Heat Gain Coefficient (SHGC):

$$Q_{\text{solar}} = \text{SHGC} \cdot A_{\text{win}} \cdot I_{\text{tot}}$$

where:

Q_{solar}—solar heat gain transmitted through glazing [W],

SHGC—Solar Heat Gain Coefficient; dimensionless coefficient describing the fraction of incident solar radiation that passes through glazing (–),

A_{win}—glazed surface area; the effective area of the window through which solar radiation is transmitted [m²],

I_{tot}—total incident solar irradiance on the glazing, including direct, diffuse, and reflected components [W/m²].

This equation quantifies the amount of solar heat transmitted through a glazing system by linking the total solar irradiance incident on the window with the physical properties of the glazing. The Solar Heat Gain Coefficient expresses the fraction of incoming solar energy that is admitted into the building, meaning that higher SHGC values correspond to greater solar heat gains and an increased risk of overheating. The glazed area determines how much of the building façade is exposed to incoming radiation, while the total irradiance reflects the combined direct, diffuse, and reflected solar components reaching the surface. By multiplying these three terms, the equation provides a direct measure of the solar load imposed on the interior environment, which is essential for predicting cooling demand, evaluating shading performance, and assessing the thermal implications of window size and orientation. It therefore forms a key element of building energy modeling and is central to the analysis of passive and active strategies aimed at reducing unwanted heat gains.

8.1.4 Reduction of Solar Heat Gains through Shading Systems

The efficiency of shading systems is quantified using the Shading Coefficient (SC):

$$SC = \frac{Q_{\text{shaded}}}{Q_{\text{unshaded}}}$$

where:

SC—shading coefficient (–), representing the ratio of solar heat gain with shading to solar heat gain without shading,

Q_{shaded}—solar heat gain through the glazing when shading is applied [W],

Q_{unshaded}—solar heat gain through the glazing without shading; reference condition used for comparison [W].

This equation defines the shading coefficient as the ratio between solar heat gains with shading applied and the corresponding gains under unshaded conditions. Because it directly compares two measurable energy flows, the formulation provides a clear indicator of how effectively a shading device reduces incident solar energy before it enters the building. A lower value of SC signifies more effective shading, as a smaller proportion of the unshaded heat gain reaches the interior. The equation therefore enables the evaluation of different shading systems, glazing types, and façade configurations by quantifying their influence on solar heat transmission. It also supports design decisions regarding overhang depth, louvre geometry and external shading placement, since variations in these elements directly alter the amount of radiation that reaches the glazing. As a result, the shading coefficient is fundamental in assessing overheating risk and in optimizing passive solar control strategies in both dynamic simulations and practical design applications.

To incorporate shading effects into the building energy balance, an adjusted SHGC is used:

$$SHGC_{\text{eff}} = SC \cdot SHGC_{\text{glass}}$$

where:

$SHGC_{\text{eff}}$—effective Solar Heat Gain Coefficient of the glazing when shading is applied (–),

SC—shading coefficient (–), representing the shading performance,

$SHGC_{\text{glass}}$—Solar Heat Gain Coefficient of the bare glazing, without shading (–).

The effective Solar Heat Gain Coefficient represents the combined influence of the glazing properties and the applied shading system on the amount of solar energy transmitted indoors. By multiplying the inherent SHGC of the glazing by the shading coefficient, the equation quantifies how much of the original solar transmission remains once shading elements have reduced the incoming radiation. A lower value of the effective SHGC indicates that shading is performing efficiently, significantly limiting heat gains and thereby reducing the cooling burden on the building. This relationship is central to evaluating the thermal performance of façades, since it allows designers to compare different glazing–shading combinations and select strategies that best control solar loads for a given climate, orientation or window configuration. It also supports energy modeling and helps assess the benefits of adaptive shading systems, which modulate solar transmission dynamically throughout the day.

8.1.5 Impact of Shading on the Reduction of Heat Gains and Cooling Loads

Shading systems significantly reduce the cooling demand of buildings by lowering solar heat gains. The cooling load reduction due to shading can be estimated as:

$$Q_{\text{cooling, reduced}} = Q_{\text{unshaded}} - Q_{\text{shaded}}$$

where:

$Q_{\text{cooling, reduced}}$—reduction in cooling load due to shading [W],
Q_{unshaded}—solar heat gain through glazing without shading (reference condition) [W],
Q_{shaded}—solar heat gain through glazing with shading applied [W.]

By comparing the solar heat gains under unshaded and shaded conditions, this equation provides a direct measure of how much cooling demand can be reduced through the use of shading systems. The difference between the two values reflects the amount of heat that would have entered the space if shading were absent, which highlights the thermal benefit delivered by external or internal shading devices. Because cooling demand is strongly influenced by solar loads, even relatively modest reductions in incident radiation can translate into substantial energy savings, improved thermal comfort and lower peak indoor temperatures. The expression is therefore particularly important in climates with strong solar exposure, and it assists designers in quantifying the performance of different shading strategies, evaluating façade configurations and estimating the potential impact on mechanical cooling requirements. In building simulations, the term plays a central role in assessing how architectural design decisions influence energy performance under varying solar conditions..

8.1.6 Estimation of Indoor Temperature Reduction Due to Shading

To assess the impact of shading on indoor temperature, a simplified heat balance model can be used:

$$\Delta T = \frac{Q_{\text{solar, net}}}{m \cdot c_p}$$

where:

ΔT—change in indoor air temperature resulting from net solar heat gain [K],
$Q_{\text{solar,net}}$—net solar heat gain entering the indoor space after shading and other reductions [W],
m—mass of the indoor air volume considered in the heat balance [kg],
c_p—specific heat capacity of air [J/(kg·K)].

At its core, this equation links the net solar heat entering a space with the resulting rise in indoor air temperature by relating heat gain to the thermal capacity of the enclosed air mass. The expression shows that a given amount of solar energy will produce a smaller temperature increase when the air volume is large or when the specific heat capacity is high, since more energy is required to raise the temperature of a greater thermal mass. Conversely, lightweight indoor air volumes respond more quickly to solar inputs, which can lead to rapid temperature fluctuations and elevated overheating risks. By quantifying this relationship, the equation provides a basis for evaluating the dynamic thermal response of rooms under solar exposure, assessing the benefits of shading, and estimating the sensitivity of indoor temperature to changes in glazing properties or solar loads. It is also widely used in simplified heat-balance models and early-stage design tools to approximate indoor temperature evolution without requiring full dynamic simulation.

8.1.7 Simulation Tools for Shading System Analysis

The performance of shading systems is most effectively assessed through computational simulation, which allows designers to evaluate solar exposure, daylight penetration, and shading effectiveness under a wide range of climatic and geometric conditions. Several established tools support this type of analysis, each offering distinct capabilities that contribute to the design and optimization of solar-control strategies.

EnergyPlus, one of the most widely used building performance simulation tools, includes detailed algorithms for the representation of shading devices, daylight-responsive controls, and solar heat-gain calculations. Its ability to model dynamic shading systems—such as operable blinds or automated louvres—makes it suitable for studies requiring integration between solar-control performance and whole-building energy behavior. Complementary to this, Radiance provides high-resolution daylighting and solar-access simulations. Through physically based ray tracing, Radiance enables rigorous evaluation of glare, luminance distributions, and shading geometry, particularly when precise assessment of visual comfort is required.

For designers working in more interactive modeling environments, platforms such as IES-VE and Ecotect have long been used to conduct shading studies that explore solar paths, façade exposure, or the seasonal performance of overhangs and screens. These tools offer intuitive, geometry-driven interfaces that make them well suited for early-stage design exploration. In more recent practice, parametric modeling environments have become increasingly important. Tools such as DIVA for Rhino and the Ladybug Tools suite integrate shading, daylighting, and solar-radiation workflows into Grasshopper's visual programming environment. This enables iterative design optimization, sensitivity analysis, and rapid comparison of alternative shading geometries across multiple climatic scenarios.

Collectively, these simulation environments allow for a more nuanced understanding of how shading systems influence thermal and visual performance. They support studies that consider the interaction between shading and building orientation, material reflectance, thermal mass, glazing properties, and occupancy schedules. Crucially, they also facilitate analysis across different seasons, enabling designers to investigate strategies that block excess summer radiation while preserving beneficial winter solar gains.

The use of simulation tools is therefore essential for developing shading systems that respond effectively to local climatic conditions, façade configurations, and evolving performance requirements. By integrating solar, daylighting, and thermal analysis into a unified design workflow, these tools support more informed decision-making and contribute to the development of environmentally responsive, energy-efficient building envelopes.

8.1.8 Building Heat Balance with Shading Systems

To assess the impact of shading systems on indoor temperature regulation, the building heat balance equation is applied:

$$\frac{dQ}{dt} = Q_{\text{solar}} - Q_{\text{shade}} - Q_{\text{loss}} - Q_{\text{ventilation}}$$

where:

dQ/dt—rate of change of internal thermal energy in the indoor environment [W],
Q_{solar}—solar heat gains entering the indoor space [W],
Q_{shade}—reduction of solar heat gains due to shading devices [W],
Q_{loss}—heat losses through the building envelope by conduction and radiation [W],
$Q_{\text{ventilation}}$—heat removed or gained through ventilation and air exchange [W].

From a thermal-dynamics perspective, this equation describes how the net heat content of a building zone evolves over time as competing heat flows enter and leave the space. Solar gains increase the internal energy, while shading, envelope losses, and ventilation all act as pathways through which heat is removed. The balance among these terms determines whether the indoor environment warms up, stabilizes, or cools down under given external conditions. Because each component represents a distinct physical mechanism, the equation makes it possible to analyse how changes in façade design, shading strategies, ventilation rates, or insulation levels influence the indoor thermal response. It also forms the backbone of simplified heat-balance models used in building performance simulations, supporting evaluations of thermal comfort, peak overheating risk, and the effectiveness of passive cooling measures across a range of climatic situations.

8.1.9 Impact of Solar Incidence Angle on Shading Efficiency

The angle of solar radiation incidence θ affects shading efficiency, and the shaded surface can be calculated as:

$$A_{\text{shade}} = A_{\text{window}} \cdot (1 - \cos\theta)$$

where:

A_{shade}—effective shaded area on the window surface [m^2],

A_{window}—total glazed window area [m^2],
θ—solar incidence angle measured between the sun's rays and the normal to the glazing [rad or degrees].

In this equation, the effective shaded area is expressed as a function of the window geometry and the solar incidence angle, illustrating how the position of the sun determines the extent of shading provided by a fixed overhang or similar device. As the solar altitude increases, the cosine term becomes smaller, resulting in a larger portion of the glazing being protected from direct radiation. Conversely, when the sun is low in the sky, shading coverage decreases, and a greater share of the incident solar energy reaches the window surface. The equation therefore captures the geometric mechanism through which shading devices modulate solar exposure throughout the day and across different seasons. It also offers designers a clear way to evaluate how changes in overhang depth, window height or façade orientation alter shading effectiveness and, by extension, influence indoor thermal conditions and cooling performance.

The principal equations governing solar radiation control and shading performance are summarized in Table 8.1, together with the definition of variables and units.

TABLE 8.1
Summary of Equations and Variables Used in the Theoretical Foundations of Passive Cooling

Equation	Description	Variables
$I_{tot} = I_{dir} + I_{diff} + I_{refl}$	Total solar irradiance on a surface	I_{dir}: direct irradiance [W/m^2]; I_{diff}: diffuse irradiance [W/m^2]; I_{refl}: reflected irradiance [W/m^2]
$Q_{solar} = SHGC \cdot A_{win} \cdot I_{tot}$	Solar heat gain through glazing without shading	SHGC: solar heat gain coefficient (–); A_{win}: glazed area [m^2]; I_{tot}: total irradiance [W/m^2]
$SC = Q_{shaded}/Q_{unshaded}$	Shading coefficient	Q_{shaded}, $Q_{unshaded}$: solar gains with and without shading [W]
$SHGC_{eff} = SC \cdot SHGC_{glass}$	Effective SHGC with shading	SC: shading coefficient (–); $SHGC_{glass}$: glazing coefficient (–)
$\Delta Q_{cool} = Q_{unshaded} - Q_{shaded}$	Cooling load reduction due to shading	Solar gains as above [W]
$\Delta T_{air} \approx (Q_{solar,net\ \Delta t})/(m_{air}\ c_p)$	Indoor air temperature change due to solar gain	m_{air}: air mass [kg]; c_p: specific heat of air [J/(kg · K)]; Δt: time step [s]; $Q_{solar,net}$: net solar gain [W]
$C_{th}\ dT_{in/dt} = Q_{solar,trans} + Q_{int} - Q_{loss} - Q_{vent}$	Lumped heat balance	C_{th}: thermal capacitance [J/K]; Q_{int}: internal gains [W]; Q_{loss}: conduction/radiation losses [W]; Q_{vent}: ventilation losses [W]
$\alpha_c = \arctan(d/h_v)$	Cut-off solar altitude (overhang)	d: overhang depth [m]; h_v: vertical distance window–overhang [m]

(Continued)

TABLE 8.1 *(Continued)*
Summary of Equations and Variables Used in the Theoretical Foundations of Passive Cooling

Equation	Description	Variables
$Q = -k_{eff} A \Delta T/\Delta x$	Fourier's law (steady conduction)	k_{eff}: conductivity [W/(m · K)]; A: area [m^2]; ΔT: temperature difference [K]; Δx: thickness [m]
$\rho c_p (\partial T/\partial t) = k_{eff} (\partial^2 T/\partial x^2)$	Fourier–Kirchhoff (transient conduction)	ρ: density [kg/m^3]; c_p: specific heat [J/(kg · K)]; T: temperature [K]; t: time [s]; k_{eff}: conductivity [W/(m·K)]
$dQ/dt = Q_{in} - Q_{out} + Q_{phase}$	Energy balance with PCM	Q_{in}: incoming sensible heat [W]; Q_{out}: outgoing sensible heat [W]; Q_{phase}: latent heat [W]
$\varphi = t_{Ti,max} - t_{Te,max}$	Time lag (peak indoor vs outdoor); adjust by 24 h if negative	$t_{Ti,max}$: indoor peak time [h]; $t_{Te,max}$: outdoor peak time [h]
$f = (T_{i,max} - T_{i,min}) / (T_{e,max} - T_{e,min})$	Decrement factor (amplitude damping)	$T_{i,max}$, $T_{i,min}$: indoor surface extremes [K]; $T_{e,max}$, $T_{e,min}$: outdoor surface extremes [K]
$Q = h A (T_s - T\infty)$	Convective heat transfer	h: convective coefficient [W/(m^2 · K)]; A: area [m^2]; T_s: surface temperature [K]; T∞: ambient temperature [K]
$Q_{rad} = \sigma \varepsilon A(T_s^4 - T_{amb}{}^4)$	Radiative heat transfer	σ: Stefan–Boltzmann constant ($5.6\,7 \times 10^{-8}$ W/(m^2 · K^4)); ε: emissivity (–); A: area [m^2]; T_s: surface temp [K]; T_{amb}: ambient temp [K]
$\Delta P = P_{wind} + P_{stack}$	Driving pressure for ventilation	P_{wind}: wind-induced pressure [Pa]; P_{stack}: stack-induced pressure [Pa]
$P_{wind} = ½ \rho v^2 C_p$	Wind-induced pressure	ρ: air density [kg/m^3]; v: wind speed [m/s]; C_p: pressure coefficient (–)
$P_{stack} = g \rho h (1/T_{in} - 1/T_{out})$	Stack-induced pressure	g: gravity [9.81 m/s^2]; ρ: air density [kg/m^3]; h: height difference [m]; T_{in}, T_{out}: temperatures [K]
$Q = C_d A \text{ sqrt}(2 \Delta P / \rho)$	Airflow through opening	C_d: discharge coefficient (–); A: opening area [m^2]; ΔP: pressure difference [Pa]; ρ: air density [kg/m^3]
$Q_{stack} = C_d A \text{ sqrt}(2 g h (T_{in} - T_{out})/T_{in})$	Stack-driven airflow	C_d: discharge coefficient (–); A: opening area [m^2]; g: gravity [9.81 m/s^2]; h: height [m]; T_{in}, T_{out}: temperatures [K]
$ACH = (Q / V) \times 3600$	Air changes per hour	Q: airflow [m^3/s]; V: room volume [m^3]
$dU/dt = Q_{solar} + Q_{int} - Q_{vent} - Q_{loss}$	Heat balance with ventilation	U: internal energy [J]; Q_{solar}: solar gains [W]; Q_{int}: internal gains [W]; Q_{vent}: ventilation losses [W]; Q_{loss}: conduction/radiation losses [W]

8.2 THERMAL BUFFERING WITH HIGH THERMAL MASS BUILDING MATERIALS. THEORETICAL FOUNDATIONS

8.2.1 Introduction

The ability of a building envelope to moderate indoor temperature depends largely on the thermal properties of the materials from which it is constructed. High thermal mass materials—such as concrete, brick, adobe, rammed earth, and natural stone—play a central role in this process by absorbing, storing, and releasing heat over extended periods. In conventional systems, this heat is stored primarily as sensible heat, meaning that energy absorption is reflected in an increase in the material's temperature. The magnitude of this effect is governed by density, specific heat capacity, thermal conductivity, and thickness, which together determine both the rate at which heat flows through a material and the extent to which it can buffer indoor conditions against outdoor fluctuations.

Thermal mass exerts its influence not simply by storing energy, but by doing so in a time-dependent manner. When subjected to daily temperature cycles, massive envelope components delay and attenuate the transmission of heat into the occupied spaces, creating a characteristic time lag between outdoor and indoor temperature peaks and reducing the amplitude of indoor temperature swings. These mechanisms underpin many passive cooling strategies and are the reason why buildings constructed from heavy materials often maintain more stable indoor climates during hot periods. Understanding these effects requires a dynamic view of heat transfer, extending beyond steady-state simplifications to the transient behavior captured by the heat diffusion equation and related formulations.

Phase Change Materials (PCM) introduce an additional dimension to thermal energy storage. Unlike conventional materials, which rely solely on sensible heat, PCM can absorb or release substantial quantities of latent heat during melting and solidification. This enables the material to moderate indoor temperature over a narrower range, as the phase transition process temporarily stabilizes surface temperatures even under variable external loads. By combining sensible and latent storage mechanisms, PCM-enhanced assemblies offer a powerful means of increasing the thermal buffering capacity of lightweight or hybrid constructions.

To analyse these processes rigorously, it is essential to draw on the fundamental physical equations that govern heat conduction, dynamic thermal storage, convective and radiative heat exchange, and phase-change phenomena. The subsections that follow present the theoretical foundations needed to model heat transfer through building envelopes, starting with Fourier's Law of conduction, progressing to the transient Fourier–Kirchhoff equation, and extending to thermal energy balance formulations, time lag and decrement factor calculations, and the role of convection and radiation. Together, these principles form the analytical basis for understanding the performance of thermal mass materials and PCM, and for applying simulation tools that support design decisions aimed at improving indoor thermal stability and passive resilience.

8.2.2 Heat Conduction in Building Envelopes—Fourier's Law

The primary mechanism of heat transfer in building materials is conduction, which is described by Fourier's Law:

$$Q = -k_{\text{eff}} \cdot A \cdot \frac{\Delta T}{\Delta x}$$

where:

Q—heat transfer rate through the material layer [W],
k_{eff}—effective thermal conductivity of the material [W/(m·K)],
A—area through which heat is conducted [m^2],
ΔT—temperature difference across the material layer [K],
Δx—thickness of the material layer [m].

In practical terms, Fourier's Law shows that the magnitude of heat flow through a solid depends not only on the temperature gradient and the geometry of the material layer but also on its thermal conductivity. Materials with a high value of k_{eff}, such as metals, allow heat to pass through rapidly, which can result in sharp and immediate temperature changes across the surface. In contrast, thermal insulation materials with low values of k_{eff}, such as mineral wool or polystyrene, strongly resist heat flow and therefore reduce the rate at which thermal energy is transferred. This behavior highlights the fundamental distinction between conductive materials and insulating materials in building applications and illustrates why increasing the thickness of insulation or selecting materials with lower conductivity is essential for limiting heat losses, improving energy efficiency, and maintaining stable indoor temperatures.

At the same time, the equation underlines the importance of considering material conductivity not only for winter performance but also for summer conditions, when excessive heat gains through poorly insulated envelopes can significantly increase indoor temperatures. By capturing these effects in a simple mathematical form, Fourier's Law serves as a foundational tool for evaluating envelope performance, informing material selection, and supporting the design of buildings that respond effectively to climatic conditions.

8.2.3 Transient Heat Conduction—Fourier-Kirchhoff Equation

For time-dependent temperature variations, the transient heat conduction equation is used:

$$\rho \cdot c_p \cdot \frac{\partial T}{\partial t} = k_{\text{eff}} \cdot \frac{\partial^2 T}{\partial x^2}$$

where:

ρ—density of the material [kg/m^3],
c_p—specific heat capacity of the material [J/(kg·K)],

T—temperature [K],
t—time [s],
k_{eff}—effective thermal conductivity of the material [W/(m·K)].

As expressed in the Transient Heat Conduction Equation, the temperature within a material changes over time due to the combined effects of its heat-storage capacity and its ability to conduct thermal energy. The term on the left-hand side represents the material's thermal inertia, indicating how density and specific heat together determine the resistance of the temperature field to rapid fluctuations. The right-hand side describes the diffusive flow of heat through the material, captured by the second spatial derivative of temperature and governed by the effective thermal conductivity.

This formulation, often referred to as the Fourier–Kirchhoff or heat diffusion equation, reveals how thermal disturbances propagate through building components and why massive materials respond slowly to external temperature variations, whereas lightweight materials react more quickly. It provides essential insight into how walls, roofs, and floors moderate indoor temperature through thermal lag and amplitude damping, which are key mechanisms in passive thermal stability. By characterizing the time-dependent interaction between heat storage and heat flow, the equation forms a cornerstone of dynamic thermal modeling and supports the evaluation of materials, construction solutions, and passive design strategies intended to improve indoor comfort and reduce energy demand.

8.2.4 Thermal Energy Balance of Building Envelopes

To evaluate how a building envelope manages heat, the thermal energy balance equation is applied:

$$\frac{dQ}{dt} = Q_{\text{in}} - Q_{\text{out}} + Q_{\text{phase change}}$$

where:

dQ/dt—rate of change of thermal energy stored in the material or system [W],
Q_{in}—incoming heat flow entering the material or control volume [W],
Q_{out}—outgoing heat flow leaving the material or control volume [W],
$Q_{\text{phase change}}$—latent heat absorbed or released during the phase change of PCM [W].

In this transient heat balance formulation, the change of stored thermal energy within a material or control volume is expressed as the difference between incoming and outgoing sensible heat flows, supplemented by an additional term representing latent heat exchange during phase transitions. The equation captures the fact that heat storage is not solely a function of fluid or material temperature changes but can also be strongly influenced by the absorption or release of latent heat when phase-change materials melt or solidify. This feature allows PCM systems to buffer indoor temperature swings by shifting significant amounts of energy without large changes in temperature.

By representing these processes in a unified way, the equation shows how PCM-enhanced envelopes or thermal storage layers can moderate temperature fluctuations, delay peak thermal loads and reduce reliance on mechanical cooling. It also highlights the dynamic nature of heat transfer in systems that include latent storage, where the timing and magnitude of heat flows depend on both the boundary conditions and the state of the phase-change medium. As a result, the transient heat balance with a phase-change term provides a rigorous basis for simulating PCM performance, evaluating its contribution to passive cooling and informing decisions related to material selection and envelope design.

8.2.5 Time Lag and Temperature Amplitude Damping

Time Lag ϕ—Shift in Peak Temperature

$$\phi = t_{i,max} - t_{e,max}$$

$$\phi = \phi + P \quad \text{if } \phi < 0$$

where:

ϕ—time lag between the occurrence of the maximum outdoor and indoor temperatures [h],

$t_{i,\max}$—time at which the indoor surface (or indoor air) temperature reaches its daily maximum [h],

$t_{e,\max}$—time at which the outdoor surface (or outdoor air) temperature reaches its daily maximum [h],

P—period of the temperature cycle, typically 24 hours [h].

As defined by this time lag relationship, the delay between the outdoor and indoor temperature peaks reflects the ability of a building element to slow down the propagation of heat through its thickness. The first expression quantifies the raw difference between the hour of the indoor maximum and the hour of the outdoor maximum, while the second expression corrects this value when the indoor peak occurs after midnight, ensuring that the result remains within a single diurnal cycle. Together, these two equations capture the fundamental mechanism through which thermal mass shifts heat flow in time, allowing interior spaces to remain cooler during the hottest hours of the day.

This behavior is central to passive design because time lag directly indicates how effectively a wall, roof or floor can buffer indoor conditions against rapid external temperature changes. Materials with high heat capacity and sufficient thickness typically generate a large time lag, meaning that heat absorbed during the day is released only after outdoor temperatures have started to decline. Conversely, lightweight constructions exhibit a small time lag and therefore transmit heat almost immediately to the interior. By quantifying this temporal shift, the time lag equation supports the evaluation of different material assemblies, informs the selection of thermal mass strategies, and provides critical insight into the resilience of buildings during heatwaves and other extreme thermal conditions.

Temperature Amplitude Damping (Decrement Factor, f)

$$f = \frac{T_{i,max} - T_{i,min}}{T_{e,max} - T_{e,min}}$$

where:

f—decrement factor (–), describing how much indoor temperature amplitude is reduced relative to the outdoor temperature amplitude,
$T_{i,max}$—maximum indoor surface (or indoor air) temperature during the daily cycle [°C or K],
$T_{i,min}$—minimum indoor surface (or indoor air) temperature during the daily cycle [°C or K],
$T_{e,max}$—maximum outdoor temperature during the daily cycle [°C or K],
$T_{e,min}$—minimum outdoor temperature during the daily cycle [°C or K].

By expressing the ratio between indoor and outdoor temperature amplitude, the decrement factor equation provides a direct measure of how effectively a building element moderates the intensity of external temperature fluctuations. The numerator reflects the difference between the indoor maximum and minimum temperatures over a diurnal cycle, while the denominator represents the corresponding outdoor temperature range. When the resulting value is small, it indicates that the building envelope significantly dampens the outdoor temperature swings, producing a more stable indoor environment even under highly variable external conditions.

This behavior is a hallmark of materials with substantial thermal inertia, such as concrete or stone, which absorb and release heat slowly and thereby limit the amplitude of indoor temperature variations. Lightweight materials, by contrast, exhibit a much higher decrement factor because they transmit outdoor temperature changes to the interior with little attenuation. The equation therefore serves as an essential tool for comparing wall and roof assemblies, evaluating the effectiveness of thermal mass strategies, and identifying constructions suited to climates with pronounced daily temperature cycles. In design practice, the decrement factor supports decisions about material selection, envelope composition, and passive cooling approaches aimed at improving indoor thermal stability during periods of intense heat.

8.2.6 Convection and Radiation in Heat Exchange

Heat transfer in building envelopes is influenced not only by conduction but also by convection and radiation.

For convective heat transfer, Newton's Law of Cooling is applied:

$$Q_{\text{conv}} = h \cdot A \cdot (T_s - T_\infty)$$

where:

Q_{conv}—convective heat transfer rate at the surface [W],
h—convective heat transfer coefficient [W/(m^2·K)],

A—surface area over which convection occurs [m^2],
T_s—surface temperature of the material [K],
$T\infty$—temperature of the surrounding fluid (air) far from the surface [K].

From a heat-transfer perspective, this equation illustrates how convection carries thermal energy away from a surface as a function of the temperature difference between that surface and the surrounding air. The convective heat transfer coefficient represents the combined influence of airflow velocity, fluid properties, and surface orientation, while the surface area term captures the extent of the physical interface through which heat exchange occurs. When the temperature difference is large or when convection is enhanced by wind or mechanical airflow, the rate of heat removal increases substantially.

Understanding this relationship is essential for analysing how exterior walls, roofs and glazing surfaces release heat to the environment, particularly during night-time cooling or under conditions of forced ventilation. It also helps explain why smooth or sheltered surfaces experience weaker convective cooling, whereas wind-exposed façades dissipate heat more efficiently. In building design, Newton's convective law informs decisions regarding material selection, façade geometry and passive cooling strategies, enabling designers to predict thermal behavior under varying airflow conditions and to optimize heat loss pathways as part of a broader thermal management approach.

For thermal radiation exchange, the Stefan-Boltzmann Law is used:

$$Q_{\text{rad}} = \sigma \cdot \varepsilon \cdot A \cdot \left(T_s^4 - T_{\text{ambient}}^4\right)$$

where:

Q_{rad}—radiative heat transfer rate [W],
σ—Stefan–Boltzmann constant (5.67×10^{-8} W/($m^2 \cdot K^4$)),
ε—surface emissivity (–),
A—radiating surface area [m^2],
T_s—surface temperature [K],
T_{ambient}—ambient (surrounding) air temperature [K].

This equation quantifies how a surface emits thermal energy in proportion to the fourth power of its absolute temperature and how this emission is tempered by the temperature of the surrounding environment. The Stefan–Boltzmann constant establishes the strength of thermal radiation, while the emissivity term accounts for the material's ability to radiate heat relative to an ideal blackbody. Because radiative heat transfer increases rapidly with temperature, small changes in surface conditions can produce significant differences in heat emission.

This formulation is especially important for building surfaces exposed to clear night skies, where radiative cooling can lower roof or façade temperatures below ambient air temperature and enhance passive cooling. It also plays a role during intense solar heating, as elevated surface temperatures intensify outgoing long-wave radiation. By capturing these radiative exchanges, the equation helps designers understand how materials, coatings, and surface temperatures influence net heat transfer, supporting decisions related to passive cooling, surface emissivity optimization, and the thermal performance of exterior building components.

8.2.7 Simulation Tools for PCM and Thermal Mass Modeling

The practical evaluation of thermal mass and phase change materials increasingly relies on numerical simulation tools capable of resolving the transient, multi-physics nature of heat transfer in building envelopes. Because PCM systems combine both sensible and latent storage mechanisms, and because masonry and concrete elements exhibit strong time-dependent thermal responses, computational modeling provides the most reliable means of predicting how these materials will behave under real climatic conditions.

Dynamic building performance simulation engines form the backbone of PCM and thermal mass analysis at the whole-building scale. EnergyPlus, widely considered the standard in this domain, incorporates detailed heat-balance algorithms and dedicated PCM modules that model phase transitions using enthalpy-based methods. Through its layer-by-layer conduction routines, it can assess how PCM-enhanced assemblies interact with diurnal temperature cycles, internal gains, and ventilation strategies. TRNSYS, another established tool for transient thermal simulation, is frequently used in research contexts, especially when PCM behavior must be examined in combination with system-level interactions, such as solar collectors or thermal storage units.

For applications requiring detailed, spatially resolved heat-transfer modeling, finite element platforms such as COMSOL Multiphysics provide the ability to represent conduction, convection, radiation, and latent heat processes within complex geometries. COMSOL enables high-fidelity modeling of PCM melting and solidification fronts, making it suitable for material-level analysis, prototype evaluation, or layered-envelope studies. In cases where airflow plays a significant role—such as ventilated façades, double-skin envelopes, or PCM-integrated air channels—computational fluid dynamics tools like ANSYS Fluent allow designers to capture the combined effects of buoyancy-driven flow, forced convection, and phase-change behavior.

Several simulation tools bridge the gap between research-level modeling and architectural design practice. DesignBuilder, which serves as a user-friendly interface for EnergyPlus, allows architects and engineers to explore the impact of PCM integration and thermal mass strategies within detailed building geometry. IES-VE provides similar capabilities and is widely used in consultancy contexts, particularly where passive cooling strategies, dynamic shading, or mixed-mode ventilation are being evaluated. Meanwhile, WUFI Pro offers a specialized approach for hygrothermal analysis, enabling the examination of coupled heat-and-moisture transport in PCM-enhanced or mass-intensive envelopes—an important consideration for assessing durability and indoor moisture conditions alongside thermal performance.

Collectively, these tools support the simulation of time lag, temperature damping, phase-transition dynamics, and the overall thermal buffering capacity of heavy or PCM-enhanced constructions. Their use enables designers to explore how specific material combinations respond to diurnal and seasonal cycles, how PCM activation interacts with climate conditions, and how thermal mass contributes to overheating mitigation or passive survivability. By integrating theoretical principles with simulation-based evaluation, these tools form an essential component of contemporary building design workflows.

For clarity, the governing equations relevant to conduction, transient heat diffusion, latent energy exchange, and damping behavior are consolidated in Table 8.1 along with the associated symbols introduced throughout this section.

8.3 THEORETICAL FOUNDATIONS OF NATURAL VENTILATION

8.3.1 Introduction

Natural ventilation is one of the most established passive design strategies used to enhance indoor air quality, regulate thermal conditions, and improve overall energy efficiency in buildings. It relies on naturally occurring pressure differences—generated by wind forces, temperature-induced buoyancy, or a combination of both—to drive air exchange without the need for mechanical equipment. As buildings increasingly aim to reduce reliance on energy-intensive cooling and ventilation systems, natural ventilation remains central to low-carbon and climate-responsive design.

Three primary mechanisms govern natural ventilation behavior in buildings. The first is single-sided ventilation, which occurs when air enters and leaves a space through openings located on the same façade. This airflow pattern is typically driven by short-term fluctuations in local wind pressure, turbulence, and small-scale eddies, resulting in relatively modest ventilation rates that depend heavily on window geometry and incident wind direction.

Cross ventilation, by contrast, arises when openings are located on opposing façades, allowing a pressure gradient to establish a more robust flow path through the building. When the incoming and outgoing openings are aligned or strategically positioned, cross ventilation can deliver significantly higher air-exchange rates and is particularly effective under moderate to strong wind conditions. Its performance is also influenced by outdoor obstructions, building orientation, and façade porosity.

The third mechanism is buoyancy-driven airflow, commonly referred to as the stack effect. In this case, differences in air density between warmer indoor air and cooler outdoor air create vertical air movement, with warm air rising and escaping through high-level openings while cooler air is drawn in from lower inlets. This process can operate independently of wind and is especially valuable at night or during periods of calm weather. Stack ventilation becomes more effective as the vertical distance between inlets and outlets increases, making it well suited for tall spaces, atria, stairwells, and ventilation shafts.

A comprehensive understanding of these mechanisms requires an examination of the underlying physical principles that govern airflow, including pressure differentials, momentum transfer, thermal stratification, and the interaction between indoor and outdoor boundary conditions. The following sections introduce the fundamental equations describing ventilation rates, driving pressures, flow coefficients, and heat exchange, establishing the analytical basis for the performance assessment and design optimization of naturally ventilated buildings.

8.3.2 Airflow and Pressure Differences in Natural Ventilation

The movement of air in naturally ventilated spaces is influenced by wind-driven forces and buoyancy-driven forces (stack effect). The total pressure difference driving ventilation can be expressed as:

$$\Delta P = P_{\text{wind}} + P_{\text{stack}}$$

where:

ΔP—total pressure difference driving natural ventilation [Pa],

P_{wind}—wind-induced pressure acting on the building façade or openings [Pa].

$$P_{\text{wind}} = \frac{1}{2}\rho v^2 C_p$$

ρ—air density [kg/m^3],

v—wind velocity at the opening or façade location [m/s],

C_p—wind pressure coefficient determined by wind direction and façade orientation (–),

P_{stack}—buoyancy-induced (stack effect) pressure is generated by indoor–outdoor temperature differences [Pa].

$$P_{\text{stack}} = g\rho h\left(\frac{1}{T_{\text{in}}} - \frac{1}{T_{\text{out}}}\right)$$

where:

g—gravitational acceleration (9.81 m/s^2),

ρ—air density [kg/m^3],

h—vertical distance between the inlet and outlet openings [m],

T_{in}—indoor absolute air temperature [K],

T_{out}—outdoor absolute air temperature [K].

By combining wind-induced and buoyancy-induced pressure components, this equation provides a comprehensive measure of the forces that drive natural airflow through building openings. The wind term reflects the dynamic pressure created by external air movement, while the buoyancy term captures the contribution of temperature-driven density differences between indoor and outdoor air. Together, these two mechanisms determine both the direction and magnitude of ventilation flow paths.

Understanding this combined pressure difference is essential for optimizing window placement, determining the required size of openings, and designing ventilation stacks or atrium ventilation routes. Because wind speed, wind direction, and indoor–outdoor temperature gradients vary throughout the day, the equation helps ensure that natural ventilation strategies remain effective under a wide range of environmental conditions. By quantifying the physical drivers of airflow in a single relationship, it provides a strong foundation for performance evaluation and for designing resilient, climate-responsive ventilation systems.

8.3.3 Ventilation Rate Calculation

The airflow rate through an opening, determined by the pressure difference across it, can be estimated using:

$$Q = C_d\, A\sqrt{\frac{2\,\Delta P}{\rho}}$$

where:

Q—volumetric airflow rate through the opening [m^3/s],
C_d—discharge coefficient accounting for flow contraction and friction losses,
A—effective area of the opening [m^2],
ΔP—pressure difference across the opening [Pa],
ρ—air density [kg/m^3].

This equation helps determine the required size and configuration of windows and ventilation openings in order to achieve the desired airflow rate under specific operating conditions. By linking the pressure difference across an opening with the resulting volumetric flow, the formulation enables designers to assess whether a given opening provides sufficient ventilation capacity or whether its area must be increased to meet comfort or air quality targets. It also supports performance evaluations under varying environmental conditions, such as differing wind speeds, building orientations or indoor–outdoor temperature gradients, ensuring that natural ventilation strategies remain effective and reliable.

8.3.4 The Navier-Stokes Equation and Its Role in Natural Ventilation

In complex ventilation scenarios, particularly those involving turbulent flow, wake formation, and interaction with building geometry, airflow is governed by the Navier-Stokes equation, which describes the motion of fluids, including air:

$$\rho\left(\frac{\partial \boldsymbol{u}}{\partial t} + \boldsymbol{u}\cdot\nabla\boldsymbol{u}\right) = -\nabla P + \mu\nabla^2\boldsymbol{u} + \rho\boldsymbol{g}$$

where:

ρ—air density [kg/m^3],
u—airflow velocity vector [m/s],
$\partial u/\partial t$—local (temporal) acceleration term [m/s^2],
$u \cdot \nabla u$—convective acceleration term representing changes in velocity along a flow path [m/s^2],
∇P—pressure gradient driving the flow [Pa/m],

μ—dynamic viscosity of air [Pa·s],
$\nabla^2 u$—viscous diffusion term accounting for momentum dissipation [m/s²],
g—gravitational acceleration vector [m/s²].

The Navier–Stokes equation forms the foundation of Computational Fluid Dynamics simulations used to analyse airflow behavior in naturally ventilated buildings. Because the equation describes the motion of air under the combined influence of pressure gradients, inertial effects, and viscous forces, CFD tools can resolve detailed flow structures within and around architectural spaces. This includes turbulence, eddies, and wake formation, all of which have a strong influence on the performance of natural ventilation systems. Such flow features cannot be captured by simplified ventilation rate equations, yet they play a critical role in determining how air enters, moves through, and exits a building. As a result, the Navier–Stokes formulation is essential when designing ventilation towers, atriums, and complex wind-driven ventilation pathways, where a clear understanding of localized pressure variations and airflow patterns is necessary to ensure robust and efficient passive ventilation performance.

8.3.5 Stack Effect and Buoyancy-Driven Ventilation

The ventilation rate due to stack effect is given by:

$$Q_{\text{stack}} = C_d A \sqrt{2gh\left(\frac{T_{\text{in}} - T_{\text{out}}}{T_{\text{in}}}\right)}$$

where:

Q_{stack}—airflow rate driven by the stack effect [m³/s],
C_d—discharge coefficient accounting for flow contraction and friction losses,
A—effective area of the opening [m²],
g—gravitational acceleration (9.81 m/s²),
h—height difference between the lower and upper openings [m],
T_{in}—indoor absolute air temperature [K].
T_{out}—outdoor absolute air temperature [K].

This equation is crucial for the design of solar chimneys, double-skin façades, and passive ventilation shafts because it quantifies the buoyancy-driven airflow generated by temperature differences between indoor and outdoor air. By expressing how variations in height, temperature, and density influence the resulting pressure difference, the formulation allows designers to predict the airflow capacity of vertical ventilation elements and evaluate their performance under different climatic conditions. It also helps determine the appropriate dimensions and placement of inlets and outlets so that buoyancy forces can be harnessed effectively, ensuring reliable ventilation even when wind-driven airflow is weak or absent. As a result, the equation provides a foundation for the optimization of passive ventilation components that depend on thermal uplift and stack-induced pressure gradients.

8.3.6 Air Change Rate—Evaluating Ventilation Efficiency

The air change rate (ACH) indicates how frequently the air in a room is replaced and is given by:

$$ACH = \frac{Q}{V} \times 3600$$

where:

ACH—air changes per hour, indicating how many times the air volume in a space is replaced in one hour [1/h],
Q—volumetric airflow rate supplied to or extracted from the space [m^3/s],
V—volume of the room or zone being ventilated [m^3],
3600—conversion factor from seconds to hours.

ACH is a key parameter for evaluating the adequacy of minimum winter ventilation, ensuring that even in periods of low outdoor temperatures, a building receives enough fresh air to prevent excessive humidity and the associated risks of condensation or mould growth. It is also essential in assessing night-time ventilation during heatwaves, when enhanced air exchange after sunset helps purge accumulated heat and improve thermal comfort without mechanical cooling. In addition, ACH provides an indirect measure of building airtightness, since the number of air changes per hour reflects how much uncontrolled infiltration or exfiltration occurs through the building envelope. Beyond these functions, ACH remains important for indoor air quality assessment because it indicates the system's ability to dilute indoor pollutants and maintain healthy environmental conditions. By capturing both the efficiency of intentional ventilation strategies and the performance of the building fabric, ACH serves as a central metric in ventilation design, operation and regulatory evaluation.

8.3.7 Building Heat Balance with Natural Ventilation

The thermal energy balance in a naturally ventilated building is given by:

$$\frac{dU}{dt} = Q_{\text{solar}} + Q_{\text{internal}} - Q_{\text{ventilation}} - Q_{\text{loss}}$$

where:

dU/dt—rate of change of internal energy within the building zone [W],
Q_{solar}—solar heat gains entering the building [W],
Q_{internal}—internal heat gains from occupants, lighting, and equipment [W],
$Q_{\text{ventilation}}$—heat removed or added due to ventilation and air exchange [W],
Q_{loss}—heat losses through the building envelope by conduction and radiation [W].

This energy balance equation describes how the internal energy of a building zone changes as a result of interacting heat flows. It shows that the indoor thermal state is governed by the combined influence of solar gains, internal heat sources, ventilation-driven heat exchange and conductive or radiative losses through the building envelope.

Because these four components represent the principal pathways by which energy enters and leaves a space, the equation captures the fundamental physics behind indoor temperature evolution. It also provides a framework for analysing the effect of different design strategies, such as improvements in glazing performance, optimization of shading, adjustments in ventilation rates or enhancements in envelope insulation. By quantifying the relative contribution of each term, the formulation supports assessments of thermal comfort, energy demand and building responsiveness under both steady and transient environmental conditions. In dynamic simulations, it forms the core of heat balance models used to predict indoor temperatures and evaluate the effectiveness of passive and active thermal control strategies.

8.3.8 Simulation Tools for Natural Ventilation Analysis

The analysis of natural ventilation performance increasingly relies on computational simulation tools capable of representing complex airflow patterns, buoyancy-driven movement, pressure-driven exchange, and their interactions with building geometry and environmental boundary conditions. Because natural ventilation is sensitive to wind conditions, façade porosity, thermal stratification, and the transient coupling between indoor and outdoor airflows, numerical modeling provides a rigorous framework for evaluating ventilation effectiveness across different design strategies and seasonal conditions.

Computational Fluid Dynamics (CFD) is among the most powerful approaches for studying natural ventilation at the detailed spatial scale. Software environments such as ANSYS Fluent provide high-resolution modeling of airflow trajectories, turbulence behavior, temperature distribution, and thermal stratification. These simulations are particularly valuable when evaluating cross-ventilation effectiveness, atrium-driven airflow, double-skin façades, or localized comfort conditions associated with drafts and thermal asymmetry. OpenFOAM, an open-source CFD framework, offers similar capabilities and is widely used in academic and advanced design contexts due to its flexibility and ability to integrate custom turbulence models or boundary conditions.

At the whole-building scale, building energy simulation engines represent natural ventilation using airflow network models or simplified pressure-based algorithms. EnergyPlus, for example, includes robust modules for simulating wind-driven and buoyancy-driven flows and can couple ventilation performance with thermal mass effects, shading operation, and internal gains. TRNSYS provides comparable functionality and is frequently used in research settings where natural ventilation interacts with other dynamic building systems. DesignBuilder, acting as a graphical interface to EnergyPlus, allows architects and engineers to examine natural ventilation scenarios within detailed building geometries, while IES-VE integrates airflow modeling into its thermal performance workflow to support passive ventilation optimization and mixed-mode control strategies.

For applications focused on airflow pathways, infiltration, pollutant transport, or multi-zone ventilation networks, specialized tools such as CONTAM provide a complementary analytical framework. CONTAM is widely used for evaluating indoor air quality, ventilation rates across complex floor plans, and the interaction between natural ventilation, mechanical systems, and occupant behavior.

Together, these simulation environments offer a comprehensive platform for assessing natural ventilation strategies, enabling designers to evaluate ventilation effectiveness under different climatic conditions, quantify airflow rates and thermal interactions, and explore the implications of façade geometry, aperture configuration, and passive design interventions. The equations governing pressure differentials, buoyancy forces, airflow resistance, and thermal balance are consolidated in Table 8.1, providing a theoretical foundation for the simulation methods applied in this section.

8.4 SUMMARY OF PASSIVE COOLING STRATEGIES AND INTEGRATED PERFORMANCE IMPLICATIONS

The three passive cooling strategies examined in this chapter—solar control through shading, thermal buffering with high thermal mass and phase change materials, and natural ventilation—form a coherent framework for reducing indoor heat stress without mechanical systems. Each strategy operates through distinct physical mechanisms, yet their combined performance determines the building's ability to maintain stable indoor conditions under diurnal temperature swings, seasonal variations, and increasing occurrences of extreme heat.

Shading systems constitute the building's first layer of defence against overheating. By limiting direct and diffuse solar gains at the façade, shading reduces the magnitude of thermal loads that must be absorbed, stored or removed by subsequent cooling strategies. The effectiveness of shading depends on geometry, orientation and seasonal sun paths, and is governed by principles of solar obstruction and irradiance reduction. By mitigating peak solar exposure at the source, shading directly lowers surface temperatures, delays heat penetration into the envelope, and helps preserve indoor thermal stability.

Thermal mass and phase change materials form the second component of passive cooling, moderating heat transfer through time-dependent storage processes. Massive materials absorb and release heat slowly, introducing time lag and amplitude damping that decouple indoor conditions from short-term outdoor fluctuations. PCM extend this buffering capacity by adding latent heat storage, stabilizing temperatures during peak hours. The equations governing heat conduction, transient heat diffusion, thermal energy balance and phase-change dynamics (summarized in Table 8.1) collectively explain how envelope elements redistribute heat throughout the day and prevent rapid indoor temperature rise.

Natural ventilation acts as the third and final mechanism, enabling accumulated heat to be expelled from the building. Whether driven by wind-induced pressure differences or buoyancy forces, ventilation supports heat removal, promotes night flushing of thermal mass and restores indoor temperatures to favourable levels before the next diurnal cycle. Its performance depends on opening geometry, vertical separation, façade porosity and climate-specific wind conditions, making it highly sensitive to building form and outdoor boundary conditions.

When these three mechanisms operate together, the building exhibits an integrated passive cooling behavior far more effective than any single strategy applied in isolation. Shading suppresses heat gains, allowing thermal mass to operate within a reduced thermal load. Thermal mass buffers and delays heat flows, creating conditions under which natural ventilation can effectively purge stored energy. Ventilation, in turn, resets the thermal state of the building overnight, preventing cumulative heating across consecutive hot days. This *block–buffer–purge* sequence represents the core logic of passive thermal design and is indispensable in highly insulated or cooling-limited buildings.

The consolidated overview presented in Table 8.1 provides a reference to the principal physical equations underlying these three strategies. It unifies the analytical tools for assessing conductive, latent, convective, and radiative heat flows, and serves as a technical foundation for evaluating how shading, thermal mass, and natural ventilation interact within an integrated passive cooling system. Together, these principles support performance-based design approaches capable of enhancing indoor comfort, improving energy efficiency, and strengthening resilience to extreme heat events.

9 Solar Shading Principles in Architecture
Climatic Logics and Built Examples

9.1 OVERHANGING ROOFS AND DEEP EAVES AS PASSIVE SHADING STRATEGIES

9.1.1 Introduction

Overhanging roofs and deep eaves represent one of the most enduring passive shading strategies in both vernacular and contemporary architecture. Their effectiveness arises from geometric modulation of solar access: in summer they block high-angle sunlight, reducing façade temperature and limiting conductive heat transfer; in winter, depending on roof pitch and depth, they admit low-angle solar radiation that contributes to passive heating. This selective seasonal control makes them a fundamental component of climate-responsive design.

Beyond shading, overhangs protect façades from precipitation, reduce wetting–drying cycles, and extend the durability of external materials. Their thermal performance is reinforced by material properties. For example, thatch provides natural insulation due to its high porosity and low thermal conductivity, which enhances the shading effect of deep overhangs. Clay tiles and timber shingles also reduce heat transfer more effectively than lightweight membranes.

Across climates—from humid maritime regions and temperate continental zones to hot, arid, and subtropical areas—overhanging roofs remain a versatile and robust element of passive environmental control.

9.1.2 Cold and Temperate Climates

In cold and temperate climates, overhanging roofs and deep eaves serve a dual role: they prevent summer overheating while maintaining access to beneficial winter sunlight. Traditional European buildings exhibit wide variation in overhang geometry, reflecting local temperature patterns, moisture exposure, and winter solar angles.

A representative example from a temperate continental climate is shown in Figure 9.1, which presents a thatched villa in Heelsum, the Netherlands. This example illustrates how vernacular material and contemporary geometry can be combined to control seasonal solar gains.

The sharply profiled thatch provides partial shading for balconies and façades during summer while still allowing low-angle winter sun to reach interior spaces.

DOI: 10.1201/9781003645573-9

FIGURE 9.1 A thatched villa in Heelsum, Netherlands, integrating vernacular roofing with modern architectural geometry.

This differs from the deeply rounded thatch typical of maritime British cottages, where uninterrupted shading, moisture protection, and thermal inertia dominate due to milder winters and lower winter irradiance.

Although overhanging roofs appear across Europe, their use was far less common in northern, western, central, and eastern regions. The primary reason is climatic: buildings in these zones depend heavily on winter solar gains, and deep eaves would block the low-angle sunlight required for passive heating during long, cold seasons. High precipitation and snow loads also necessitated steep roofs with minimal horizontal projection to support rapid snow shedding and reduce structural risk. For these reasons, pronounced overhangs remained relatively uncommon despite their technical advantages in warmer climates.

9.1.3 Coastal Architecture

In coastal regions, roof overhangs are essential for protecting façades from intense solar radiation, rainfall, and strong winds. Their geometry mitigates heat gains throughout the day while creating ventilated zones beneath the roof surface.

The climatic adaptation of this strategy is illustrated in Figure 9.2, which shows a modern house near Biarritz with a deep overhanging roof.

A second example from the same region is presented in Figure 9.3, demonstrating how extended roof planes function as climatic filters in an oceanic environment.

Both buildings employ wide overhangs that shield façades particularly during midday and afternoon hours, reducing solar heat gains without obstructing daylight

FIGURE 9.2 Contemporary house near Biarritz with a deep roof overhang providing solar protection.

FIGURE 9.3 Detached house near Biarritz with a large tiled roof and extended eaves.

or natural ventilation. While thatched roofs in temperate continental climates adjust shading seasonally, coastal tiled roofs emphasize continuous environmental buffering throughout the year.

9.1.4 Hot and Subtropical Climates

In hot and subtropical regions, the primary objective of roof overhangs is continuous exclusion of solar heat. Buildings in these climates rely on persistent shading rather than balancing between heating and cooling seasons. As a result, roof geometry tends to be more uniform and horizontal, with extended eaves providing constant façade protection.

This strategy is exemplified in Figure 9.4, showing a residential building in Arguineguín, Gran Canaria.

FIGURE 9.4 Single-family house in Arguineguín, Gran Canaria, using flat projecting eaves for continuous shading.

In such environments, buildings often lack the heavy thermal mass typical of temperate regions. Instead, they rely on a combination of reflective surfaces, compact geometry, and multi-layered shading to limit heat accumulation. Continuous overhangs reduce direct solar exposure on external walls and windows, lower the mean radiant temperature, and allow windows to remain open for natural ventilation even during peak solar hours. This complementarity between shading and airflow is essential for maintaining comfort in hot-dry and hot-humid climates.

9.2 SOLAR PROTECTION THROUGH FAÇADE GEOMETRY IN HOT AND TEMPERATE CLIMATES

Façade geometry plays a decisive role in controlling solar exposure, reducing heat gains, and stabilizing indoor thermal conditions. Overhangs, terraces, deep recesses, and loggia-like volumes modulate sunlight through spatial configuration rather than mechanical devices. In regions with high solar intensity—or with moderate climates where façades face multiple orientations—form-based shading becomes a central passive strategy. This section discusses how façade projections and recesses mediate solar access across different climatic contexts.

9.2.1 Cornices, Balconies, and Recessed Openings in Temperate Urban Contexts

Layered façades composed of cornices, projecting balconies, pilasters, and recessed glazing create effective shading in temperate and transitional climates. Their geometric depth breaks direct solar exposure and reduces the mean radiant temperature of external surfaces.

This principle is illustrated in Figure 9.5, showing a historic building in Bilbao where horizontal and vertical façade components contribute to shading throughout the day.

The projecting cornices interrupt high-angle sunlight, while balconies provide shading to the windows below. Pilasters add vertical structure to the façade and shade glazing from low-angle morning and afternoon sun. Combined with masonry mass and recessed window placement, these elements form a continuous thermal buffer.

9.2.2 Deep Arched Recesses and Loggia-Based Shading in Hot Climates

In hot Mediterranean and North African climates, façade recesses often replace exterior shading devices. Deep loggias and arched niches reduce direct solar access on multiple sides of an opening, enabling thermal buffering without movable components.

This approach is demonstrated in Figure 9.6, which presents a building in Ronda, Spain, where deep arched recesses function as volumetric sunshades.

A more compact but highly repetitive example appears in Marrakesh, where loggias form the primary thermal interface of hotel façades. As shown in Figure 9.7, the depth of the arches, the thickness of masonry, and reduced aperture size limit solar penetration and stabilize indoor temperatures during hot afternoons.

FIGURE 9.5 Historic multi-storey building in Bilbao using cornices, pilasters, and recessed windows for passive solar protection.

FIGURE 9.6 Building in Ronda, Spain, with deep arched recesses providing full façade shading.

FIGURE 9.7 Hotel El Andalous, Marrakesh, Morocco, integrating deep arched loggias into a compact façade.

9.2.3 Sculpted Loggias and Thermal-Mass Façades in Subtropical Contexts

In subtropical island climates, loggias are often combined with topography and thermal mass to create deeper shading volumes. The façade becomes a carved surface, where geometry, mass, and airflow interact to regulate heat.

This principle is illustrated in Figure 9.8, showing the Gloria Palace Royal Hotel and Spa on Gran Canaria.

Here, the wavy balcony edges and deep sidewalls reduce sky view and exposure to direct sunlight, lowering mean radiant temperature in a high-solar, low-humidity environment. The loggias' mass absorbs heat during the day and releases it slowly at night, contributing to diurnal thermal stability.

FIGURE 9.8 Gloria Palace Royal Hotel & Spa, Gran Canaria, using deep organic loggias carved into a thermal-mass façade.

9.2.4 Façade Reduction: Minimizing Aperture Size to Limit Solar Gain

Another geometric strategy involves limiting and recessing openings. Buildings with perforated or highly controlled façade apertures can achieve substantial reductions in solar heat gain without additional shading devices.

This approach is shown in Figure 9.9, depicting a student housing building in Bordeaux characterized by a densely perforated façade.

The geometry filters daylight while preserving thermal inertia and privacy. It demonstrates how solar exclusion can be achieved through façade restraint rather than added devices.

9.2.5 Fixed Horizontal and Vertical Louvers as Spatial Solar Filters

In climates with high summer irradiance and strong directional sunlight, fixed louvers offer a form-based shading strategy that enhances thermal comfort while allowing light diffusion.

This principle appears in Figure 9.10, showing a circular residential building in Bordeaux wrapped with fixed horizontal louvers.

FIGURE 9.9 University housing building in Bordeaux using a perforated façade with deep-set window openings.

An even more articulated example is found at the Bordeaux INP campus, shown in Figure 9.11, where two contrasting façade systems function together as static solar filters.

9.3 ADAPTIVE SHADING STRATEGIES

Adaptive shading systems complement the fixed geometric measures described earlier by allowing real-time or seasonal regulation of solar exposure. Their purpose is to provide occupants with control over daylight, glare, and heat gains while maintaining visual comfort and natural ventilation. These systems range from simple manually operated shutters to more complex multi-layered façade assemblies. In many climates, especially those with pronounced seasonal or diurnal variation, adaptive shading offers a level of flexibility that fixed elements cannot achieve.

Adaptive devices can be grouped into two categories:

1. Manually operated systems such as shutters, blinds, and operable screens.
2. Hybrid or layered systems that combine fixed shading with adjustable components, enabling fine-tuned modulation of solar radiation.

In vernacular architecture, these systems emerged from practical needs—privacy, daylighting, and thermal comfort—long before contemporary environmental design formalized their function. Today, they remain essential across European, Mediterranean, and subtropical climates.

FIGURE 9.10 Circular apartment block in Bordeaux with fixed horizontal louvers forming a continuous façade screen.

9.3.1 Operable Shading Devices: Shutters, Blinds, and Exterior Screens

In temperate regions, adaptive solar control typically relies on simple manually operated systems. Their advantage lies in the ability to adjust shading instantaneously in response to weather, time of day, or interior needs. Unlike fully automated systems, manual operation reflects user awareness and often results in more energy-efficient behavior.

This principle is illustrated in Figure 9.12, which shows a semi-detached house in Wrocław equipped with two distinct external shading systems.

Roller shutters represent the simplest type of adaptive shading. They allow full closure of the window opening, significantly reducing solar heat gains during hot periods. In their perforated configuration, they admit partial daylight while maintaining ventilation gaps.

FIGURE 9.11 Bordeaux INP campus building with vertical fins and horizontal cladding used as fixed solar filters.

FIGURE 9.12 Semi-detached house in Wrocław with external roller shutters and exterior venetian blinds.

Exterior venetian blinds offer finer control through adjustable slat angles. They allow occupants to regulate the amount and direction of incoming light, reduce glare, and maintain outward visibility while moderating heat gains. This increased flexibility makes them particularly effective for living rooms, offices, or other spaces where both daylight and thermal comfort are critical.

These devices exemplify user-driven microclimate regulation. Manual systems often achieve superior performance compared with automated mechanisms because they reflect real, immediate occupant needs rather than pre-programmed assumptions.

A traditional counterpart to these systems appears in the vernacular architecture of Bayonne, as shown in Figure 9.13.

Here, wooden shutters remain essential, offering full or partial shading depending on daylight and thermal conditions. They reflect a long-standing tradition of occupant interaction with the building envelope.

In Mediterranean climates with more intense sunlight, adaptive systems are often layered. Figure 9.14 presents an example from Mijas, Spain, where esparto blinds and a retractable awning operate together to filter and block sunlight as needed.

Esparto blinds diffuse sunlight and allow ventilation, while the upper awning offers full shading during peak irradiance. This combination demonstrates climate-responsive layering that adapts to diurnal and seasonal variation.

At an architectural scale, adaptive shading can be integrated into lighter façade systems. Figure 9.15 shows a compact public building in Las Palmas wrapped in a ventilated double skin composed of mesh and fixed horizontal louvers.

A more dense version of this strategy appears in Figure 9.16, where a multi-storey building in Las Palmas uses tightly packed horizontal louvers that generate a kinetic façade effect—expressive in appearance despite being non-operable.

FIGURE 9.13 Townhouse façades in Bayonne with traditional wooden shutters.

FIGURE 9.14 Combined esparto blinds and retractable awning in Mijas, Spain.

FIGURE 9.15 Public building in Las Palmas with ventilated façade composed of fixed horizontal louvers.

FIGURE 9.16 Multi-storey building in Las Palmas using dense horizontal louvers as a kinetic façade system.

These examples illustrate the continuum from occupant-controlled devices to integrated façade systems that provide shading, ventilation, and daylight modulation without mechanical actuation.

9.3.2 Vegetated Shading

Vegetation-based shading offers a dynamic, climate-responsive alternative to material-based shading systems. Through evapotranspiration and foliage density, plants reduce surface temperatures, filter sunlight, and cool microclimates. Their performance varies seasonally, yet this natural adaptability often enhances both thermal comfort and aesthetic integration.

In temperate climates with sufficient rainfall, vegetation can operate as a low-maintenance external shading layer. Figure 9.17 presents a historic villa in Wrocław where ivy covers the façade during warm months.

In humid oceanic climates such as southwestern France, vegetation often becomes an integral part of architectural identity. The example in Figure 9.18 shows residential buildings near Biarritz where climbing plants and small garden trees contribute to solar control.

Vegetation also extends to rooftops, where green roofs mitigate solar exposure and reduce heat transmission through the upper building envelope. Figure 9.19 illustrates an example from Bordeaux.

In hot and dry climates, extensive vegetated surfaces can also function as public infrastructure. Figure 9.20 shows the large-scale green roof of the Torre Sevilla complex in Spain.

FIGURE 9.17 Historic villa in Wrocław covered in ivy for passive solar protection.

FIGURE 9.18 Residential buildings near Biarritz featuring climbing vegetation and garden trees used for solar protection.

FIGURE 9.19 Business school building in Bordeaux with an accessible green roof and planted courtyards.

FIGURE 9.20 Torre Sevilla building with large-scale green roof integrated into public landscape.

9.3.3 Urban-Scale Shading Structures

Urban morphology significantly influences microclimate, particularly in hot or arid regions where shading at street level is essential for outdoor comfort. Dense building arrangements create self-shading, while overhead structures such as fabric canopies and wooden coverings extend protection to pedestrian zones.

Narrow alleys in Mediterranean and Middle Eastern cities illustrate this well. Although the following figures appear later in the chapter (Section 9.4), their logic is relevant here: shading is achieved not only by individual buildings but also by urban geometry itself. Overhead textile canopies—common in Andalusian cities—create seasonal shading that reduces radiant load on streets, while wooden coverings in North African souks provide permanent microclimatic buffering.

Vegetation-based shading also extends to urban environments in the form of pergolas and climbing plants, which create cooled pedestrian corridors through evapotranspiration and filtered sunlight.

9.4 Whitewashed Walls and High-Reflectivity Surfaces as Solar Control Strategies

In hot and high-solar climates, surface reflectivity becomes a critical parameter shaping the thermal response of building envelopes. Unlike overhangs, loggias, or geometric recesses, which regulate solar exposure through form, reflective surfaces operate at the material level by reducing the proportion of absorbed solar radiation. With high-albedo coatings—whether limewash, acrylic paint, or reflective glazing—external walls and roofs absorb substantially less heat, keeping surface temperatures lower and reducing conductive heat transfer into interior spaces.

Whitewashed surfaces are among the oldest and most geographically widespread passive cooling strategies. Their performance depends on both optical and hygrothermal properties: high solar reflectance reduces radiant heating, while lime-based coatings often support moisture buffering and antimicrobial performance. In dense, sun-exposed urban fabrics, these coatings also help reduce local heat accumulation and mitigate urban heat island effects.

9.4.1 Whitewashed Architecture in Hot Mediterranean and Arid Regions

Whitewashed lime coatings appear throughout hot-summer Mediterranean regions, where their high reflectance counters prolonged and intensive solar exposure. They are particularly effective when applied to thick masonry walls that can store coolness overnight and delay heat transfer during daytime.

The spatial and thermal logic of whitewashed settlements is illustrated in Figure 9.21, showing Setenil de las Bodegas in southern Spain.

In Setenil—where many houses are partially carved into rock—the combination of reflective lime coatings and thermal mass creates a strong buffering effect. Even under intense summer irradiance, the dwellings remain relatively cool because reflected radiation reduces daytime heat accumulation, while the surrounding rock and masonry slow heat transfer into interior spaces.

FIGURE 9.21 Whitewashed townscape of Setenil de las Bodegas, Spain.

9.4.2 High-Reflectivity Façades in Contemporary Architecture

Modern buildings employ reflective surfaces through glass, metal, and advanced coatings. While the traditional logic remains the same—reducing absorbed solar energy—contemporary materials allow higher precision in optical performance and a more integrated aesthetic.

An example of this approach appears in Figure 9.22, illustrating a contemporary hotel in Bilbao that uses reflective glazing panels as a passive solar protection strategy.

Reflective glazing functions as both a visual and thermal filter. By rejecting a significant portion of direct solar radiation, it reduces internal heat gain without requiring external shading devices—even in moderate climates where cooling loads are typically lower. In west- or south-facing hotel rooms, this strategy can substantially improve thermal comfort and reduce reliance on mechanical cooling systems during summer peaks.

9.4.3 Performance Considerations

Reflective surfaces are most effective when combined with:

- high thermal mass, which delays heat transfer,
- ventilated wall assemblies, which disperse heat from outer layers,
- shaded exterior zones, which reduce diffuse and reflected radiation.

FIGURE 9.22 Contemporary hotel in Bilbao with reflective glazing panels.

Their performance is reduced when surface coatings deteriorate, accumulate dust, or are replaced by low-albedo materials. For this reason, limewashed façades in many southern regions are regularly renewed—often as part of seasonal maintenance traditions that support both thermal comfort and cultural identity.

9.5 URBAN SHADING STRUCTURES

Urban morphology plays a decisive role in shaping microclimatic conditions, particularly in regions with high solar exposure, limited vegetation, and compact historic fabrics. While individual buildings may employ overhangs, recesses, or reflective surfaces, the broader thermal environment is governed by street geometry, shading continuity, and overhead structures that limit radiant load at pedestrian level. Narrow streets, arcaded walkways, textile canopies, and wooden coverings constitute essential passive strategies for regulating outdoor comfort in hot-arid and hot-summer Mediterranean climates.

These urban shading systems provide not only thermal protection but also structure social interaction and circulation patterns. Their effectiveness depends on geometry, materiality, and the degree to which they integrate with prevailing wind patterns.

9.5.1 Compact Urban Blocks and Narrow Streets

Dense street networks common in Mediterranean and Middle Eastern cities create natural self-shading conditions. Reduced street widths and increased building

FIGURE 9.23 Aerial view of Seville showing compact urban blocks and dense street network.

heights decrease sky view factor, significantly limiting direct solar radiation on façades and pavements.

The spatial logic of this strategy is illustrated in Figure 9.23, showing the compact fabric of Seville.

Closely spaced buildings cast mutual shade, lowering radiant temperatures during peak hours. Irregular alley geometries enhance wind funneling and natural ventilation, which helps dissipate accumulated heat. This contrasts with open-plan developments, where broader exposures result in higher thermal loads and greater dependence on mechanical cooling.

9.5.2 Shaded Alleys and Connecting Arches

In many historic cities, shading is intensified by architectural elements that span across streets. Connecting arches and overhead masonry structures further reduce solar penetration while preserving airflow and visibility.

This principle is shown in Figure 9.24, depicting a narrow alley in Évora, Portugal.

Such structural connections are common in Southern Europe and Islamic cities, where dense urban forms and material mass create persistent shading. They function as transitional environments that lower daytime heat stress and contribute to the thermal stratification of public spaces.

Further intensification of shading occurs in fully or semi-covered passages, such as those in Marrakesh. Figure 9.25 presents a semi-covered passage framed by a pointed arch and thick earthen walls.

FIGURE 9.24 Narrow street with connecting arches in Évora, Portugal.

In denser zones of the medina, passages become fully vaulted, creating enclosed microclimates with stable temperatures throughout the day.

Such structures represent a long-established response to extreme solar conditions, where shaded circulation spaces are essential for daily function and pedestrian comfort.

9.5.3 Textile Canopies and Seasonal Overhead Structures

In hot-summer Mediterranean climates, lightweight textile canopies provide seasonal shading for streets and plazas. Their flexibility allows installation during high-irradiance months and removal during cooler seasons.

FIGURE 9.25 Semi-covered passage in Marrakesh, Morocco.

This strategy is shown in Figure 9.26, depicting semi-permanent textile shading structure in pedestrian street in Mijas.

Beyond textile systems, traditional markets in North Africa employ wooden or bamboo coverings as permanent shading solutions. Figure 9.27 illustrates such a canopy from the souks of Marrakesh.

9.5.4 Vegetation-Based Shading in Urban Environments

Vegetation-based shading provides an adaptive and visually coherent complement to structural shading. Pergolas, climbing plants, and green corridors reduce solar exposure through foliage density and evapotranspiration, creating perceptibly cooler pedestrian environments.

FIGURE 9.26 Pedestrian street in Mijas with semi-permanent textile shading structure.

This is illustrated by Figures 9.28 and 9.29, which show vegetation-based shading in Puerto de Mogán, Gran Canaria, and Marrakesh.

Such systems contribute not only to thermal comfort but also to the sensory and spatial identity of urban environments, integrating ecological performance with cultural character.

9.6 SUMMARY

This chapter provides an integrated examination of passive shading strategies applied across building and urban scales, highlighting how geometry, material properties, vegetation, and adaptive systems influence solar exposure and thermal performance

FIGURE 9.27 Traditional wooden canopy in the Marrakesh souk.

in diverse climates. The examples demonstrate that effective control of solar gains results from the combined action of fixed architectural forms, adjustable shading devices, reflective surfaces, and urban morphology.

At the building scale, overhanging roofs and deep eaves proved to be one of the most enduring passive shading measures. Their geometry allows summer protection while, in some climates, enabling winter solar gains. The strategy is strongly climate-dependent: extended eaves are common in warm and coastal regions, whereas in northern and eastern Europe, their use has historically been limited by the need for winter sunlight and by high snow loads that require steeper roof profiles.

Façade geometry constitutes the second major category of passive shading. Elements such as cornices, balconies, recessed windows, arched loggias, and sculpted thermal-mass façades regulate solar access through spatial depth and selective exposure. These solutions are especially effective in Mediterranean and subtropical climates, where reducing the sky view factor and shading glazing surfaces significantly improves indoor thermal stability.

Adaptive shading systems introduce flexibility into the envelope. Manually operated shutters, roller blinds, and exterior venetian blinds allow occupants to respond to changing solar conditions, while hybrid solutions combine diffusing materials with retractable elements to achieve layered protection. Vegetated shading expands this adaptability through seasonal foliage density and evapotranspiration, providing

FIGURE 9.28 Vegetation-based shading in semi-arid urban environments in Puerto de Mogán.

both cooling and microclimatic regulation. Green roofs and climbing plants demonstrate how biological strategies can complement material-based shading.

Reflective surfaces, including whitewashed walls and mirrored glazing, operate at the material level by reducing the proportion of absorbed solar energy. Lime-based coatings traditionally used in southern Europe and North Africa offer high reflectivity and moisture buffering, while contemporary metal and glass façades achieve similar effects through engineered optical properties. These strategies are most effective when paired with thermal mass or ventilated assemblies that dissipate absorbed heat.

At the urban scale, narrow streets, compact block structures, overhead arches, textile canopies, and wooden coverings illustrate how entire districts can function as

FIGURE 9.29 Vegetation-based shading in semi-arid urban environments in Marrakesh.

shading systems. These configurations reduce direct solar exposure, improve pedestrian comfort, and enhance natural ventilation, particularly in hot-arid and hot-summer Mediterranean climates. Vegetation in public spaces further supports cooling by filtering sunlight and reducing radiant temperatures.

Taken together, the strategies presented in this chapter demonstrate that effective solar protection requires coordination across multiple spatial layers: roof geometry, façade depth, adjustable devices, vegetation, surface reflectivity, and urban form. Their integration enables buildings and cities to reduce heat stress, stabilize indoor conditions, and lower reliance on mechanical cooling, forming a comprehensive framework for climate-responsive design.

10 Thermal Mass Principles
Climate, Materiality, and Architectural Examples

10.1 INTRODUCTION

Thermal mass is one of the fundamental passive mechanisms used to stabilize indoor environmental conditions. Its performance is based on absorbing heat during periods of warming and releasing it gradually during cooling, which reduces diurnal temperature swings, delays peak indoor temperatures, and improves overall thermal stability. In traditional European and Mediterranean architecture, this principle has long served as protection against both winter cold and summer overheating, demonstrating sustained effectiveness across a wide range of climatic contexts.

Recent simulation studies show that even moderate differences in the heat capacity of external building envelope components can significantly influence the trajectory of indoor operative temperatures. Pajek et al. (2017) demonstrated that massive walls combined with night ventilation reduce daytime indoor temperatures by 1.8–2.8 K and stabilize night-time values through stored coolth. Mantesi et al. (2019), in a full-scale two-storey building, found that increasing envelope mass reduces indoor temperatures by approximately 2 K over the entire diurnal cycle, with even stronger effects when floor mass is increased. Jakovics et al. (2015) observed reductions of 3.5°C–4.0°C in maximum indoor temperatures under rapidly rising outdoor conditions. Brambilla and Jusselme (2017) showed that combining thermal mass with intensive night ventilation lowers peak temperatures by around 3°C and reduces diurnal amplitudes even under moderate outdoor conditions.

The importance of thermal mass becomes especially evident in studies where its buffering capacity is intentionally limited. Tink et al. (2018) reported that applying internal insulation to heavy walls increases mean indoor temperatures by 1.5°C and maximum temperatures by 2°C–2.5°C. Grynning et al. (2019) found that replacing a timber floor with a concrete one reduces peak temperatures during heatwaves by approximately 2.5 K, with differences exceeding 3 K in extreme conditions.

In a full-scale experiment carried out in Zielona Góra, Kuczyński and Staszczuk (2020) compared two identical residential buildings differing only in thermal mass (B1: 400 kJ/m^2K; B2: 192 kJ/m^2K) during a 14-day heatwave. The massive building reached a maximum operative temperature of 26.5°C, compared with 29.3°C in the lightweight version. The mean difference was 2.8°C, increasing to 3.4°C during peak hours. A TM59 analysis showed a reduction in overheating from 18.6 days above 28°C in the lightweight building to only 8 hours in the massive one.

DOI: 10.1201/9781003645573-10

Broader implications of thermal mass are demonstrated in the work of Johra et al. (2019), who highlighted its role in increasing building energy flexibility. Higher thermal mass lengthens the time constant of the building and enables more effective shifting of thermal loads. Their results show that thermal mass has the greatest influence in lightweight and medium-mass buildings, while additional mass produces smaller gains in already heavy structures. Internal mass contributed by furniture and finishes can increase the effective inertia of lightweight buildings by up to 40 percent, with the strongest impact occurring during prolonged heat extremes.

Theoretical analyses by Reilly and Kinnane (2017) underline that the benefits of thermal mass depend strongly on climate and occupancy patterns. In cold climates, particularly under intermittent heating, high thermal mass may increase energy use due to delayed thermal response. Using the metrics Transient Energy Ratio and Effective U-value, the authors showed that dynamic thermal behavior can raise the effective heat transfer of envelope components. In hot climates, however, thermal mass remains advantageous when paired with strategies that limit solar gains and ensure sufficient night-time regeneration. These findings highlight the need to evaluate thermal mass in relation to climate, envelope insulation levels, and building use.

Against this analytical background, the examples presented in the following sections illustrate how different European and neighboring regions—from cold climates to temperate and hot zones—have adapted thermal-mass-based strategies to local environmental and urban conditions. The cases confirm that thermal mass remains central to climate-resilient design and derives its effectiveness from the interaction of materials, geometry, shading, ventilation, and seasonal temperature rhythms. Given the increasing frequency of heatwaves and prolonged thermal extremes, the strategic role of thermal mass is once again emerging as a foundation of passive thermal comfort and building resilience.

10.2 HIGH THERMAL MASS IN URBAN MASONRY CONSTRUCTION: CENTRAL AND WESTERN EUROPE

The role of thermal mass in temperate and cold-temperate regions is especially evident in the historic centers of Central Europe, where stone and brick construction enabled buildings to buffer temperature swings throughout the year. Massive walls, arcades, and reduced window-to-wall ratios served as passive stabilizers of indoor conditions long before these principles were described in modern thermal theory.

The first example, shown in Figure 10.1, illustrates this logic in Bamberg, Germany.

These buildings were constructed primarily from stone and brick, with wall thicknesses often exceeding 50 cm. Their high thermal inertia moderates seasonal and diurnal temperature fluctuations by absorbing solar gains and internal heat during cold months and delaying heat penetration during summer. The stabilization effect significantly reduces dependence on mechanical heating and cooling. In cities such as Munich or Nuremberg, pastel- and white-colored façades further reduce solar absorption, while arcades at street level provide shading, improving pedestrian comfort and buffering ground-floor interiors.

FIGURE 10.1 Historic streetscape in Bamberg, Germany, illustrating high-mass masonry architecture.

10.3 LIGHTWEIGHT TIMBER-FRAMED ARCHITECTURE: FACHWERK AND WATTLE-AND-DAUB

In contrast to the high-mass masonry traditions of Central Europe, many regions developed lightweight timber-framed systems that rely on rapid thermal response rather than long-term heat storage. These structures warm up and cool down quickly, making them suitable for transitional climates and daily temperature variability but requiring additional measures to prevent heat loss in winter and overheating in summer.

10.3.1 Fachwerk Construction Technique

The Fachwerk system integrates a load-bearing timber frame with infill panels made from brick, clay, plaster, or other composite materials. This method significantly reduces thermal mass while offering structural flexibility and efficient use of local materials. The architectural and thermal principles of this strategy are illustrated in Figure 10.2.

Compared with the stone masonry of Bamberg, Fachwerk buildings exhibit low heat capacity and limited thermal inertia. Their lightweight envelopes respond rapidly to changing outdoor temperatures, which can be advantageous during shoulder seasons but leads to faster heat loss in winter. As a result, many Fachwerk buildings require internal insulation to maintain acceptable thermal comfort during colder months. Overhanging upper stories, typical of this typology, provide passive shading for lower façade levels and optimize interior space in narrow medieval streets.

FIGURE 10.2 Traditional timber-framed (Fachwerk) buildings in Braunschweig, Germany.

10.3.2 Wattle-and-Daub Construction Technique

A second traditional low-mass system is wattle-and-daub, employed particularly in areas where stone was scarce. The technique combines a woven wooden lattice (wattle) with a clay-based mixture reinforced with straw or other organic fibers (daub). This system produces walls with moderate thermal inertia and excellent humidity regulation.

Figure 10.3 presents an example from Honfleur, Normandy.

Wattle-and-daub walls balance thermal mass and breathability. The porous daub mixture regulates indoor humidity, supports moisture buffering, and reduces condensation risk. Combined with steeply pitched roofs, which improve rain protection and promote convective cooling, these buildings achieve stable indoor conditions despite their low structural mass. Although mechanically less durable than masonry, their reparability, low embodied energy, and reliance on local materials supported their use for centuries across Western Europe.

Together, Fachwerk and wattle-and-daub illustrate an alternative to the high-mass approach: low thermal inertia compensated by adaptive shading, ventilation patterns, and user operation. These systems form a crucial historical reference for contemporary lightweight construction in temperate climates.

10.4 HYBRID MASONRY AND TIMBER STRATEGIES: CENTRAL EUROPEAN ADAPTATIONS

Hybrid structures combining massive masonry with lightweight timber-framed upper floors constitute an intermediate approach between high thermal inertia and low-mass construction. This configuration, found widely in Central Europe,

FIGURE 10.3 Traditional half-timbered house in Honfleur, Normandy, France, built with wattle-and-daub infill.

responds to climatic, economic, and urban conditions by distributing material functions across different parts of the building. Heavy ground floors stabilize indoor temperatures, while lighter upper levels reduce construction costs and allow faster thermal response in residential spaces.

An example of this strategy is shown in Figure 10.4, illustrating a traditional hybrid building in Braunschweig, Germany.

In this typology, the masonry ground floor—often used historically for storage, workshops, or commercial activity—benefits from high thermal mass. The walls absorb and release heat slowly, maintaining stable conditions throughout the year and reducing reliance on mechanical conditioning. In contrast, the upper floors constructed in the Fachwerk tradition have low thermal mass, allowing rapid adjustment to daily temperature fluctuations. This duality enhances functional efficiency: durable masonry supports stable indoor environments, while lightweight timber improves constructability and adapts to varied residential needs.

Cantilevered façades, common in such buildings, provide additional passive shading. The projecting upper floors limit direct solar exposure on lower façades and reduce heat gain on street-facing elevations. The arrangement also increases usable interior space without expanding the building footprint, a valuable adaptation in the dense medieval fabrics where these structures originated.

Hybrid masonry-timber buildings illustrate how historical construction practices integrated material performance with climate-responsive design. By balancing thermal buffering and rapid thermal response, they demonstrate a layered approach to environmental regulation that remains relevant to contemporary strategies seeking to combine mass, insulation, shading, and efficient material use.

FIGURE 10.4 Hybrid masonry and timber-framed building in Braunschweig, Germany.

10.5 BASQUE COUNTRY AND ATLANTIC TRANSITIONAL REGIONS

Moving from colder continental climates toward the Atlantic coast of France and the western Iberian Peninsula, the role of thermal mass shifts from predominantly winter buffering to a dual function: moderating summer heat and stabilizing humidity. Although diurnal temperature swings are milder than in continental Europe, they remain sufficient for heavy envelopes to provide short-term damping of indoor thermal fluctuations.

The first example, shown in Figure 10.5, comes from Bordeaux, where most eighteenth- and nineteenth-century residential buildings were constructed from local limestone with wall thicknesses typically between 60 and 80 cm.

The façade is relatively flat yet incorporates deep window recesses that reduce solar gains and slow conductive heat transfer. Combined with traditional wooden shutters, the limestone mass forms an effective passive system. When shutters are closed during the day and windows opened at night, the building retains cool air and delays daytime warming, a favorable strategy for Bordeaux's warm-summer oceanic climate.

A contrasting but related example appears in Bayonne, further southwest within the same climatic zone but with slightly higher summer irradiance and lower ventilation due to denser street canyons. As shown in Figure 10.6, the arcaded building at the corner of Rue des Tanneurs and Rue d'Espagne employs limewashed, rendered masonry over substantial wall mass.

FIGURE 10.5 Residential limestone building on Rue Sainte-Catherine, Bordeaux.

While the masonry core provides moderate-to-high thermal inertia, the lime-washed external render increases reflectivity and reduces surface heating. Deep-set windows and operable shutters mitigate solar gains, and the arcades shade the ground-floor façade. The building illustrates a calibrated response to a climate requiring both buffering and controlled ventilation.

A more explicit juxtaposition of thermal strategies appears in the narrow street shown in Figure 10.7, where high-mass stone buildings stand adjacent to lightweight timber-frame structures.

In this configuration, heavy masonry buildings act as thermal capacitors, cooling slowly and reducing indoor temperature peaks, while lighter timber structures depend more on daily shading routines and ventilation management. Notably, nearly all buildings—regardless of material—use external shutters, which remain essential for mitigating summer solar loads in this climatic region.

Further west, in San Sebastián, traditional masonry townhouses continue this logic. Their compact form, reduced window-to-wall ratios, and shaded street

FIGURE 10.6 Limewashed masonry building with arcades in central Bayonne.

geometry contribute to effective moderation of summer heat. The progression of this strategy becomes even clearer in Bilbao, where high-mass institutional buildings respond to rising summer temperatures through combined use of material mass, façade depth, and controlled ventilation.

The Jesuit College building in Bilbao, shown in Figure 10.8, demonstrates these principles.

Constructed from solid stone and clad in rendered façades, the building provides very high thermal mass suitable for stabilizing interior conditions in a climate that increasingly experiences summer heatwaves. Thick walls and small windows reduce solar gains and prolong heat transfer to interior spaces. Deep arcades enhance shading, while the building's symmetrical plan and riverside location support cross-ventilation during cooler night-time hours.

Across Bordeaux, Bayonne, San Sebastián, and Bilbao, thermal mass emerges as an adaptable strategy that changes in emphasis according to climate, material tradition, and urban density. In the Atlantic transitional zone, it is neither uniformly

FIGURE 10.7 Stone-masonry and timber-frame buildings coexisting within a narrow street in Bayonne.

dominant nor absent; instead, it coexists with lightweight systems, shading devices, and user-driven ventilation practices. This coexistence reflects an ongoing negotiation between mass-based buffering and dynamic solar control in regions facing increasingly variable summer temperatures.

10.6 HOT-SUMMER MEDITERRANEAN REGIONS: MADRID AND ANDALUSIA

As the climatic transition continues southward toward the interior and southern Iberian Peninsula, thermal mass becomes not only advantageous but also essential. In hot-summer Mediterranean climates, prolonged periods of high solar radiation, limited nocturnal cooling, and extended heatwaves require envelope strategies that can store heat during the day, delay its penetration, and release it gradually

FIGURE 10.8 Jesuit College (Colegio de los Jesuitas), Bilbao, with high-mass stone walls and recessed façades.

overnight. Across Madrid and Andalusia, heavy masonry construction, compact urban morphology, and controlled solar exposure form a coherent response to these demanding conditions.

In Madrid, where daytime summer temperatures frequently exceed 35°C, thermal mass serves both functional and symbolic roles, underpinning civic architecture designed for stability, permanence, and environmental moderation. The Banco de España building, shown in Figure 10.9, exemplifies this approach.

Constructed primarily from granite or limestone, the building features exceptionally thick walls, recessed openings, and a restrained glazing ratio. These characteristics provide strong thermal inertia, delaying heat transfer to interior spaces and reducing cooling loads. Arcades, cornices, and roof overhangs contribute to self-shading, while the symmetrical floor plan and tall ceilings enhance airflow and facilitate night-time cooling. The building's longevity illustrates the enduring relevance of high thermal mass in climates where overheating risk is severe.

Moving further south into Andalusia, the role of thermal mass intensifies due to longer summers, narrower diurnal temperature ranges, and intermittent humidity increases. Cities such as Córdoba have refined passive cooling strategies over centuries, combining heavy materials, shaded façades, and compact urban form into a coherent thermal system. A representative example is shown in Figure 10.10.

The building's massive masonry envelope absorbs heat slowly during the day and releases it overnight. The low window-to-wall ratio and deep recesses limit solar penetration, while the light-colored plaster increases reflectivity and reduces

FIGURE 10.9 Banco de España, Madrid: late nineteenth-century civic building with high-mass stone walls.

FIGURE 10.10 Historic masonry building in central Córdoba with deeply recessed openings and light-colored plaster.

surface heating. Traditional Andalusian buildings of this type often include enclosed patios and thick interior partitions, which further support thermal buffering and convective airflow. Set within a compact urban fabric and supplemented by temporary shading devices, the structure demonstrates a layered adaptation to extreme summer heat.

A complementary example from Córdoba reveals how contemporary architecture can reinterpret vernacular strategies while maintaining thermal performance. The neo-mudejar Hostal El Triunfo, presented in Figure 10.11, applies high-mass walls, deep façade articulation, and flexible shading in a stylized form.

Although modern in construction, the building relies on proven passive principles: moderate-to-high thermal inertia, controlled solar exposure, and compact siting within a dense urban block. The recessed windows maintain self-shading throughout the day, while fixed awnings and retractable blinds allow for gradual modulation of solar gain. The result is a contemporary expression of traditional Andalusian thermal logic, adapted to the region's increasingly demanding climate.

Across Madrid and Andalusia, high thermal mass remains a central determinant of thermal resilience. Massive envelopes delay and dampen heat ingress, while solar protection, façade depth, and controlled ventilation address the strong radiative and convective forces characteristic of the Mediterranean interior. The examples demonstrate that traditional heavy construction provides a robust foundation for contemporary passive cooling strategies, particularly in regions facing intensifying climatic extremes.

FIGURE 10.11 Neo-mudejar building on Plaza del Triunfo, Córdoba, with passive shading elements and recessed façades.

10.7 CANARY ISLANDS: SUBTROPICAL ATLANTIC CONDITIONS

In the dry-subtropical climates of the Canary Islands, high solar radiation, extended dry seasons, and pronounced diurnal temperature swings create conditions in which thermal mass becomes a primary passive cooling resource. Traditional architecture in inland areas of Gran Canaria demonstrates a clear evolution in material use, façade configuration, and building form, shifting from purely functional thermal buffering toward hybrid strategies that combine mass, reflectivity, and controlled ventilation.

One of the clearest examples is found in Mogán Pueblo, a small inland settlement characterized by steep terrain and intense solar exposure. Two adjacent vernacular dwellings, shown in Figure 10.12, illustrate early and later stages of thermal-mass-based adaptation.

The older building on the right uses irregular volcanic fieldstone coated with thick earth render, producing wall thicknesses of approximately 60–80 cm. The heavy envelope stores heat during the day and releases it after sunset, stabilizing interior conditions despite strong radiation. The newer, whitewashed house on the left incorporates deeper reveals and a small shaded porch, improving solar control while maintaining substantial mass. Together, the two structures reveal long-standing reliance on thick masonry, low glazing ratios, and bright exterior finishes as primary defense against heat stress.

A more developed iteration of this vernacular logic appears in the exposed-stone dwelling presented in Figure 10.13.

Built likely in the late nineteenth or early twentieth century, the house employs larger volcanic stones bonded with visible mortar. Wall thicknesses still exceed 70 cm,

FIGURE 10.12 Paired vernacular dwellings in Mogán Pueblo, Gran Canaria.

FIGURE 10.13 Traditional unrendered stone house in Mogán Pueblo with small recessed windows.

ensuring strong thermal inertia and slow heat ingress. Small, recessed windows with shutters provide user-controlled ventilation and shading, while the unrendered stone façade blends with the surrounding rocky landscape. The structure maintains cool interiors through the combined effect of mass, limited glazing, and delayed heat transmission.

In contrast, contemporary adaptations in Mogán often prioritize visual continuity over true thermal inertia. Figure 10.14 presents such a case.

Although visually aligned with traditional architecture, the building is likely constructed using hollow block or concrete-frame systems with stone applied as cladding. Its thermal mass is therefore moderate, and indoor comfort depends more on reflective surfaces, orientation, and shading devices than on inherent material inertia. The example illustrates a broader shift toward aesthetic historicism in contemporary construction throughout the region.

Further insights into thermal mass adaptation come from the historic center of Agüimes, where pre-modern and modern buildings coexist within a compact urban fabric exposed to high solar loads. The first example, shown in Figure 10.15, demonstrates the traditional Canarian approach.

The structure is built from local volcanic stone bonded with lime and clay mortar, with wall thicknesses exceeding half a meter. Its deep reveals, heavy envelope, and high-reflectance render limit solar gains and stabilize indoor conditions. Minimal glazing and compact form further restrict heat ingress. The building exemplifies the continued relevance of thermal mass in inland areas with extreme radiation.

The neighboring contemporary example, shown in Figure 10.16, applies a similar volumetric logic but uses lightweight construction.

FIGURE 10.14 Modern house in Mogán with decorative stone façade and tiled roof.

FIGURE 10.15 Traditional masonry building in Agüimes, Gran Canaria, with deep window recesses and ochre render.

FIGURE 10.16 Contemporary cubic dwelling in Agüimes referencing traditional forms but built with lightweight construction.

Although it echoes historic proportions and colors, the structure's lightweight envelope results in low thermal inertia. It is more sensitive to temperature fluctuations and depends on solar orientation, reflective finishes, and potential mechanical cooling to maintain comfort. The contrast between Figures 10.15 and 10.16 highlights the performance gap between traditional mass-based cooling and modern lightweight construction in dry subtropical climates.

Across Mogán and Agüimes, the evolution of architectural form illustrates ongoing negotiation between thermal mass, surface reflectivity, shading, and user operation. Despite changes in construction techniques, the region's vernacular heritage demonstrates the persistent value of heavy envelopes and controlled apertures in mitigating severe heat loads—principles that remain central to contemporary passive design in subtropical environments.

10.8 EARTH-SHELTERED ARCHITECTURE: PRINCIPLES, TYPOLOGIES, AND CASE STUDIES

Earth-sheltered buildings constitute one of the most effective passive strategies for thermal stabilization in regions with high diurnal temperature variation or extreme seasonal heat. Their performance derives from the thermal properties of soil, which acts as a large-capacity buffer capable of absorbing and releasing heat with significant delay. By embedding building volumes fully or partially within the ground, temperature fluctuations in interior spaces can be greatly reduced, often eliminating the need for mechanical heating or cooling during large parts of the year.

10.8.1 Typologies of Earth-Sheltered Systems

Earth-sheltered architecture exists in three principal configurations that differ in their degree of integration with the surrounding soil. These typologies, illustrated schematically in Figure 10.17, correspond to distinct climatic and spatial requirements:

1. Fully underground buildings: Volumes entirely surrounded by soil with minimal contact with outdoor air.
2. Partially buried buildings: Structures embedded in a slope or berm, with at least one façade exposed for daylight and ventilation.
3. Above-ground high-mass earth constructions (e.g., earthbag, rammed earth): Buildings that achieve similar thermal buffering through thick earthen walls without being physically buried.

Traditional examples can be found in Meymand (Iran), Sacromonte (Spain), and Setenil de las Bodegas (Spain), where living spaces are surrounded on three or four sides by geological mass. Modern approaches combine bermed walls, earth roofs, and recessed courtyards to introduce daylight and facilitate natural ventilation while retaining thermal stability.

10.8.2 Ground Temperature Profile and Thermal Mechanisms

Soil temperature varies seasonally and diurnally, but fluctuation amplitude decreases substantially with depth. Near the surface, the temperature curve resembles outdoor air patterns, yet at approximately 1 m depth, fluctuations are significantly damped. Between 1 and 8 m, ground temperature becomes nearly constant and approximates

FIGURE 10.17 Typological scheme of three earth-sheltered systems: (a) fully underground, (b) partially embedded, (c) above-ground earthbag construction. (From Khaksar et al. 2022.)

the annual mean outdoor temperature. Below 8–10 m, only a slight geothermal gradient is observed.

This stability enables earth-sheltered buildings to function as large-mass heat exchangers. In summer, heat is transferred from the building to cooler subsurface layers; in winter, heat flows from the relatively warmer soil to the structure. Consequently, the temperature difference between interior and exterior becomes much smaller than in above-ground buildings, reducing heat flux through envelope components and decreasing heating and cooling demand by up to 40 percent.

10.8.3 Field and Simulation Study: Meymand, Iran

Khaksar et al. (2022) conducted a detailed analysis of 22 earth-sheltered structures in the historic settlement of Meymand. Buildings differed in geometry, orientation, and depth of embedding. Temperature and humidity were recorded at 10-minute intervals during summer and winter using Testo 175-H2 loggers. Numerical models were developed in EnergyPlus and Ladybug Tools, incorporating measured thermal properties of volcanic rock, soil moisture, and coupled convective–radiative heat transfer.

The thermal equilibrium depth was found at approximately 1.5 m, where temperature variation was minimal. Phase shift between outdoor air temperature and subsurface conditions reached 4–5 hours in summer, ensuring that indoor spaces remained cool during peak heat.

Measurements showed indoor temperatures between 19.6°C and 28.4°C, while outdoor temperatures ranged from 20.5°C to 35.7°C, corresponding to an average reduction in diurnal amplitude of around 6.4 K. Square-plan buildings performed better (±2.9°C variation) than elongated rectangular layouts (±5.5°C), indicating that plan geometry influences temperature uniformity.

Orientation also proved critical. South-east-oriented buildings exhibited the most stable temperatures and the highest percentage of annual comfort hours, reaching nearly 60 percent according to ASHRAE 55 adaptive criteria. Increased embedding depth enhanced thermal stability but raised relative humidity, which reached up to 58 percent in the deepest rooms.

Model validation demonstrated high agreement with field data ($R^2 = 0.76$; RMSE = 0.6 K). Using a genetic algorithm, six design variables were optimized: length, width, height, orientation, window-to-wall ratio, and shading depth. The final configuration achieved 90.5 percent annual comfort hours compared to 82.6 percent in the best-performing existing building-without HVAC use.

10.8.4 Limitations and Emerging Directions

The primary limitation of earth-sheltered systems is ventilation and moisture control. Restricted air exchange can lead to condensation and high humidity, particularly in deeply buried spaces. Traditional solutions include recessed courtyards and open vestibules enabling natural air movement. Contemporary designs increasingly incorporate vertical ventilation shafts or solar chimneys, which can increase air-exchange efficiency by 60–80 percent.

Mihalakakou et al. (2024) highlight future advances including integration with phase-change materials, automated ventilation control driven by environmental sensors, and machine-learning models that predict soil temperature and moisture for proactive microclimate management. The authors emphasize the need to combine thermal modelling with daylighting and acoustic analyses to achieve holistic performance optimization.

10.8.5 Case Studies: Andalusian Cave Architecture

Earth-sheltered principles are well illustrated by cave architecture in southern Spain, particularly in Guadix, Sacromonte, and Setenil de las Bodegas. These typologies use geological mass for thermal buffering and minimize exposure to solar radiation.

The first case, shown in Figure 10.18, is a carved dwelling in Sacromonte, where interiors are excavated directly into soft limestone and marl.

A different variant appears in Setenil de las Bodegas, where buildings are constructed under natural rock overhangs rather than excavated. The example in Figure 10.19 illustrates this hybrid condition.

A larger residential and commercial structure positioned fully beneath a limestone ledge is shown in Figure 10.20. Although partly modernized, the building benefits from continuous geological shielding.

Sacromonte and Setenil de las Bodegas differ mainly in geological conditions. Soft limestones and marls in Sacromonte are easy to excavate, allowing multi-room dwellings fully embedded within the hillside. Harder limestones in Setenil de las Bodegas limit excavation potential, resulting in hybrid systems where masonry façades support natural overhangs. Despite these differences, both typologies rely on the same principle: reducing solar exposure and leveraging geological mass to achieve stable thermal environments.

FIGURE 10.18 Carved hillside dwelling in Sacromonte, Granada.

FIGURE 10.19 Semi-troglodytic dwelling under a natural rock overhang, Setenil de las Bodegas.

FIGURE 10.20 Residential and commercial building beneath a limestone ledge, Setenil de las Bodegas.

10.9 SUMMARY

This chapter examines the role of thermal mass across a wide climatic gradient, from cold-temperate Central Europe to the hot-summer Mediterranean and dry-subtropical Atlantic environments. The examples demonstrate that thermal mass is not a uniform solution but a climate-responsive strategy whose performance depends on material properties, envelope configuration, shading, ventilation patterns, and the magnitude of diurnal temperature swings.

In cold and temperate regions, high-mass masonry buildings moderate seasonal and daily temperature fluctuations, reducing dependence on mechanical heating and cooling. Thick stone and brick walls act as thermal reservoirs, absorbing heat during the day and releasing it slowly at night. Historic districts in Germany and England illustrate this approach through compact forms, limited glazing, and arcaded ground floors that enhance shading and environmental stability.

Lightweight systems such as timber-framed Fachwerk and wattle-and-daub construction offer a contrasting condition. Their low thermal inertia results in rapid thermal response, which is beneficial during transitional seasons but requires careful management of insulation and shading. These typologies depend more on occupant behavior—adjusting shutters and ventilation pathways—than on envelope mass.

Hybrid masonry–timber buildings combine the advantages of both strategies. Heavy ground floors stabilize storage or commercial spaces, while lighter upper stories allow faster heating and cooling in residential areas. The cantilevered façades common in these structures provide self-shading and optimize interior space in dense medieval settings.

In Atlantic transitional climates, the role of thermal mass becomes more nuanced. Limestone buildings in Bordeaux, rendered masonry in Bayonne, and high-mass institutional structures in Bilbao all rely on envelope inertia to dampen summer heat while maintaining acceptable indoor conditions during warm periods. Lightweight postwar buildings in the same cities compensate for lower thermal mass through shading devices, reflective surfaces, and controlled ventilation.

In the hot-summer Mediterranean, thermal mass becomes essential. Buildings in Madrid and Córdoba demonstrate deeply recessed openings, limited glazing, reflective plaster, and compact morphology. These features delay heat ingress and support night-time re-cooling, forming a reliable passive cooling system in climates with prolonged high temperatures and limited nocturnal cooling potential. Contemporary adaptations, such as neo-mudejar façades, maintain these thermal principles while reinterpreting them stylistically.

In the dry-subtropical Atlantic climate of Gran Canaria, thermal mass coexists with strategic solar control and high reflectance. Vernacular houses in Mogán and Agüimes employ thick volcanic stone walls, minimal apertures, and whitewashed surfaces to buffer daytime heat loads. More recent lightweight buildings mimic traditional forms visually but rely on shading and orientation rather than envelope mass to maintain comfort.

Earth-sheltered systems represent the most advanced expression of passive thermal buffering. Fully or partially buried structures—such as those documented in Meymand, Sacromonte, and Setenil de las Bodegas—use geological mass to stabilize indoor temperatures and reduce thermal amplitude by more than 6 K. Their efficiency, however, depends on careful control of ventilation and moisture. Recent studies highlight opportunities to integrate such systems with phase-change materials, automated ventilation, and machine-learning-based environmental prediction.

Across all regions, the performance of thermal mass is shown to depend not only on material selection but also on its interaction with form, shading, ventilation, and surface reflectivity. This chapter demonstrates that thermal mass remains a central, adaptable, and resilient strategy for passive environmental control, retaining relevance in contemporary climates marked by intensifying heatwaves and increasing temperature variability.

REFERENCES

Brambilla, A., & Jusselme, T. (2017). Preventing overheating in offices through thermal inertial properties of compressed earth bricks: A study on a real-scale prototype. *Energy and Buildings*, *156*, 281–292. https://doi.org/10.1016/j.enbuild.2017.09.070

Grynning, S., Nocente, A., Gullbrekken, L., & Skjeggeru, K. (2019). Thermal mass and thermal comfort in offices–experimental studies of a concrete floor. *MATEC Web of Conferences* 282, 02087, CESBP 2019. https://doi.org/10.1051/matecconf/201928202087

Jakovics, A., Gendelis, S., & Bandeniece, L. (2015). Energy efficiency and sustainability of different building structures in Latvian climate. *IOP Conference Series: Materials Science and Engineering*, *96*(1), 012032. https://doi.org/10.1088/1757-899X/96/1/012032

Johra, H., Heiselberg, P., & Dréau, J. L. (2019). Influence of envelope, structural thermal mass and indoor content on the building heating energy flexibility. *Energy and Buildings*. https://doi.org/10.1016/j.enbuild.2018.11.012

Khaksar, A., Tabadkani, A., Mofidi Shemirani, S. M., Hajirasouli, A., Banihashemi, S., & Attia, S. (2022). Thermal comfort analysis of earth-sheltered buildings: The case of Meymand village, Iran. *Frontiers of Architectural Research*, *11*(6), 1214–1238. https://doi.org/10.1016/j.foar.2022.04.008

Kuczyński, T., & Staszczuk, A. (2020). Experimental study of the influence of thermal mass on thermal comfort and cooling energy demand in residential buildings. *Energy*, *195*, 116984. https://doi.org/10.1016/j.energy.2020.116984

Mantesi, E., Hopfe, C. J., Mourkos, K., Glass, J., & Cook, M. (2019). Empirical and computational evidence for thermal mass assessment: The example of insulating concrete formwork. *Energy and Buildings*, *188–189*, 314–332. https://doi.org/10.1016/j.enbuild.2019.02.021

Mihalakakou, G., Paravantis, J. A., Nikolaou, P., Menounou, P., Tsangrassoulis, A. E., Malefaki, S., Fotiadi, A., Papadaki, M., Giannakopoulos, E., Romeos, A., Escobar-Hernandez, H. U., & Souliotis, M. (2024). Earth-sheltered buildings: A review of modeling, energy conservation, daylighting, and noise aspects. *Journal of Cleaner Production*, *472*, 143482. https://doi.org/10.1016/j.jclepro.2024.143482

Pajek, L., Hudobivnik, B., Kunič, R., & Košir, M. (2017). Improving thermal response of lightweight timber building envelopes during cooling season in three European locations. *Journal of Cleaner Production*, *156*, 939–952. https://doi.org/10.1016/j.jclepro.2017.04.098

Reilly, A., & Kinnane, O. (2017). The impact of thermal mass on building energy consumption. *Applied Energy*, *198*, 108–121. https://doi.org/10.1016/j.apenergy.2017.04.024

Tink, V., Porritt, S., Allinson, D., & Loveday, D. (2018). Measuring and mitigating overheating risk in solid wall dwellings retrofitted with internal wall insulation. *Building and Environment*, *141*, 247–261. https://doi.org/10.1016/j.buildenv.2018.05.062

11 Natural Ventilation Principles

Airflow Mechanisms and Climatic Applications

11.1 INTRODUCTION

Natural ventilation constitutes one of the most enduring and technically sophisticated passive strategies in architectural design. Long before the development of mechanical ventilation systems, buildings relied on wind pressure, buoyancy, and the geometric configuration of openings to regulate indoor air quality, temperature, and humidity. These mechanisms were neither accidental nor secondary. Across climates and cultures, they shaped building typologies, spatial organization, and entire urban morphologies. Courtyards, windcatchers, atria, and double-skin façades (DSFs) represent different expressions of a shared aerodynamic logic that has evolved over centuries in response to environmental pressures.

In the context of contemporary climate change, natural ventilation regains strategic importance. Rising temperatures, more frequent heatwaves, and the increasing energy cost of mechanical cooling make passive airflow systems essential for maintaining thermal comfort and reducing operational energy demand. At the same time, new technologies—such as adaptive façades, advanced sensors, and renewable-energy integration—allow traditional ventilation principles to be reinterpreted within high-performance building envelopes and dense urban environments.

This chapter examines natural ventilation across multiple scales. At the building scale, it begins with the three fundamental mechanisms that govern all passive airflow—cross-ventilation, single-sided ventilation, and buoyancy-driven stack effect—and analyses how each has been adapted to cold, temperate, and Mediterranean climates. It then explores courtyard systems, atria, and windcatchers, illustrating how these typologies amplify or refine basic aerodynamic forces. At the façade scale, the chapter discusses how DSF and ventilated façade form advanced interfaces that combine shading, buffering, airflow regulation, and, increasingly, photovoltaic integration. Finally, at the urban scale, it addresses ventilation corridors as part of performance-based planning frameworks that enhance air movement, reduce heat accumulation, and mitigate the urban heat island effect.

Viewed together, these systems demonstrate that natural ventilation is not a single technique but a continuum of climate-responsive strategies operating across building layers and spatial hierarchies. Their physical principles are universal, yet their architectural forms are highly differentiated, shaped by regional climates, material

DOI: 10.1201/9781003645573-11

traditions, and evolving technological capabilities. This integrative perspective underscores the ongoing relevance of natural ventilation for resilient, low-energy, and health-supportive built environments.

11.2 MECHANISMS AND CLIMATIC LOGICS OF NATURAL VENTILATION

11.2.1 Universal Physical Mechanisms of Airflow

Natural ventilation has long served as one of the primary passive strategies for regulating indoor air quality, thermal comfort, and humidity in traditional architecture. Before the advent of mechanical systems, buildings relied entirely on the interaction of wind pressure, temperature differences, and air-density gradients to generate airflow through openings and internal passages. When supported by appropriate building geometry and spatial organization, these physical mechanisms enabled effective ventilation without energy input.

Across climatic regions, vernacular architecture developed sophisticated and often highly localized methods for directing air movement and moderating indoor temperatures. In temperate and Mediterranean climates, cross-ventilation and courtyard-based airflow formed the core of passive cooling. Deep-set windows, arcades, and shaded patios facilitated air circulation and reduced heat accumulation while maintaining daylight access. Similar principles appear in the Jiangnan region of China, where sky-wells, narrow wind corridors, and water–air interaction along canal networks improved airflow within dense urban districts.

In hot and humid climates, such as Southeast Asia, stilt houses use elevated floors, lightweight materials, and stack-effect ventilation to reduce heat storage and manage high relative humidity. In hot-arid regions, Iranian windcatchers and Indian stepwells provide notable examples of passive strategies that harness air-pressure differentials and evaporative cooling to maintain comfortable indoor conditions. These systems demonstrate refined environmental logic based on aerodynamic control, microclimatic buffering, and the strategic use of water surfaces.

By examining these diverse systems, this chapter highlights how natural ventilation in vernacular architecture has been optimized across cultures and climates. Despite the emergence of mechanical ventilation, the principles embedded in these traditional solutions remain fundamental for reducing energy demand, enhancing resilience during heatwaves and supporting adaptive comfort in contemporary design.

Natural ventilation remains one of the most fundamental passive strategies for controlling indoor air quality and thermal comfort. Although the physical principles governing airflow are universal, the ways in which these mechanisms manifest in buildings vary significantly across climatic regions. Temperature gradients, prevailing winds, humidity, diurnal variability, and cultural patterns of occupancy create distinctive ventilation logics that have shaped vernacular and contemporary architecture alike. In cold, temperate, and Mediterranean climates, natural ventilation operates through three primary airflow mechanisms: cross-ventilation driven by wind-pressure differences, single-sided ventilation induced by local turbulence, and stack ventilation driven by buoyancy forces. These mechanisms, illustrated

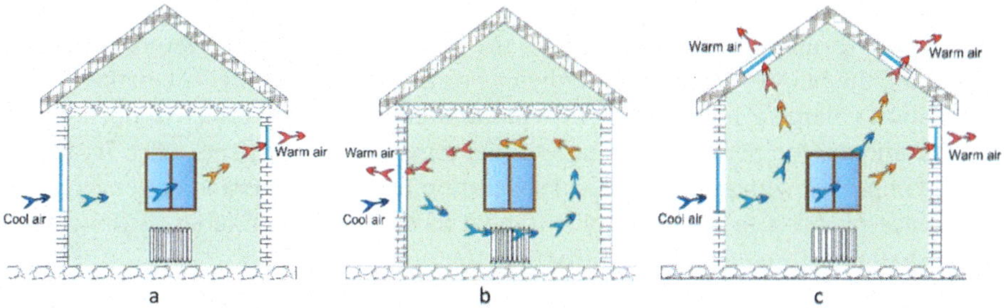

FIGURE 11.1 Mechanisms of natural ventilation: (a) cross-ventilation, (b) single-sided ventilation, (c) stack-driven ventilation. (From Myroniuk et al., 2026.)

schematically in Figure 11.1, form the aerodynamic foundation of passive ventilation strategies applied across a wide range of building types.

11.2.2 Cross-Ventilation: Wind-Driven Air Exchange

Cross-ventilation relies on the creation of pressure differentials between two façades exposed to different wind conditions. When wind impinges on the windward façade, it generates a zone of positive pressure. On the leeward façade, a region of negative pressure arises as air accelerates around the building's edges. Air flows from the high-pressure side to the low-pressure side, traversing interior spaces and providing rapid air exchange. This mechanism is particularly effective during periods when indoor–outdoor temperature differences are small because buoyancy plays only a limited role. In summer, especially in temperate and Mediterranean climates, cross-ventilation provides one of the most effective methods of passive cooling by removing heat accumulated in indoor air and in the upper layers of thermal mass.

11.2.3 Single-Sided Ventilation: Turbulence-Driven Flow

In contrast, single-sided ventilation depends on a more subtle interplay of aerodynamic effects. When both the inlet and outlet are located on the same façade, airflow is generated not by large pressure differences but by small-scale turbulence, vortex shedding, and local eddies induced by wind interacting with the building surface. Temperature differences within the room may contribute additional buoyancy-driven movement, although the resulting flow remains intermittent and pulsating. Because the driving forces are weaker, the effective depth of a single-sided ventilated room is limited, typically to two to two-and-a-half times the opening height. This mechanism is therefore common in narrow residential buildings, terraced houses, classrooms, and offices with limited façade exposure.

11.2.4 Stack Ventilation: Buoyancy-Driven Airflow

Stack ventilation, the third fundamental mechanism, operates primarily on temperature differences. Warm indoor air becomes less dense and rises, creating an upward flow toward high-level openings. At the same time, cooler and denser outdoor air is drawn in through lower openings. As the indoor–outdoor temperature difference increases, the buoyancy force strengthens, producing a stable vertical airflow. Stack ventilation is largely independent of wind and is particularly effective during winter and mid-season periods when substantial temperature gradients exist. Its performance improves with greater vertical separation between inlets and outlets: a height difference of 2.5–3 m is generally sufficient to maintain continuous airflow, while taller shafts or stairwells can amplify the effect.

11.2.5 Climatic Variations in Ventilation Logic

Although these three mechanisms are universal, their relative importance changes across climatic regions. In cold climates, natural ventilation is carefully controlled to avoid excessive heat loss, and stack ventilation plays a dominant role because large temperature differences exist for prolonged periods of the year. In temperate climates, airflow patterns shift seasonally, with cross-ventilation providing summer cooling and stack ventilation supporting winter air renewal. In Mediterranean climates, cross-ventilation becomes the principal strategy, often combined with night-flush cooling of thermal mass. To understand how these mechanisms function in practice, it is necessary to examine how cultural, geometric, and material factors interact with climatic drivers in different regions.

11.2.5.1 Cold and Subarctic Climates

In cold and subarctic climates, natural ventilation is defined by the need to balance air quality with stringent thermal protection. During winter, temperature differences of 20°C–40°C between indoors and outdoors are common. Under these conditions, stack ventilation is extremely powerful: the vertical pressure gradient generated by buoyancy can easily exceed that of wind forces, especially in sheltered or snow-covered environments. Traditional architecture in northern Europe and the circumpolar regions reflects this logic. Semi-subterranean dwellings in Scandinavia and Iceland used earth as both insulation and thermal mass, while smoke vents at high level acted as outlets for rising warm air. In Inuit winter dwellings, compact forms with low openings limited infiltration losses, and the minimal ventilation that occurred was governed almost entirely by buoyancy. Contemporary Scandinavian buildings continue to use stack-driven ventilation through stairwells, service shafts, and mechanical-assist chimneys, though with controlled openings that minimize heat loss during cold periods.

11.2.5.2 Temperate Climates

In temperate climates such as Central Europe, North America's northern states, and much of East Asia, natural ventilation follows a seasonal rhythm. During winter, stack ventilation dominates due to pronounced temperature differences, while cross-ventilation is used sparingly to avoid rapid heat loss. Buildings constructed in the nineteenth and early twentieth centuries often incorporated vertical shafts in kitchens and bathrooms to maintain continuous background ventilation, supplemented by brief airing of living rooms. As temperatures rise in spring and autumn, mixed ventilation patterns emerge: buoyancy provides a baseline upward flow, while moderate winds enhance cross-flow through operable windows. Brick and stone buildings with high thermal mass benefit from night cooling, in which cool outdoor air is drawn through the structure after sunset, dissipating heat accumulated during the day.

During summer in temperate climates, cross-ventilation becomes essential. Moderate breezes of 2–4 $m \cdot s^{-1}$ can produce substantial indoor air movement when facilitated by correctly placed openings on opposing or adjacent façades. The geometry of such openings is critical. Rooms deeper than 6–8 m often experience stagnant zones unless equipped with additional interior openings or transoms. Buildings designed with two or more orientations perform best, as airflow can follow different wind directions throughout the day. In cities such as Wrocław or Vienna, nineteenth-century tenement buildings with windows on two adjacent façades exemplify this approach. Their combination of large operable windows, high ceilings, and thick walls creates favorable conditions for both cross-ventilation and night-flush cooling.

11.2.5.3 Mediterranean Climates

In Mediterranean climates, cross-ventilation forms the backbone of passive cooling strategies. Summers are characterized by high solar radiation, modest indoor–outdoor temperature differences by day and significant cooling at night, conditions under which wind-driven flow becomes the most effective mechanism for temperature regulation. Buildings in regions such as Andalusia, southern France, coastal Italy, and the Aegean islands typically feature operable windows on at least two orientations: balconies or shutters that modulate solar exposure and internal layouts that minimize airflow obstructions. Courtyards play a particularly important role, acting as cool reservoirs of air that enhance cross-flow through surrounding rooms. During the hottest hours, shutters block direct sunlight while still permitting airflow. At night, the temperature drop accelerates air exchange, cooling both indoor air and thermal mass.

Mediterranean cross-ventilation also benefits from local wind systems. Sea breezes, valley winds, and nightly downslope flows can be channeled through streets and courtyards, improving indoor airflow even in dense urban fabrics. The interplay of shading, high thermal mass, and ventilation enables buildings to maintain comfortable temperatures without mechanical cooling. Thick stone or adobe walls store coolness during the night and release it slowly during the day, while cross-ventilation removes the heat accumulated in upper boundary layers. This combined shading-ventilation-mass effect forms one of the most robust passive cooling logics found in traditional architecture.

11.2.5.4 Hot-Arid Climates

In hot-arid climates, natural ventilation relies on a more complex interaction between wind pressure and buoyancy. Large diurnal temperature swings, sometimes exceeding 20°C, create strong night-time buoyancy forces that support ventilation. During the day, however, intense solar radiation and high outdoor temperatures limit the usefulness of cross-flow, as outdoor air may be hotter than indoor air. Traditional architecture in regions such as Iran, Iraq, Egypt, and parts of North Africa addresses this through the use of windcatchers, or badgir systems, which harness both wind-pressure gradients and buoyancy forces. Tall towers capture prevailing winds at height, where temperatures are lower and airflow is stronger, directing them downward into interior rooms or underground cooling channels. The vertical shafts simultaneously act as stack chimneys, exhausting warm indoor air through buoyancy. These combined systems extend the principles of basic stack and cross-ventilation into a more advanced climatic adaptation.

11.2.5.5 Tropical and Monsoon Climates

Tropical and monsoon climates present yet another variation. High humidity and minimal indoor–outdoor temperature differences suppress buoyancy forces, rendering stack ventilation largely ineffective. Instead, buildings rely predominantly on cross-ventilation and high-level exhaust openings that promote moisture removal and enhance thermal comfort through air movement rather than cooling. Traditional houses in Southeast Asia, South India, and coastal West Africa feature porous envelopes, elevated floors, and extended eaves that facilitate airflow even under heavy rainfall. Large openings on opposite façades allow breezes to pass freely through interior spaces, while roof vents or clerestories remove warm air from upper levels. Because thermal mass can become a liability in humid climates, lightweight construction complements ventilation strategies by enabling rapid heat dissipation.

11.2.6 Synthesis: Ventilation as Part of an Integrated Passive System

Across all climates, the effectiveness of natural ventilation depends on an interplay of environmental conditions, building geometry, and the thermal properties of materials. Ventilation alone rarely suffices for comfort; instead, it forms part of a larger passive system that includes shading, insulation, thermal mass, and airflow pathways. In cold climates, stack ventilation provides necessary air renewal without excessive heat loss, while in temperate climates a seasonal alternation between cross- and stack ventilation optimizes performance. In Mediterranean climates, cross-ventilation combined with night cooling offers a powerful means of passive thermal regulation. These principles form the aerodynamic foundation for the advanced strategies discussed later in this chapter, including atrium-driven ventilation, solar chimneys, courtyard ventilation, and wind towers

11.3 Atrium-Based Ventilation in Mediterranean and North African Architecture

11.3.1 Introduction

Courtyards and patios constitute one of the most characteristic elements of architecture in hot regions, and their functioning is based on physical mechanisms that systematically regulate microclimatic conditions within the building. The literature highlights that courtyards influence both the air temperature in semi-open spaces and the thermal parameters of adjacent rooms. Their operation relies on a combination of shading, natural convection, stack effect, energy buffering by massive external partitions, and evaporative cooling.

In Mediterranean and North African architecture, patios are central spaces open to the sky that support effective natural ventilation and thermal regulation. The microclimate within patios is shaped by local convective flows, temperature differences between shaded and sunlit zones, and the presence of vegetation and water surfaces. These elements work together to provide conditions that support thermal comfort while insulating interior spaces from overheated outdoor air and street noise. Although these systems originate from Islamic and Iberian traditions, similar principles were employed in many other regions. In traditional Chinese siheyuan dwellings, central courtyards enable cross-ventilation and moderate the microclimate through vegetation shading and the presence of water features. On the Indian subcontinent, spaces such as chowks and verandas function as thermal buffers in which trees and fountains reduce heat gains. In the Sahel and Sub-Saharan Africa, central courtyards in traditional earthen compounds support air exchange, retention of coolness, and community life adapted to climatic conditions. These examples confirm the universality of courtyard design principles regardless of cultural context.

These relationships are reflected in empirical and numerical analyses. Sánchez de la Flor et al. (2021) demonstrated that courtyards in Mediterranean climates reduce air temperature within the open space by 10°C–12°C during extreme days and lower cooling demand in adjacent rooms by more than 7 percent. This effect is linked to the geometric proportions of the courtyard, the height of surrounding walls, the ratio of openings to floor area, and the orientation of façades relative to cardinal directions.

Chidiadi and Taki (2025) emphasize the role of the courtyard as an aerothermal system in which local vortices, shear flows, and air stratification determine the intensity of natural ventilation. The authors note the significant influence of vegetation and water elements, which, through evaporation, reduce air temperature and increase relative humidity. This effect is particularly important in hot dry and hot humid climates.

Research by Torres-González et al. (2025) confirmed that courtyards also act as microclimatic buffers that support the preservation of heritage materials. Stabilization of temperature and humidity reduces daily amplitudes, delays the occurrence of extremes, and decreases thermal stresses responsible for the degradation of plaster, timber, and ceramics. It was shown that even during heatwaves the environmental parameters within the patio remain more favorable than in open urban space.

The importance of convective and evaporative processes is evident in numerical analyses. Sun et al. (2024) demonstrated that evaporative cooling in courtyards is particularly effective in single-sided ventilation systems, in which airflow depends primarily on courtyard geometry and pressure losses within the atrium. In cross-ventilated systems the effect is weaker because dominant inflow streams originate from outside the courtyard and limit the impact of evaporative cooling.

In another study, Sun et al. (2025) showed that the shape of the courtyard covering is a key factor determining cooling performance. A comparison of ten roof types revealed that streamlined forms, such as domed or single-pitched roofs, generate favorable low-pressure zones and higher airflow velocities, which translate into intensified passive cooling within interior spaces. Flat roofs and parapet configurations, by contrast, result in air stagnation and higher operative temperatures.

Collected research findings indicate that the courtyard should be interpreted as an integrated passive cooling system whose effectiveness depends on the interaction of spatial geometry, thermal mass, local convective flows, evaporative processes, and the aerodynamic properties of the roof. These principles are reflected across a wide spectrum of architectural traditions in which the courtyard serves as a fundamental environmental regulation component. The following section of the chapter presents case studies from Morocco and Spain that illustrate how different geometric configurations, proportions, vegetation, and function influence the performance of courtyard-based ventilation systems.

11.3.2 Courtyard-Based Environmental Strategies in Marrakesh and Seville

Figure 11.2 shows a traditional riad in the Marrakesh medina, characterized by a fully open central patio with no external windows. All rooms face the courtyard, allowing for one-sided ventilation and efficient air circulation within a highly compact urban fabric. The narrow depth of rooms, approximately 2.5–3.5 m, ensures full penetration of ventilating air from the patio.

FIGURE 11.2 Open-air patio of a traditional riad in Marrakesh, providing one-sided ventilation to all surrounding rooms.

The high, whitewashed courtyard walls function as a vertical thermal shaft: during the day, warm air rises and escapes through the open top, while at night, the cooled walls draw fresh air downward into the interior. The ground floor contains semi-open living areas (Figure 11.2), which facilitate convection-driven airflow and enhance the integration of indoor and outdoor spaces.

Figure 11.3 depicts a modern hotel atrium in Seville featuring partial glazing at the top, which provides rain protection while maintaining vertical ventilation. The overall geometry supports natural upward airflow through gaps and openings, assisted by well-ventilated side corridors.

Figure 11.4 presents a commercial courtyard in Marrakesh, repurposed for retail activities. Here, retractable fabric canopies provide sun protection while preserving upward airflow. Despite the change in use, the courtyard retains its environmental performance, thanks to its height, enclosure, and open sky exposure. These architectural qualities ensure effective removal of heat build-up even during high occupancy and intensive daytime use.

FIGURE 11.3 Partially glazed atrium in Hotel Palacio Alcázar, Seville, balancing rain protection with vertical airflow.

FIGURE 11.4 Commercial courtyard in Marrakesh equipped with retractable fabric canopies for solar control.

This example underscores the adaptability of traditional patio configurations across changing functions, as long as their spatial and climatic logic remains intact.

Figures 11.5 and 11.6 depict formal palace courtyards in the Real Alcázar of Seville, designed for both aesthetic and climatic performance. The use of long, narrow reflecting pools (Figure 11.6) enhances evaporative surfaces, while symmetrically spaced trees offer solar protection for the paved surfaces and seating areas.

Large colonnades create thermal buffers, reducing direct exposure to solar radiation while supporting lateral air exchange between courtyard and surrounding rooms. Although monumental in character, these courtyards represent highly functional passive cooling systems based on geometric control, material mass, water features, and plant-assisted microclimate regulation.

These case studies illustrate how courtyard-based ventilation strategies are deeply embedded in the architectural heritage of the Mediterranean and North Africa. Despite differences in scale, materials, and spatial articulation, they all employ a

FIGURE 11.5 Garden courtyard in the Real Alcázar of Seville, featuring dense vegetation and shaded arcades.

FIGURE 11.6 Long reflecting pool in the Real Alcázar, Seville, flanked by arcades and shaded vegetation.

shared climatic logic that integrates shading, evaporative cooling, and thermal buffering to maintain indoor comfort.

Although courtyard-based systems are most widespread in the Mediterranean Basin and in warm regions of Asia, similar environmental principles can also be found in temperate climates. An example is the atrium of the Tabakalera Cultural Center in San Sebastián (Figure 11.7), located in the oceanic climate of northern Spain. The former tobacco factory, now converted into a cultural complex, incorporates a large central courtyard covered with a glazed roof that functions as a naturally ventilated atrium. Warm air accumulates beneath the glass canopy and rises through the stack effect, while cooler air from the lower levels is drawn upward, ensuring continuous air renewal. This passive air circulation moderates temperature fluctuations and reduces the need for mechanical ventilation. Vegetation placed

FIGURE 11.7 Central atrium of the Tabakalera Cultural Center, San Sebastián, Spain. The glazed roof structure supports passive ventilation through the stack effect, while vegetation and reflective surfaces enhance microclimatic stability within the restored industrial courtyard.

along balconies contributes to humidity control and enhances the thermal buffering capacity of the space.

In this adaptation, the traditional Mediterranean courtyard logic, originally grounded in air movement, shading, and the interplay of light and mass, is reinterpreted within the context of a temperate climate.

The atrium of Tabakalera demonstrates how the physical principles of passive ventilation and thermal moderation can be successfully applied to rehabilitated industrial buildings in northern Europe, maintaining comfort while minimizing energy use.

11.4 WIND TOWERS PAST AND PRESENT: TECHNOLOGICAL DEVELOPMENT AND CONTEMPORARY APPLICATIONS

(based on Almazmumi et al., 2025; Jomehzadeh et al., 2017, 2020; Li et al., 2025)

11.4.1 Introduction: The Badgir as an Archetype of Natural Ventilation

The idea of natural ventilation and passive cooling is one of the oldest and most advanced examples of interaction between architecture and the environment. In Middle Eastern cultures, particularly in Iran, Egypt, Bahrain, and the United Arab Emirates, badgir systems, or wind towers, formed a fundamental component of climatic architecture. Their purpose was to capture wind at roof level and direct it into building interiors to improve thermal comfort and indoor air quality. As shown in Figure 11.8, in cities such as Yazd, Meybod, and Kerman, the dense network of

FIGURE 11.8 Traditional windcatchers in Iran and the Gulf region: (a) one-sided windcatchers in Meybod, Iran; (b) two-sided windcatchers of the water cistern in Dowlat-Abad Garden, Yazd, Iran; (c) four-sided windcatchers of the water cistern on Kish Island, Iran; (d) six-sided windcatchers in Yazd, Iran; (e) eight-sided windcatcher in Dowlat-Abad Garden, Yazd, Iran; (f) cylindrical windcatcher in Dubai. (From Jomehzadeh et al., 2017.)

badgirs formed a distinctive architectural landscape in which built form was directly shaped by climatic conditions.

Badgirs functioned as self-regulating systems. Warm air rose and was exhausted through the tower, while cooler masses entered the interior, producing a natural convective loop. They were often connected to underground qanat channels, which supplied cool water and further cooled incoming air. The result was a complex passive system operating through synergy between thermal mass, water, and airflow. In Iranian architecture, the badgir symbolized a balance between technique and nature, serving functional, aesthetic, and symbolic purposes.

11.4.2 Contemporary Applications: Roof Windcatchers in Public Architecture

In the second half of the twentieth century, as modern building technologies advanced, traditional natural ventilation systems began to decline and were replaced by mechanical ventilation. However, rising energy costs and the need to reduce CO_2 emissions renewed interest in passive solutions. Since the 1990s, the concept of the badgir has re-emerged in the form of modern windcatcher systems installed in public buildings, schools, and shopping centers in the United Kingdom, Canada, and Western Europe. As shown in Figure 11.9, contemporary towers preserve the operational principle of traditional badgirs but include louvres, attenuators, and diffusers that allow precise airflow control.

Studies by Jomehzadeh et al. (2017) demonstrated that contemporary windcatchers can provide up to 57 air changes per hour with minimal energy use. They achieve this by exploiting pressure differences between the windward and leeward sides of the building. In some projects, hybrid ventilation systems have been implemented in

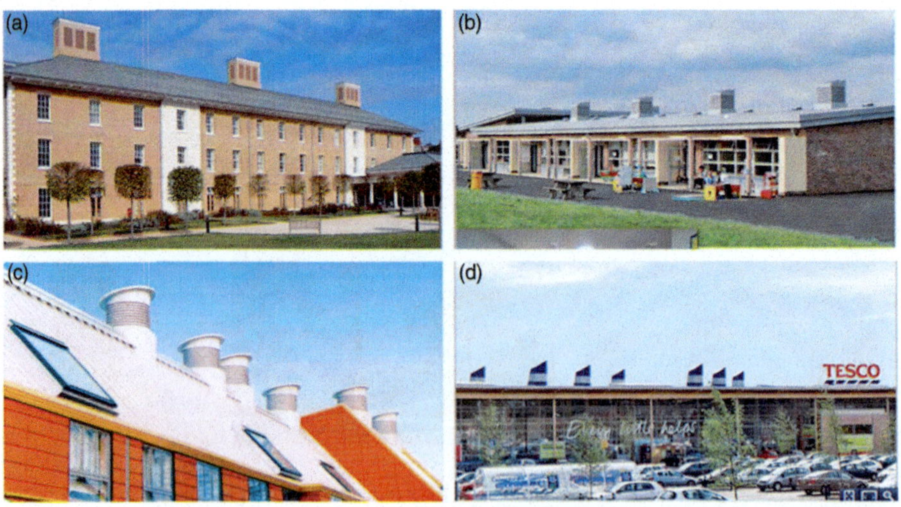

FIGURE 11.9 Examples of Monodraught windcatcher installations in the United Kingdom: (a) Royal Chelsea Hospital, London; (b) Seaside School, Lancing; (c) Addey and Stanhope School, Deptford; (d) Tesco Eco-store, Cheetham Hill. (From Jomehzadeh et al., 2017.)

which the windcatcher operates alongside mechanically assisted exhaust, providing optimal airflow even under calm conditions. These systems combine tradition and modernity and align with the idea of passive and adaptive architecture.

11.4.3 Operating Principle and Integration with Passive Systems

The operating principle of a modern windcatcher is based on a simple yet highly effective aerodynamic effect. The tower uses the pressure difference between the windward and leeward sides of the building to direct air indoors and then exhaust it outward. This generates natural circulation without the need for electricity. In practice, the system integrates two mechanisms: the wind-driven effect, produced by airflow over façades and roofs, and the stack effect, arising from temperature differences between indoor and outdoor air. During the day, warm air rises and is removed through upper openings, while at night cooler outdoor air enters and reduces indoor temperature.

As shown in Figure 11.10, the geometry of the windcatcher head is a key factor in operational performance. A venturi-shaped configuration produces local acceleration of airflow and stronger pressure gradients, increasing air supply even at low wind speeds. This arrangement limits turbulence and improves directional flow.

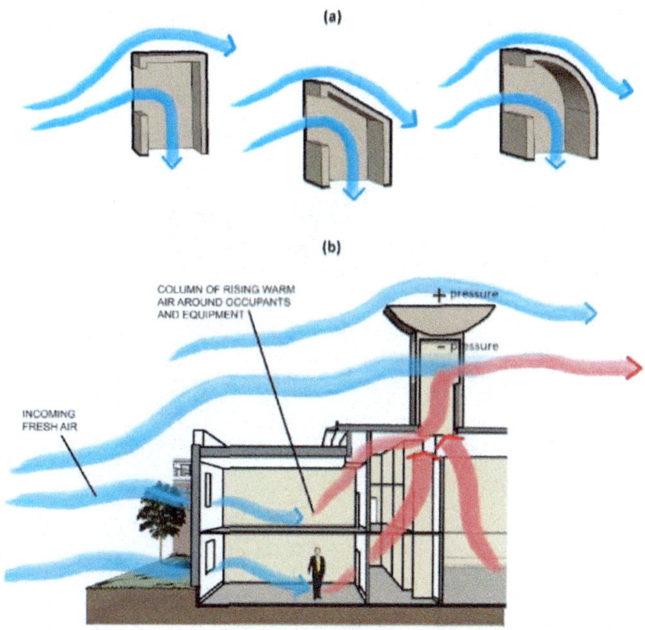

FIGURE 11.10 Schematic illustration of roof-mounted windcatchers showing the influence of head geometry on airflow performance: (a) comparison of flat, inclined, and curved windcatcher head configurations; (b) building-scale schematic of a windcatcher integrated with a venturi-shaped roof, illustrating pressure differences and buoyancy-driven airflow. (From Jomehzadeh et al., 2020.)

System efficiency depends on many factors, including channel proportions, tower height, and the angle of inlet openings. Jomehzadeh et al. (2020) found that an optimal height-to-width ratio ranges from 3:1 to 5:1, which ensures stable airflow and minimizes pressure losses. Best performance is achieved with polygonal cross-sections such as hexagonal or octagonal forms, which capture wind from multiple directions.

Modern windcatchers are multi-layered systems. Supply and exhaust channels are separated, and airflow is controlled by directional louvres and volume dampers. Sensors measuring CO_2 concentration and temperature allow the system to automatically adjust supply airflow to occupancy and environmental conditions. This arrangement is illustrated in Figure 11.11.

Integration of automation with aerodynamic layers enables the development of a smart windcatcher system in which control is continuous and dynamic. Sensor data are used to regulate louvre positions and damper openings in real time, maintaining a balance between supply and exhaust. This mechanism ensures thermal comfort with minimal energy consumption. The control logic is shown in Figure 11.12.

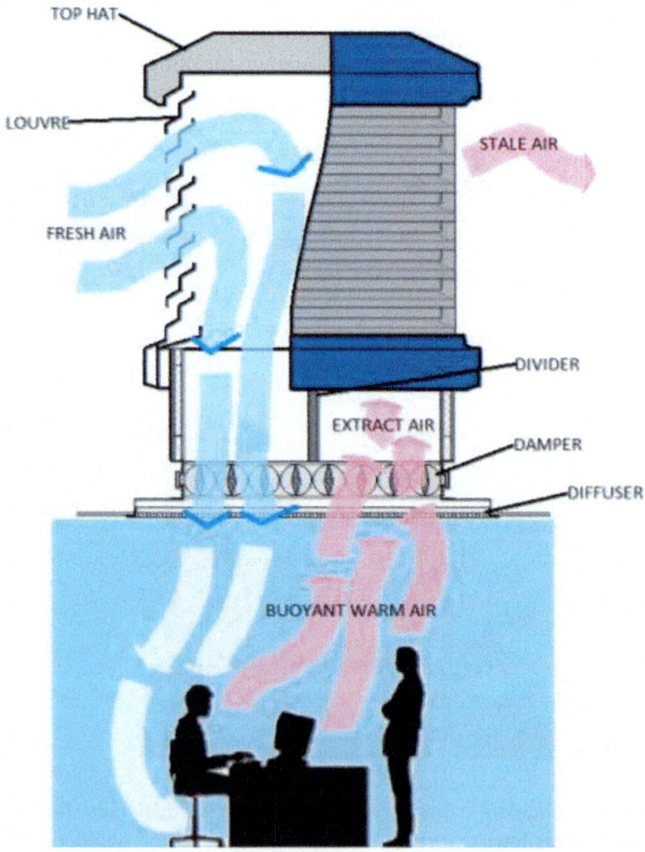

FIGURE 11.11 Modern windcatcher with louvres, dampers, and sensors. (From Jomehzadeh et al., 2020.)

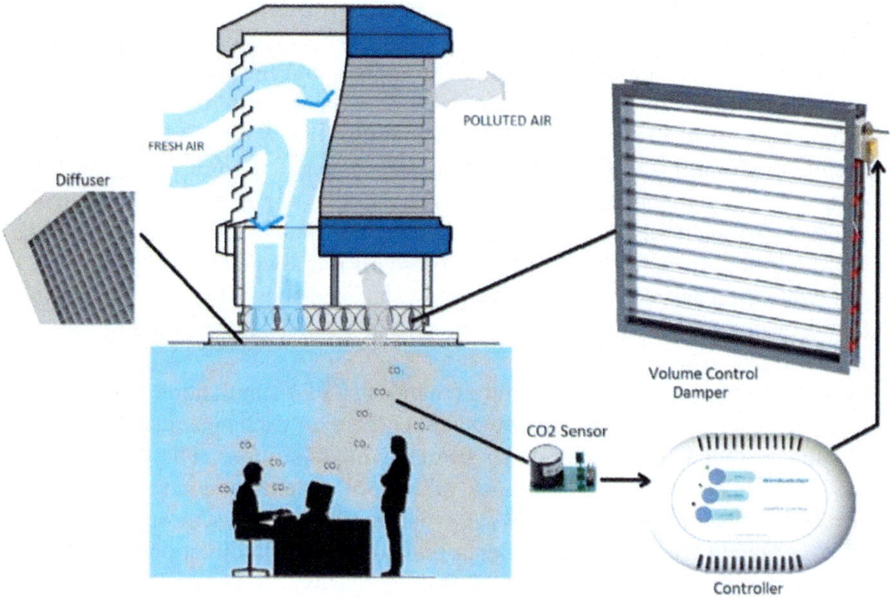

FIGURE 11.12 Smart windcatcher system integrated with automated volume control. (From Jomehzadeh et al., 2020.)

Automatic flow control allows the system to maintain a high level of comfort while reducing energy use. The ventilation module can switch to a night mode that relies on passive cooling through outdoor temperature reduction or to a daytime mode in which airflow is restricted to prevent overheating. These solutions form a bridge between passive operational logic and low-power active technologies, combining traditional aerodynamic principles with intelligent control.

Experimental studies have confirmed that integration of windcatchers with renewable systems significantly increases annual performance. Solar chimneys maintain ventilation during calm wind periods, while evaporative cooling reduces supply air temperature by 5°C–7°C in dry climates. Thermal buffers and photovoltaic panels allow sensors and actuators to operate independently from the electrical grid.

Next-generation windcatchers are increasingly integrated with thermal energy storage and phase change materials (PCMs). As a result, supply air can be pre-cooled or pre-heated depending on weather conditions, enabling daily operation without external energy input. These solutions become the foundation of micro-autonomous architecture in which each building is capable of self-regulating its environmental and energy conditions.

11.4.4 Innovations of the Twenty-First Century: The Wall Windcatcher (WWC)

Rapid development of digital technologies and the need to reduce energy consumption in dense urban environments have led to the emergence of a new wind-driven system: the Wall Windcatcher (WWC). Its concept shifts the function of the traditional roof-mounted tower to the building façade. This enables effective ventilation of multiple storeys simultaneously, eliminating the main limitation of classical roof windcatchers, which operate most effectively only on upper floors.

Studies by Almazmumi et al. (2025) show that the WWC meets the needs of contemporary multi-storey buildings where roof space is limited and airflow is obstructed by neighboring structures (Figure 11.13). Air is drawn through inlets located at various levels of the façade and then distributed through a vertical shaft divided into supply and exhaust sections. Fresh air is delivered to each floor individually, while used air is directed upward and exhausted above the roof. This configuration provides full independence of storeys and ensures stable airflow throughout the building.

According to Computational Fluid Dynamics (CFD) simulations and field measurements, the WWC provides 35–40 percent higher air exchange rates compared to conventional roof windcatchers under similar wind conditions. The greatest differences occur on middle and lower floors, where roof systems generate stagnant zones and reduced flow velocities. The WWC ventilates these areas through distributed inlets that benefit from façade turbulence and variable wind directions.

A major advantage of the WWC is its modularity. As shown in Figure 11.14, the vertical shaft can be installed entirely on the exterior surface, which simplifies retrofitting in existing buildings. External mounting eliminates the need for internal ductwork and preserves usable floor area. This approach is especially valuable for office and residential buildings without pre-existing ventilation shafts.

Experiments by Almazmumi et al. (2025) showed that the division of the shaft into independent flow zones maintains stable operation even at low wind speeds

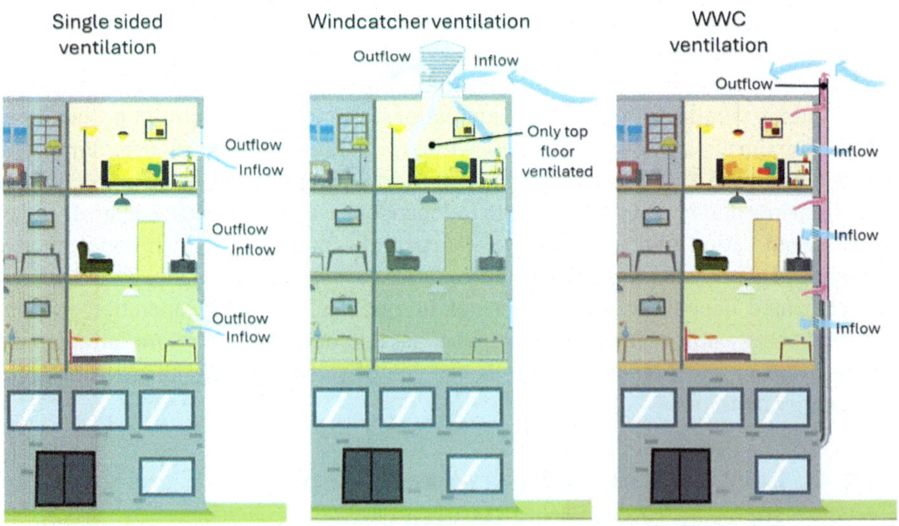

FIGURE 11.13 Comparison of single-sided, roof-mounted, and wall windcatcher ventilation. (From Almazmumi et al., 2025.)

of 1.5–2 m·s^{-1}. Flow reversal, typical for lower floors in multi-level configurations, was eliminated through optimized duct cross-sections and volume regulators. Simulations also demonstrated that a façade equipped with a WWC improves the building's aerodynamic balance. The channels function as stabilizing elements and reduce pressure gradients between windward and leeward sides.

The WWC can also be integrated with air quality sensors, wind sensors, and building management systems, providing adaptive operation. During periods of increased heat load or higher CO_2 concentration, the control system opens dampers and increases airflow; during cooler periods or at night, it restricts flow to maintain comfort without draughts. This capability makes the WWC an intelligent system responsive to microclimatic conditions.

Constructively, the new tower type offers an additional advantage. It can incorporate photovoltaic panels and solar radiation sensors that power the control mechanisms and allow environmental monitoring. As a result, the WWC can operate independently of the electrical grid and forms part of micro-autonomous building strategies.

Aesthetically, the system introduces a new architectural language in which ventilation becomes a visible façade element. Vertical windcatcher modules can be arranged rhythmically along glazed surfaces or integrated with shading systems, giving the building a technological character while remaining rooted in climatic responsiveness.

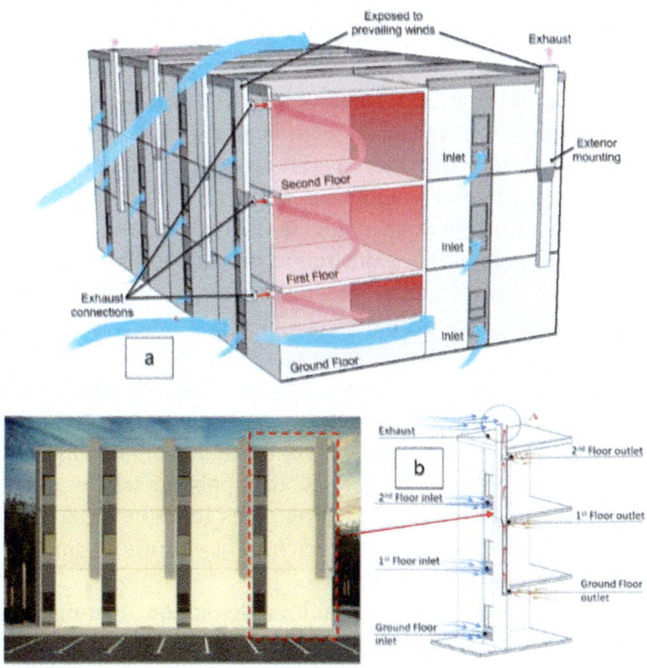

FIGURE 11.14 (a) Integration of a wind-driven wall cavity (WWC) system in a multi-storey building; (b) schematic illustration of its operating principle. (From Almazmumi et al., 2025.)

Findings by Almazmumi et al. (2025) confirm that the WWC significantly increases natural ventilation performance in multi-storey buildings and reduces reliance on mechanical systems. Its ease of installation, adaptability, and compatibility with intelligent building management systems make it a new stage in the evolution of the badgir concept, from desert tower to modern façade-based environmental module.

11.4.5 Windcatchers in Hot and Cold Climates: New Directions of Integration

Although the windcatcher originated in desert climates, its contemporary applications are no longer confined to hot and dry regions. A recent review by Li et al. (2025) shows that advancements in passive and low-power technologies have enabled wind-driven ventilation systems to be used across nearly all climate zones, from tropical to temperate and even cold regions. In a study of 147 scientific analyses, the authors demonstrated that modern windcatchers, when integrated with heat recovery, earth-to-air heat exchangers, evaporative cooling, and PCMs, can function as year-round environmental regulators, providing cooling, heating, and dehumidification.

In hot and dry climates, the most effective configuration remains the combination of a windcatcher with evaporative cooling and a solar chimney. Incoming air is cooled by water evaporation, while the solar chimney strengthens the stack effect and maintains airflow even during weak winds. Field studies in Yazd, Kerman, and Abu Dhabi recorded reductions of indoor temperature by 10°C–15°C during the day, along with increases in relative humidity of 20–30 percent. This approach is particularly effective in low-humidity regions where adiabatic cooling can be used without condensation risk.

In temperate climates, systems combining windcatchers with earth-to-air heat exchangers have become increasingly significant. These systems use the stable ground temperature at several meters depth to pre-heat supply air in winter and cool it in summer. In practice, supply air temperature remains within the range of 16°C–22°C for most of the year, reducing energy demand in heating and cooling systems by 25–30 percent. Demonstration projects in the United Kingdom and Germany showed that a windcatcher integrated with an earth heat exchanger maintained thermal comfort with minimal mechanical assistance, while a building management system regulated airflow by season and time of day.

In cold and transitional climates, integration with heat recovery systems is crucial. Li et al. (2025) describe applications of passive heat pipes and rotary thermal wheels that transfer energy from exhaust to supply air. Studies in the United Kingdom and Scandinavia reported increases of 2°C–4°C in supply-air temperature, sufficient to maintain comfort without additional heating. This solution is particularly valuable in buildings with high ventilation losses such as schools, offices, and libraries, where natural ventilation is required year-round.

For very low winter temperatures, traditional solutions can result in significant heat losses. To prevent this, the authors propose pairing windcatchers with local renewable sources such as solar thermal collectors and small-scale biomass systems. Waste heat from combustion or heating systems can be fed into ventilation channels to warm incoming air. Pilot projects in Finland and Canada demonstrated that a windcatcher combined with a heat recovery unit and a solar thermal collector maintained indoor temperatures of 20°C at outdoor temperatures of approximately –5°C without mechanical heating.

Li et al. (2025) also emphasize the potential of integrating windcatchers with PCMs, which act as thermal energy storage. During the day, PCM absorbs excess heat through melting and at night releases it during solidification, stabilizing indoor temperature fluctuations. Laboratory studies showed that PCM integrated into windcatcher channels maintained supply-air temperature for approximately 7 hours without external energy. In continental and arid climates, this strategy supports both daytime cooling and mild night-time heating.

Another development direction is the incorporation of desiccant systems, particularly important in humid and tropical climates. Desiccant wheels containing hygroscopic materials such as silica gel reduce humidity in supply air, improving air quality and reducing mold risk. Pilot projects in Malaysia and Singapore reported that integrating a windcatcher with a passive desiccant wheel maintained indoor relative humidity below 60 percent while reducing cooling energy use by approximately 35 percent.

Deployment of integrated systems across different climates indicates that modern windcatchers can act as year-round environmental comfort systems. In summer they provide cooling and ventilation, while in winter they contribute to heat recovery and energy storage, reducing the demand for full mechanical heating, ventilation, and air-conditioning (HVAC) operation. At the urban scale, this approach supports the development of low-energy neighborhoods where buildings share local sources and thermal storage, forming micro-environmental networks.

Li et al. (2025) conclude that the future of windcatcher technology lies in hybridization: combining passive strategies with intelligent control and renewable energy sources. Unlike historical systems, which relied solely on wind forces, contemporary windcatchers function as integrated micro-autonomous modules capable of operating in all weather conditions. Consequently, they should be regarded not merely as cooling devices but as comprehensive energy and air-management systems, synthesizing tradition and technology, architecture, and environment.

11.5 DOUBLE AND VENTILATED FAÇADES AS ELEMENTS OF NATURAL VENTILATION IN CONTEMPORARY ARCHITECTURE

11.5.1 Introduction and Functional Logic of Double-Skin Façades

(based on Tahmasbi et al., 2025)

Contemporary energy-efficient architecture increasingly relies on systems that function as active mediators between the external environment and the building interior. The façade is understood not as a barrier but as a dynamic interface that enables the exchange of air, light, and energy. DSF systems serve not only as energy-saving solutions but also as integral components of natural ventilation strategies and microclimate adaptation in contemporary buildings. In this context, double and ventilated façades constitute one of the most technologically advanced passive strategies. As noted by Tahmasbi et al. (2025), they are part of the broader category of adaptive façades that respond to changing environmental conditions while combining shading, ventilation, and energy-generation functions.

Although these systems were initially developed in the context of energy efficiency, their significance extends considerably further, including the improvement of microclimatic quality and support for natural ventilation in buildings with extensive glazed areas. DSFs combine the characteristics of a thermal buffer, a ventilation system, and a layer of solar protection. With appropriate control of airflow and material transparency, they can also warm incoming ventilation air during winter, making them adaptable and suitable for year-round operation (Alkhatib et al., 2025).

Within the framework of passive strategies discussed in this chapter, double and ventilated façades represent a key link between classical natural ventilation systems and modern intelligent building envelopes. Their operation is based on convection and pressure differences that enable autonomous microclimate regulation without mechanical energy. In this sense, DSF can be interpreted as an architectural extension of natural airflow principles used in atria, solar chimneys, or vertical ventilation shafts. The double-layer structure, however, allows precise control of heat transfer and airflow, thereby integrating ventilation with the management of solar-radiation energy.

To situate DSF within the broader landscape of adaptive façade technologies, it is useful to consider the conceptual framework illustrated in Figure 11.15, adapted from a recent review of innovative façade systems. The diagram maps two parallel developments shaping contemporary façade practice: increasing performance demands (user comfort, cost, integration, energy and dependency) and accelerating

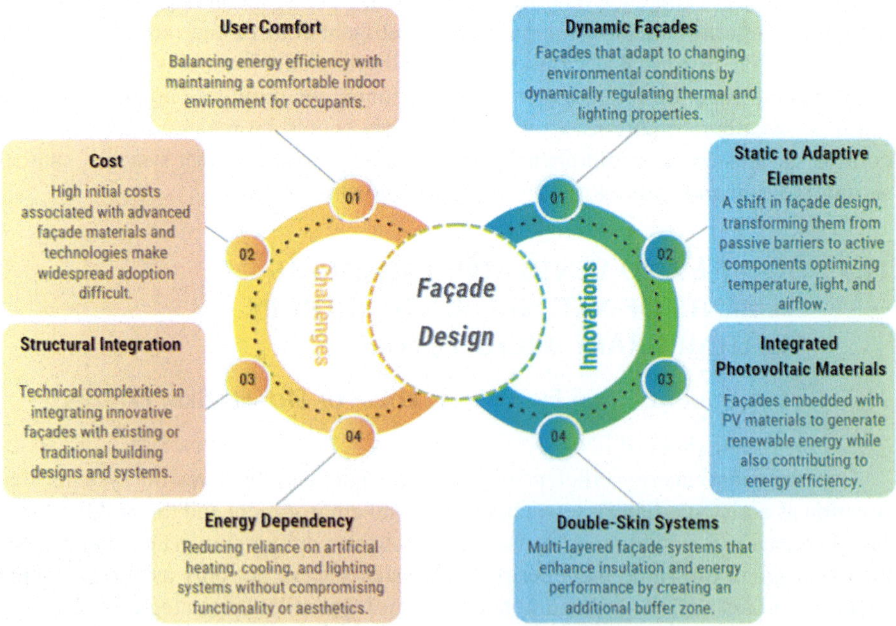

FIGURE 11.15 Key challenges and innovations in façade design. (From Tahmasbi et al., 2025.)

technological innovations (dynamic façades, adaptive elements, photovoltaic integration, and double-skin systems). Together, they reflect a shift from passive envelope design toward multi-layered façades that regulate microclimate, enhance ventilation, and contribute to building-wide energy strategies. This relationship between performance demands and emerging façade innovations is illustrated in Figure 11.15, which synthesizes the principal challenges and technological directions shaping contemporary envelope design.

11.5.2 Typologies and Climatic Adaptation Strategies

A DSF consists of two glazed layers separated by an air cavity, typically ranging from several to several dozen centimeters in width. Within this intermediate space, air moves either by natural convection or with fan assistance, providing a stabilizing buffer between the exterior climate and the occupied interior. In passive configurations, the cavity functions as a vertical thermal channel in which the stack effect is exploited: solar radiation heats the air in the cavity, causing it to rise and exit through upper openings while cooler outdoor air is drawn in from lower inlets. This buoyancy-driven circulation contributes to temperature moderation and supports natural ventilation without mechanical intervention (Alkhatib et al., 2025).

The typology of DSFs can be broadly understood through the framework introduced by Gratia and De Herde (2007), who identified four principal configurations that differ in their geometric arrangement and airflow logic. The first is the box-window façade, where each room is equipped with its own independent cavity module, allowing decentralized control of air circulation. The second configuration, the corridor façade, extends the air cavity horizontally or vertically along the entire building front, creating a shared ventilation layer serving multiple spaces. A third type, the multi-storey façade, spans several floors with a continuous cavity that acts as a single large ventilation shaft. The fourth configuration, the shaft-box façade, combines elements of these systems and incorporates local constrictions within the cavity to intensify the stack effect and increase air velocity.

Recent studies further extend this typological catalogue by introducing variants tailored to specific climatic conditions. In hot and humid regions, spray-cooled façades have been developed to enhance evaporative effects within the cavity and limit overheating under strong solar exposure. In temperate climates, phase-change materials embedded in the cavity or glazing assembly provide additional thermal buffering by absorbing and releasing latent heat, thereby reducing peak temperatures and improving stability across diurnal cycles (Li et al., 2024; Wang et al., 2022). These developments illustrate a broader trend toward hybridized systems in which DSFs are adapted to the thermal characteristics of their climatic context rather than applied as uniform technological solutions.

The appropriate selection of façade type depends on building height, façade orientation, shading context, and interior functional requirements (Poirazis, 2004). Tall buildings often benefit from multi-storey or shaft-box arrangements that promote continuous vertical airflow, while low-rise structures may favor box-window or corridor configurations that allow more granular control of ventilation. Contemporary design frequently employs mixed systems in which different sections of a single building envelope adopt distinct DSF typologies to create zones with differentiated microclimatic functions. Such combined strategies support year-round performance by aligning façade behavior with variable solar exposure, seasonal airflow patterns, and the thermal demands of individual interior spaces.

11.5.3 Integration with Photovoltaic and Renewable Technologies

Contemporary energy-efficient architecture increasingly relies on façade systems that mediate heat, light, and airflow in a controlled and adaptive manner. Within this transition, DSF systems have emerged as one of the most effective hybrid strategies, operating simultaneously as thermal buffers, ventilation channels, and solar-control layers. Their defining characteristic is the transformation of the façade from a static barrier into a dynamic environmental interface capable of modulating airflow and heat exchange without mechanical assistance.

As shown in Figure 11.16, energy-efficient façade systems can be divided into four main technological groups: smart glazing, photovoltaic façades, double-skin systems, and dynamic façades. Each group operates through a distinct but complementary mechanism to improve the building's energy balance by means of insulation, storage, or active control of solar radiation and airflow. Integrating photovoltaics with DSFs creates a dual-function system: the outer layer acts as an energy generator

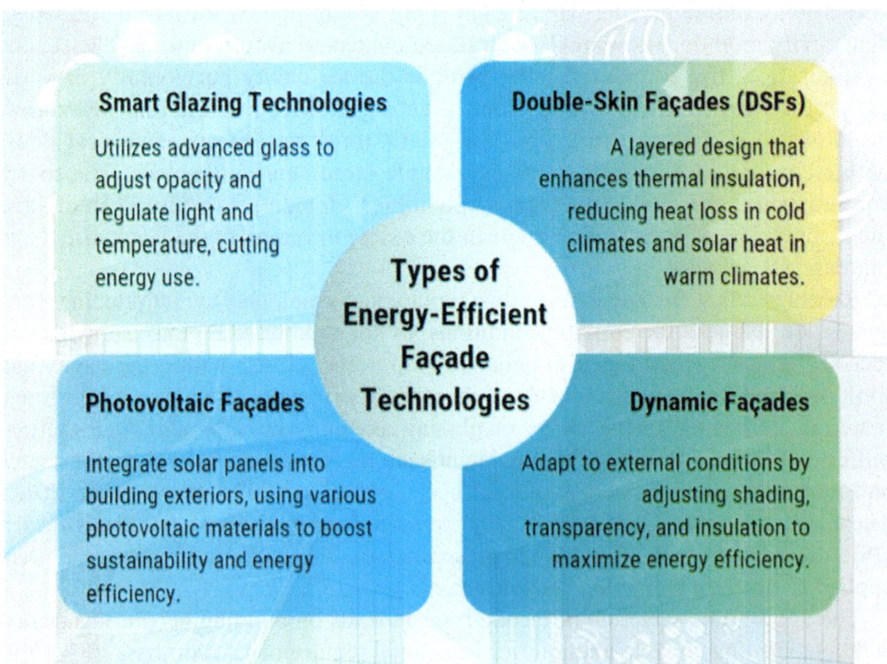

FIGURE 11.16 Types of energy-efficient façade technologies, including photovoltaic and double-skin systems. (From Tahmasbi et al., 2025.)

and solar-shading device, while the inner layer serves as a buffering element that stabilizes heat transfer between indoor and outdoor environments.

The integration of DSFs with building-integrated photovoltaics (BIPV) represents one of the most advanced developments in energy-efficient architectural design. According to Tahmasbi et al. (2025), the outer layer of the façade may serve a hybrid role, combining electricity generation with control of solar gains and daylight. Semi-transparent photovoltaic (STPV) modules allow simultaneous energy generation and selective solar-radiation filtering, thereby reducing cooling loads while maintaining indoor daylight.

The effectiveness of BIPV systems depends largely on the orientation and tilt of the modules relative to the sun, which is particularly important in tall buildings with limited rooftop exposure (Barone et al., 2022; Xiang & Matusiak, 2022). Studies indicate that façade-integrated photovoltaic systems not only reduce energy use but also improve the thermal characteristics of the building envelope and provide a more even distribution of solar gains. Appropriate mounting angles and materials with controlled light transmission ensure visual comfort and reduce overheating during summer.

DSFs may operate in different seasonal modes, utilizing both the stack effect and passive convective cooling. When combined with a photovoltaic layer, the DSF-BIPV system forms a multi-layer envelope with controllable transparency and variable thermal-transmittance properties, enabling optimization of energy exchange throughout the year. The seasonal adaptability of DSF-BIPV systems is presented in Figure 11.17, which illustrates how airflow patterns, glazing states, and thermal behavior shift across heating and cooling periods.

11.5.4 Smart and Adaptive Materials in Façade Design

Contemporary façades increasingly incorporate materials with variable optical and thermal properties, including electrochromic glazing, phase-change materials, hydrogels, and aerogels (Tahmasbi et al., 2025). Their use enables dynamic adjustment of light transmission, reflectivity, and heat storage depending on solar exposure.

Research cited by Tahmasbi et al. (2025) shows that electrochromic glazing can reduce energy consumption by up to 60 percent, thermochromic materials by 7–46 percent, and PCM by 2–3 percent annually, depending on climatic conditions and façade orientation. When combined with natural ventilation, these materials form hybrid systems that integrate passive and active processes. Aerogels provide high insulation while maintaining transparency and light transmission, making them particularly useful for façades with extensive glazing (Pizzatto et al., 2025). The application of hydrogels and intelligent coatings within the cavity further supports microclimatic self-regulation, especially in high-humidity regions.

11.5.5 Artificial Intelligence and Predictive Control

Modern DSFs increasingly incorporate artificial intelligence tools for predictive control of airflow, louvre position, and glazing transparency. Temperature, solar-radiation, and air-quality sensors transmit data to control systems that learn from

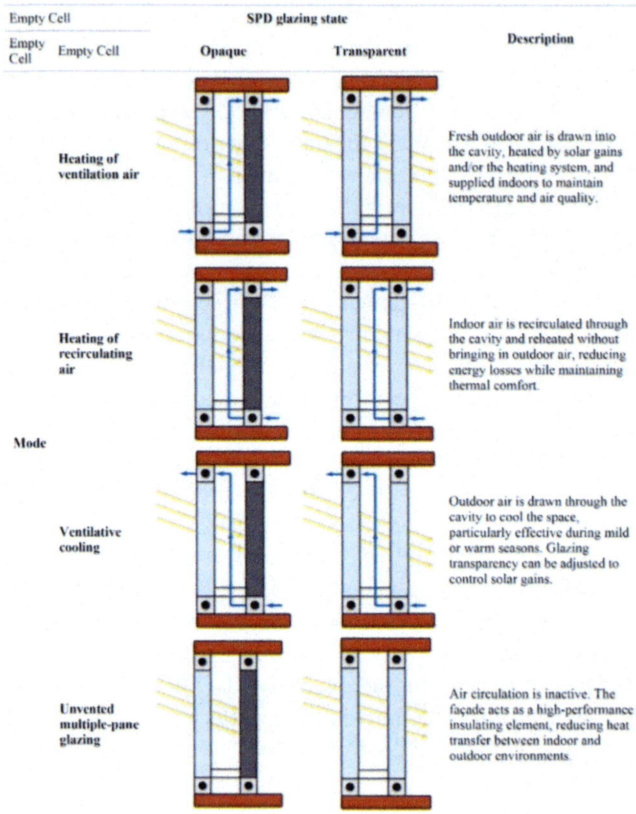

FIGURE 11.17 Seasonal operating modes of an adaptive double-skin façade with suspended particle device (SPD) glazing integration. (From Tahmasbi et al., 2025.)

weather and occupancy patterns. According to Tahmasbi et al. (2025), façades equipped with XGBoost and Random Forest algorithms can adjust operating parameters every 90 seconds, responding in real time to changes in solar exposure and temperature.

Machine-learning algorithms enable dynamic optimization of thermal conditions with minimal energy use. In this way, the façade becomes part of an intelligent building envelope that reacts in real time to external and internal stimuli.

11.5.6 Design Challenges and Future Perspectives

Experimental studies by Alkhatib et al. (2025) and reviews by Pizzatto et al. (2025) and Tahmasbi et al. (2025) confirm that DSFs are a key component of sustainable architectural design. Their ability to combine natural ventilation with renewable-energy technologies makes them a bridge between passive and active design strategies.

Despite their many advantages, DSF systems are characterized by high investment costs and require precise design of airflow cavities. Common issues include uneven vertical temperature distribution and the accumulation of pollutants within the cavity (Gratia & De Herde, 2007). High-rise office buildings require particular attention regarding airflow control and cleaning of glazed surfaces.

High installation and maintenance costs, especially in BIPV and advanced DSF systems, remain a major barrier to wider adoption. This indicates the need for financial support and standardized design procedures (Tahmasbi et al., 2025).

11.5.7 Conclusion: Toward Integrated Passive Ventilation Systems

Within the broader context of passive strategies, double and ventilated façades can be understood as a form of hybrid natural ventilation that combines physical convection processes with intelligent microclimate control. Their function extends beyond thermal buffering to include active regulation of airflow, light, and humidity, and in the most recent solutions, energy generation. In this sense, DSF systems form a bridge between traditional passive ventilation systems and integrated environmental management technologies. In contemporary architecture, they represent a spatial manifestation of natural ventilation concepts adapted to the demands of dense urban environments, sustainable development, and user comfort.

11.6 PERFORMANCE-BASED URBAN VENTILATION CORRIDORS

(based on Yin & He, 2025)

In the previous sections, ventilation systems at the building scale were analyzed, including natural, cross-, and atrium-based ventilation, all of which contribute to improved thermal comfort with limited energy use. However, maintaining stable and safe temperatures in the built environment does not end at the level of individual structures. An equally important component of thermal balance and urban livability is city-scale ventilation, which determines whether accumulated heat and pollutants can be effectively dispersed and whether fresh air can reach densely built-up areas.

In this context, Yin & He (2025) propose a performance-based urban ventilation corridor (PB-UVC) system in which design decisions are evaluated not through declarative criteria but through actual microclimatic outcomes. This represents a shift from static planning principles toward a dynamic, data-driven model based on wind direction and speed, urban morphology, surface roughness, and air-exchange indicators. The authors emphasize that contemporary cities need not only projects that appear sustainable in principle but also systems that can be quantitatively assessed in terms of measurable environmental improvements. From this perspective, UVCs constitute one of the most significant tools of climate-sensitive planning, integrating aerodynamic, morphological, and environmental analysis within a unified design framework (Figure 11.18).

Airflow begins in source zones, often river valleys, green belts, or open periurban areas, where masses of cool, clean air are generated. As this air moves toward urbanized zones, its flow is stabilized by surfaces with high aerodynamic permeability such as vegetation, water bodies, and open spaces. These constitute the inflow zone, where convection is supported by cool ground surfaces and soil moisture. Yin and He emphasize that this is where the thermal exchange cycle between natural and urban landscapes begins.

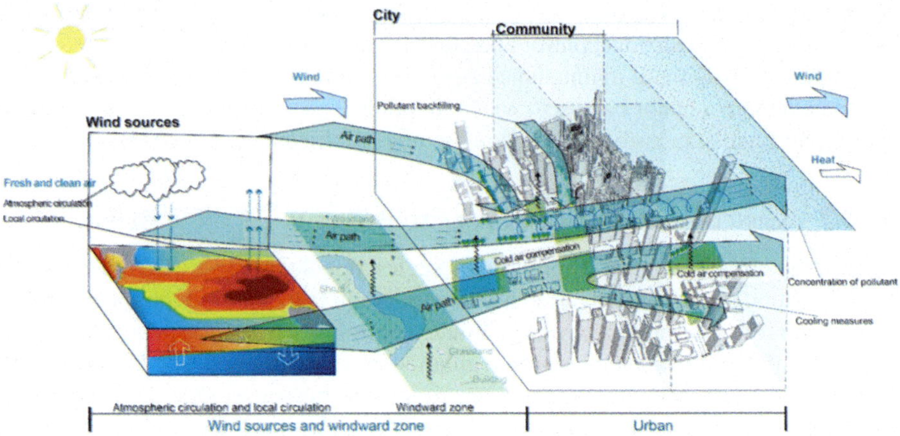

FIGURE 11.18 General mechanism of air circulation within an urban ventilation corridor system. (From Yin & He, 2025.)

In the urban center, airflow encounters a varied building structure. Differences in height and density lead to compression, turbulence, and recirculation. It is under these conditions that the importance of ventilation corridors becomes clear. They act as channels that guide cool air masses, counteracting local overheating and enabling pollutant dispersion. UVCs therefore function like ecological air-transport systems: cooler air advances from open landscapes while warmed air rises in the central area, forming a vertical exchange pattern that reduces the intensity of the urban heat island (Figure 11.19).

The diagram shows how climatic, spatial, and morphological data are integrated into a single decision-making model. Its core consists of four groups of parameters: meteorological conditions, building morphology, topography, and surface characteristics such as vegetation coverage and materials with high thermal capacity. Each factor influences aerodynamic resistance and ventilation performance.

The authors introduce the concept of urban resistance, which can be quantitatively assessed through CFD. Designing ventilation corridors becomes an iterative process in which spatial variants are evaluated and optimized until the desired microclimatic outcomes are achieved.

The PB-UVC system not only visualizes airflow but also enables its quantitative assessment. Indicators such as the Ventilation Potential Index, Surface Temperature Drop, and Air Exchange Rate allow planners to measure the effectiveness of urban interventions, positioning ventilation corridors as tools for managing the urban environment (Figure 11.20).

The model illustrates a hierarchical network of natural ventilation corridors in which primary pathways form the structural skeleton, while secondary corridors and local connections distribute air deeper into the built environment. Each spatial type has a distinct microclimatic function: green areas and river valleys act as sources of cool air, plazas and intersections serve as exchange zones, and low-rise streets function as airflow channels.

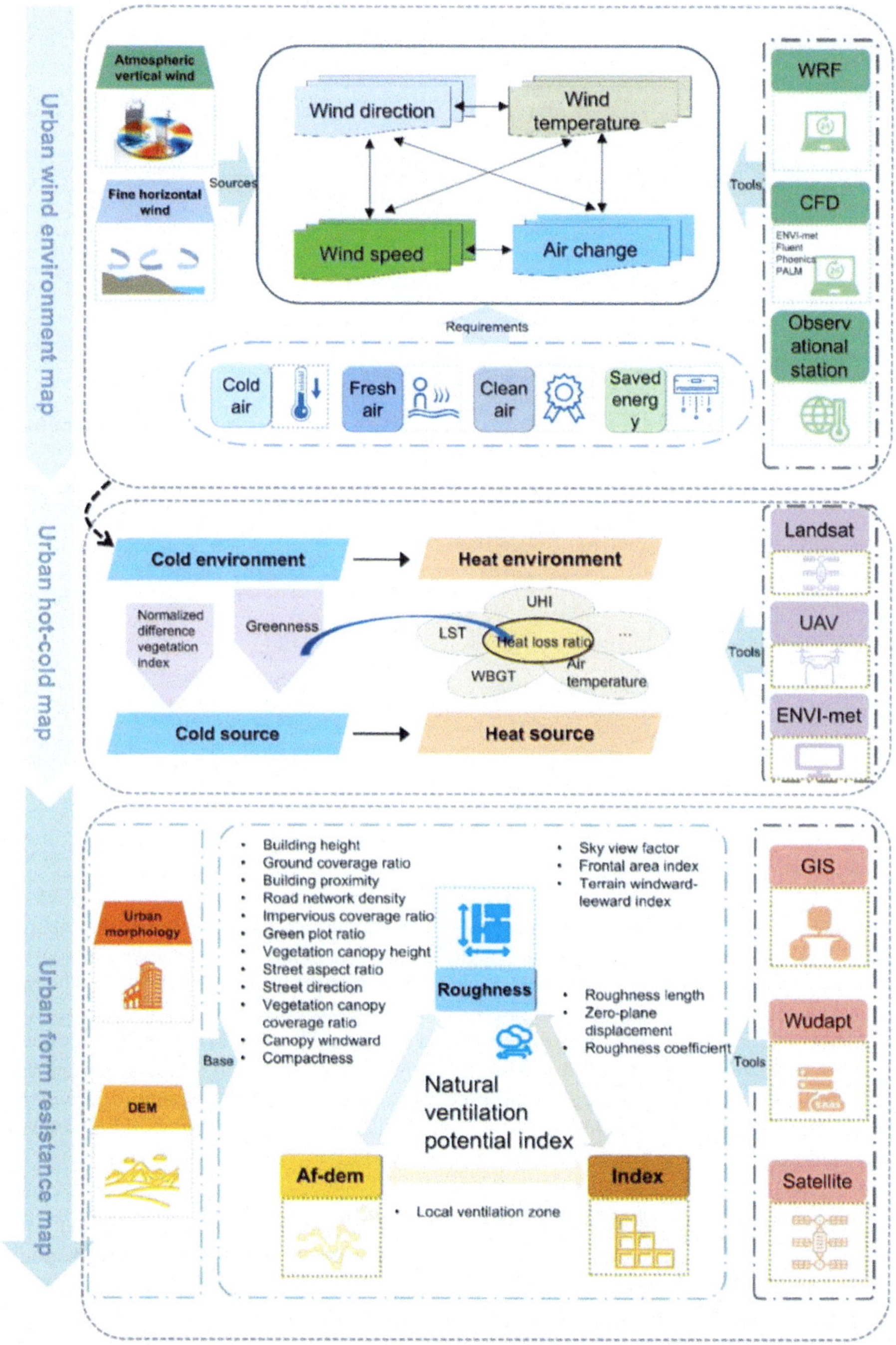

FIGURE 11.19 Analytical structure of the PB-UVC (performance-based urban ventilation corridor) system. (From Yin & He, 2025.)

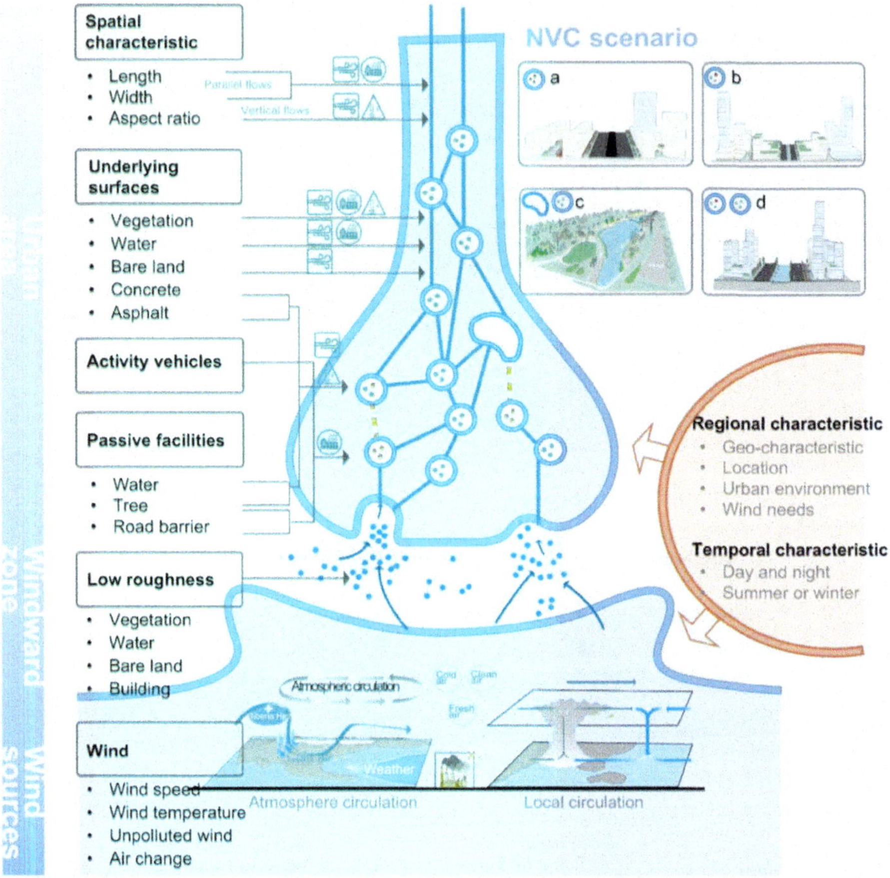

FIGURE 11.20 Scenario of a natural ventilation corridor (NVC) illustrating the theoretical PB-UVC system and typical urban applications. The diagram integrates wind sources, urban morphology, surface characteristics, and ventilation performance indicators. The inset scenarios show different spatial configurations of ventilation corridors: (a) a street-based ventilation corridor aligned with the prevailing wind direction, (b) a high-rise urban canyon with limited ventilation efficiency due to building height and density, (c) a blue-green corridor combining water bodies and vegetation to enhance cooling and airflow, and (d) a widened urban corridor with reduced roughness and improved air exchange through building setbacks and open spaces. (From Yin & He, 2025.)

Simulation results show that implementing the PB-UVC system can reduce average surface temperatures by 1.8°C–2.3°C and increase pedestrian-level wind speeds by 0.6 m·s^{-1}. The reduction in urban heat island intensity reached 15–20 percent, with the greatest improvements occurring where the ventilation network was linked to green infrastructure.

Analysis of the three diagrams reflects the full logic of the approach proposed by Yin and He, from regional-scale airflow mechanisms (Figure 11.17) to the integration of environmental data within the PB-UVC model (Figure 11.8) and its implementation in the urban structure (Figure 11.19). Understood in this manner, ventilation corridors constitute a performance-based design strategy that connects urban climate science with practical planning. They enable the creation of more breathable, heat-resilient, and health-supportive urban environments.

11.7 SUMMARY

This chapter examined natural ventilation across multiple spatial scales, from building-level airflow mechanisms to district-wide ventilation corridors. The examples highlighted how vernacular and contemporary strategies regulate air movement, moderate heat gains, and support thermal comfort under diverse climatic conditions.

At the building scale, three fundamental mechanisms—cross-ventilation, single-sided ventilation, and stack-driven ventilation—were presented as complementary systems whose effectiveness depends on wind pressure, temperature gradients, and building geometry. Cross-ventilation dominates in Mediterranean climates due to its ability to harness both wind and night-time cooling. Single-sided ventilation serves shallow-plan buildings, while buoyancy-driven ventilation remains essential in cold climates where controlled airflow with minimal heat loss is required.

Courtyard- and atrium-based systems formed the second category of ventilation strategies. Their performance arises from the interaction of shading, thermal mass, convective currents, and evaporative cooling. Case studies from Marrakesh, Seville, and San Sebastián showed how geometric proportions, vegetation, and water features generate stable microclimates and support natural ventilation in both historic and adapted modern contexts.

Windcatchers represented a third system, ranging from traditional badgirs to contemporary sensor-driven devices. Vernacular towers capture wind and promote buoyancy-driven airflow, often enhanced by evaporative cooling from qanats. Modern windcatchers incorporate adjustable louvers, automated dampers, and hybrid ventilation modes. Wall-mounted windcatchers demonstrated how multi-storey buildings can achieve uniform natural ventilation, overcoming airflow limitations typical of roof-only systems.

DSFs were discussed as adaptive interfaces capable of integrating natural ventilation with solar control and thermal buffering. Their performance depends on cavity geometry, seasonal operating modes, and the use of smart materials or photovoltaic glazing. These systems function as hybrid climatic devices that mediate air, light, and heat within a single envelope.

The chapter then expanded to the urban scale through the PB-UVC framework. This methodology integrates wind data, morphology, topography, and surface characteristics into an outcome-focused design process. Hierarchical ventilation corridors—primary, secondary, and local—were shown to reduce surface temperatures and increase pedestrian-level airflow, supporting adaptation to urban heat stress.

Together, the systems presented in this chapter show that natural ventilation remains a fundamental component of climate-responsive design. Whether implemented through openings, courtyards, windcatchers, DSFs, or urban corridor networks, these strategies reduce energy use, enhance comfort, and support resilience across a wide range of climatic and cultural contexts.

REFERENCES

Alkhatib, H., O'Neill, N., & Norton, B. (2025). Month-by-month prioritization of heating and ventilation from an adaptive mechanically-ventilated double-skin façade. *Energy Reports*, *14*, 2627–2641. https://doi.org/10.1016/j.egyr.2025.09.038

Almazmumi, S., Sun, H., Liu, M., Calautit, J., & Jimenez-Bescos, C. (2025). A novel wall windcatcher (WWC) natural ventilation system evaluated through CFD and experimental field testing: A solution to single-sided ventilation in multi-story buildings? *Building and Environment*, *283*, 113307. https://doi.org/10.1016/j.buildenv.2025.113307

Barone, G., Zacharopoulos, A., Buonomano, A., Forzano, C., Giuzio, G. F., Mondol, J., Palombo, A., Pugsley, A., & Smyth, M. (2022). Concentrating PhotoVoltaic glazing (CoPVG) system: Modelling and simulation of smart building façade. *Energy*, *238*(Part B), 121597. https://doi.org/10.1016/j.energy.2021.121597

Chidiadi, E., & Taki, A. (2025). Examining the impact of multilevel courtyards in hot-dry and humid climates. *Energies*, *18*(10), 2425. https://doi.org/10.3390/en18102425

Gratia, E., & De Herde, A. (2007). Greenhouse effect in double-skin façade. *Energy and Buildings*, *39*(2), 199–211. https://doi.org/10.1016/j.enbuild.2006.06.004

Jomehzadeh, F., Hussen, H. M., Calautit, J. K., Nejat, P., & Ferwati, M. S. (2020). Natural ventilation by windcatcher (Badgir): A review on the impacts of geometry, microclimate and macroclimate. *Energy and Buildings*, *226*, 110396. https://doi.org/10.1016/j.enbuild.2020.110396

Jomehzadeh, F., Nejat, P., Calautit, J. K., Yusof, M. B. M., Zaki, S. A., & Hughes, B. R., Yazid, M. N. A. W. M. (2017). A review on windcatcher for passive cooling and natural ventilation in buildings, Part 1: Indoor air quality and thermal comfort assessment. *Renewable and Sustainable Energy Reviews*, *70*, 736–756. https://doi.org/10.1016/j.rser.2016.11.254

Li, J., Calautit, J., Jimenez-Bescos, C., Song, W., Riffat, S., & Chen, Q. (2025). Climate-adaptive windcatcher natural ventilation integrated with passive and low-energy technologies: A review of current and future developments. *Building and Environment*, *284*. https://doi.org/10.1016/j.buildenv.2025.113436

Li, Y., Fu, S., Li, Y., Peng, Y., Yang, C., Tao, W., & Wang, H. (2024). Seasonal thermal performance of a macro-encapsulated phase change material blind integrated double skin façade system: An experimental study. *Energy and Buildings*, *325*, 114952. https://doi.org/10.1016/j.enbuild.2024.114952

Myroniuk, K., Zhelykh, V., Furdas, Y., Jesson, M., & Mitoulis, S.-A. (2026). Energy-saving of passive ventilation systems in thermally modernized residential buildings – A review. *Energy and Buildings*, *350*, 116681. https://doi.org/10.1016/j.enbuild.2025.116681

Pizzatto, S. M., dos, S., Pizzatto, F., Raupp-Pereira, F., Arcaro, S., Angioletto, E., & Montedo, O. R. K. (2025). Ventilated facade system: A review/Sistema de fachada ventilada: una revisión. *Boletín de la Sociedad Española de Cerámica y Vidrio*, *64*(3), 100443. https://doi.org/10.1016/j.bsecv.2025.100443

Poirazis, H. (2004). Double Skin Façades for Office Buildings 2004, Division of Energy and Building Design, Department of Construction and Architecture, Lund Institute of Technology Lund University, Report EBD-R–04/3.

Sánchez de la Flor, F. J., Ruiz-Pardo, Á., Diz-Mellado, E., Rivera-Gómez, C., & Galán-Marín, C. (2021). Assessing the impact of courtyards in cooling energy demand in buildings. *Journal of Cleaner Production*, *320*, 128742. https://doi.org/10.1016/j.jclepro.2021.128742

Sun, H., Bensalem, R., Dik, A., Tao, Z., Wang, Z., Jimenez-Bescos, C., & Calautit, J. K. (2025). The impact of courtyard roof shape on adjacent building natural ventilation and passive cooling. *Journal of Building Engineering*, *111*, 113331. https://doi.org/10.1016/j.jobe.2025.113331

Sun, H., Zhong, H., Dik, A., Ding, K., Jimenez-Bescos, C., & Calautit, J. K. (2024). Numerical investigation of evaporative cooling strategies on the aero-thermal performance of courtyard buildings in hot-dry climates. *Building and Environment, 258*, 111588. https://doi.org/10.1016/j.buildenv.2024.111588

Tahmasbi, F., Khdair, A. I., Aburumman, G. A., Tahmasebi, M., Thi, N. H., & Afrand, M. (2025). Energy-efficient building façades: A comprehensive review of innovative technologies and sustainable strategies. *Journal of Building Engineering, 99*, 111643. https://doi.org/10.1016/j.jobe.2024.111643

Torres-González, M., Rodríguez-Antuña, L., Bienvenido-Huertas, D., Alducin-Ochoa, J. M., León-Muñoz, M., & Rubio-Bellido, C. (2025). Courtyards as passive climate buffers: *Enhancing* thermal comfort and preventive conservation in Mediterranean climates. *Energy and Buildings, 336*, 115496. https://doi.org/10.1016/j.enbuild.2025.115496

Wang, Y., He, F., Lv, Z., Yang, W., & Wang, Q. (2022). Experimental investigations on thermal performance of spray cooling double-skin façade in hot humid climate region. *Energy and Buildings, 277*, 112605. https://doi.org/10.1016/j.enbuild.2022.112605

Xiang, C., & Matusiak, B. S. (2022). Façade integrated photovoltaics design for high-rise buildings with balconies, balancing daylight, aesthetic and energy productivity performance. *Journal of Building Engineering, 57*, 104950. https://doi.org/10.1016/j.jobe.2022.104950

Yin, M., & He, B.-J. (2025). Performance-based urban ventilation corridor for actual achievement of environmental benefits. *Building and Environment, 270*, 112536. https://doi.org/10.1016/j.buildenv.2025.112536

12 Integrated Passive Cooling in Vernacular Architecture

Synergies of Shading, Thermal Mass, and Natural Ventilation

12.1 INTRODUCTION

The preceding chapters examined the three principal passive strategies for heat mitigation: shading, the use of thermal mass, and natural ventilation. Each was considered in theoretical terms. The analysis demonstrated that, while each strategy can effectively reduce the thermal load on a building, their greatest potential emerges when they are implemented in combination, forming a coherent environmental system.

Vernacular architecture provides numerous precedents for such integration. Developed under diverse climatic and cultural conditions, traditional building forms frequently merge shading, thermal mass, and ventilation within arrangements in which individual components fulfil multiple roles: protecting against excessive solar radiation, moderating indoor temperatures, enabling air movement, and supporting moisture control. In many cases, these strategies are embedded not only in the design of individual buildings but also in the organization of entire settlements, influencing street orientation, the placement of open spaces, and the relationship to landscape features.

This chapter presents case studies that illustrate the integrated use of these three strategies in vernacular architecture. The examples, drawn from Mediterranean, arid, monsoonal, and humid subtropical climates, show how traditional settlements transformed environmental constraints into assets. They demonstrate how climate-responsive design was achieved through the synergy of form, materiality, and microclimatic control, offering enduring lessons for the development of low-energy, high-resilience architectural solutions today.

12.2 CASE STUDY: VERNACULAR PASSIVE COOLING STRATEGIES ACROSS THREE CLIMATIC ZONES IN IRAN

(based on Ganjimorad et al., 2024)

Iran's geographical diversity encompasses a wide range of climatic zones, from humid subtropical coasts to temperate-mountainous valleys and the hyper-arid

DOI: 10.1201/9781003645573-12

deserts of the central plateau. This diversity has generated distinct vernacular building traditions, each adapted to its own environmental demands while employing the universal passive strategies of shading, thermal mass, and natural ventilation. Understanding these climatic contexts is essential for appreciating the sophistication of the solutions developed over centuries, as illustrated in Figure 12.1, which maps the three locations examined: Rasht, Hajij, and Yazd, each representing a unique climatic zone and architectural response.

The first example, Rasht, demonstrates adaptations to a humid subtropical climate with high rainfall and persistent atmospheric moisture.

12.2.1 Rasht: Passive Cooling Strategies in a Humid Subtropical Climate

Rasht, located in the Gilan province of northern Iran near the Caspian Sea, experiences a humid subtropical climate characterized by year-round precipitation, high relative humidity often exceeding 80 percent, and mild winters with hot, humid summers. These conditions demand an architectural response that simultaneously limits solar heat gain, promotes continuous ventilation, and mitigates moisture accumulation.

One of the most recognizable features of Rasht's vernacular houses is the extended roof overhang, designed to protect walls, doors, and windows from both direct solar radiation and heavy rainfall. The deep eaves shade upper façades during

FIGURE 12.1 Map of Iran highlighting the cities selected as case studies. (From Ganjimorad et al., 2024.)

FIGURE 12.2 Traditional house in Rasht with wide eaves, preventing rainwater infiltration and limiting solar heat gain. (From Ganjimorad et al., 2024.)

peak summer hours, significantly reducing wall surface temperatures, while directing rainwater away from the building base to avoid dampness in structural elements. The use of light-colored clay plaster or lime on façades reflects incident solar radiation, further reducing heat absorption. An example of this can be seen in Figure 12.2, which shows a traditional house in Rasht with wide eaves protecting against both excessive heat and moisture infiltration.

Another defining adaptation is the Pileh Khaneh system, where buildings are elevated on wooden stilts or brick platforms. This creates an air cavity beneath the structure, enhancing underfloor ventilation and preventing ground moisture from entering interior spaces. This arrangement also facilitates convective cooling of the floor during hot days, as cooler air circulates underneath. Cross-ventilation is carefully orchestrated through opposing window and door openings, positioned to capture prevailing north-easterly breezes. The deliberate orientation and shallow interior layouts ensure air passes completely through each space. This airflow logic is illustrated in Figure 12.3, which diagrams the cross-ventilation effect in traditional Rasht houses, showing how elevated structures and opposing openings enhance passive cooling.

Material selection further supports thermal regulation and moisture management. Adobe bricks provide thermal mass to moderate diurnal temperature changes, while timber elements enable vapor permeability. Roofs made from rice-straw thatch offer lightweight insulation that resists heat gain yet allows moisture to escape. In some cases, double-layered walls with internal air cavities reduce thermal conductivity.

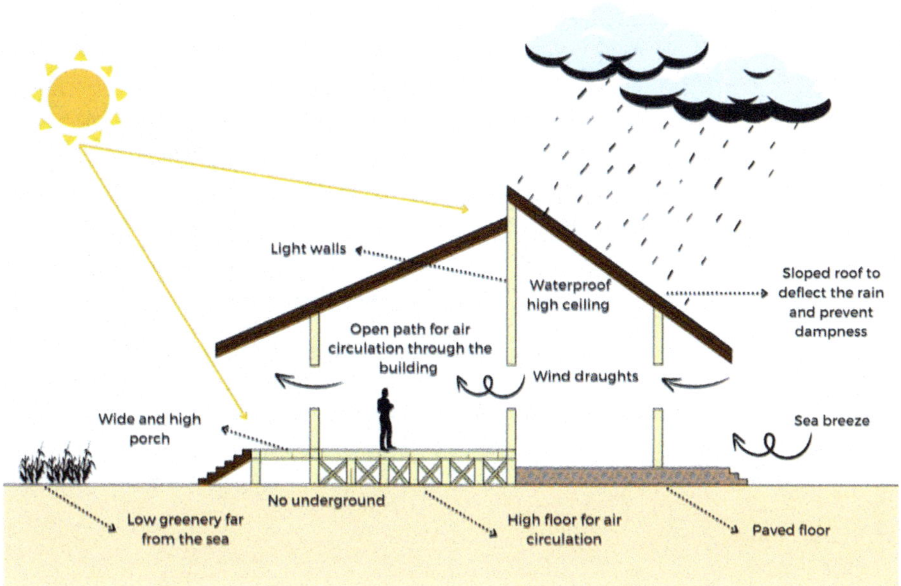

FIGURE 12.3 Diagram illustrating the cross-ventilation effect in traditional Rasht houses, where opposing windows and elevated structures improve passive cooling. (From Ganjimorad et al., 2024.)

Urban planning reinforces these strategies. Buildings are spaced to allow airflow between them, avoiding wind stagnation, while surrounding vegetation and gardens contribute to evaporative cooling and shade. Together, these measures create a comfortable indoor climate with reduced reliance on mechanical cooling despite persistent humidity and seasonal extremes.

12.2.2 Hajij: Passive Strategies in a Temperate-Mountainous Climate

Hajij, situated in the mountainous Kermanshah province of western Iran, lies within a temperate-mountainous climate characterized by cold winters, mild summers, high diurnal temperature variation, and significant seasonal shifts in solar angle. Precipitation is concentrated in the cooler months, while summers are comparatively dry. Architectural responses in this region focus on retaining warmth during winter, moderating heat gain in summer, and managing airflow in steep, terraced settlements.

The urban morphology follows the natural slope, producing a cascading, terraced settlement in which the roof of one dwelling often forms the courtyard or access terrace for the house above. This arrangement reduces heat loss in winter by limiting exposed façades and provides mutual shading in summer. The stepped settlement and its microclimatic benefits are visible in Figure 12.4, which depicts the interlocked stone-built forms on the mountain slope.

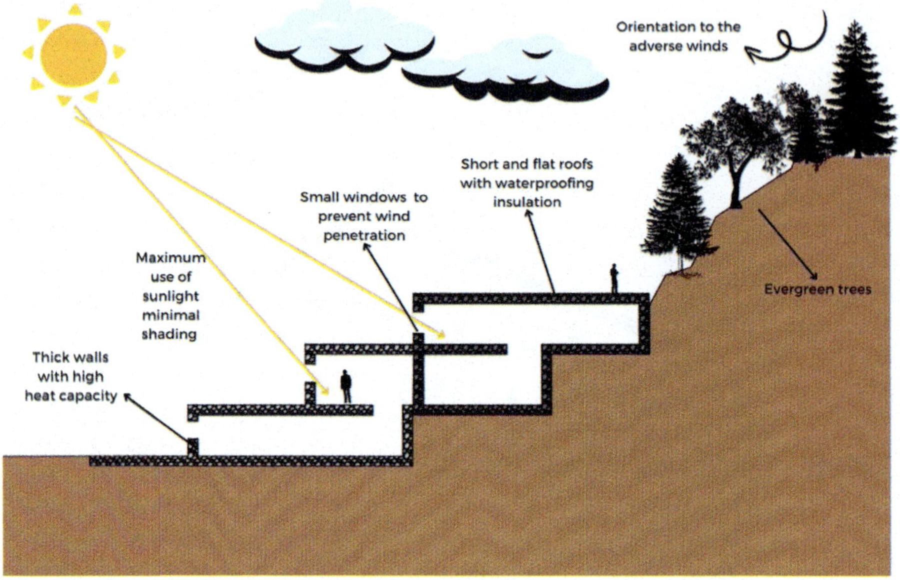

FIGURE 12.4 Vernacular terraced houses in Hajij, Kermanshah province. (From Ganjimorad et al., 2024.)

Construction materials are locally sourced stone and timber, chosen for both availability and performance. The thick stone walls, often exceeding 50 cm, provide substantial thermal mass, absorbing heat during the day and releasing it at night. This is critical in a climate with large daily temperature swings. A closer look at this building fabric is shown in Figure 12.5, which combines a wider view of Hajij's terraced setting with a detailed image of a single-stone building, highlighting the thick masonry, recessed windows, and compact form that together control heat transfer.

Building details further reflect climatic logic: small, recessed window openings limit winter heat loss while controlling solar gain in summer. Timber floors and frames aid vapor exchange, helping prevent moisture build-up from snowmelt and rain. Compact plan forms and clustered arrangements shelter from prevailing winter winds while maximizing winter solar access. Stepped terraces also act as microclimatic platforms, warming quickly in the morning sun and transferring heat to spaces below.

Seasonal adaptability is evident in daily life. In summer, upper terraces and rooftops are used for cooking, food preparation, and drying produce. In winter, activities move indoors to retain heat, exemplifying the integration of environmental response into both built form and social practice.

12.2.3 Yazd: Passive Cooling and Climate Adaptation in a Hot-Arid Environment

Yazd, on the central plateau, is one of the most celebrated examples of architectural adaptation to a hyper-arid desert climate. Summers are long and intensely hot, with

FIGURE 12.5 Shape and material of a traditional building of cold mountainous regions, Hajij village. (From Ganjimorad et al., 2024.)

daytime temperatures regularly exceeding 40°C, while winters can be cold. Annual precipitation is minimal, and relative humidity is typically below 30 percent. In such conditions, design must protect against solar radiation, mitigate diurnal temperature swings, and ensure comfort with minimal mechanical systems.

A defining element of Yazd's skyline is the badgir, or windcatcher—shown in Figure 12.6—which captures prevailing winds and channels them into the building.

FIGURE 12.6 Badgir (wind tower) system in Yazd. (From Ganjimorad et al., 2024.)

Many are multi-directional, ensuring ventilation under variable conditions. In traditional houses, windcatchers often work with subterranean spaces, cooling air via evaporation over water channels before it enters living spaces.

The second major element is the hozkhaneh, a courtyard integrated with water features and vegetation. High walls provide full shading for much of the day, reducing mean radiant temperature (MRT). Pools and fountains supplied by the qanat system enable continuous evaporative cooling. This arrangement is illustrated in Figure 12.7, which shows a shaded courtyard with a central water pool and surrounding vegetation that creates a cool microclimate and supports airflow.

Yazd's shading and façade strategies also support passive cooling. Narrow, winding streets create continuous shade and channel breezes, while limestone plaster reflects solar radiation. Small, deeply recessed windows with mashrabiya screens filter sunlight while permitting ventilation. Domed or vaulted roofs minimize direct solar exposure and allow hot air to rise and dissipate.

Thermal mass plays a central role: adobe brick walls and vaulted roofs absorb heat slowly and release it at night, smoothing diurnal temperature swings. External lime or gypsum plaster enhances reflectivity and durability.

At the urban scale, the compact, maze-like street network shades streets and accelerates airflow through venturi effects. Seasonal zoning within houses places summer living spaces near courtyards and water, while winter rooms are oriented for maximum solar exposure.

The synergy between windcatchers, courtyards, thermal mass, and shading ensures that indoor temperatures remain far lower than outdoor peaks in summer and retain warmth in winter. In Yazd's hyper-arid context, these features work as a multi-layered environmental mechanism, each component serving multiple functions.

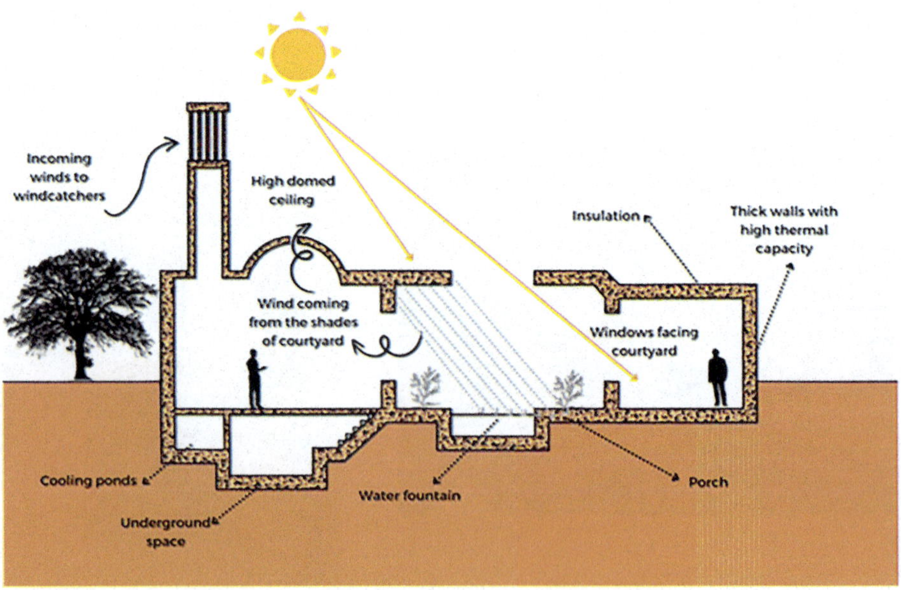

FIGURE 12.7 Traditional courtyard house in Yazd. (From Ganjimorad et al., 2024.)

12.2.4 Summary and Contemporary Relevance

The three Iranian case studies—Rasht in the humid subtropical north, Hajij in the mountainous west, and Yazd in the arid central plateau—demonstrate the adaptability of vernacular architecture when shading, thermal mass, and natural ventilation are applied in different climatic contexts. In Rasht, openness, airflow, and moisture control dominate; in Hajij, compact terracing and stone massing balance seasonal extremes; in Yazd, all three strategies operate together in a cohesive system.

Across all, the principles are consistent but locally calibrated, and as the mapped locations in Figure 12.1 remind us, each case reflects an integrated response to its distinct climate. For contemporary design, these systems offer lessons in low-energy, climate-resilient architecture that works with, rather than against, environmental forces.

12.3 SUNKEN COURTYARDS AS A VARIATION OF THE PASSIVE COURTYARD STRATEGY

(based on Wang et al. (2022))

12.3.1 Introduction

The courtyard typology demonstrates remarkable adaptability across climates. In Marrakesh and Córdoba, courtyards operate as inward-facing climatic regulators, combining shading, evaporative cooling, and thermal mass to stabilize conditions during the hottest periods of the year. In Seville, partial glazing and vertical greenery illustrate how the same principle can be adapted to mixed climates, balancing protection from rain with effective stack ventilation. These examples confirm that courtyards in the Mediterranean and North Africa act as self-regulating systems where air, water, vegetation, and masonry work together to maintain comfort.

The principle of mass and enclosure is taken further in the cave dwellings of southern Spain. Here, geological strata replace masonry walls as the main thermal buffer, creating stable environments through near-total exclusion of solar gain and reliance on the earth's inertia. Both the courtyard house and the cave dwelling show how vernacular forms integrate ventilation, shading, and mass to moderate extreme climates.

A related strategy appears in China's Loess Plateau, where sunken courtyards, or Dikengyuan, exploit the insulating properties of soil. By lowering the central space below ground, these dwellings reduce solar exposure and create powerful thermal buffering, while maintaining ventilation through carefully placed openings. The subterranean courtyard thus represents a further evolution of the same environmental logic that underlies Mediterranean patios and Iberian cave houses, demonstrating the versatility of earth and void as complementary tools of passive cooling.

12.3.2 Thermal Regulation and Energy Efficiency

Unlike traditional above-ground courtyards, sunken courtyards leverage the insulating properties of soil to regulate indoor temperatures more effectively. The surrounding earth acts as a thermal buffer, reducing heat gain in summer and heat loss in winter. Studies such as those by Wang et al. (2022) highlight that the interior of sunken courtyards remains 3°C–5°C cooler during hot periods while maintaining 2°C higher temperatures in winter compared to above-ground structures. These thermal advantages lead to a 23–35 percent reduction in energy consumption, significantly decreasing reliance on mechanical heating and cooling systems.

The thermal regulation mechanism in sunken courtyards is closely linked to their spatial configuration and interaction with the surrounding environment. As demonstrated in Figure 12.8a, the subterranean layout minimizes exposure to direct solar radiation, while simultaneously allowing walls to function as thermal storage masses that absorb heat during the day and release it at night. The depth of the courtyard influences the effectiveness of this thermal buffering, with deeper courtyards providing better insulation but requiring additional ventilation strategies to prevent excessive humidity accumulation.

The layout of a sunken courtyard also influences its ability to regulate temperature effectively. As illustrated in Figure 12.8a, a typical plan includes a central courtyard, surrounding living spaces, and auxiliary areas. The courtyard's placement and depth determine how much solar radiation it receives and how well it retains warmth during colder periods.

The perspective drawing in Figure 12.8c provides a three-dimensional visualization of the spatial organization of a sunken courtyard, showcasing the excavated earth surrounding the living areas and how it contributes to thermal insulation.

The real-life application of these principles can be observed in Figure 12.8d, which presents a photographic example of an existing sunken courtyard. The image illustrates the physical structure of the courtyard, including the earthen walls, shaded interior, and traditional materials. This photograph provides a direct reference to the practical effectiveness of passive cooling strategies applied in vernacular architecture.

12.3.2.1 Microclimate Regulation and Heat Exchange

The process of microclimate regulation in sunken courtyards is further detailed by Zhu et al. (2023), who emphasize the dynamics of heat exchange between the soil and the indoor space. As illustrated in Figure 12.9, during the day, solar radiation is absorbed by the loess walls and soil mass, preventing excessive heat gain in interior spaces. At night, the stored energy is gradually released, maintaining a relatively stable indoor temperature. The insulating properties of the soil limit longwave radiation losses, ensuring that heat is retained during colder periods.

The absence of direct solar exposure to interior courtyard surfaces contributes to a reduced MRT, which is critical for maintaining thermal comfort. The depth and orientation of the courtyard also influence its ability to regulate temperature and humidity, requiring careful design to prevent overheating in summer and excessive cooling in winter.

12.3.2.2 Ventilation and Airflow Dynamics

As depicted in Figure 12.9b, warm air rises from sunlit courtyard surfaces during the day, creating an upward flow that is replaced by cooler air from shaded areas or openings at the perimeter. This effect is reversed at night, as the stored heat is released from the walls and soil, sustaining mild air movement that prevents thermal stagnation.

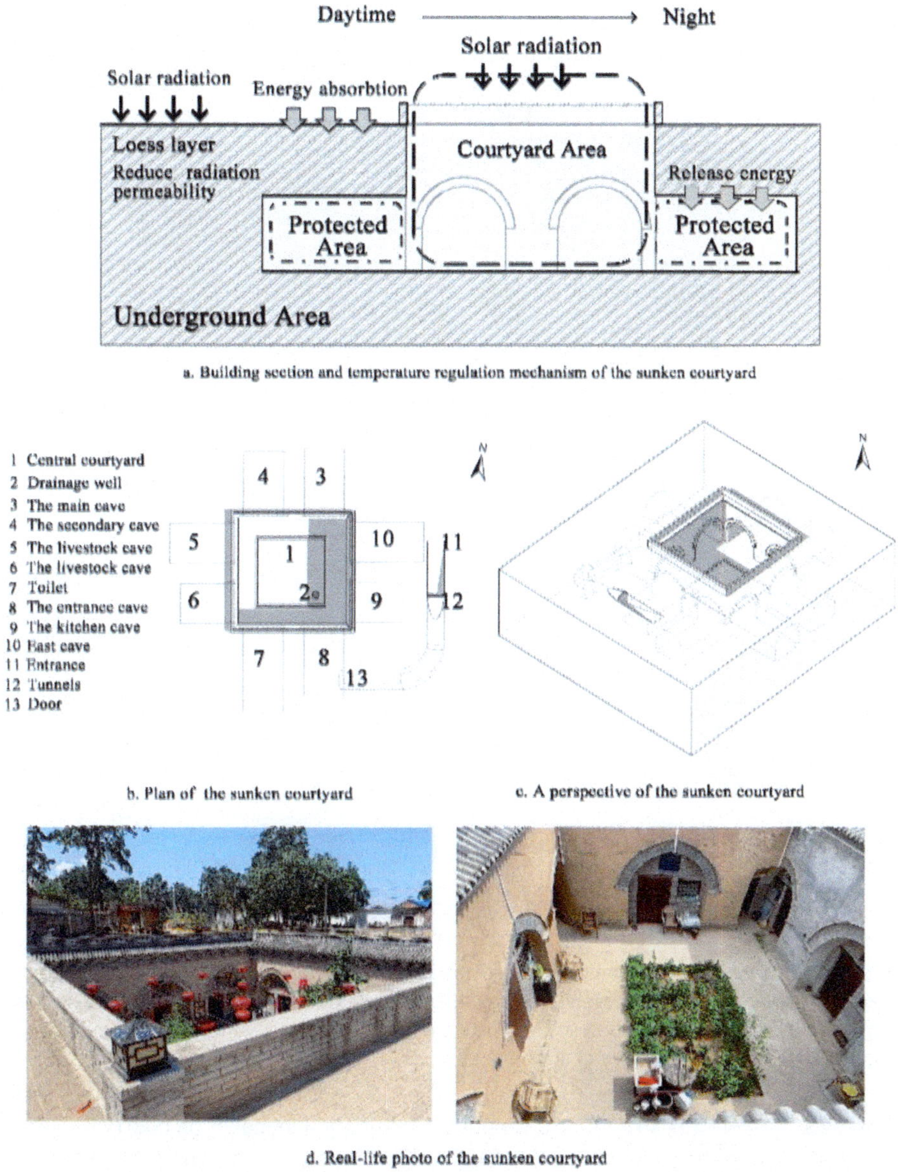

FIGURE 12.8 (a) Building section illustrating the temperature regulation mechanism of the sunken courtyard through reduced radiation permeability and thermal storage in loess; (b) plan of a typical sunken courtyard showing the spatial arrangement of caves and functional zones; (c) three-dimensional perspective view of the sunken courtyard system; (d) real-life photograph of an inhabited sunken courtyard demonstrating its spatial form and use. (From Wang et al., 2022)

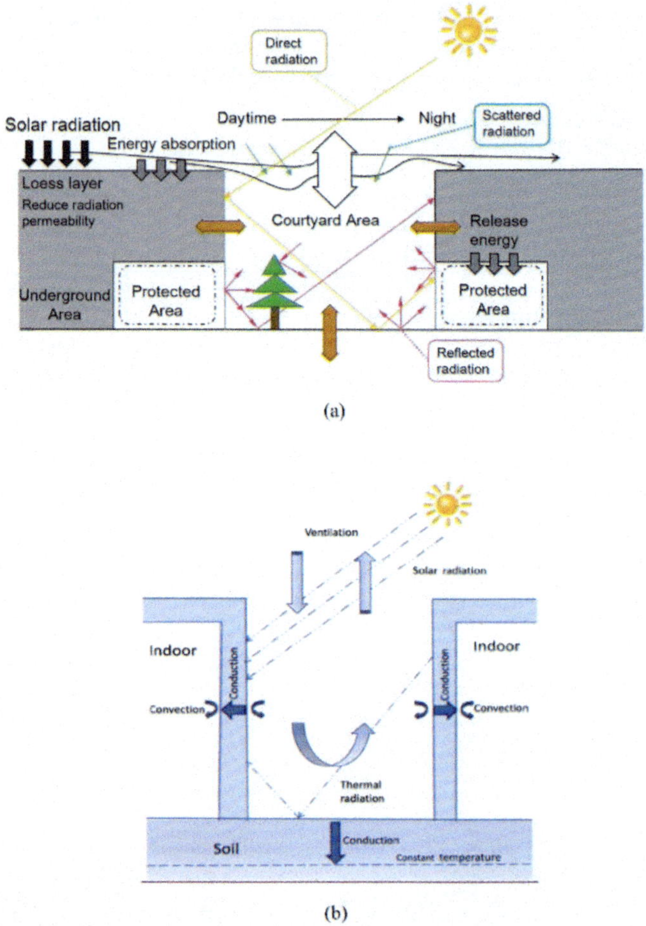

FIGURE 12.9 (a) Main components governing the microclimate of a sunken courtyard, including solar radiation, thermal storage, and energy release mechanisms; (b) schematic illustration of ventilation processes and heat exchange in a courtyard, highlighting conduction, convection, and radiative heat transfer.

However, the efficiency of natural ventilation is highly dependent on courtyard geometry, orientation, and the presence of openings. Deeper courtyards may experience restricted airflow, necessitating additional ventilation strategies such as high-level vents or thermal chimneys to enhance air circulation and mitigate potential humidity issues.

The efficiency of natural ventilation in sunken courtyards is also influenced by seasonal variations, wind direction, and external temperature fluctuations. During hot summer months, well-designed openings and the thermal mass of the surrounding soil can create pressure-driven ventilation, encouraging cooler air to enter at ground level while allowing warm air to exit through higher openings. In contrast, during colder months, the airflow pattern must be carefully regulated to prevent excessive heat loss, as uncontrolled air exchange could lead to discomfort in winter conditions. In addition to ventilation openings, incorporating evaporative cooling elements such as shallow water features or vegetation within the courtyard can further enhance cooling performance by increasing air humidity and promoting temperature differentials that drive airflow. Studies suggest that the integration of vegetation in sunken courtyards not only improves air quality but also contributes to heat absorption and delayed release, further stabilizing indoor temperatures.

When properly designed, sunken courtyards serve as efficient climate-responsive architectural solutions, leveraging a combination of thermal mass, shading, and airflow management to reduce energy demand and improve passive thermal regulation. By fine-tuning courtyard depth, optimizing the size and placement of ventilation pathways, and incorporating adaptive design strategies, these structures can serve as a valuable precedent for modern passive cooling solutions in hot and arid climates.

12.3.2.3 Comparative Perspective: Courtyard-Based Climate Strategies

Across the Mediterranean and North Africa, the courtyard has long served as the organizing element of vernacular dwellings, simultaneously providing light, air, and a thermally moderated core. In Marrakesh, riads use tall enclosing walls, shaded arcades, and vegetation to regulate heat and channel airflow vertically through the patio. In Córdoba and Seville, water features and planted courtyards introduce evaporative cooling, reinforcing the courtyard's role as a self-regulating microclimate.

The same environmental logic is visible in the cave dwellings of southern Spain, where geological mass rather than masonry walls provides the thermal buffer. Here, the reliance on earth as an envelope minimizes solar exposure and stabilizes indoor temperatures throughout the year. Although very different in form, both courtyard and cave house exploit mass, shading, and selective ventilation to temper extreme climates.

This continuity extends to the sunken courtyards of the Chinese Loess Plateau. By lowering the courtyard below ground level, these dwellings intensify the buffering effect of soil, reduce exposure to wind and solar radiation, and achieve remarkable thermal stability in climates with large diurnal swings. Despite variations in geography and construction, the principle remains the same: the courtyard, whether above ground, cut into rock, or excavated into loess soil, functions as a climatic regulator that integrates shading, ventilation, and mass into a coherent system of passive cooling.

12.4 CASE STUDY: THE FIRST GRANARY OF XUAN'EN COUNTY—PASSIVE BUFFERING AND VENTILATION IN A HUMID SUBTROPICAL CLIMATE

(based on Xu et al., 2024a)

The First Granary of Xuan'en County represents a distinct case compared with the other vernacular dwellings examined in this chapter. Unlike domestic architecture, granaries were designed for storage rather than habitation, yet their environmental performance is highly relevant because they operate within temperature and humidity ranges comparable to those required for human comfort. This case therefore broadens the climatic scope of vernacular strategies while demonstrating the effectiveness of passive regulation in hot-humid mountain conditions.

12.4.1 Climatic Context and Site Layout

Xuan'en County in Hubei province lies at an elevation of about 845 m and is characterized by a humid subtropical monsoon climate with abundant rainfall and marked seasonal variation. Summers are hot and humid, with mean highs around 31.8°C in August, while winters are cold and damp, with average lows near 5°C. In this environment, grain storage requires quasi-low temperature conditions that do not exceed 25°C throughout the year.

The granary complex is located within a valley and oriented to capture prevailing southeast winds while remaining sheltered by surrounding mountains. The long axis of the main buildings follows the contour lines, optimizing exposure to valley air circulation. The aerial photograph in Figure 12.10 shows the compact layout of the facility, positioned directly adjacent to Jiaoshi Provincial Road and surrounded by forested slopes that reinforce its climatic protection.

12.4.2 External Form and Seasonal Operation

The granary buildings are rectangular in plan, with double-hipped roofs and thick rendered masonry walls. Openings are small and fitted with wooden shutters, a

FIGURE 12.10 Location of the First Granary of Xuan'en County in Xuan'en, Hubei province. (From Xu et al., 2024a.)

FIGURE 12.11 External views of the First Granary of Xuan'en County. (a) Summer operation with shutters open for ventilation. (b) Winter operation with shutters closed for insulation. (From Xu et al., 2024a.)

design that limits direct solar gain while allowing controlled ventilation when required. Seasonal photographs in Figure 12.11 highlight the adaptability of this envelope: during summer the shutters are opened outward to permit airflow, while in winter they are tightly closed, reinforcing insulation and protecting the interior from damp and cold.

12.4.3 Internal Organization and Passive Strategies

Inside, the building reveals a sophisticated combination of passive mechanisms. Storage chambers are elevated on stone piers, creating a ventilated underfloor cavity that reduces moisture infiltration and carries away excess heat. Buffer aisles of 1.5 m separate the storage bins from the external walls, providing an intermediate zone that dampens external fluctuations. The tall central hall works as a vertical shaft: warm air rises through this volume and escapes through clerestory windows, while cooler air is drawn in through vents at the base. In addition, the walls and roofs form a double–skin envelope with air cavities that improve insulation and encourage airflow. The sequence of interior photographs in Figure 12.12 illustrates these elements, from narrow aisles with slit openings to raised storage bins, the lofty ventilation shaft, and the roof structure with timber trusses and clerestory vents.

The airflow pathways are clarified in Figure 12.13, where the sectional diagram traces the operation of cross- and stack ventilation. Cool air enters at ground level, passes through underfloor cavities and buffer aisles, then rises through the central hall to exit via clerestory openings.

12.4.4 Results

Field monitoring demonstrated the robustness of the passive design strategies employed in the First Granary of Xuan'en County. During the hottest summer periods, when outdoor temperatures reached 33°C and fluctuated by up to 7°C within a single day, the indoor environment remained below 25°C with a maximum

FIGURE 12.12 Internal construction features of the First Granary of Xuan'en County. (a) Buffer aisle with ventilation slits. (b) Raised storage chambers supported on stone piers. (c) Tall central hall with vertical ventilation shaft. (d) Roof structure with timber trusses and clerestory vents. (From Xu et al., 2024a.)

variation of only ±2°C. This narrow thermal band illustrates the buffering capacity of the combined wall mass, air cavities, and underfloor ventilation. The interior conditions therefore complied with the quasi-low temperature standard required for grain storage, a threshold that closely parallels the comfort needs of human occupants.

Relative humidity measurements further confirmed the effectiveness of the system. While external humidity swung widely between 40 and 80 percent in response to rainfall and daytime heating, interior levels stabilized within a narrow band of 70–80 percent. Although high compared with recommended comfort levels for human habitation, the stability of this range demonstrates the system's ability to

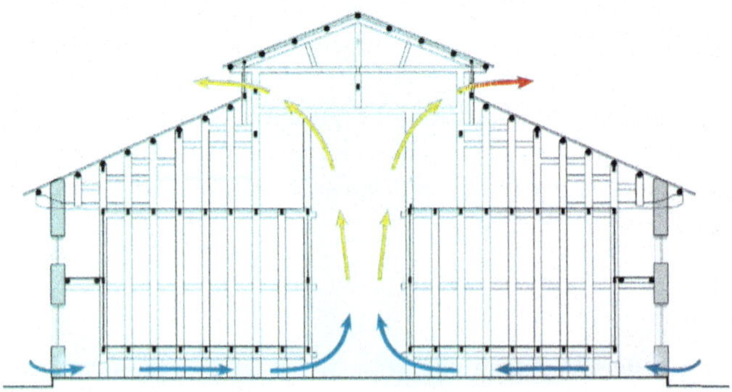

FIGURE 12.13 Sectional diagram of the granary showing airflow dynamics. (From Xu et al., 2024a.)

prevent the sharp fluctuations that accelerate spoilage. The constancy of indoor humidity, combined with low thermal variation, provides a clear picture of how the architecture modulates a highly unstable external climate.

In winter, the building continued to moderate external conditions. Interior temperatures remained on average 2°C higher than outdoors, reducing the risks of frost and condensation. Humidity was consistently lower inside than outside, particularly during cold, damp periods, thereby reducing the potential for mold growth in stored grain. The seasonal measurements together reveal a year-round capacity for stabilization, achieved entirely through passive means.

12.4.5 Lessons and Broader Implications

The evidence from Xuan'en underscores how vernacular builders were able to create a resilient passive environment in a humid subtropical climate by integrating multiple strategies into a coherent system. Although originally designed for grain storage, the building's thermal performance demonstrates principles that are directly transferable to human comfort. Thick masonry walls provide high thermal inertia, delaying heat transfer and smoothing temperature peaks. Buffer aisles reduce the direct impact of external swings by creating an intermediate microclimate zone. Raised floors both separate the structure from ground moisture and enable airflow beneath the storage bins. Finally, sequenced ventilation—combining underfloor inlets, buffer corridors, and clerestory outlets—creates a layered air circulation system that operates effectively across seasons.

The broader implication is that comfort and resilience can be achieved without mechanical intervention, even in regions with hot, humid summers and cold, damp winters. The quasi-low temperature achieved inside the granary mirrors the desirable thresholds for human living spaces, confirming that these passive techniques are not limited to storage architecture but can inspire contemporary housing design. The concept of a double-skin envelope with ventilated cavities, for instance, has direct application in modern low-carbon construction, offering both thermal insulation and dynamic air regulation.

Equally important is the reliance on local materials and construction methods. Stone piers, clay masonry, lime render, and timber trusses were all sourced and fabricated with minimal external input, lowering the embodied energy of construction. The introduction of a double-layered roof and cavity walls illustrates how vernacular architecture could innovate within the limits of available resources to enhance performance. This approach points toward a design logic that combines environmental responsiveness, economic efficiency, and material circularity.

Taken together, the lessons of the First Granary of Xuan'en County expand the geographic and climatic scope of vernacular adaptation strategies. They show how passive design can stabilize environments within the narrow temperature and humidity ranges required for sensitive storage, while also demonstrating principles applicable to human comfort. For contemporary sustainable architecture, the case illustrates that effective passive design emerges not from isolated devices but from the integration of form, material, and operation into a comprehensive climatic system.

12.5 CASE STUDY: VERNACULAR DWELLINGS IN JIANGNAN AS A COUPLED WIND–THERMAL REGULATION SYSTEM

(based on Min & Zhang, 2025)

12.5.1 Climate

The Jiangnan region of southeastern China lies within the northern subtropical humid monsoon climate zone, defined by extreme seasonal contrasts, including hot, humid summers with intense rainfall and cold, damp winters. These conditions create one of the most thermally challenging environments at this latitude, with average summer highs reaching 33°C (and peaks exceeding 42°C) and winter lows dropping to –10°C. Architectural responses in this region must therefore address the full spectrum of climatic demands, including cooling, ventilation, moisture control, and winter heat retention.

12.5.2 Urban Morphology and Village Form

Traditional Jiangnan dwellings form part of a dense urban fabric interwoven with water channels. This configuration reflects a long evolution of vernacular bioclimatic adaptation. Built from white lime-rendered masonry and gray-tiled roofs, the structures maximize thermal inertia while minimizing solar gain.

A defining feature of Jiangnan settlements is the "double chessboard" structure of interlaced river channels and narrow alleys. The rivers lower ambient temperature through evaporative cooling, while perpendicular alleys funnel prevailing southeast summer winds through the settlement, expelling trapped humidity and heat (Figures 12.14 and 12.15).

12.5.3 Wind–Thermal Coupling via Sky-Wells

At the building scale, the sky-well (or tianjing) is the primary climate-regulating element. It functions as a vertical shaft for airflow and light, while also providing thermal stratification. Openings facing the sky-well have a large average opening coefficient of 0.5–0.65, allowing for dynamic seasonal adjustment (Figure 12.16).

The sky-well also acts as a thermal buffer, with self-shading and evaporative cooling from vegetation and water features. In summer, cool air is drawn through shaded ground-level spaces, while heated air escapes vertically, a system enhanced by thermal pressure and night-time pre-cooling effects.

12.5.4 Passive Cooling and Ventilation Strategies

Jiangnan dwellings incorporate a sophisticated passive design logic for climate regulation:

- Guided and induced ventilation: Horizontal cooling alleys and north–south ventilation corridors support summer airflow, while vertical shafts like sky-wells enable thermal draft systems. These facilitate both day-time air

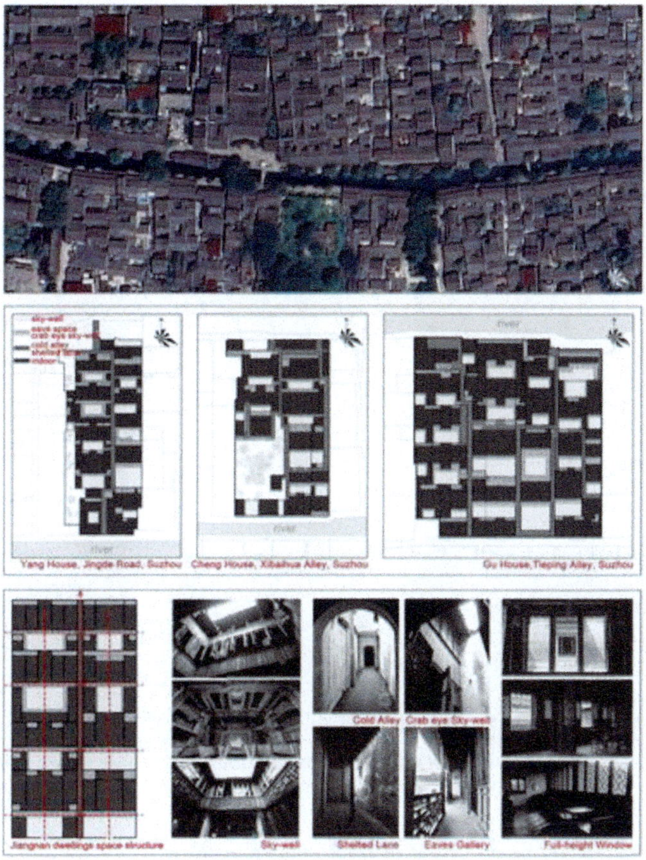

FIGURE 12.14 Spatial configuration from village scale to building section showing the breathable thermal structure. (From Min & Zhang, 2025.)

extraction and night-time ventilation, supported by thermal mass storage effects.

- Evaporative pre-cooling: Water features and vegetation within courtyards help lower incoming air temperature and reduce indoor heat gain.
- Differentiated opening strategies: Openings on east- and west-facing façades are minimized to approximately 2–3 percent of wall area and are positioned high to reduce direct solar exposure. In contrast, north-facing façades may include larger openings, up to 6 percent, to promote cross-ventilation during summer months.
- Thermal regulation interfaces: Internal courtyard façades use operable full-height windows with high opening ratios (50–65 percent), enabling convective cooling while maintaining spatial privacy and shading.
- Self-shading and geometric buffering: Sky-wells are dimensioned with height-to-width ratios of 1.0–1.5, and multiple shafts can be combined to

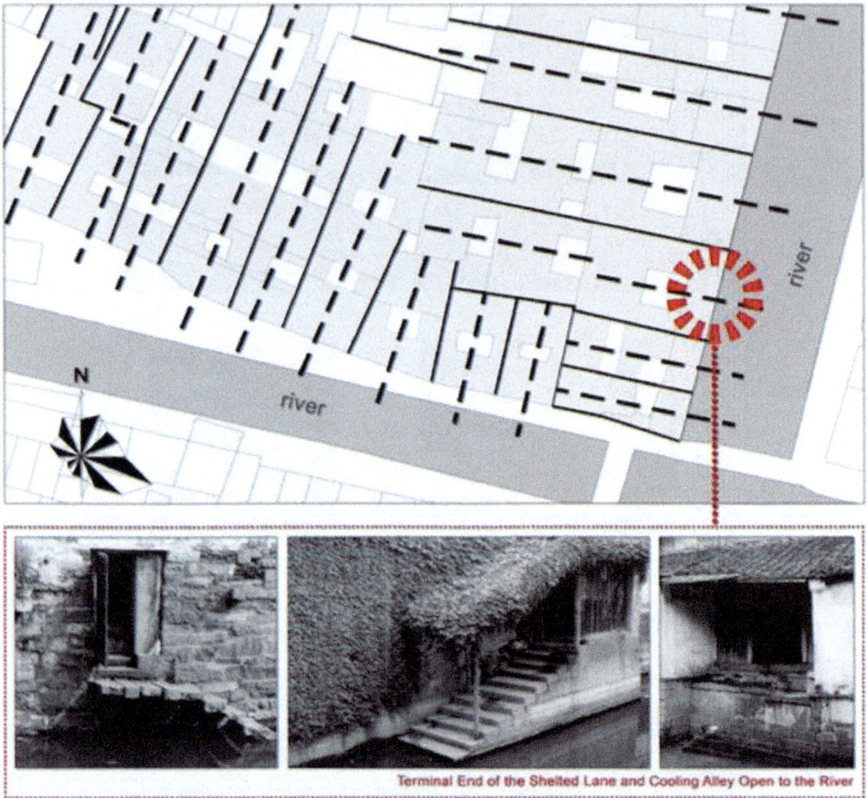

FIGURE 12.15 River-based layout of dwellings and cooling alleys. (From Min & Zhang, 2025.)

FIGURE 12.16 Full-height windows used on sky-well-facing thermal regulation interfaces. (From Min & Zhang, 2025.)

increase self-shading. Secondary "summer sky-wells" positioned on the north side enhance ventilation and mitigate heat.

- Shading layers and spatial gradient: Eaves corridors, overhangs, and recessed entries form microclimatic buffer zones. Proper aspect ratios (e.g., shading depth equal to 0.45 times window height) ensure full shading during peak radiation periods.
- Opening height modulation: Window height varies by floor to regulate neutral pressure zones and improve airflow dynamics. Lower intakes are placed at ground level, while upper-level outlets enhance exhaust.

These interlinked measures form a coherent multi-layered passive cooling and ventilation system, operating from the scale of urban morphology down to window detailing. The full set of strategies, along with their key features, is summarized in Table 12.1, providing a concise reference to the design logic applied in Jiangnan dwellings.

The external envelope employs white lime plaster (solar absorption coefficient 0.48) and gray roof tiles (0.52) to balance solar reflectivity and thermal inertia. These materials reduce daytime heat gain and support warmth retention in winter. The spatial logic of opening distribution differentiates between climate-exposed façades and those facing internal courts.

TABLE 12.1
Summary Table of Passive Strategies

Strategy Type	Key Features
Guided ventilation	North–south cooling alleys enhance summer airflow in narrow urban fabrics
Induced ventilation	Sky-wells use thermal pressure gradients for vertical airflow, day and night
Evaporative pre-cooling	Water and vegetation in courtyards lower air temperature through evaporation
Differentiated opening strategy	Opening sizes and placement adapted by façade orientation to balance solar protection and airflow performance
Thermal regulation interface	Courtyard façades with large operable openings (50–65 percent) for convective heat dissipation
Self-shading through sky-wells	Shaft depth provides solar protection and thermal gradient buffering
Summer sky-well	North-facing auxiliary shaft enhances ventilation and blocks seasonal heat
Shading and spatial buffering	Eaves, overhangs, and corridors form microclimatic buffer zones
Opening height modulation	Floor-dependent intake/exhaust strategy improves pressure-driven airflow
Thermal mass and envelope materials	Lime plaster and gray tiles provide heat reflection and seasonal thermal inertia

12.5.5 Summary and Implications

Min and Zhang (2025) demonstrate that Jiangnan vernacular dwellings embody a climatically intelligent, multi-scalar system. At its core is a feedback relationship between morphology, climate, and occupant adaptation. The combination of thermal mass, evaporative cooling, targeted solar control, and guided ventilation creates a passive strategy resilient to seasonal extremes.

This case illustrates how future architecture in hot-humid and cold-humid regions can draw from vernacular models by:

- integrating sky-lit cavities and shaded courtyards as ventilation engines;
- employing microclimatic buffering layers through water, vegetation, and narrow urban grain; and
- designing building envelopes with differentiated opening logic for external versus internal interfaces.

The Jiangnan case underscores that regional thermal comfort can be achieved without reliance on mechanical systems when spatial logic, environmental forces, and user practices are properly aligned.

12.6 CASE STUDY: VERNACULAR DWELLINGS AND ALLEY MICROCLIMATES IN ZHONGSHAN OLD TOWN

*(based on Xu et al., 2024*b*)*

12.6.1 Introduction

Zhongshan Old Town in Chongqing is a historic riverine settlement where timber dwellings are arranged in two parallel rows: the west row pressed against the hillside and the east row partially elevated above the river. Narrow alleys run between these rows, creating semi-outdoor spaces that function simultaneously as circulation corridors, public gathering places, and microclimatic regulators. Field measurements conducted in summer reveal how airflow, shading, evaporative cooling from the river, and the buffering effect of soil-adjacent structures combine to stabilize the thermal environment of houses.

12.6.2 Climate Context of Zhongshan Old Town

The region has a hot summer and cold winter climate. Summers are humid and intensely hot, with daily averages of 26°C–29°C and extreme maxima approaching 39°C. Relative humidity often exceeds 80 percent, creating sultry conditions with limited wind. Winters are damp and cold, but the main thermal challenge is summer overheating. These climatic pressures explain the reliance on passive strategies: narrow alleys to guide breezes, eaves and overlapping roofs to provide shading, river orientation to capture evaporative cooling, and soil-adjacent structures to buffer heat.

12.6.3 Settlement Morphology and Alley Typologies

The spatial organization of Zhongshan Old Town shows how settlement form and microclimatic function are inseparable. As illustrated in Figure 12.17, the dwellings are pressed between hillside and river, with narrow alleys weaving through the fabric. These passages are not only central to community life but also act as environmental regulators, offering shaded retreats during hot summer afternoons. Building form, terrain, and alley space together define the living microclimate.

The typological variations of these alleys further explain their thermal role. Figure 12.18 demonstrates the progression from unilateral-open alleys, most exposed to solar radiation, through top-open alleys with partial sky exposure, to top-closed alleys, where overhanging eaves create deep shade and the strongest buffering effect. This sequence highlights how modest shifts in enclosure and shading produce measurable differences in thermal performance, reducing air temperatures by up to 3.6°C at peak heat.

12.6.4 Structural Features: Chuandou Construction

Nearly all dwellings in Zhongshan adopt the Chuandou timber-frame system, in which purlins rest directly on columns and horizontal elements interlock to form a stable lattice. Timber board or bamboo-mud walls are inserted into this skeleton, producing a lightweight, flexible structure. Wall thickness ranges from 30 to 50 mm with heat transfer coefficients of 2.3–2.8 W/(m^2·K), indicating poor insulation. Roofs are covered with small sintered clay tiles without insulation, with a coefficient of 4.9 W/(m^2·K). Floors are either soil-cement or timber planks. These envelope properties are characterized by low thermal mass, high permeability, and limited airtightness, making the buildings highly dependent on spatial strategies such as alleys and soil adjacency for thermal regulation.

12.6.5 Split-Level Logic and Soil-Adjacent Spaces

Because of the steep topography, floor plans are elongated rectangles (3–5 m wide, >8 m deep) with varying heights along the depth. This creates split-level interiors where some rooms are partially subterranean or adjacent to soil-retaining walls. These soil-adjacent structures function as thermal sinks: their large mass and stable temperature dampen diurnal variations. Field measurements confirmed that rooms in contact with soil, such as storage spaces with stone retaining walls, exhibited the lowest temperature fluctuations, while elevated rooms without soil adjacency were most vulnerable to overheating.

12.6.6 Airflow and Ventilation Pathways

Air movement in Zhongshan is shaped by the steep terrain, the river corridor, and the alley geometry. During daytime, heated air rises along the hillside, while cooler river air enters lower alleys. Unilateral-open alleys allow cross-ventilation but expose rooms to direct heat. Top-closed alleys, with flanking buildings and connected eaves, channel airflow more slowly yet maintain stable temperatures.

FIGURE 12.17 Zhongshan Old Town: settlement, topography, and alley use. (a) overall view of Zhongshan Old Town showing timber stilted buildings arranged along the riverbank, with the settlement embedded between the hillside and the water; (b) schematic section illustrating the spatial relationship between hill, buildings, central alley, and river, highlighting differences in elevation and airflow paths; (c) photograph of a narrow shaded alley used as a public activity space, where residents gather and remain during hot periods; (d) photograph of a covered alley with strong enclosure, showing social use of shaded space and passive cooling supported by reduced solar exposure. (From Xu et al., 2024b.)

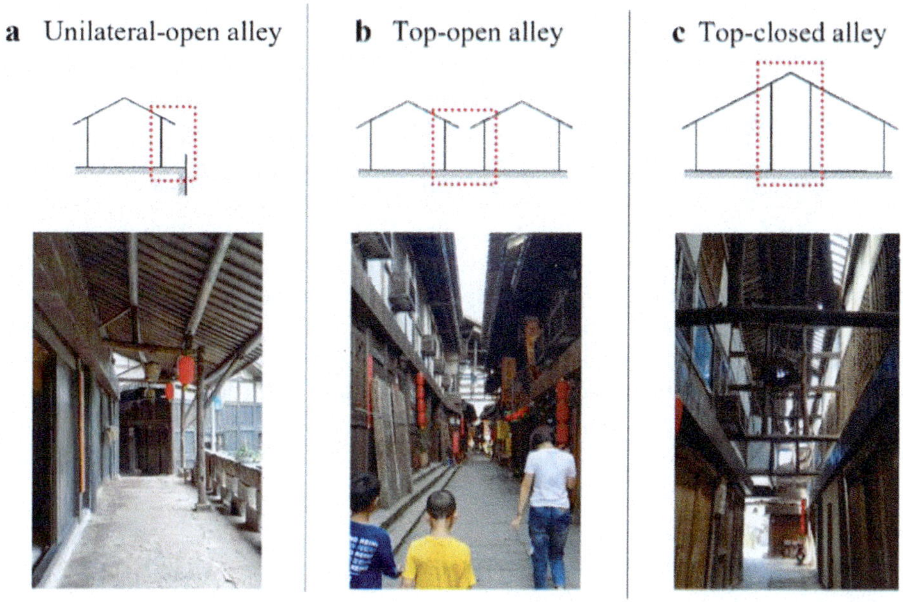

FIGURE 12.18 Morphological types of alleys in Zhongshan Old Town. (a) unilateral-open alley, with buildings located on one side only, allowing lateral openness and partial exposure to solar radiation; (b) top-open alley, with buildings on both sides and non-overlapping eaves, permitting vertical openness and controlled daylight penetration; (c) top-closed alley, where overlapping or connected eaves create continuous shading of the ground and walls, resulting in strong thermal protection and enhanced passive cooling. (From Xu et al., 2024b.)

Measurements in the living room adjacent to the alley (R1) showed that natural ventilation and infiltration contributed 40 percent of daily heat gain, illustrating the critical role of airflow between indoor and semi-outdoor spaces. Opening doors and windows in morning and evening hours increased the air change rate up to 7.5 times/h, creating significant cooling potential. At night, when river breezes are cooler, ventilation flushed accumulated heat, underlining the importance of night-time airflow as a passive strategy.

12.6.7 Shading and Solar Control

Shading is a defining factor in the performance of the alleys. Top-closed alleys, with overlapping eaves at heights of 8 m, provided the most complete shadowing, transforming them into "cold alleys." Thermal images showed surface temperature differences of 3.8°C between unilateral-open and top-closed alleys during peak sun. Wall surfaces in shaded alleys remained around 30–32°C compared to 36–38°C in exposed alleys.

For dwellings, deep eaves and bamboo-mud infill walls provided partial solar control. Split-level arrangements reduced the solar exposure of lower rooms. However, lightweight envelopes alone offered little protection; it was the integration of architectural shading with alley morphology that produced resilient thermal conditions.

12.6.8 Evaporative Cooling and the River Influence

The river plays a decisive role in shaping the summer microclimate of Zhongshan. Dwellings along its eastern edge are partially elevated on timber stilts, allowing cool air currents from the water surface to circulate beneath the floors. At night, when the temperature gradient between land and water is greatest, these breezes become especially effective, lowering indoor air temperatures and flushing out accumulated heat.

Alleys oriented towards the river act as natural ventilation corridors. During heat peaks above 35°C, they maintained air temperatures 1–2°C lower than enclosed alleys further inside the settlement. This effect, achieved through a combination of evaporative cooling and channeled airflow, demonstrates how the presence of the river enhances the resilience of the urban fabric, creating semi-outdoor spaces that support both comfort and social activity.

12.6.9 Thermal Mass and Soil-Adjacent Structures

While envelopes were lightweight and poorly insulated, soil-adjacent structures introduced crucial thermal inertia. The monitored dwelling included a storage room (R4) embedded against the hillside with two stone retaining walls. These surfaces remained stable at 28.5°C with only 0.6°C daily variation, acting as powerful heat sinks. By contrast, the riverside bedroom (R3), with no soil contact, experienced wall surface peaks above 42°C.

Heat balance analysis of R1 showed that walls, ceiling, and ventilation provided net gains, while the ground was the only net heat sink, absorbing –3393 W per day. The proportion of heat gain through envelope surfaces was 34 percent from walls and 26 percent from the ceiling, but this was offset by ground losses. These findings demonstrate that thermal stability in Zhongshan dwellings relies heavily on mass provided by soil contact, compensating for lightweight timber and clay-tile construction.

12.6.10 Cold Alleys and Thermal Buffering

Beyond typology and orientation, certain alleys operate as cold alleys, drawing in cooler air from the hillside and river. Unlike similar examples in the Middle East or North Africa, where massive stone walls act as heat sinks, in Zhongshan the ground and surrounding topography are the dominant cooling elements, as timber walls have very low thermal mass.

As illustrated in Figure 12.19, cold air flows from the hill and river into shaded alleys, displacing warm air and reducing indoor heat gain in adjacent rooms. Residents often occupy these alleys in the afternoon, benefiting from their reduced air temperature. This integration of terrain, water, and built form underscores how vernacular settlements harness natural flows to produce microclimates that stabilize comfort conditions.

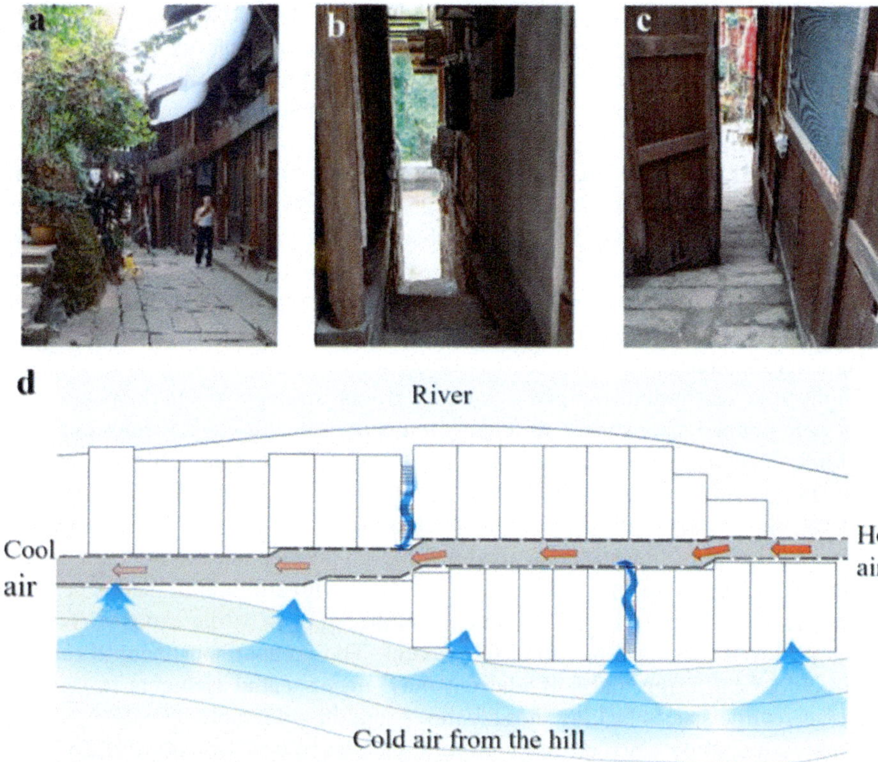

FIGURE 12.19 Cold alleys and terrain-driven airflow in Zhongshan Old Town. (a) alley located close to the hill, benefiting from proximity to the terrain; (b) alley opening towards the river, enabling the inflow of cooler air from the water body; (c) alley opening towards the hill, allowing cold air to descend into the built fabric; (d) schematic diagram illustrating how alleys take advantage of cold air from both the hill and the river, forming linear ventilation paths that displace warmer air. (From Xu et al., 2024b.)

Adaptive comfort analysis based on Chinese GB/T 50785-2012 and ASHRAE 55-2020 standards showed that rooms adjacent to alleys (R1) or soil (R4) achieved comfort conditions on over 90 percent of monitored days. Intermediate spaces (R2) achieved 85 percent, while the exposed riverside room (R3) achieved only 58 percent. These results underline the significance of semi-outdoor spaces and earth-contacted walls in ensuring resilience against summer heat stress without reliance on active cooling.

12.6.11 Lessons and Implications

The results of the Zhongshan study provide clear evidence of how vernacular design strategies operate under extreme summer conditions. Heat balance analysis revealed that the main sources of indoor heat gain were external walls, ceilings, and infiltration, contributing 34, 26, and 40 percent, respectively. In contrast, the ground consistently acted as a heat sink, absorbing excess energy throughout the day. This duality illustrates the vulnerability of lightweight envelopes, which readily transmit heat, and the stabilizing role of earth contact, which provides reliable thermal buffering.

The thermal buffering capacity of alleys was shown to follow distinct thresholds. When outdoor daily averages were below 25.6°C, alleys had little effect. Between 25.6°C and 31.1°C, however, they actively absorbed heat from adjacent rooms, accelerating cooling. Above 31.1°C, the alley ceased to cool but still reduced further heat gain, effectively acting as a shield. This progression demonstrates that semi-outdoor spaces are not static features but adaptive systems whose influence strengthens with rising external stress.

Night-time ventilation emerged as another crucial strategy. When doors and windows were opened after sunset, air change rates increased up to 7.5 times per hour, providing substantial cooling and flushing of accumulated heat. This highlights a simple but powerful synergy: airtightness by day to block excessive infiltration, and deliberate night-time ventilation to restore comfort.

Thermal comfort analysis confirmed the uneven performance of different spatial conditions. Rooms adjacent to alleys or in contact with soil maintained acceptable comfort levels on more than 90 percent of monitored days, while exposed riverside rooms achieved this on fewer than 60 percent. The comparison underscores how semi-outdoor buffering and soil-adjacent mass are essential elements in securing resilience, while spaces lacking these features remain vulnerable.

Finally, the study clarified the hierarchy of alley types. Unilateral-open alleys provided the least protection, top-open alleys offered intermediate buffering, and top-closed alleys achieved the most pronounced cooling effect. This typological sequence demonstrates how incremental increases in shading and enclosure translate directly into improved microclimatic performance. The logic of layered shading found in Zhongshan can inspire modern urban design, suggesting that networks of semi-outdoor passages, properly proportioned and oriented, could serve as low-energy climate regulators in dense settlements.

12.7 CASE STUDY: PASSIVE STRATEGIES IN ANATOLIA: THE VERNACULAR SETTLEMENT OF MARDIN

(Based on Sözen & Oral, 2019*)*

The architectural heritage of Anatolia presents a wide range of passive design strategies adapted to different climatic conditions. In hot-arid regions, such as Mardin in southeastern Turkey, vernacular architecture has evolved to mitigate extreme temperatures, optimize shading, and enhance natural ventilation. The study by Sözen and Oral (2019) provides a detailed parametric analysis of these strategies, focusing on urban canyons and courtyards as key elements in outdoor thermal comfort regulation. Through microclimate simulations and empirical observations, their research highlights the effectiveness of high-density settlement patterns, shading elements, and thermal mass utilization in reducing heat stress and improving livability.

12.7.1 High-Density Stone Built Urban Fabric and Shading

The dense, terraced layout of Mardin follows the contours of the Mazı Mountain, creating an intricate network of narrow, irregular streets and stepped passages that respond to the region's steep topography and hot-arid climate. The settlement's compact urban fabric is characterized by high-density, stone-built structures that provide mutual shading and protection from intense solar radiation. This arrangement minimizes heat accumulation within the urban space while promoting natural ventilation through temperature-driven airflows facilitated by the varying elevations.

12.7.2 Urban Canyons and Thermal Comfort Optimization

As illustrated in Figure 12.20, the cityscape consists of modular housing blocks arranged in a cascading pattern, where each level of buildings serves as a foundation for the next, ensuring that homes have access to daylight and prevailing winds while remaining shielded from direct sun exposure. The interweaving of winding alleyways, stepped pathways, and covered passages creates a layered microclimate, where air circulation is enhanced through venturi effects within the narrow urban canyons. This design helps regulate ambient temperatures by reducing heat island effects, maintaining cooler shaded zones throughout the day.

Additionally, Mardin's street orientation and aspect ratio play a crucial role in thermal comfort. The east-west orientation of the main streets, coupled with high aspect ratios (H/W: 2 to 4), effectively reduces solar heat gain by ensuring that

FIGURE 12.20 Top view of the vernacular settlement of Mardin, illustrating the terraced urban form, modular housing pattern, and integration of shaded streets and courtyards. (From Sözen and Oral, 2019.)

streets remain shaded for extended periods. The integration of semi-open spaces such as revaks (colonnaded galleries) and eyvans (vaulted reception areas) within the built fabric enhances thermal buffering, providing gradual transitions between outdoor and indoor environments.

12.7.3 Courtyards as Climatic Modifiers

This combination of high-density terraced construction, interconnected courtyards, and shaded walkways contributes to a passively cooled environment, where thermal inertia, mutual shading, and natural ventilation work in harmony to mitigate extreme summer temperatures. These features, refined over centuries, exemplify the resilience of vernacular architecture in arid climates, offering valuable insights into contemporary sustainable urban design. As shown in Figure 12.20, the east–west orientation of the main streets, combined with high aspect ratios (H/W: 2 to 4), effectively minimizes solar exposure and reduces radiant heat gain.

Simulations conducted by Sözen and Oral (2019) reveal that higher aspect ratios lead to lower MRT and physiological equivalent temperature (PET) values, particularly in north–south oriented streets. At peak afternoon hours, PET values in high-aspect-ratio streets were up to 13°C lower than in wider streets with lower building heights. The analysis also demonstrated that asymmetrical street sections, where the southern-facing buildings are lower, can dissipate heat toward the southern edge of the street, contributing to overall thermal balance.

Another key passive feature in Mardin's streetscape is the use of galleries (revak), which are shaded colonnades integrated into commercial streets. These galleries create cooler microclimates, reducing PET values by up to 10°C in shaded areas and offering comfortable outdoor spaces during peak summer heat (Figure 12.21).

In addition to urban canyons and revaks, courtyards play a crucial role in passive climate adaptation in Mardin. The study analyzed multiple courtyard configurations to assess their impact on temperature regulation, air movement, and shading effectiveness.

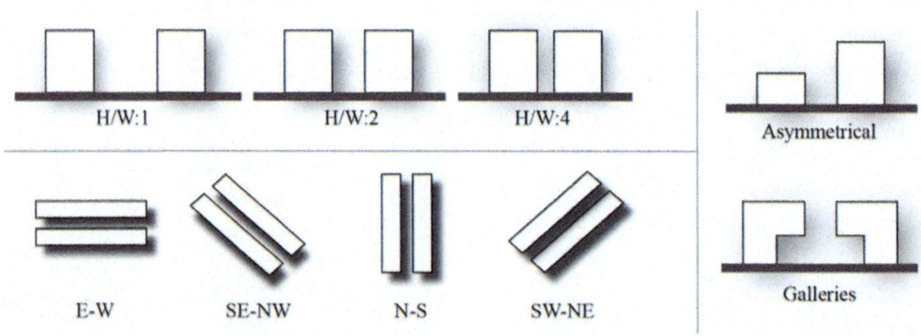

FIGURE 12.21 Revak (colonnaded gallery) and eyvan (vaulted reception space) as key features in Mardin's traditional courtyard houses, providing shaded areas that enhance thermal comfort. (From Sözen and Oral, 2019.)

Courtyard Plan Ratios

- Courtyards with a 1:2 aspect ratio (elongated north-south) performed best in thermal comfort, with PET values up to 8°C lower than those with a 2:1 ratio (elongated east–west).
- These courtyards benefited from enhanced shading from surrounding walls and lower solar exposure in the morning and afternoon, leading to better thermal regulation.

Shading and Semi-Open Spaces

- A defining feature of Mardin's traditional courtyard homes is the revak, a colonnaded semi-open gallery that extends from the house into the courtyard.
- The eyvan, a vaulted reception area open to the courtyard, is another key element that improves thermal comfort.
- As illustrated in Figure 12.22, these semi-open spaces provide shaded transition areas, reducing PET values and enhancing overall usability.

Surface Materials and Vegetation

- Courtyards paved with natural stone or soil exhibited lower surface temperatures compared to those with concrete flooring, reducing PET values by up to 4.5°C.
- The integration of trees further improved thermal comfort, with shading effects reducing PET by up to 15.6°C during peak afternoon hours.
- Figure 12.22 highlights the eyvan and courtyard configuration, demonstrating how shaded areas enhance thermal performance throughout the day.

The findings of Sözen and Oral (2019) demonstrate that high-density settlement patterns, optimized shading strategies, and natural ventilation elements are key passive strategies in Anatolia's hot-arid climate. The integration of narrow, high-aspect-ratio streets, shaded galleries, and well-proportioned courtyards significantly

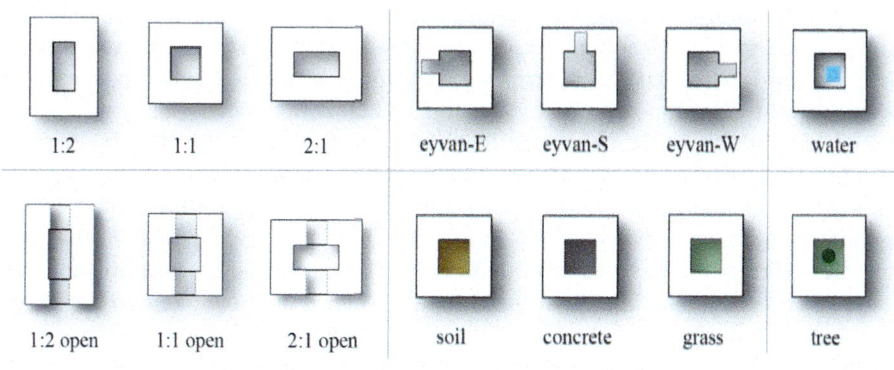

FIGURE 12.22 Eyvan and courtyard layout in a traditional Mardin house, showing the integration of semi-open spaces and shaded areas to reduce heat stress. (From Sözen and Oral, 2019.)

improves outdoor thermal comfort, reducing solar heat gain and enhancing airflow regulation. The study reinforces the relevance of vernacular design principles in modern sustainable architecture, particularly in adapting to contemporary challenges of urban heat stress and energy efficiency.

12.7.4 Comparative Synthesis and Design Implications

The case studies presented in this chapter illustrate the remarkable breadth and adaptability of vernacular passive cooling strategies. Despite differences in geography, material resources, and cultural traditions, a consistent environmental logic emerges: shading, thermal mass, and natural ventilation form the foundation of resilient climatic adaptation. These elements rarely operate in isolation. Instead, they are combined into layered systems in which form, material, and spatial configuration reinforce one another to buffer extremes, regulate airflow, and stabilize comfort conditions.

In hyper-arid regions such as Adrar, compact urban form, thick adobe walls, and sequenced ventilation enabled communities to withstand summer maxima approaching 50°C. The lessons from Ksar Tamentit underline the capacity of earthen mass and controlled night-time flushing to align material properties with daily behavioral rhythms. Marrakesh riads and the vernacular houses of Gran Canaria translated the same logic into different material registers. In Marrakesh, rammed earth and brick inclusions provided thermal inertia, reinforced by shaded passages and covered souks. On Gran Canaria, volcanic stone walls 60–80 cm thick combined with reflective lime renders and narrow streets to achieve year-round buffering under constant solar radiation.

A more radical interpretation appears in the cave dwellings of southern Spain, where geological strata replaced masonry altogether. Here, the ground itself served as structure, envelope, and thermal buffer, producing stable interiors of 18°C–22°C without mechanical input. This principle of earth sheltering finds a parallel in the sunken courtyards of China's Loess Plateau, where subterranean design intensified thermal buffering by reducing solar exposure and leveraging soil's stable temperature profile. Both traditions demonstrate that embedding living space directly into the ground provides exceptional resilience against diurnal and seasonal variation.

Courtyards across the Mediterranean and North Africa highlight the synergy of shading, ventilation, and evaporative cooling. Whether in the riads of Marrakesh, the patios of Córdoba, or the palaces of Seville, courtyards served as climatic engines, combining vertical stack effects, vegetative shading, and water features to produce self-regulating microclimates. In Zhongshan Old Town, narrow alleys functioned as parallel devices, channeling airflow and moderating solar exposure, while adjacency to soil and river breezes enhanced buffering. These semi-outdoor spaces not only extended comfort into circulation zones but also shaped interior conditions, confirming that settlement form and microclimate regulation are inseparable.

Windcatchers in Iran and the Middle East represent another archetype of vernacular ingenuity. By capturing high-level winds and combining them with evaporative cooling from qanat-fed courtyards, they ensured both ventilation and temperature reduction in some of the harshest desert environments. Modern reinterpretations have shown how aerodynamic refinement and sensor-driven control can extend this logic into dense urban contexts, maintaining resilience under variable and polluted conditions.

At the humid subtropical margins, such as Jiangnan and Xuan'en County, vernacular systems responded to the dual challenges of hot, wet summers and cold, damp winters. Jiangnan dwellings combined sky-wells, alley corridors, shading, and evaporative pre-cooling from rivers and vegetation to form coupled wind–thermal regulation networks. The First Granary of Xuan'en County, though built for grain storage, demonstrated similar principles through buffer aisles, raised floors, double-skin walls, and sequenced stack ventilation. Despite its utilitarian purpose, the granary maintained temperatures below 25°C in summer and reduced humidity fluctuations, operating within ranges directly relevant to human comfort. Its performance illustrates how passive design can deliver both thermal stability and resilience in one of the world's most demanding climatic contexts.

Taken together, these cases underscore three fundamental lessons. First, effective passive cooling arises not from single devices but from integrated systems in which mass, shade, and airflow are combined at multiple scales. Second, successful adaptation requires calibration to local climate and cultural practice, whether through behavioral routines in Adrar, water management in Seville, or seasonal shutter operation in Xuan'en. Third, the enduring effectiveness of these strategies demonstrates their relevance for contemporary design. Vernacular precedents provide not only empirical evidence of long-term performance but also conceptual models for sustainable construction.

The synthesis across climates reveals a continuum of solutions: compact urban canyons, massive earthen or stone envelopes, subterranean excavation, courtyard microclimates, wind-driven ventilation, and soil buffering. Each represents a variation on the same principle: the deliberate harnessing of natural forces to create comfort without mechanical intervention. For modern architecture, these lessons affirm that low-energy, climate-responsive design must be systemic, adaptive, and rooted in the interplay of environment, form, and material.

REFERENCES

Ganjimorad, M., Fernandez, J. D., & Heiranipour, M. (2024). Impact of wind in urban planning: a comparative study of cooling and natural ventilation systems in traditional Iranian architecture across three climatic zones. *Architecture Papers of the Faculty of Architecture and Design STU*, *29*(4), September. https://doi.org/10.2478/alfa-2024-0020

Min, T., & Zhang, T. (2025). "Coupling adaptation mode and regulation strategy for the wind-thermal climate environment of vernacular dwellings in Jiangnan, China based on regional bioclimatic evaluation. *Journal of Asian Architecture and Building Engineering*. https://doi.org/10.1080/13467581.2025.2454615

Sözen, İ., & Oral, G. K. (2019). Outdoor thermal comfort in urban canyon and courtyard in hot arid climate: A parametric study based on the vernacular settlement of Mardin. *Sustainable Cities and Society*, *48*, 101398. https://doi.org/10.1016/j.scs.2018.12.026

Wang, F., Wang, S., Cheng, B., & Wang, W. (2022). To inhabit, retain or abandon? Adaptive utilization of energy-efficient sunken buildings by rural households in Shanzhou, China. *Energy & Buildings*, *255*, 111668. https://doi.org/10.1016/j.enbuild.2021.111668

Xu, W., Wang, Q., Deng, H., & Zhu, Z. (2024a). Research on the sustainable design strategies of vernacular architecture in Southwest Hubei: A case study of the First Granary of Xuan'en County. *PLoS ONE, 19*(12), e0316518. https://doi.org/10.1371/journal.pone.0316518

Xu, Y., Yang, Z., Zheng, X., Ren, J., Feng, C., & Tang, M. (2024b). Thermal environments of vernacular dwellings and the adjacent alley in summer: An experimental study in Southwest China. *Building and Environment, 261*, 111634. https://doi.org/10.1016/j.buildenv.2024.111634

Zhu, J., Feng, J., Lu, J., Chen, Y., Li, W., Lian, P., & Zhao, X. (2023). A review of the influence of courtyard geometry and orientation on microclimate. *Building and Environment, 236*, 110269. https://doi.org/10.1016/j.buildenv.2023.110269

13 Research in Passive Cooling Technologies

13.1 GLAZING, GREENERY, AND PASSIVE COOLING

Shading strategies and vegetative systems play a crucial role in reducing solar heat gains and enhancing thermal comfort in buildings, particularly in warm and humid climates. This section reviews key scientific studies on the effectiveness of static and dynamic shading devices, smart glazing technologies, and vegetated systems in mitigating indoor overheating and reducing cooling energy demand.

The discussion begins with an evaluation of shading devices for glazed facades, including fixed and dynamic systems, assessing their impact on energy savings, visual comfort, and return on investment. Studies on electrochromic and thermochromic glazing follow, highlighting their adaptive capabilities in regulating solar heat gain while maintaining daylight availability. The section then explores the thermal and energy performance of double-skin green facades, vertical greenery systems (VGS), and green roofs, emphasizing their role in reducing façade temperatures and improving urban microclimates.

Comparative analyses of these strategies demonstrate their varying effectiveness based on climate conditions, building typologies, and economic feasibility. By integrating shading and vegetative solutions, buildings can achieve significant cooling energy reductions while improving occupant comfort. The findings offer practical guidance for architects and engineers seeking sustainable, climate-responsive design solutions.

13.1.1 External Shading Devices in Hot-Humid Office Retrofits: Case Studies from Chennai, India

Chandrasekaran and Madhumathi (2023) investigated the effectiveness of external shading devices (ESDs) as a retrofitting strategy for fully glazed office buildings in Chennai, India. The study focused on two multi-story buildings, Lancor Westminster (Sample A) and Raheja Towers (Sample B), representing typical glass-dominated façades constructed in the early 2000s. The aim was to assess the impact of both fixed and dynamic ESDs on cooling energy demand, daylight access, and payback period under tropical hot-humid conditions. The general layout, geometric configuration, and façade characteristics of the two case studies are illustrated in Figure 13.1, which shows their location, floor plans, and external views.

13.1.1.1 Methodology

The analysis used annual simulations in E-Quest (for energy performance) and Revit Insight (for daylighting), comparing eighteen fixed shading strategies, including

DOI: 10.1201/9781003645573-13

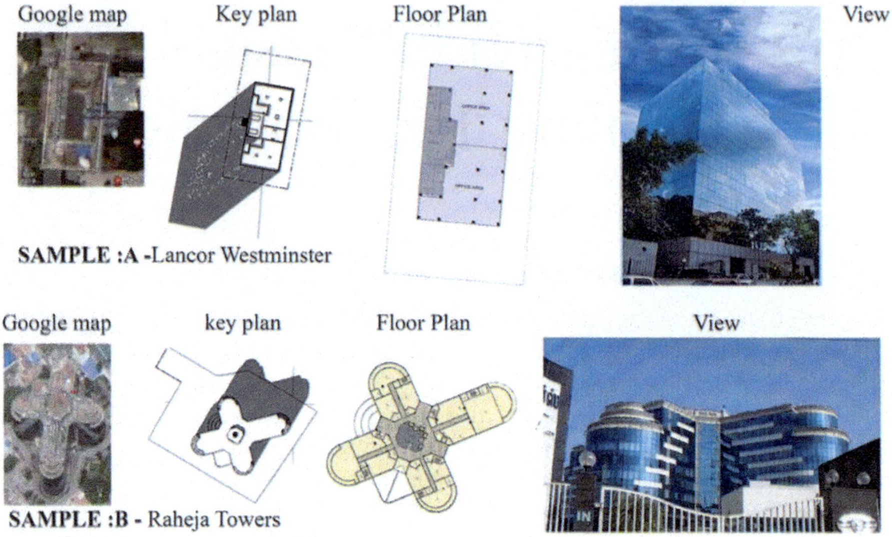

FIGURE 13.1 Location, layout, and façade view of the two case study buildings in Chennai. Sample A (Lancor Westminster) features a compact east–west-oriented rectangular form, while Sample B (Raheja Towers) is a large cruciform structure with complex solar exposure patterns. (Chandrasekaran & Madhumathi, 2023.)

variations in glazing type, window-to-wall ratio, horizontal/vertical fins, and fabric solar screens, with a base case. The most effective fixed solution (perforated fabric covering 100% of glazing) was then tested as a dynamic shading system using Dynamo scripts in Revit, responding to hourly solar angles over key seasonal days.

Cooling setpoints, occupancy schedules, and envelope parameters were standardized across both buildings to ensure comparability. The evaluation focused on three criteria: cooling energy savings, daylight availability (SDA and ASE), and economic payback based on Indian construction cost data.

13.1.1.2 Results

Among fixed strategies, the use of a perforated solar fabric screen covering the entire glazed surface yielded the best performance. This solution reduced cooling energy use by 22.7% in Sample A and 17.8% in Sample B, with acceptable daylighting (SDA > 56%) and ASE below 15%. Payback periods ranged from 5 to 7 years, making it both effective and cost-efficient.

Fixed aluminum fins also improved performance, particularly in Sample B, where vertical fins with 0.6 m depth achieved 14.8% cooling savings. However, these systems had longer payback periods and posed greater installation complexity.

Dynamic shading with the fabric system achieved even greater energy savings, up to 24% in Sample A and 18% in Sample B, particularly during summer months and on east- and west-facing façades. Nonetheless, annual daylight simulations could not be performed for the dynamic system, and its implementation costs remain difficult to quantify.

13.1.1.3 Conclusion

The case study confirms that external shading, particularly lightweight and easily deployable systems such as fabric screens, can significantly reduce cooling energy demand in fully glazed office buildings located in hot and humid climates. Dynamic systems offer additional performance gains, but their cost-effectiveness must be further explored. These findings support the adoption of envelope-level retrofitting strategies that improve thermal resilience and daylighting while remaining affordable in developing country contexts.

13.1.2 Electrochromic and Thermochromic Glazing for Energy Efficiency: A Study in Athens

Detsi et al. (2024) analyzed how electrochromic (EC) and thermochromic (TC) glazing influence energy demand and cost efficiency in a Mediterranean climate, focusing on a multi-story office building in Athens. The study examined how dynamic glazing technologies affect cooling and heating loads, particularly in buildings with extensive glazing areas and variable solar exposure.

Figure 13.2 provides an overview of the examined building. The first part (Figure 13.2a) presents an external perspective of the structure, highlighting its general architectural form and surroundings. The second part (Figure 13.2b) offers a top–down view of the building model, illustrating its spatial layout and relationship with the surrounding environment.

Figure 13.3 details the digital model used for simulations. The first part (Figure 13.3a) shows an overview of the building model developed in SketchUp, capturing key architectural features used in energy simulations. The second part (Figure 13.3b) presents the front view, emphasizing the extensive glazing on the south-facing facade, which played a crucial role in evaluating the impact of dynamic glazing technologies. The final part (Figure 13.3c) displays the rear elevation, highlighting window distribution and other construction elements affecting the building's energy balance.

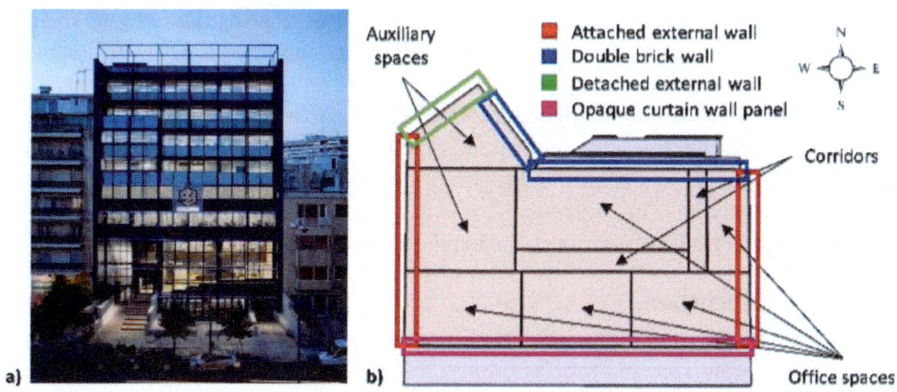

FIGURE 13.2 Overview of the examined office building, illustrating its general structure and spatial layout. (Detsi et al., 2024.)

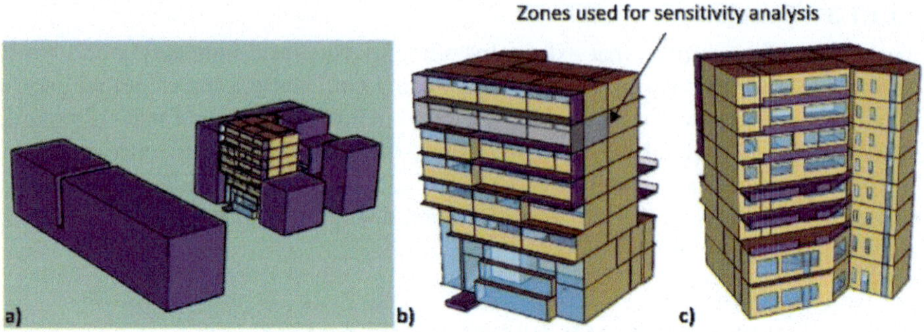

FIGURE 13.3 SketchUp model illustrating the building geometry and key simulation parameters. (Detsi et al., 2024.)

13.1.2.1 Methodology

The study employed dynamic energy simulations in EnergyPlus, testing multiple glazing configurations under varying climate and building conditions. The analysis included electrochromic glazing in triple- and quadruple-pane systems filled with argon (T-EC-Ar) or krypton (Q-EC-Kr) gases, thermochromic glazing with temperature-responsive solar transmittance, and hybrid electrochromic-thermochromic systems. Additionally, different insulation levels and window-to-wall ratios (WWR) were considered to determine optimal energy performance.

13.1.2.2 Results and Key Findings

Smart glazing reduced primary energy demand by 7–16%, with the most significant savings occurring in warmer climates and buildings with WWR $\geq$ 60%. Quadruple electrochromic glazing with krypton (Q-EC-Kr) achieved the highest energy savings, particularly in office buildings with extensive glazed surfaces. Triple electrochromic glazing with argon (T-EC-Ar) was the most cost-effective solution, offering a fast return on investment while maintaining significant energy reductions. The introduction of thermochromic layers to electrochromic glazing did not lead to notable improvements, indicating that electrochromic technology alone is sufficient for optimizing energy performance.

Economic analysis confirmed that NPC could be reduced by up to 18% when using optimized dynamic glazing instead of standard low-emissivity (Low-E) glass. While Q-EC-Kr glazing required a higher initial investment, it resulted in the greatest long-term savings. In contrast, T-EC-Ar provided a more affordable alternative with a balanced trade-off between cost and efficiency.

13.1.2.3 Conclusion

The study demonstrated that electrochromic glazing effectively reduces cooling demand in Mediterranean office buildings. The findings highlight the importance of selecting glazing configurations based on climate conditions and budget constraints, with electrochromic systems offering a viable balance between energy efficiency and cost-effectiveness.

13.1.3 Electrochromic Glazing for Energy and Comfort in Tropical Climates: A Study in Singapore

Li et al. (2023) conducted an experimental evaluation of electrochromic glazing (EC) in a tropical environment, assessing its impact on thermal comfort and cooling energy demand in Singapore. Unlike simulation-based studies, this research was based on real-world measurements, using two identical climate chambers constructed on the rooftop of the National University of Singapore. One chamber was fitted with double-glazed electrochromic windows, while the other featured conventional single-pane glazing, commonly used in buildings in the region.

13.1.3.1 Methodology

As illustrated in Figure 13.4, the experimental setup included two identical climate chambers positioned side by side, enabling direct comparison of thermal and energy performance. The aerial view in Figure 13.4a provides an overview of the site, while

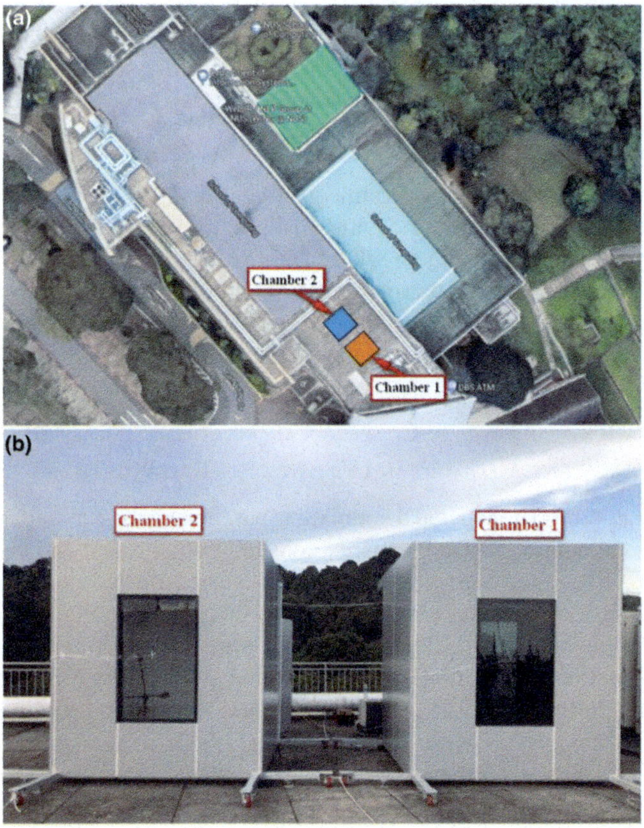

FIGURE 13.4 (a) Aerial view of the experimental chambers' location on a satellite map, (b) External views of the climate chambers at the National University of Singapore. (Li et al., 2023.)

Figure 13.4b shows the external appearance of the chambers, highlighting their structural similarities and glazing differences.

The chambers, with lightweight construction, were exposed to direct solar radiation for several weeks. During the experiment, the researchers monitored glazing surface temperatures, indoor thermal conditions, and air-conditioning energy consumption under different EC states: fully darkened, fully clear, and intermediate transitions.

13.1.3.2 Results

The findings demonstrated that electrochromic glazing significantly reduced the internal surface temperature of windows, limiting heat transfer into the building. In fully darkened mode, the inner glass surface was 4.4°C cooler than in the reference chamber, while in clear mode, the difference was 3.0°C. This effect improved thermal comfort, as indicated by the Predicted Mean Vote (PMV) index, which improved by 50% in fully darkened mode and by 37% in partially clear mode, effectively reducing heat stress for occupants.

Energy savings were also significant. The chamber with electrochromic glazing required 19% less cooling energy than the one with conventional glazing. The energy reduction was more pronounced on sunny days, while on overcast days, differences were minimal. A cost analysis estimated annual cooling cost savings in Singaporean office buildings at 9.12 SGD/m^2 when using EC windows.

13.1.3.3 Conclusion

The study confirmed that electrochromic glazing is an effective strategy for improving indoor comfort and reducing cooling loads in tropical buildings. The ability to dynamically adjust solar heat gain without completely blocking daylight allows for a balance between visual and thermal comfort. However, the authors note that fully darkened states may necessitate additional artificial lighting, which could offset some of the energy savings.

13.1.4 Static and Dynamic Glazing Combinations: Impact on Energy and Comfort in Lightweight Buildings

Es-sakali et al. (2024) conducted a simulation-based study on the impact of combining static and dynamic glazing on energy consumption, thermal and visual comfort, and operational costs in lightweight buildings located in the semi-arid climate of Benguerir, Morocco. The research analyzed seven types of static glazing alongside dynamic thermochromic (TC) and electrochromic (EC) coatings, both separately and in combination. Simulations were performed using EnergyPlus and Radiance, incorporating thermal comfort indices (PMV and PPD) and daylight performance metrics, including glare assessment.

Figure 13.5 illustrates the developed research methodology, outlining the workflow used to evaluate glazing configurations under different environmental and operational conditions.

13.1.4.1 Methodology

The study assessed the energy performance of different glazing types through a series of dynamic simulations. Static glazing configurations included single, double,

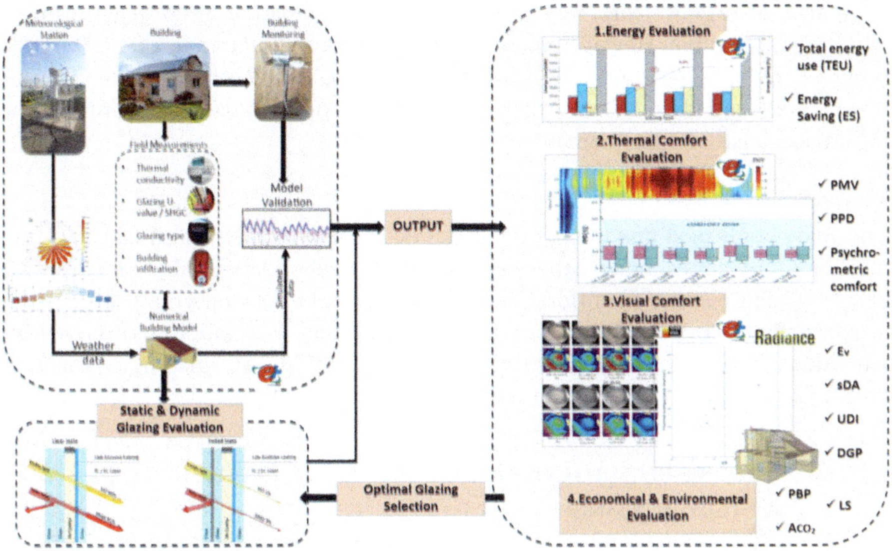

FIGURE 13.5 Developed research methodology illustrating the workflow for assessing the impact of glazing configurations on energy performance, comfort, and emissions reduction. (Es-sakali et al., 2024.)

and triple-pane Low-E glazing with various gas fillings, while dynamic glazing solutions involved TC and EC coatings. Their effectiveness was measured in terms of annual energy demand, CO_2 emissions reduction, and indoor comfort. The analysis also examined visual comfort, evaluating daylight autonomy and glare control.

13.1.4.2 Results

The use of double-pane Low-E glazing with an air-filled cavity provided the best balance between energy savings (21% reduction in annual energy consumption), thermal comfort, and visual performance. Compared to single glazing, it lowered CO_2 emissions by 1522.38 kg/year, with a payback period of 12.86 years. While triple-pane Low-E glazing improved energy efficiency further, it led to increased reliance on artificial lighting, increasing overall energy demand.

Integrating dynamic coatings yielded additional benefits. Electrochromic glazing reduced energy consumption by 6.3% compared to the optimal static glazing, while thermochromic glazing provided a smaller reduction of 3%. The highest savings were achieved by combining TC and EC, leading to a 26% decrease in annual energy demand. This configuration was most effective in preventing overheating during hot periods while maintaining comfortable indoor temperatures in colder months.

Visual comfort analysis indicated that dynamic coatings improved the balance between daylight levels and glare reduction. Electrochromic glazing provided the highest daylight autonomy (sDA), whereas thermochromic coatings were more effective in mitigating excessive light intensity. The TC-EC combination offered the optimal compromise, reducing overheating while ensuring adequate daylight levels.

Economic analysis revealed that the high initial cost of dynamic glazing remains a significant barrier. The payback period for TC coatings was approximately

28 years, for EC 44 years, and for the TC-EC combination 58.65 years, limiting widespread adoption. However, these technologies contributed to further CO_2 reductions—1723.05 kg/year (TC), 1909.07 kg/year (EC), and 1928.83 kg/year (TC-EC). The authors emphasized that cost reductions and improved availability are essential for broader market implementation.

13.1.4.3 Conclusion

The study confirmed that integrating dynamic glazing with an optimized static system can significantly enhance the energy efficiency of buildings in dry climates while ensuring high occupant comfort. The results highlight the importance of selecting appropriate glazing configurations based on building orientation and daylighting conditions, demonstrating that the combination of TC and EC glazing provides the most effective energy savings but requires further cost optimization.

13.1.5 Cooling Potential of Double-Skin Green Facades: A Study in Shanghai

Yang et al. (2018) investigated the cooling performance and energy-saving potential of a double-skin green facade (DSGF) on a recently renovated five-story administrative building located on a university campus in Shanghai. The study aimed to evaluate how a vertical greening system (VGS) influences the thermal environment of a building envelope under hot-humid climatic conditions.

Figure 13.6 provides a site plan and perspective view of the examined building, illustrating its orientation and spatial context. Figure 13.7 presents a comparative view of the south-facing facade, showing both the DSGF-covered and the original bare masonry facade, which served as a reference in the study.

13.1.5.1 Methodology

The research was conducted as a field study, measuring the thermal effects of a DSGF installed on a Southwest-South-facing facade, with a Northeast-North-facing facade serving as a reference without greenery. The building, originally constructed in the

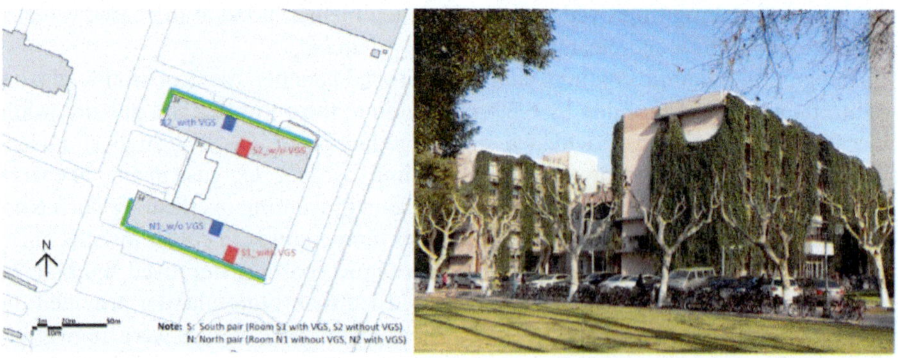

FIGURE 13.6 Site plan and perspective of the analyzed office building, illustrating its orientation and spatial context. (Yang et al., 2018.)

FIGURE 13.7 Comparative view of the south-facing facade, highlighting the DSGF-covered section vs the original masonry façade. (Yang et al., 2018.)

1980s, has 300 mm-thick brick and mortar walls with an additional VGS frame supporting evergreen climbing plants (Mucuna sempervirens "Hemsl"). The leaf area index (LAI) ranged from 3.0 to 3.5, ensuring consistent vegetation coverage.

Two office rooms on the fourth floor were selected for monitoring, each measuring 4.2 m × 3.6 m × 3.6 m. Continuous 15-day measurements were carried out in summer, during which the rooms remained unoccupied, with windows and doors closed to maintain free-running thermal conditions. The study recorded: Surface temperatures (Ts) of the outer vegetated skin and masonry facade, Cavity air temperature (Ta) and relative humidity (RH) between the VGS and the masonry wall, Indoor parameters: air temperature (Ta), globe temperature (Tg), wind velocity (WV), and operative temperature (Top), Weather conditions using a rooftop meteorological station on a nearby mid-rise building.

13.1.5.2 Results

The analysis confirmed that the DSGF created a distinct microclimate, lowering facade temperatures and improving indoor comfort. On the Southern facade, the air temperature in the cavity was on average 0.4°C lower, with peak reductions reaching 5.5°C, while the Northern facade experienced a smaller cooling effect, averaging 0.2°C with a maximum of 3.3°C.

The Southern facade's exterior surface was, on average, 1.5°C cooler, with peak reductions of 9°C, while the interior surface was 1.2°C lower on average and 2°C at maximum. The Northern facade showed reductions of 0.5°C (exterior) and 1.3°C (interior).

Indoor air temperature in the South-facing office was on average 1.2°C lower, with a maximum reduction of 2.7°C, while the North-facing office had reductions of 0.6°C on average and 1.9°C at peak.

13.1.5.3 Conclusion

The study confirmed that DSGF effectively reduces cooling demand in high-density urban environments. The results suggest that integrating VGS into building retrofits can improve thermal comfort and reduce energy consumption, particularly for South-facing facades exposed to direct solar radiation.

13.1.6 Vertical Greenery and Green Roofs: Effects on Indoor Temperature in Air-Conditioned Rooms

Hao et al. (2020) analyzed the impact of vertical greenery systems (VGS) and green roofs (GR) on indoor operative temperature in air-conditioned spaces. The study focused on two experimental rooms constructed in Xiangtan, China, a city characterized by a hot summer and cold winter climate. One room was equipped with VGS and GR, while the other served as a control room without greenery. The objective was to evaluate the extent to which these passive strategies influence thermal comfort and energy consumption.

Figure 13.8 presents a photograph of the experimental setup, showing both the VGS-GR-equipped room and the reference room for direct comparison.

13.1.6.1 Methodology

The study was conducted using two identical experimental rooms, each measuring 3.0 m × 2.5 m × 3.0 m. The VGS system consisted of a living wall with Schefflera octophylla, planted in modular planters supported by a steel frame, with an 80 mm air gap between the greenery and the wall. The green roof (GR) was installed using modular trays with Sedum linear vegetation. Both systems were irrigated using an automated watering system.

The rooms were air-conditioned and controlled at an identical set-point temperature of 26°C. Data collection was performed over an eight-day period (July 24–August 1, 2018), measuring both indoor and outdoor thermal conditions. The recorded parameters included:

- Indoor operative temperature (Top), assessed using black-globe thermometers,
- Mean radiant temperature (MRT), derived from dry-bulb and black-globe temperature readings,

FIGURE 13.8 Photograph of the experimental setup, showing the room equipped with VGS and GR alongside the reference room without greenery. (Hao et al., 2020.)

- Air temperature and relative humidity, continuously monitored,
- Surface temperatures of walls and roof in both rooms.

To complement the field study, EnergyPlus simulations were performed to assess long-term seasonal impacts and evaluate energy savings potential.

13.1.6.2 Results

The presence of VGS and GR significantly affected the indoor operative temperature. The VGS-GR room exhibited an average reduction of 0.4°C in Top, with a maximum difference of 2.1°C compared to the control room. The cooling effect was more pronounced during the daytime, with an average operative temperature decrease of 0.6°C between 7:00 AM and 7:00 PM.

The VGS and GR systems also reduced indoor temperature fluctuations, leading to a 1.1°C decrease in operative temperature oscillation, creating a more stable and comfortable indoor environment.

Energy simulations showed that the set-point temperature of the VGS-GR room could be raised by 0.76–0.84°C without compromising thermal comfort, leading to a 7–8% reduction in daytime cooling energy consumption.

13.1.6.3 Conclusion

The study confirmed that VGS and GR contribute to reducing indoor operative temperature, particularly during peak cooling hours. While the cooling effect was smaller in an air-conditioned setting compared to free-running buildings, the results indicate potential energy savings by allowing for a higher air-conditioning set-point while maintaining thermal comfort.

13.1.7 Passive Cooling Roofs for Climate Resilience: A Case Study in Camas, Spain

Palma et al. (2024) examined the impact of a double-skin roof system on indoor thermal comfort and overheating mitigation in a social housing complex in Camas, Seville. The study aimed to assess the effectiveness of passive cooling strategies, including night ventilation and evaporative cooling, in reducing indoor temperature fluctuations and thermal discomfort in naturally ventilated buildings.

Figure 13.9 provides an urban context view of the Camas social housing district before renovation, illustrating the arrangement of the buildings and their exposure to environmental conditions.

Figure 13.10 presents the double-skin roof system, detailing its key components and operational mechanism.

13.1.7.1 Methodology

The study was conducted on a multi-family social housing complex comprising six buildings with a total of 36 dwellings, each ranging from 38 to 80 m^2. Two buildings were selected for detailed monitoring—one retrofitted with a conventional roof and another with a double-skin passive cooling roof. The buildings had identical geometry, boundary conditions, and occupancy rates, ensuring comparability of results.

FIGURE 13.9 Urban context of the Camas social housing district before renovation, illustrating the building layout and environmental exposure. (Palma et al., 2024.)

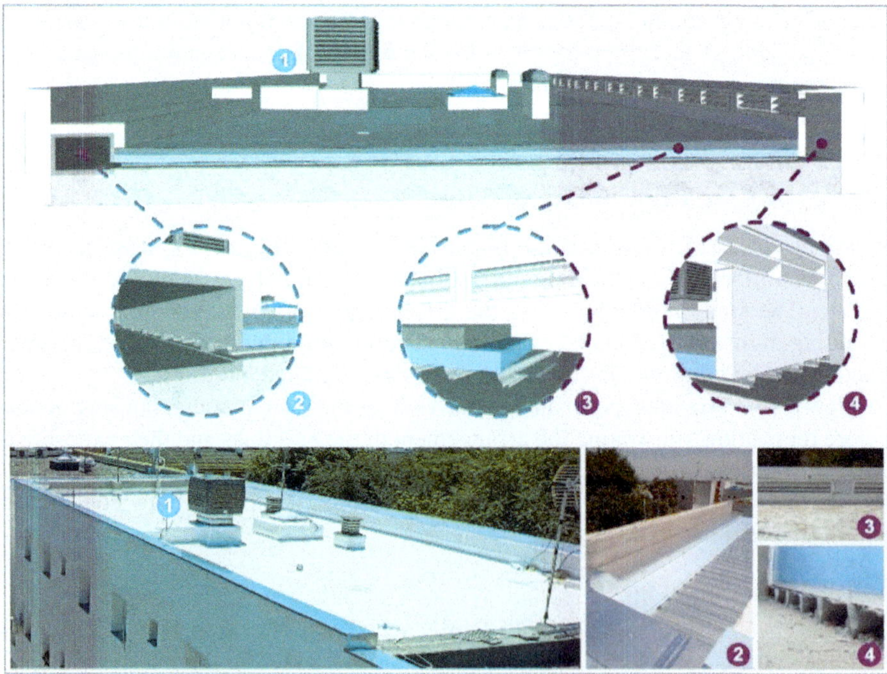

FIGURE 13.10 Schematic overview of the double-skin roof system, illustrating its components and airflow mechanisms. (Palma et al., 2024.)

The double-skin roof system was designed with an air cavity of 56 mm, a high-reflectivity outer skin (87%), and a ventilation system that integrated evaporative cooling. The roof operated in three distinct modes, alternating between passive insulation during the day, night ventilation when outdoor temperatures allowed for natural cooling, and evaporative-assisted cooling when temperatures exceeded 22°C. A fully automated control system regulated airflow based on real-time environmental conditions.

To assess the performance of the retrofit, a monitoring campaign was conducted over three years, collecting real-time data from both buildings. Measurements included indoor air temperature, ceiling surface temperature, and relative humidity, recorded at 10-minute intervals. Outdoor environmental parameters, including air temperature, humidity, wind speed, and solar radiation, were continuously monitored using a local weather station. The study also incorporated a simulation-based approach, using calibrated Building Energy Models (BEMs) developed in EnergyPlus. The models were validated against field data and used to assess seasonal variations in thermal performance and potential energy savings.

13.1.7.2 Results

The double-skin roof significantly improved thermal conditions by maintaining lower ceiling temperatures and reducing overheating. Compared to the conventional roof, ceiling temperatures remained consistently below indoor air temperature, preventing excessive radiant heat exposure. The daily difference between ceiling and indoor air temperatures was reduced by 54%, with a peak cooling effect of 3.9°C. Overheating discomfort hours decreased by 94.1%, and weighted discomfort degree-hours were reduced by 76.9% according to the Fanger model and 94.1% according to the ASHRAE adaptive model.

Energy simulations indicated that the set-point temperature for air conditioning could be increased by 0.76–0.84°C without compromising comfort. This adjustment resulted in a 7–8% reduction in daytime cooling energy consumption. Water and electricity consumption analysis showed that the operational cost of the system represented less than 3% of the minimum monthly income in the social housing district. The evaporative cooling system used an average of 35 L of water per day, with peak consumption of 80 L during extreme heat waves.

13.1.7.3 Conclusion

The study confirmed that a double-skin roof system effectively mitigates overheating in naturally ventilated buildings, particularly in social housing where mechanical cooling is often unavailable. The system is cost-effective, seamlessly integrates into existing structures, and enhances building resilience against future climate challenges.

13.1.8 Renovation Strategies for Shanghai Residential Districts: Microclimate and Energy Simulations

Yu et al. (2023) investigated the impact of renovation strategies on the air-conditioning load and microclimate of existing residential districts in Shanghai. The study assessed how four key strategies—increasing the greening rate, improving pavement reflectivity and permeability, increasing wall and roof reflectivity, and upgrading window

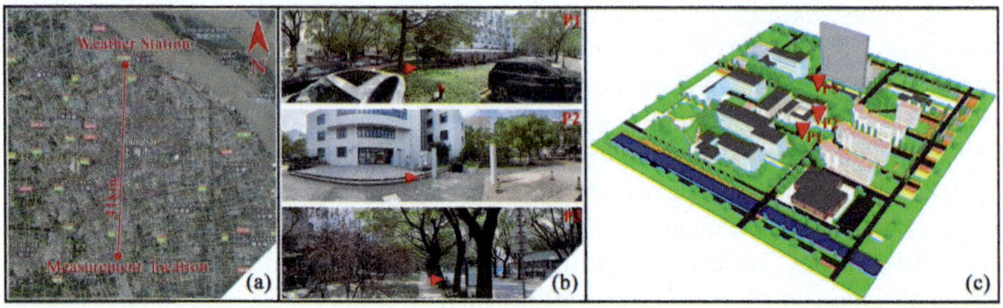

FIGURE 13.11 (a) Map indicating the measurement site and its distance (approx. 4.1 km) from the reference weather station used to define simulation boundary conditions; (b) photographs of the three on-site measurement points (P1–P3) at 1.5 m height, representing grassed, paved, and shaded pedestrian locations; (c) three-dimensional model of the study area applied in the microclimate simulation, showing buildings, vegetation, and the marked measurement points.

thermal performance—affect building cooling and heating demand through a coupled simulation approach using ENVI-met and EnergyPlus.

Figure 13.11 presents an overview of the studied residential district and the ENVI-met simulation model, illustrating the layout of buildings and the urban context.

Figure 13.12 provides a 3D representation of the target building used in EnergyPlus simulations, highlighting key parameters for energy performance evaluation.

13.1.8.1 Methodology

The study focused on a typical residential district model consisting of 12 buildings oriented north–south, each measuring 48 m in length, 12 m in width, and 18 m in height. The buildings were separated by 20 m (north–south) and 12 m (east–west), ensuring a representative urban layout for microclimate assessment.

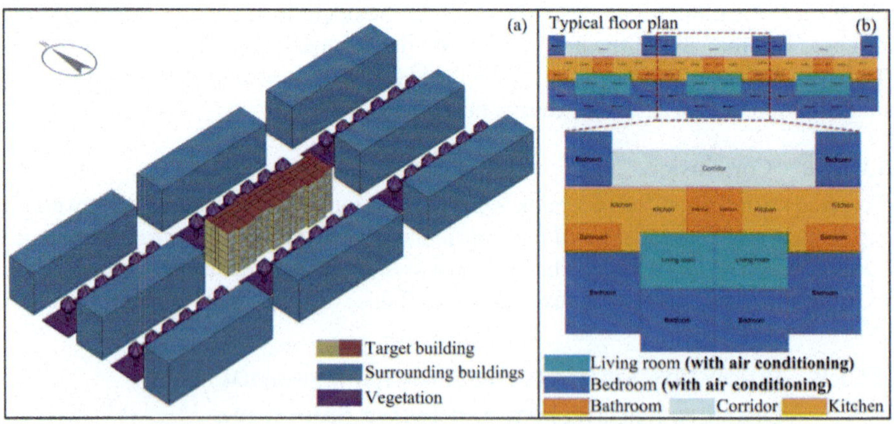

FIGURE 13.12 (a) Three-dimensional representation of the EnergyPlus model showing the target building in relation to surrounding buildings and vegetation; (b) typical floor plan of the target building used for energy consumption simulation, including functional zoning and air-conditioned spaces.

Microclimate conditions were simulated using ENVI-met 5.1.1, integrating meteorological data from the Baoshan District Weather Station in Shanghai. Air temperature, relative humidity, wind speed, and solar radiation were applied as boundary conditions, and the simulation domain covered 408 m × 276 m × 54 m with a 4 m × 4 m × 1.5 m resolution grid.

The energy consumption analysis was conducted using EnergyPlus 22.1, where weather files were modified to include microclimate data from ENVI-met. The target building was simulated under various renovation scenarios, adjusting key envelope properties, solar reflectivity, and vegetation coverage. The performance of different strategies was evaluated based on their effect on cooling and heating loads, comparing results obtained using conventional weather station data and microclimate-adjusted inputs.

13.1.8.2 Results

The study evaluated the impact of various renovation strategies on building energy demand, with window upgrades (W) and reflective roofs (R4) demonstrating the most significant reductions in both cooling and heating loads. The integration of high-performance windows (W3), high-reflectivity roofs (R3), and enhanced wall coatings (C3) resulted in a 25.1% decrease in cooling demand, representing the most effective combination. While green roofs (R4) contributed to cooling by lowering roof surface temperatures, their influence was less pronounced compared to reflective coatings. Similarly, vegetation (G3) exhibited a limited effect on cooling demand despite its substantial role in modifying local air temperature.

The renovation measures also influenced heating loads, with high-performance windows (W3) reducing heating demand by 9.2%, marking the most effective intervention for winter energy savings. Green roofs (R4) further provided a 4.2% reduction in heating loads, enhancing insulation performance. However, the application of high-reflectivity coatings on walls and roofs led to increased heating demand, emphasizing the trade-off between cooling efficiency and winter thermal retention.

The integration of microclimate-adjusted weather data in energy simulations revealed notable discrepancies when compared to conventional weather station inputs. Cooling load reductions were underestimated by up to 12.6% when using standard datasets, while heating load reductions were overestimated. These findings underscore the importance of incorporating localized microclimate effects into energy modeling to improve the accuracy of performance assessments.

13.1.8.3 Conclusion

The study confirmed that building envelope improvements, particularly window upgrades and reflective roofing, significantly reduce cooling and heating energy demand. While greening strategies improve the urban microclimate, their direct impact on building energy consumption is limited. The findings underscore the importance of integrating microclimate data into energy simulations to enhance the accuracy of renovation strategy evaluations.

FIGURE 13.13 Overview of the experimental setup, illustrating the three tested cubicles with the green wall (GW), double-skin green facade (GF), and reference system. (Coma et al., 2017.)

13.1.9 Green Walls vs Green Facades: A Comparative Study on Energy Savings

Coma et al. (2017) investigated the thermal performance and energy-saving potential of vertical greenery systems (VGS) by comparing two configurations: a double-skin green facade (GF) with deciduous climbing plants and a green wall (GW) with evergreen shrubs. The study assessed the effectiveness of these systems during both cooling and heating periods using experimental house-like cubicles under Mediterranean continental climate conditions.

Figure 13.13 presents an overview of the experimental setup, showing the three tested cubicles, including the green wall system (GW), the double-skin green facade (GF), and the reference cubicle without greenery.

13.1.9.1 Methodology

The experiment was conducted in Puigverd de Lleida, Spain, using three identical 3 × 3 × 3 m cubicles, each with the same wall and roof construction. The reference cubicle had no greenery, while the GF system was installed with a 25 cm air gap and supported by a wire mesh, allowing climbing plants to form a double-skin. The GW system consisted of polyethylene modules filled with coconut fiber, supporting dense evergreen shrubs with an integrated micro-irrigation system.

The study assessed thermal performance and energy savings by applying two distinct testing conditions. In the controlled temperature mode, an HVAC system maintained indoor conditions at predefined setpoints, with temperatures of 18°C, 21°C, or 24°C during cooling and 22°C during heating. This approach ensured a stable indoor environment, allowing for precise evaluation of energy demand under regulated conditions. In contrast, the free-floating mode operated without HVAC intervention, enabling the analysis of passive cooling effects and thermal lag responses under naturally fluctuating indoor temperatures.

The monitored parameters encompassed indoor air temperature, wall surface temperature, air temperature within the greenery layers, relative humidity, global solar radiation, and HVAC energy consumption. These variables provided a comprehensive dataset for assessing the thermal behavior of the building under both active and passive conditions.

13.1.9.2 Results

The study demonstrated the significant cooling benefits of the Vertical Green System (VGS), with the Green Wall (GW) system achieving the highest energy savings. In terms of cooling performance, the GW system reduced cooling energy demand by up to 58.9%, while the Green Façade (GF) system achieved a 33.8% reduction. Peak energy savings were observed during periods of high solar irradiance, confirming that shading and evapotranspiration played a crucial role in cooling efficiency. Additionally, the GW air chamber temperature was found to be 6°C lower than that of the GF system, highlighting the enhanced insulation effect provided by dense vegetation and substrate layers.

The impact on heating performance varied between the two systems. The GF system, featuring deciduous plants, did not increase heating demand, as it allowed solar radiation to reach the façade during winter months. Meanwhile, the GW system contributed to slight heating energy savings (4.2%), primarily due to its insulative effect at night, which helped reduce radiative heat losses.

In terms of thermal lag and stability, both GF and GW systems effectively lowered wall surface temperatures, with reductions reaching up to 21.5°C on the south-facing façade for the GW system. Furthermore, the GW system extended thermal lag by 9.7 hours, helping to delay peak heat loads into the evening, thus enhancing overall thermal stability.

13.1.9.3 Conclusion

The study confirmed that VGS significantly improves thermal performance, particularly in hot summer climates where shading and evapotranspiration enhance cooling efficiency. The GW system outperformed the GF system in terms of energy savings and thermal stability, while the GF system provided winter benefits by maintaining solar exposure. These findings highlight the potential of VGS as a passive cooling and insulation solution for energy-efficient buildings.

13.1.10 Roofing Mitigation Strategies: Impact on Temperature and Energy in Chicago during a Heat Wave

13.1.10.1 Introduction

Tan et al. (2023) conducted a regional-scale modeling study to evaluate the impact of three widely used roofing strategies-cool roofs, green roofs, and photovoltaic (PV) panel roofs-on urban climate and cooling energy demand. The study focused on the Chicago Metropolitan Area (CMA) during a heat wave event, utilizing a high-resolution regional climate model coupled with a multilayer building energy model (BEP + BEM in WRF v4.3.1). The study aimed to assess reductions in near-surface temperature, urban heat island intensity (UHII), and cooling energy demand under different roofing scenarios. The conceptual framework of the study and the modeling system setup are illustrated in Figure 13.14.

13.1.10.2 Methodology

The study utilized the Weather Research and Forecasting (WRF) model, integrated with a multilayer urban energy model (BEP + BEM), to evaluate the impact of roofing mitigation strategies on a city-wide scale. Simulations were conducted for a

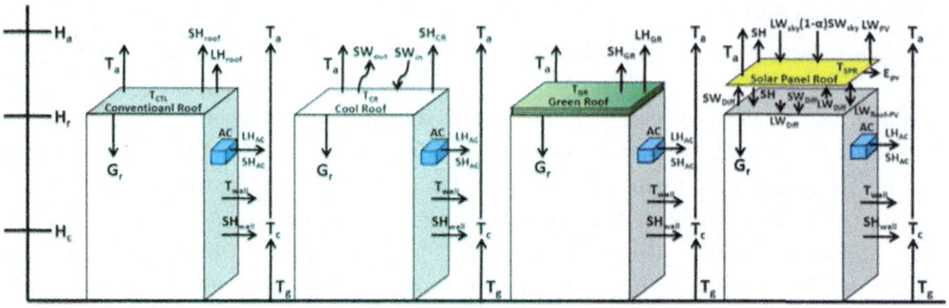

FIGURE 13.14 Schematic representation of urban grid cells in WRF model.

six-day heat wave event in August 2021, during which peak air temperatures reached 35°C. The analysis focused on three distinct roofing strategies:

1. Cool Roofs (CR)—Increased surface reflectivity (albedo = 0.8) to minimize absorbed solar radiation.
2. Green Roofs (GR)—Enhanced evapotranspiration through a 0.3 m soil and vegetation layer, improving rooftop cooling.
3. Solar Panel Roofs (SPR)—Photovoltaic panels mounted 80 cm above rooftops, reducing direct heat absorption while simultaneously generating electricity.

To achieve high spatial accuracy, the study employed a three-level nested modeling approach, with resolutions of 4.5 km, 1.5 km, and 500 m, allowing for a detailed representation of urban heat dynamics. Urban surface classifications and land use data were derived from WUDAPT datasets, ensuring realistic urban parameters.

Model validation was conducted using observational weather station data, confirming the accuracy of temperature and wind speed predictions. These results underscore the effectiveness of high-resolution urban climate modeling in assessing mitigation strategies at the city scale.

13.1.10.3 Results

The study demonstrated that all three roofing strategies effectively reduced near-surface temperature and cooling energy demand, with cool roofs achieving the greatest overall mitigation effects. Cool roofs lowered near-surface temperature by 1.5°C, green roofs by 1.2°C, and PV panel roofs by 0.6°C. The maximum daytime reduction in urban heat island intensity (UHII) was 3.9°C for cool roofs, 2.5°C for green roofs, and 2.3°C for PV panel roofs. However, green roofs caused nighttime warming due to retained heat in soil layers, while cool roofs had minimal impact on nighttime temperatures.

In terms of cooling energy demand, cool roofs reduced city-wide air-conditioning consumption by 16.6%, green roofs by 14.0%, and PV panel roofs by 7.6%. If all electricity generated by PV panels were exclusively used for cooling, potential savings could reach 46.7%. Cool roofs exhibited the highest efficiency in reducing daytime cooling loads, while PV panels provided long-term energy benefits by offsetting electricity demand.

Regarding the effects of urbanization on mitigation strategies, cool roofs were the only approach that maintained cooling effectiveness as urbanization increased, whereas green roofs led to higher humidity levels, limiting their efficiency. Additionally, higher urban fractions amplified the impact of cool roofs, reinforcing their role as the most effective urban heat mitigation strategy.

13.1.10.4 Conclusion

The study concluded that cool roofs are the most effective and cost-efficient roofing strategy for mitigating urban heat and reducing cooling energy demand. They require no additional water consumption and are cheaper than green roofs and PV systems. However, their performance may degrade over time due to material aging and dirt accumulation. Green roofs enhance urban greenery and air quality benefits but increase water consumption and humidity levels, reducing their efficiency in extreme heat. PV panel roofs offer long-term energy savings but are less effective in directly reducing urban temperatures unless the generated electricity is exclusively used for cooling.

13.1.11 Summary

The reviewed studies assess passive cooling measures across a wide range of climates and building types, using both field measurements and dynamic simulation. Test beds include highly glazed office and administrative buildings in hot-humid and Mediterranean contexts, rooftop and outdoor experimental chambers, house-scale cubicles, social housing, and residential districts. Three families of strategies dominate the evidence base: façade shading and smart glazing, vegetated envelopes and roofs, and composite roof or district retrofits that bundle several passive measures. Reported outcomes are expressed in energy savings, air and surface temperature reductions, comfort metrics, and daylight quality, with frequent use of Spatial Daylight Autonomy (SDA) and Annual Sunlight Exposure (ASE) to balance thermal and visual performance.

13.1.11.1 Façade Shading and Smart Glazing

External shading on highly glazed façades in hot-humid Chennai was consistently effective. Full-coverage perforated fabric screens reduced cooling energy by about 18–23% in fixed configurations and 18–24% with simple control, while maintaining acceptable daylight (Spatial Daylight Autonomy [SDA] above 56%) and low glare risk (Annual Sunlight Exposure [ASE] below 15%). Payback periods were typically 5 to 7 years. Fixed aluminum fins also reduced cooling energy, commonly up to about 15%, although with higher installation complexity and longer payback.

Smart glazing delivered reliable, medium-sized gains. In Mediterranean simulations for Athens, electrochromic systems lowered primary energy by roughly 7 to 16%, with quadruple units offering the largest absolute reductions and triple units providing the best cost balance. In tropical Singapore, rooftop chamber measurements showed that Electrochromic glazing (EC) reduced cooling energy by about 19%, lowered inner glass surface temperature by 3 to 4.4°C, and improved comfort. In semi-arid Benguerir, an optimized static baseline of double low-emissivity glazing reduced cooling demand by about 21%. Adding Electrochromic glazing (EC) and Thermochromic coatings (TC) raised total savings to roughly 26%, although long payback periods remain a barrier for dynamic coatings.

13.1.11.1.1 Vegetated Envelopes and Roofs

Field data from Shanghai indicate that double-skin green façades create a cooler façade microclimate and reduce indoor air temperature by about 1.2°C on average, with peaks near 2.7°C. In Xiangtan, pairing a vertical greenery system with a green roof in air-conditioned rooms allowed the thermostat set-point to rise by about 0.8°C, enabling 7–8% daytime cooling energy savings and reducing operative temperature oscillations. Experimental cubicles in Spain reported upper-bound cooling savings for dense green walls up to 58.9% and about 33.8% for green façades. These results illustrate the potential under controlled conditions, while studies in occupied buildings typically find smaller but still meaningful reductions in indoor temperature and peak loads.

13.1.11.2 Composite Roofs and District Retrofits

A double-skin passive cooling roof in social housing in Camas, Seville, reduced overheating hours by about 94% and lowered peak ceiling-adjacent temperatures by up to 3.9°C. By enabling a set-point increase of roughly 0.8°C, it delivered 7 to 8% daytime energy savings with modest operating costs. At the neighborhood scale in Shanghai, coupled ENVI-met and EnergyPlus simulations showed that a package comprising window upgrades, reflective roofs, and high-reflectance wall coatings reduced cooling demand by about 25%. Using microclimate-adjusted weather files was essential for accurate benefit estimation, since standard weather data underestimated cooling savings by up to 12.6%.

13.1.11.3 Cross-Study Comparability and Aggregate Effectiveness

Although climates and typologies vary, the studies share a common context of high solar gains and, frequently, high window-to-wall ratios. Metrics are broadly comparable: energy savings, indoor and surface temperature reductions, comfort indicators such as Predicted Mean Vote and hours above fixed thresholds, and daylight balance via SDA and ASE. The findings are consistent across sources. External shading provides the most repeatable energy savings on highly glazed façades, typically 15–25%, with textile screens offering favorable payback. Electrochromic (EC) glazing yields dependable energy and comfort improvements of roughly 7–19%, with cost-effectiveness dependent on climate, glazing ratio, and tariffs. Vegetated façades and roofs usually deliver modest indoor temperature reductions and permit higher cooling set-points in conditioned spaces, yielding 7–8% savings, and produce larger effects in free-running or experimental settings. Composite roof and envelope retrofits are particularly strong in cutting overheating hours and peak temperatures. Across all cases, performance assessment is most robust when building-specific microclimate inputs are used and daylight is tracked with Spatial Daylight Autonomy (SDA) and Annual Sunlight Exposure (ASE) to balance thermal and visual outcomes.

13.2 THERMAL MASS AND PHASE CHANGE MATERIALS

Thermal mass is a fundamental passive cooling strategy that mitigates indoor temperature fluctuations by absorbing and releasing heat, reducing overheating risks, and cooling energy demand. This section examines the role of traditional high thermal inertia materials and the effects of modifying building mass through insulation adjustments.

The discussion begins with an evaluation of concrete, masonry, and earth-based materials, highlighting their effectiveness in stabilizing indoor temperatures. Studies on heavyweight walls, insulated concrete formwork (ICF), and compressed earth bricks (CEB) illustrate the benefits of increased heat capacity in warm and temperate climates. Special attention is given to the impact of removing floor insulation to enhance thermal coupling with the ground, a strategy that has shown promise in stabilizing indoor temperatures during heat waves.

The section also explores phase change materials (PCM) as an emerging thermal storage solution. Research on PCM-enhanced walls and ceilings assesses their role in absorbing excess heat during peak hours and releasing it at night. While PCM can complement traditional thermal mass, its performance depends on ventilation rates, climate conditions, and phase-transition efficiency.

Findings highlight that increased thermal mass can significantly improve indoor thermal stability, reducing overheating risks and peak temperature fluctuations. The section provides guidance on optimizing thermal mass strategies for different building typologies and climates.

13.2.1 Thermal Performance of Heavy Walls in New Zealand Houses

Bellamy and Mackenzie (2001) investigated the impact of heavy exterior walls on thermal comfort and heating energy use in New Zealand houses. The study aimed to determine how wall thermal mass affects temperature stability, overheating prevention, and energy savings, using side-by-side test buildings with identical design but different wall materials.

13.2.1.1 Methodology

The study was conducted at Lincoln University, New Zealand, using two single-room test buildings, each covering a 20 m^2 floor area, designed to replicate real residential conditions. One building featured lightweight timber-framed walls with exterior insulation, while the other incorporated exterior-insulated concrete walls with significantly higher thermal mass.

The research examined four experimental configurations, each tested over a period of at least six months. The base case trial (June–December 1998) assessed heating performance without ventilation or curtains. In the ventilation trial

(January–June 1999), a controlled ventilation system was introduced, activating when indoor temperatures exceeded 27°C. The curtain trial (July–October and December 1999) utilized motorized roller shutters to simulate the thermal effects of curtains, reducing nighttime heat loss. Finally, the bare floor trial (January–June 2000) removed carpets to expose the concrete floor, influencing heat retention and temperature regulation.

Monitored parameters included air temperature, wall surface temperature, ventilation rate, solar gain, and auxiliary heating energy consumption.

13.2.1.2 Results

The findings confirmed that high thermal mass significantly improved indoor comfort, reducing overheating and yielding energy savings under specific conditions. The concrete-walled building lowered peak indoor temperatures by up to 5°C in unventilated conditions and by 3–4°C with ventilation. Additionally, the thermal mass provided nighttime warming, with pre-sunrise temperatures in the concrete building remaining 1.5°C higher than in the timber-framed structure.

In terms of heating energy savings, heavy walls reduced heating demand by more than 50% on sunny winter days. Monthly heating energy use was generally lower in the concrete-walled building, although in months with limited ventilation, thermal mass slightly increased heating demand.

The influence of ventilation and solar gain was also notable. Ventilation energy losses were reduced by up to 23.5 kWh/month in the concrete building due to enhanced heat retention. Energy savings strongly correlated with daily solar gain, reinforcing the importance of optimizing insulation and thermal mass balance.

Additional design features further influenced performance. Curtains improved heating efficiency, leading to energy savings of 5.1% compared to unshaded conditions. Meanwhile, the removal of carpets facilitated direct heat storage in the concrete slab, reducing overheating and lowering peak indoor temperatures in both buildings.

13.2.1.3 Conclusion

The study demonstrated that exterior-insulated heavy walls enhance thermal comfort by moderating temperature fluctuations and reducing overheating risks. While energy savings depend on ventilation and solar gain, heavy walls can significantly lower heating demand in sunny regions. The findings highlight the need for integrated design strategies, ensuring that insulation, window size, and solar exposure are optimized to fully exploit the benefits of thermal mass.

13.2.2 Preventing Office Overheating with Compressed Earth Bricks: A Prototype Study

Brambilla and Jusselme (2017) investigated the thermal performance of compressed earth bricks (CEBs) as a passive cooling strategy in office buildings. The study analyzed how thermal inertia (TI) can mitigate overheating, comparing a lightweight construction (wooden structure) with a heavyweight construction (CEBs with external insulation) under controlled ventilation conditions.

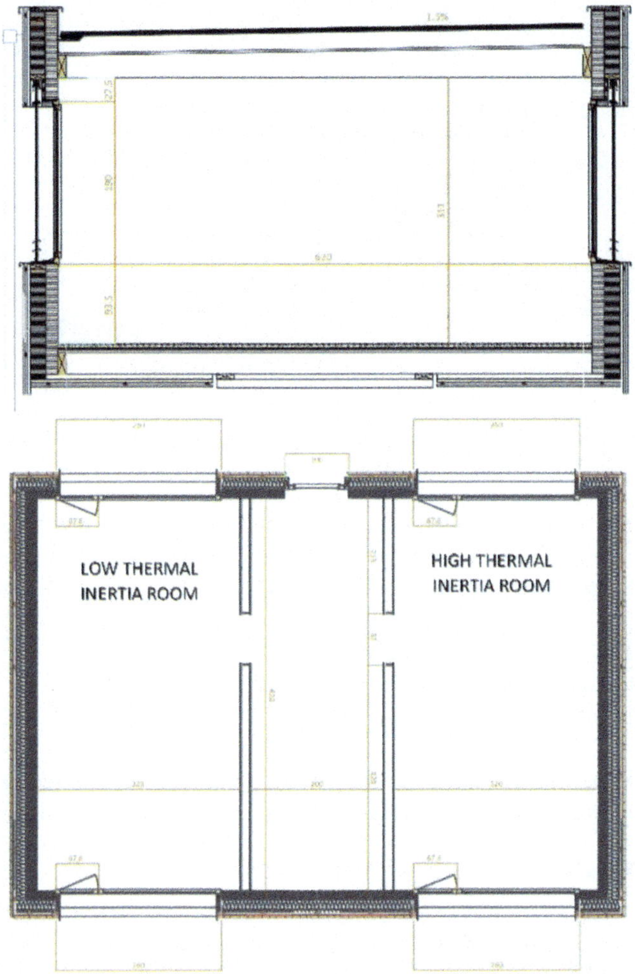

FIGURE 13.15 Plan and section of the prototype with lightweight and heavyweight test rooms. (Brambilla & Jusselme, 2017.)

Figure 13.15 presents an architectural plan and section of the prototype, illustrating the layout of the two experimental rooms. Figure 13.16 provides a visual representation of the experimental setup, showing the placement of thermal sensors and shading elements.

13.2.2.1 Methodology

The study was conducted in Fribourg, Switzerland, using two side-by-side test rooms, each 18 m^2, designed to simulate office conditions. The lightweight room (LI) featured a wooden structure insulated with 180 mm of polyurethane, while the heavyweight room (HI) had an additional 10-ton layer of CEBs to enhance thermal mass.

FIGURE 13.16 Experimental prototype with shading and sensor placement for monitoring thermal performance. (Brambilla & Jusselme, 2017.)

Both rooms were fully insulated and shaded to prevent direct solar gains, with internal heat gains simulated using electric lighting systems. The experiment lasted 21 days in August 2016, during which various ventilation strategies were tested:

- Mechanical ventilation (50 m^3/h)
- Natural ventilation through hopper windows
- No ventilation (worst-case scenario)

Monitored parameters included indoor air temperature, surface temperature at various wall layers, external climate conditions, and ventilation efficiency.

13.2.2.2 Results

The CEB room (HI) demonstrated superior thermal stability, significantly reducing indoor temperature fluctuations and delaying heat peaks. CEBs reduced maximum indoor temperatures by up to 3°C, effectively preventing overheating. The lightweight room (LI) exhibited greater temperature variability, with daily fluctuations reaching 4.4°C, compared to 3.1°C in the heavyweight room. Heat peaks in the LI room were 2.4–3.0°C higher, confirming the stabilizing effect of thermal mass.

Natural night ventilation proved to be the most effective strategy, ensuring sufficient heat dissipation from stored daytime gains. In the worst-case scenario (no ventilation), the LI room accumulated heat, leading to a continuous temperature rise over multiple days.

CEBs provided a buffering effect, maintaining more stable interior surface temperatures. Temperature differences across wall layers in the HI room were lower than in the LI room, indicating reduced heat penetration.

The thermal comfort evaluation confirmed that only 3.4% of occupied hours in the HI room exceeded comfort thresholds, compared to 7–12 times more discomfort hours in the LI room. The worst overheating conditions occurred in Scenario 4 (continuous mechanical ventilation) and Scenario 6 (no ventilation), particularly in the LI room.

13.2.2.3 Conclusion

The study confirmed that CEBs effectively mitigate overheating by enhancing thermal inertia, particularly in high-internal-gain environments like offices. The combination of high TI and night ventilation significantly reduces heat peaks and stabilizes indoor temperatures, making it an effective passive cooling strategy. These findings highlight the importance of considering thermal mass in building envelopes, especially in highly insulated structures prone to overheating.

The authors also noted that the study was conducted during a typical summer, without extreme heat waves. Given the projected increase in summer temperatures due to climate change, they suggested that future research should repeat the experiment during periods of prolonged heat waves to better assess the long-term cooling potential and cumulative overheating risks associated with CEBs in office buildings.

13.2.3 Overheating Risk in Solid Wall Dwellings with Internal Insulation

Tink et al. (2018) investigated the impact of internal wall insulation (IWI) on summertime overheating risk in solid-walled UK dwellings. The study aimed to determine whether retrofitting solid walls with internal insulation increases overheating risk and to assess the effectiveness of passive mitigation strategies such as night ventilation and shading.

Figure 13.17 presents an architectural plan and section of the test houses, illustrating the layout of the experimental setup.

13.2.3.1 Methodology

The research was conducted on a pair of semi-detached solid wall houses, built around 1910 in Leicestershire, UK. The houses were thermally matched and had identical floor plans (82 m^2), double-glazed windows, and loft insulation. One house

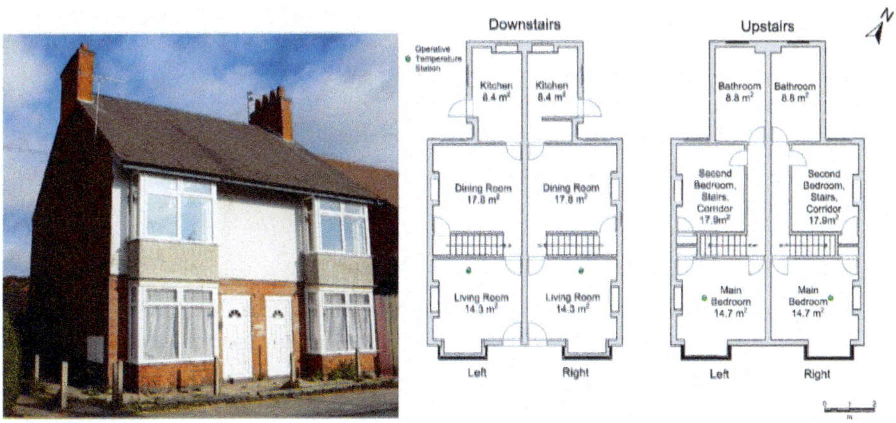

FIGURE 13.17 Architectural plan and section of the test houses, illustrating the layout of the experimental setup with the location of sensors. (Tink et al., 2018.)

was retrofitted with IWI (phenolic insulation boards), while the other remained uninsulated, serving as a control.

Three test configurations were evaluated:

1. Matched Pair Test (Baseline Condition)—Both houses were uninsulated, allowing for direct comparison.
2. Left House Insulated Test—The left house was retrofitted with IWI, while the right house remained uninsulated.
3. Mitigation Test—Passive strategies, including solar shading (closed blinds) and night ventilation (window opening), were applied to both houses.

A synthetic occupancy system simulated realistic internal heat gains, using electric heaters, lighting, and appliance loads, mimicking a two-person household. Thermal conditions were monitored at five-minute intervals, measuring operative temperature, air velocity, surface temperatures, and relative humidity.

13.2.3.2 Results

The installation of IWI significantly increased indoor operative temperatures, indicating that improved insulation reduced heat dissipation, leading to an increased risk of overheating. The insulated house consistently exhibited higher temperatures, with operative temperatures reaching 2.2°C above those in the uninsulated house in the living room and 1.5°C higher in the bedroom. Overheating thresholds were exceeded in 12.3% of occupied hours in the insulated house, compared to 4.6% in the control house.

The mitigation strategies proved effective in reducing overheating risks. Daytime shading combined with night ventilation successfully stabilized indoor temperatures, bringing nighttime conditions in the insulated house to nearly the same level as the uninsulated house (mean difference: 0.1°C). Notably, no overheating events were recorded during the mitigation test, highlighting the ability of simple passive measures to counteract overheating risks under current climatic conditions.

However, long-term climate change projections indicated increasing risks. A linear regression model forecasted a substantial rise in overheating risk after 2065, driven by increasing outdoor temperatures. By 2066–2095, passive mitigation strategies alone would no longer be sufficient to prevent overheating in living rooms on 9% of summer days, underscoring the need for additional adaptive measures in future building designs.

13.2.3.3 Conclusion

The study confirmed that IWI increases summertime overheating risk in solid wall dwellings but demonstrated that simple passive cooling measures—such as shading and night ventilation—can effectively mitigate this risk. The findings suggest that IWI remains a viable retrofit option, provided that overheating risk is addressed as part of a holistic refurbishment strategy. Future building regulations may need to mandate ventilation and shading measures to ensure thermal comfort in retrofitted homes under future climate conditions.

13.2.4 Thermal Mass, CO_2 Emissions, and Indoor Temperature in Future Climates

Hacker et al. (2008) conducted a 100-year lifecycle carbon dioxide (CO_2) emissions analysis for a two-bedroom, 65 m^2 semi-detached house in southeast England, evaluating the impact of thermal mass on energy performance under projected climate change scenarios. The study aimed to quantify how the balance between embodied and operational CO_2 emissions is affected by different levels of thermal mass and assess the long-term implications for heating, cooling, and air conditioning demand.

Figure 13.18 presents the floor plans of the case study dwelling, illustrating the internal layout and distribution of facade and window areas.

13.2.4.1 Methodology

The study analyzed four construction types with different levels of thermal mass, from lightweight timber frame to heavyweight concrete block construction. Each case had the same insulation levels, ventilation rates, and energy efficiency standards to ensure comparability. The houses were modeled using dynamic thermal

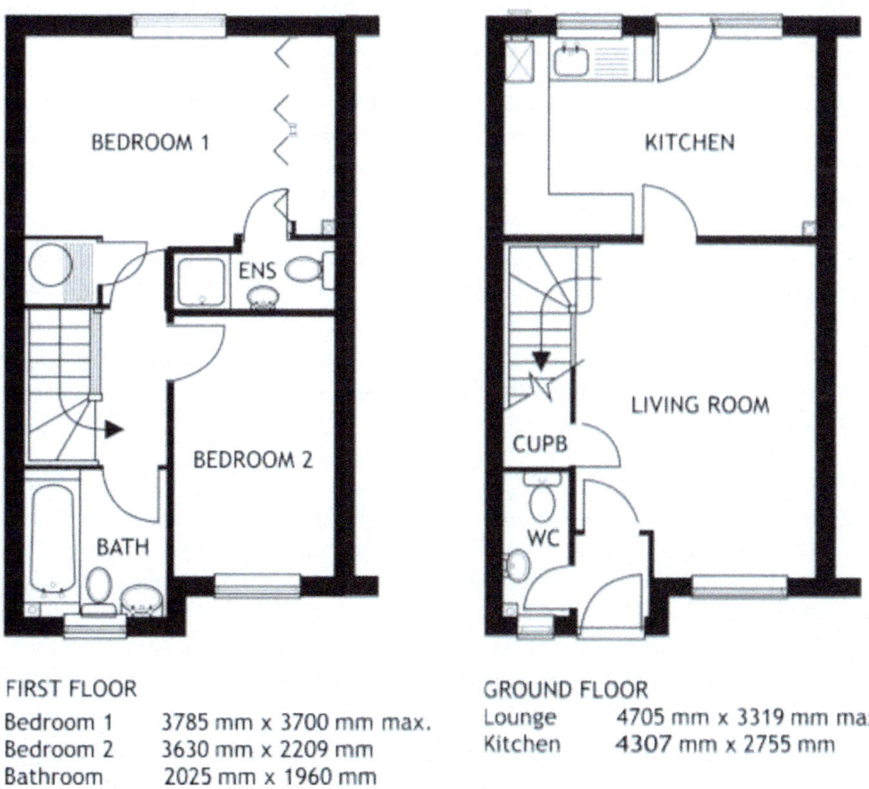

FIGURE 13.18 Floor plans of the case study dwelling, illustrating the internal layout and facade/window distribution. (Hacker et al., 2008.)

simulations under the UKCIP02 medium-high emissions climate change scenario, predicting temperature increases of up to 7°C by 2100.

The 65 m^2 semi-detached house was assumed to be occupied by two adults and a child, with space heating provided by a gas-fired condensing boiler (85% efficiency). Cooling was assessed in three modes: passive (natural ventilation at 6 ac/h), air-conditioned, and mixed-mode. Air conditioning was installed once overheating exceeded CIBSE limits for three years in a five-year period.

Operational energy modeling was performed using ENERGY 2 software, while embodied CO_2 emissions were assessed using material life-cycle data. Historical weather data from 1976 to 1995 were adjusted for climate projections up to 2100. Lifecycle CO_2 emissions were analyzed over five 20-year periods to determine the carbon payback time of heavier construction materials.

13.2.4.2 Results

The study confirmed that thermal mass significantly reduced operational CO_2 emissions by improving passive cooling and heat retention, delaying the need for air conditioning and lowering energy consumption over the building's lifetime. The heavyweight case reduced heating-related CO_2 emissions by 19–32% compared to the lightweight case, as thermal mass improved heat retention, decreasing reliance on mechanical heating during winter.

Overheating risks and cooling demand varied across construction types. The lightweight house reached overheating thresholds by 2021, leading to early air conditioning installation, while mediumweight and heavyweight houses delayed air conditioning by 20–40 years, reducing cooling-related emissions. Peak indoor temperatures in the lightweight house were up to 5°C higher than in the heavyweight case, confirming the cooling benefits of thermal mass.

Lifecycle CO_2 savings were also substantial. The heaviest construction reduced total lifecycle CO_2 emissions by up to 17% (35.7 t CO_2) over 100 years. Although it had a higher initial embodied CO_2 (ECO_2) footprint, with up to 4.93 t CO_2 more than the lightweight case, these emissions were offset within 11–25 years, depending on construction type, demonstrating the long-term sustainability benefits of incorporating thermal mass.

13.2.4.3 Conclusion

The study demonstrated that medium to heavyweight construction improves indoor thermal comfort and significantly reduces long-term CO_2 emissions, particularly in a warming climate. While heavier construction materials have a higher embodied CO_2 impact, their operational energy savings compensate for this within decades. These findings suggest that future housing should integrate thermal mass to enhance resilience against climate change and reduce lifecycle emissions.

13.2.5 Thermal Mass in Insulated Concrete Formwork: Impact on Energy and Comfort

Mantesi et al. (2019) investigated the thermal performance of Insulated Concrete Formwork (ICF) as a heavyweight construction method, analyzing its impact on

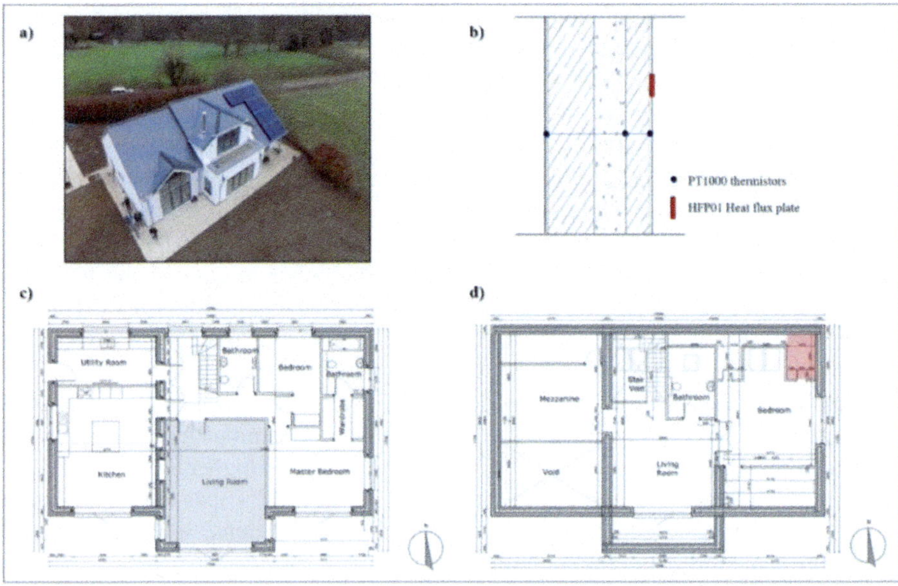

FIGURE 13.19 Southwest view and monitoring setup of the ICF case study dwelling in Guildford, UK. Panels show: (a) External perspective of the house with PV panels, (b) wall section with sensors and heat flux plate, (c) ground floor plan, and (d) first floor plan with monitored room highlighted. (Mantesi et al., 2019.)

internal air temperature, heat flux, and energy performance. The study combined empirical monitoring and calibrated simulations to evaluate the transient heat storage capacity of ICF walls and assess whether ICF functions thermally as a heavyweight or lightweight structure.

Figure 13.19 presents a southwest view of the case study building, highlighting its structural elements and monitoring setup.

13.2.5.1 Methodology

The study was conducted in a three-bedroom, two-story ICF house in Guildford, UK, designed to achieve near Passivhaus standards. The building envelope comprised ICF walls, an insulated foundation raft, a prefabricated concrete hollow-core slab, and EPS roof panels.

A 20-month monitoring campaign recorded internal air temperature, heat flux, surface temperatures, and energy consumption. Temperature sensors were placed across the ICF wall layers to evaluate heat transfer dynamics. Weather data, including solar radiation, wind speed, and humidity, were collected on-site to ensure accurate boundary conditions for simulation calibration.

The EnergyPlus 8.6 simulation model was calibrated against the measured data and validated using Root Mean Square Error (RMSE) and Mean Biased Error (MBE) metrics. To assess the thermal impact of ICF, the model compared ICF walls to low thermal mass (LTM) timber-frame walls and high thermal mass (HTM) concrete walls under identical insulation and ventilation conditions. A parametric analysis investigated the influence of floor construction on overall thermal inertia, testing lightweight, medium, and heavyweight ground floors.

13.2.5.2 Results

The study confirmed that ICF walls provide a stable internal environment, reducing heat flux variability and moderating internal temperature swings. Internal air temperature in the ICF house was 2°C lower in summer compared to the LTM house, demonstrating the benefits of thermal inertia. The concrete core of the ICF wall remained stable at 16°C, acting as a buffer to external heat gains and losses. The ICF decrement factor was similar to that of LTM walls, indicating that its thermal storage effect occurs over longer cycles rather than daily fluctuations.

Heat flux measurements confirmed that ICF walls reduced heat losses by 37% compared to LTM walls with the same insulation, confirming the concrete core's role in thermal energy storage. The LTM building exhibited higher temperature fluctuations, with heat gains and losses fluctuating by up to 12°C per day. The ground floor thermal mass significantly influenced indoor conditions, with heavyweight floors further stabilizing internal temperatures.

Sensitivity analysis revealed that the density, specific heat capacity, and thickness of the ICF concrete core were the most influential factors affecting internal air temperature. Standard energy assessment methods (SAP2012, ISO 13790) underestimate ICF thermal mass, leading to inaccurate regulatory compliance calculations.

13.2.5.3 Conclusion

The study demonstrated that ICF walls contribute to thermal mass effects, despite the presence of internal insulation. The concrete core effectively buffers heat transfer, stabilizing indoor conditions and reducing heating and cooling energy demand. While ICF walls do not provide the same short-term temperature regulation as exposed heavyweight materials, their long-term thermal storage capacity can enhance energy efficiency in well-insulated buildings. These findings highlight the need for advanced simulation models to accurately capture the thermal behavior of modern construction systems.

13.2.6 Medium-Weight vs Lightweight Walls in Passive Buildings: Experimental Study in Zielona Góra

Summary Box—Medium-Weight vs Lightweight Walls

- Masonry walls (B1) maintained indoor temperatures up to 3.4°C lower than lightweight timber walls (B2) during a prolonged heat wave.
- Overheating hours (>26°C) were reduced by more than 60% in B1.
- Cooling demand decreased by 67–75% compared to B2.
- Thermal mass demonstrated strong buffering capacity, improving comfort without mechanical cooling.

13.2.6.1 Introduction and Experimental Design

In temperate European climates, the resilience of residential buildings to summer overheating has become a growing concern. While low-energy dwellings emphasize airtightness and insulation to reduce winter heat loss, these features may exacerbate overheating during heat waves. As extreme events increase, passive measures such as increased wall mass attract renewed attention.

To investigate the buffering potential of wall mass under real conditions, Kuczyński and Staszczuk (2020) conducted a full-scale study of two single-family houses in western Poland. Both had identical floor plans, areas (≈123 m^2), orientation, roof geometry, and insulation. The sole distinction was construction: Building B1 used masonry with high storage capacity, whereas B2 employed a lightweight timber-frame with insulated walls. This side-by-side setup enabled direct comparison of thermal mass effects.

As shown in Figure 13.20, the houses were built adjacent to each other with equal solar exposure. Each had a single story with unheated attic and identical room layouts (Figure 13.21). Five temperature sensors at 1.2 m monitored key rooms, averaged to represent indoor conditions. Windows remained unshaded, ventilation was fixed at 0.6 ACH, and no internal gains were introduced.

Although wall U-values were comparable, their thermal capacity differed sharply. B1's cellular concrete with lime-sand plaster offered more than twice the heat storage of B2's mineral wool and sheathing system. Other components (roof, floor slab) were kept consistent. The layering of the lightweight assembly is illustrated in Figure 13.22, highlighting the contrast in thermal inertia between the two structures.

13.2.6.2 Results and Discussion

The monitoring campaign in July and August 2015 recorded an intense heat wave between 3 and 16 August, when outdoor temperatures exceeded 31°C for 11 consecutive days under high solar radiation and limited nocturnal cooling. From the

FIGURE 13.20 External view of the two experimental buildings constructed in Nowy Kisielin: B1 (left) with masonry walls, B2 (right) with timber-frame walls. (Kuczyński & Staszczuk, 2020.)

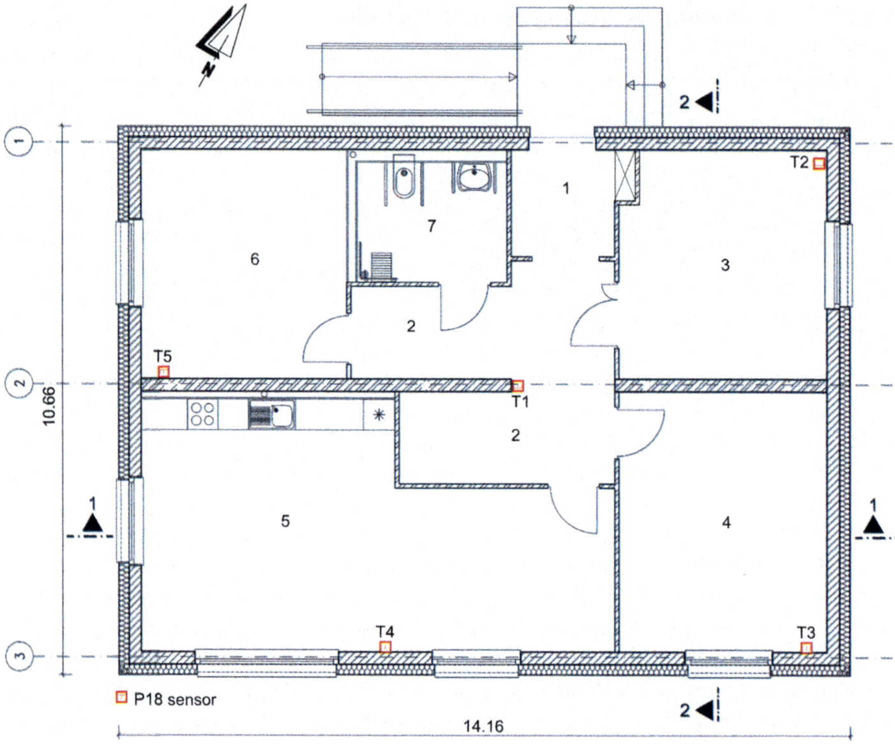

FIGURE 13.21 Ground floor plan of the experimental buildings, showing room layout and positions of five temperature sensors used for analysis. (Kuczyński & Staszczuk, 2020.)

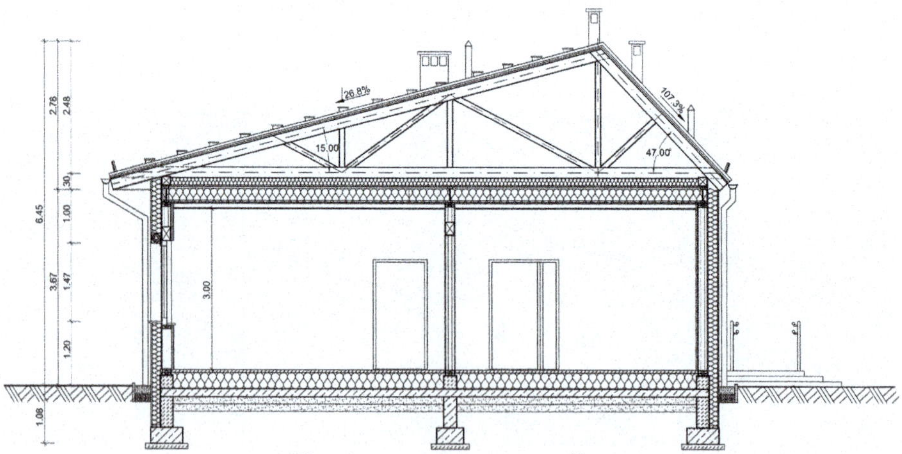

FIGURE 13.22 Vertical section of building B2, showing construction layers of walls, floor, and roof in the lightweight structure. (Kuczyński & Staszczuk, 2020.)

first days of the event, clear differences in thermal performance were observed. The masonry building B1 showed a slower and more moderate rise in indoor temperature than the lightweight timber structure B2. During the 14-day period, the average indoor temperature in B1 was 26.1°C, compared to 28.9°C in B2. At peak hours, the difference reached 3.4°C, which represents a meaningful improvement in comfort conditions. B1 also stabilized more quickly at a lower equilibrium, while B2 continued to accumulate heat and remained above 30°C for several days.

An analysis of overheating thresholds confirmed these findings. Using 25°C and 26°C as reference values in line with standards for temperate climates, B2 exceeded these limits for 1056 and 849 hours, respectively, while B1 recorded 526 and 306 hours. Moreover, B2 experienced over 18 days above 28°C, whereas B1 reached that level for a total of only eight hours. These results show that even a moderate increase in wall mass can significantly reduce thermal stress. Although nighttime differences were smaller, B1 consistently remained within acceptable comfort ranges. The gradual release of stored heat slightly slowed nocturnal cooling but was outweighed by the reduced daytime accumulation of heat.

The heavyweight structure also demonstrated greater stability. Daily indoor amplitude averaged 1.4°C in B1 and 1.8°C in B2, compared to outdoor fluctuations of 16.7°C. The decrement factor ranged between 0.07 and 0.12 in B1 and between 0.09 and 0.14 in B2. Although no significant time lag was observed, most likely due to the dominance of direct solar gains and the absence of shading, the flattened temperature curve in B1 indicated effective buffering capacity.

The thermal differences translated into substantial variations in estimated cooling demand. At setpoints of 25°C and 26°C, B2 required 1055 kWh (8.6 kWh/m^2) and 720 kWh (5.9 kWh/m^2), while B1 required 338 kWh (2.8 kWh/m^2) and 188 kWh (1.5 kWh/m^2). This represents a reduction of 67–75% achieved solely by increasing wall mass. Such savings are particularly important in Poland, where mechanical cooling is still uncommon but increasingly in demand. The ability to delay or even avoid the use of air conditioning through thermal mass has clear environmental and energy implications, especially when combined with complementary strategies such as shading or night ventilation.

These results are consistent with other studies. Brambilla and Jusselme (2017) observed indoor temperature reductions of 2–3°C using compressed earth bricks in Switzerland, while Mantesi et al. (2019) reported a 2°C moderation in a UK insulated concrete formwork test house. The present experiment not only confirms these effects under more severe conditions but also extends them by showing that in full-scale residential buildings exposed to prolonged heat waves, passive buffering provided by wall mass alone can secure substantial thermal comfort benefits without mechanical systems.

13.2.6.3 Implications for Design and Policy

This study confirms that thermal mass should not be regarded as an auxiliary feature but as a central element of climate-resilient design. Masonry structures in temperate climates can significantly reduce overheating and cooling demand under heat wave conditions, thereby supporting both comfort and energy efficiency. Integrating thermal mass with shading and ventilation strategies provides a low-cost and durable response to increasing summer heat stress in European residential buildings.

13.2.7 Ground Coupling and Indoor Thermal Performance: Modified Floor Insulation in a Masonry Building

Summary Box—Ground Coupling and Thermal Stability

- Removal of floor insulation and increased wall mass in B1 lowered summer indoor temperatures by 5.4°C compared to lightweight B2.
- B1 maintained indoor conditions below 26°C throughout the hottest summer on record in Poland.
- Overheating was eliminated in B1, while B2 exceeded comfort thresholds during more than 75% of occupied hours.
- Calibrated simulations confirmed robustness of passive buffering even with internal heat gains.

13.2.7.1 Introduction and Experimental Design

Following earlier investigations into the role of wall mass, this study examined whether coupling a building to the ground could enhance indoor thermal stability during extreme summer events. Two full-scale single-story experimental houses located in western Poland were used, with identical layouts, orientation, roof form, and glazing. Building B2 remained a lightweight timber-frame structure with insulated floors, while Building B1 was modified by removing 30 cm of floor insulation and replacing it with a solid concrete slab. This intervention increased its Thermal Mass Parameter (TMP) from 400 to 467 kJ/m^{2}K, classifying it as a heavyweight structure according to ISO 13786.

Both houses were unoccupied, with windows unshaded, identical ventilation settings, and no internal heat gains introduced. Indoor air temperatures were monitored at five points at 1.2 m height, while in B1, additional sensors were placed in the soil beneath the slab to capture ground temperature profiles. The study was conducted during the summer of 2019, the hottest ever recorded in Poland, when the mean outdoor temperature reached 20.9°C, almost 4°C above the long-term norm. Maximum values exceeded 30°C on 28 days and surpassed 32°C on 15 occasions, providing a stringent test for passive resilience.

13.2.7.2 Results and Discussion

The results revealed striking differences between the two structures. Across the summer, the average indoor temperature in B1 was 22.4°C, compared to 27.8°C in B2, a persistent gap of 5.4°C. During June, when outdoor mean temperature reached 22.7°C and daily peaks consistently exceeded 30°C, B1 maintained indoor values between 21.8 and 22.9°C. In contrast, B2 averaged 29.3°C, with highs of 30.0°C and night lows of 28.5°C. The daily difference between the two houses during this period often reached 8°C.

The July heat wave confirmed these patterns. In B1, the mean indoor temperature remained 22.7°C with fluctuations between 22.0 and 23.3°C. In B2, average indoor conditions were 27.9°C with daily maxima of 28.7°C and minima around 26.9°C. The diurnal amplitude in B2 was nearly 2°C, while in B1 it remained below 1.5°C, showing stronger damping of external loads.

The late August event provided the most severe test. While B2 reached 29.7°C indoors and averaged 28.7°C, B1 maintained a mean of 23.7°C, with peaks rarely above 24.5°C. Even during this period, when outdoor temperatures surpassed 32°C for consecutive days, B1 stayed below the 26°C comfort threshold set in European summer standards.

The decrement factor analysis supported these findings. B1 exhibited values of 0.07 in June, 0.10 in July, and 0.11 in August, while B2 ranged from 0.10 to 0.14. These modest numerical differences translated into a much greater stability of perceived comfort in B1. Nighttime recovery was particularly important. In B1, minimum night temperatures remained around 22–23°C, enabling full thermal relief. B2, by contrast, stayed above 26.9°C at night, preventing recovery and contributing to cumulative discomfort, especially in bedrooms.

A thermal comfort assessment based on CIBSE TM52 and TM59 methodologies was applied. With thresholds of 26°C for bedrooms and 28°C for the living room, B1 recorded no hours of exceedance across the entire summer. In B2, overheating was severe, occurring during more than 75% of occupied hours in all monitored rooms, reaching 82.6% in the living room. These differences illustrate the decisive impact of floor–ground coupling on passive survivability.

Dynamic simulations calibrated against measured data further validated the results. Calibration achieved high accuracy with correlation coefficients between 0.88 and 0.98. Incorporating internal heat gains of 5 W/m^2 led to average increases of 2°C in both houses, but even under these conditions, B1 remained below comfort thresholds. In B2, overheating intensified, with bedrooms exceeding 26°C during more than 77% of occupied hours and the living room surpassing 28°C for 82.6% of occupancy.

Comparisons with other studies highlight the significance of these findings. Traditional masonry dwellings in Portugal reported mean summer indoor temperatures of 23.9°C (Fernandes et al., 2015), similar to B1. Cardinale et al. (2013) observed stable indoor conditions in Matera and Alberobello, where indoor values remained below 28°C despite outdoor temperatures above 35°C. In contrast, Roque et al. (2021) found frequent overheating in a small single-room structure with uninsulated floors, underscoring that scale and envelope configuration matter.

Parallel evidence comes from retrofitting studies. Porritt et al. (2011) showed that a combination of insulation, reflective coatings, blinds, and ventilation reduced peaks by up to 4.6°C in a Victorian terraced house. O'Donovan et al. reported reductions of 2.9–3.5°C in a Budapest office using shading and night cooling. Silva et al. (2022) projected national cooling energy reductions of 84% in Switzerland from similar passive strategies. Closer to the Polish context, Kuczyński and Staszczuk (2021) found that combining thermal mass, shading, and night ventilation reduced cooling loads by more than 90% and peak temperatures by over 7°C. The present experiment confirms that even a single measure, namely ground coupling in a masonry building, can deliver comparable protection against overheating.

The climatic context reinforces these conclusions. The summer of 2019 corresponded closely to IPCC SSP 6.0 projections for late-century Central Europe, indicating that B1's performance offers a realistic vision of resilience under future climate conditions. Its total heat loss coefficient, 0.981 W/m^{2}K, was comparable to contemporary high-efficiency buildings in London (Yannas & Rodríguez-Álvarez, 2020), demonstrating that removing horizontal floor insulation, when combined with perimeter detailing, need not compromise winter efficiency.

These results support the view advanced by De Masi et al. (2021), Norouzi et al. (2023), and Yannas and Rodríguez-Álvarez (2020) that envelope performance should be assessed dynamically, focusing on time-dependent thermal behavior rather than static U-values. The experiment provides empirical evidence that even moderate changes in mass distribution and ground coupling can yield major resilience benefits in real buildings.

13.2.7.3 Implications for Design and Policy

The findings provide robust confirmation that coupling masonry structures to the ground can eliminate overheating risk under extreme summer conditions. Building B1 maintained safe indoor temperatures throughout the hottest summer on record in Poland without mechanical cooling, while B2 showed severe overheating. For practice, this suggests that moderate increases in wall mass, combined with uninsulated ground slabs and perimeter detailing, provide a cost-effective and regulation-compliant solution to improve resilience in residential construction.

The results underline the necessity of moving from static envelope criteria to dynamic performance indicators that capture how buildings behave under real heat stress. Ground coupling, combined with thermal mass and complementary strategies, should be recognized as a central element of low-carbon housing design, capable of delivering both present comfort and long-term adaptation to a warming climate.

13.2.8 Phase Change Material for Wall Applications: An Experimental Assessment

Phase change materials (PCM) have been increasingly studied for their ability to enhance thermal storage and reduce temperature fluctuations in lightweight buildings. The study conducted by Kuznik and Virgone (2009) aimed to evaluate the effectiveness of PCM wallboards in improving indoor thermal comfort and reducing overheating under different seasonal conditions. The research focused on comparing indoor temperatures in a test room with and without PCM integration, analyzing its ability to store and release heat.

13.2.8.1 Methodology

The experiment was conducted in a full-scale controlled test facility (MINIBAT) at the National Institute of Applied Sciences in Lyon (CETHIL-INSA de Lyon, France). The test setup consisted of a single experimental room measuring 3.10 m × 3.10 m × 2.50 m, enclosed on five sides by temperature-controlled air volumes simulating adjacent spaces, while the sixth side featured a glazed facade allowing controlled solar exposure. A solar simulator with twelve 1000 W metal halide lamps was used to replicate real-world radiative heat flux, ensuring repeatable test conditions. The layout of the test room and measurement setup is shown in Figure 13.23.

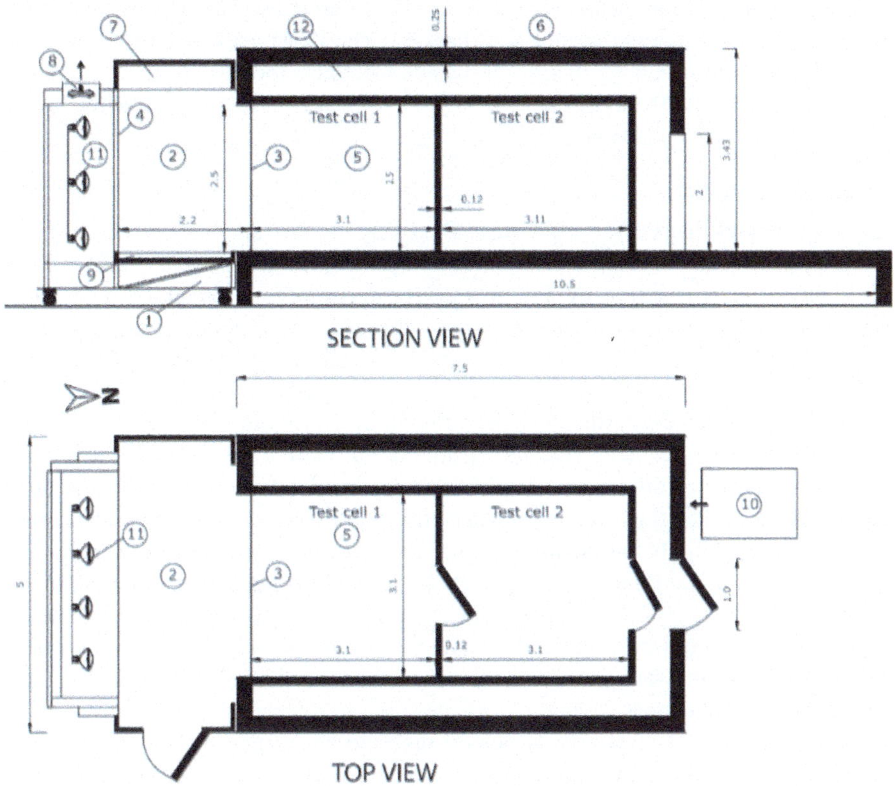

FIGURE 13.23 Layout of the MINIBAT experimental test cell at INSA Lyon, showing the test room, temperature-controlled air volumes, and glazed facade used for solar exposure. The placement of PCM wallboards and temperature sensors is also indicated.

The PCM wallboards used in the study were composed of ENERGAIN®, a composite containing 60% microencapsulated paraffin within a copolymer matrix. The material had a melting temperature range of 13.6–23.5°C, with a latent heat storage capacity of 107.5 J/g during melting and 104.5 J/g during freezing. PCM sheets were 5 mm thick, applied to the walls, and had a thermal conductivity of 0.22 W/mK (liquid phase) and 0.18 W/mK (solid phase).

The study examined three different seasonal scenarios:

- Summer conditions: External temperatures fluctuated between 15°C and 30°C, with night ventilation activated when temperatures dropped below 22.5°C to enhance heat dissipation.
- Mid-season conditions: External temperatures varied between 10°C and 18°C, representing moderate climate conditions.
- Winter conditions: External temperatures ranged from 5°C to 15°C, with a 1500 W heating system maintaining an indoor temperature of 20°C.

Temperature measurements were taken at two different heights (0.85 m and 1.70 m) inside the room to assess thermal stratification. The PCM wall surface temperature and radiative heat flux from the glazed facade were also monitored. Each test lasted three days, with data from the last two days used for analysis to minimize the impact of initial boundary conditions.

13.2.8.2 Results

The study found that PCM wallboards effectively reduced indoor temperature fluctuations, particularly during summer conditions. In the summer test, PCM reduced peak indoor temperatures by up to 4.2°C compared to a room without PCM. Additionally, the decrement factor, which measures the reduction of indoor temperature variation, ranged between 0.73 and 0.78, indicating the material's ability to buffer temperature fluctuations and improve indoor thermal stability.

PCM wallboards also influenced air circulation within the room, enhancing natural convection and reducing thermal stratification. Unlike the reference test room, where a temperature difference of 1.3°C was observed between the lower and upper measurement points, the room with PCM exhibited a more uniform temperature distribution, suggesting better air mixing and fewer localized temperature variations.

Another significant finding was that PCM delayed the occurrence of peak indoor temperatures by approximately 40 minutes, shifting the warmest period later in the day. This delay could be beneficial for passive cooling strategies, as it allows night ventilation or other cooling measures to operate more efficiently when outdoor temperatures begin to drop.

Additionally, PCM wallboards helped regulate wall surface temperatures, particularly on sun-exposed facades. The study showed that the temperature amplitude of the interior wall surface decreased by approximately 3.5°C for east and west walls and by 2.8°C for the north-facing wall. This reduction in surface temperatures contributed to better occupant comfort by limiting radiant heat exchange from the walls into the indoor space.

These findings highlight that PCM wallboards not only reduce peak indoor temperatures but also moderate temperature fluctuations, delay overheating, and improve indoor thermal uniformity, making them a valuable passive cooling strategy for lightweight buildings during summer conditions.

13.2.8.3 Conclusion

The study confirmed that PCM wallboards are an effective passive cooling strategy for reducing overheating in lightweight buildings, particularly under summer conditions. The ability of PCM to absorb heat during peak daytime hours and release it during cooler night hours helped to moderate indoor temperature fluctuations, leading to improved thermal stability and enhanced occupant comfort. The results demonstrated that peak indoor temperatures were reduced by up to 4.2°C, while the decrement factor ranged between 0.73 and 0.78, indicating the material's efficiency in dampening temperature variations.

Beyond temperature reduction, PCM wallboards also delayed the occurrence of peak indoor temperatures by approximately 40 minutes, which could be beneficial for integrating with night ventilation strategies. Additionally, the use of PCM led to a more uniform indoor temperature distribution, minimizing thermal stratification and improving natural air circulation. These effects contributed to a more comfortable indoor environment, particularly in rooms with direct solar exposure.

However, the study also highlighted that the effectiveness of PCM strongly depends on external climatic conditions and the ability to release stored heat at night through ventilation. Without sufficient cooling during nighttime hours, PCM may not undergo a full phase transition, reducing its capacity to absorb heat the following day. This limitation underscores the importance of integrating PCM with adequate night ventilation strategies to maximize its cooling potential.

13.2.9 PCM Wallboard for Thermal Comfort in a Renovated Building: In-Situ Study

13.2.9.1 Methodology

The study conducted by Kuznik et al. (2011) aimed to evaluate the real-world performance of phase change material (PCM) wallboards in improving indoor thermal comfort in a renovated office building. Unlike previous laboratory-based research, this study focused on in-situ monitoring over an extended period, providing insights into PCM effectiveness under actual building operating conditions.

The experiment was carried out in two identical office rooms located within the HELIOS building, a lightweight tertiary structure in Lyon, France, known for its significant internal heat loads and susceptibility to temperature fluctuations. Each room measured 3.55 m × 5.2 m × 2.7 m, with identical orientation and exposure to external climate conditions. One room was equipped with PCM wallboards, installed on the lateral walls and ceiling, while the other served as a reference room without PCM (Figure 13.24).

The PCM used in the study was ENERGAIN®, developed by DuPont de Nemours, consisting of 60% microencapsulated paraffin embedded in a copolymer matrix. The material had a melting temperature range of 13.6–23.5°C, allowing it to store heat during the day and release it at night, moderating indoor temperature swings. The PCM sheets were 5 mm thick, covering an approximate area of 46 m^2 per room, with a latent heat storage capacity of 107.5 J/g (melting) and 104.5 J/g (freezing).

Temperature measurements were continuously recorded using PT1000 sensors with an accuracy of ±0.6°C. Each wall and the ceiling were equipped with four temperature sensors placed at different material layer interfaces, while air temperature and humidity were monitored using additional sensors distributed throughout both rooms. A meteorological station was installed on-site to measure solar radiation on the facade and outdoor air temperature, ensuring accurate comparisons between indoor and outdoor conditions.

To maintain consistency, the study focused on weekend periods when the offices were unoccupied, ensuring that internal heat gains from occupants and equipment were minimized. The monitoring system collected data every 10 minutes via a remote connection to the research laboratory, allowing continuous observation of PCM performance under varying climatic conditions.

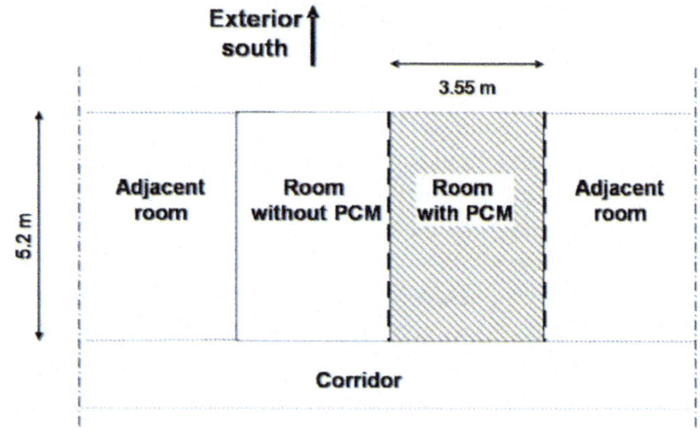

FIGURE 13.24 Layout of the monitored office rooms in the HELIOS building, Lyon, France, showing sensor placement, PCM wallboard integration, and exposure to external climate conditions.

13.2.9.2 Results

The study found that PCM significantly reduced peak indoor temperatures, particularly during warm weather conditions. The maximum air temperature in the PCM-equipped room was up to 2.2°C lower than in the reference room, confirming the material's ability to moderate overheating.

Wall and ceiling surface temperature analysis revealed that PCM actively absorbed and released heat, with wall temperatures fluctuating within the PCM's phase change range (13.6–23.5°C). This thermal behavior reduced the cooling demand during the day and limited excessive heat loss at night, creating a more stable indoor environment.

Another key observation was improved indoor thermal comfort due to reduced radiative heat exchange from wall surfaces. The mean radiant temperature in the PCM-equipped room was consistently lower, leading to a more comfortable occupant experience compared to the reference room.

Additionally, PCM contributed to enhanced air temperature uniformity, reducing vertical temperature differences between the upper and lower parts of the room. This effect minimized thermal stratification, ensuring a more even distribution of indoor temperatures.

13.2.9.3 Conclusion

The study confirmed that PCM wallboards can effectively enhance thermal comfort in lightweight renovated buildings by reducing overheating, stabilizing indoor temperatures, and improving thermal uniformity. The ability of PCM to store and release heat dynamically helped mitigate peak indoor temperatures by up to 2.2°C, offering a passive cooling solution without increasing energy consumption.

However, the effectiveness of PCM depended on outdoor conditions and the availability of nighttime cooling to allow proper heat release. Without sufficient cooling cycles, PCM may not fully transition between phases, reducing its thermal storage capacity for the next day.

These findings support the integration of PCM in building retrofits, particularly for lightweight office spaces prone to overheating. Future research should explore adaptive PCM strategies, optimized ventilation techniques, and long-term performance under extreme climate conditions to further enhance passive cooling solutions (Kuznik et al., 2011).

13.2.10 PCM in Brick Walls for Passive Cooling: An Experimental Study

13.2.10.1 Methodology

The study conducted by Castell et al. (2010) evaluated the impact of macroencapsulated phase change materials (PCM) in brick construction under real climatic conditions. The experiment was carried out in five full-scale test cubicles (2.4 m × 2.4 m × 2.4 m) located in Puigverd de Lleida, Spain, a region with a continental Mediterranean climate characterized by hot summers and significant daily temperature variations.

Two types of brick construction were tested: conventional perforated brick and alveolar brick, with and without PCM integration. The PCM materials used were RT-27 (melting at 28°C) and SP-25 A8 (melting at 26°C), encapsulated in CSM panels and installed within the walls and roofs of selected cubicles. Three cubicles were constructed using conventional brick: a reference cubicle (no insulation), a polyurethane-insulated cubicle (PU), and a PCM cubicle (RT-27 + PU). Two additional cubicles were built using alveolar brick: a reference cubicle and a PCM-enhanced cubicle (SP-25 A8 + alveolar brick).

To assess thermal performance, indoor air temperature, wall surface temperature, and heat flux through the south-facing wall were continuously monitored, along with outdoor climate conditions. Two testing approaches were applied:

1. Free-floating conditions, where no cooling system was used, allowing PCM's natural thermal behavior to be observed.

2. Controlled temperature conditions, where an air conditioning system maintained a 24°C setpoint, and energy consumption was recorded to quantify actual savings.

13.2.10.2 Results

The free-floating experiments demonstrated that PCM effectively reduced peak indoor temperatures by up to 1°C and smoothed out daily temperature fluctuations. The effect was more pronounced in alveolar brick cubicles, where the lower thermal resistance of the material allowed PCM to have a greater impact. However, some challenges were observed with PCM solidification at night, as no dedicated ventilation system was implemented to enhance cooling.

In conventional brick cubicles, the reference cubicle (without insulation) showed high temperature fluctuations, while the PU-insulated cubicle reduced these fluctuations. The PCM-integrated cubicle (RT-27 + PU) remained cooler throughout the experiment, maintaining a more stable indoor temperature profile. On days when the PCM phase change was fully activated, temperature fluctuations were almost completely eliminated.

In alveolar brick cubicles, the PCM-equipped cubicle (SP-25 A8) demonstrated superior temperature regulation, particularly after prolonged exposure to high outdoor temperatures. On days when outdoor temperatures reached 36°C, the PCM helped to maintain better indoor thermal stability, even as outdoor temperatures began to drop.

The controlled temperature experiments with a setpoint of 24°C showed that PCM cubicles consumed less energy than non-PCM cubicles. The RT-27 + PU cubicle achieved a 15% reduction in cooling energy demand compared to the PU-insulated cubicle, while the SP-25 + Alveolar cubicle reduced energy consumption by 17% compared to the standard alveolar brick cubicle.

The study also evaluated CO_2 emissions, demonstrating that PCM-integrated cubicles resulted in an annual reduction of 1–1.5 kg CO_2/m^2 due to lower cooling energy demand.

13.2.10.3 Conclusion

The study confirmed that macroencapsulated PCM can improve indoor thermal comfort by reducing peak indoor temperatures and stabilizing daily fluctuations, particularly in Mediterranean climate conditions. The PCM-integrated cubicles maintained lower maximum temperatures and required less cooling energy, proving their effectiveness as a passive cooling strategy. However, the findings also highlighted the importance of nighttime cooling for PCM to fully transition between phases. Without adequate ventilation or cooling mechanisms, solidification issues may occur, reducing the material's ability to absorb heat the following day.

The results suggest that PCM is best integrated into lightweight and low-thermal-mass structures, such as alveolar brick, where its thermal storage properties have the most significant impact. These findings reinforce the potential of PCM for energy-efficient building design, particularly in regions with high cooling demands, and highlight the need for further research into optimized ventilation strategies to maximize PCM effectiveness (Castell et al., 2010).

13.2.11 Micro-Encapsulated PCM in Building Materials: Experimental Study

13.2.11.1 Methodology

The study conducted by Schossig et al. (2005) investigated the effectiveness of micro-encapsulated phase change materials (PCM) integrated into conventional construction materials for improving thermal comfort in lightweight buildings. Unlike previous research using macroencapsulation, this study focused on microencapsulated PCM, which enables uniform distribution within materials such as plaster and gypsum board, enhancing thermal storage capacity without affecting structural integrity. The experiment took place in full-size test rooms at the Fraunhofer ISE facade testing facility. The rooms were identically constructed, oriented southward, and designed to simulate real-world indoor conditions without active cooling systems. One test room was coated with PCM-enhanced interior plaster, while the other served as a reference with standard plaster (Figure 13.25).

FIGURE 13.25 Exterior view of the Fraunhofer ISE facade testing facility, where two identical lightweight test rooms were constructed for PCM evaluation.

The PCM used had a melting temperature of 25°C and a melting range of 2K, allowing it to absorb heat during peak hours and release it at night. The micro-encapsulated PCM particles had an average diameter of 8 μm, ensuring uniform dispersion within the plaster matrix for optimal heat transfer.

A night ventilation system was implemented, operating at an air exchange rate of 4 ACH during nighttime hours to facilitate heat release. During the day, a minimum ventilation rate of 1 ACH was maintained, simulating real-world conditions without active cooling. Surface temperatures, indoor air temperatures, and wall heat fluxes were continuously monitored using a network of temperature sensors installed at different depths within the PCM and reference walls.

13.2.11.2 Results

The study found that PCM significantly reduced peak indoor temperatures and improved thermal stability. The surface temperature of PCM-enhanced walls increased more slowly compared to conventional walls, demonstrating the material's ability to absorb and store heat effectively.

The measured air temperature profiles showed that the PCM test room maintained temperatures up to 4°C lower than the reference room at peak hours, highlighting its cooling potential. Additionally, the PCM delayed peak indoor temperatures by approximately one hour, allowing passive night ventilation to be more effective in removing stored heat.

Analysis of thermal comfort indicators revealed that the PCM room experienced significantly fewer hours above 26°C, reducing overheating discomfort. The cumulative frequency analysis indicated that the reference room exceeded 28°C for more than 50 hours, while the PCM-equipped room was above this threshold for only 5 hours, confirming its effectiveness in preventing extreme indoor temperatures.

13.2.11.3 Conclusion

The study confirmed that micro-encapsulated PCM can effectively enhance the thermal performance of lightweight buildings, offering a practical and durable solution for passive cooling. By integrating PCM directly into plaster and gypsum, the research demonstrated that thermal storage capacity can be significantly increased without compromising structural integrity or requiring complex installation methods. The most significant benefits were observed when PCM was combined with night ventilation, allowing stored heat to be effectively dissipated. However, the study also emphasized that PCM integration must be carefully dimensioned based on expected thermal loads and shading conditions to maximize its performance.

13.2.12 PCM for Retrofitting: Enhancing Comfort in a Modern Home

13.2.12.1 Methodology

The study conducted by Jamil et al. (2016) assessed the effectiveness of phase change material (PCM) as a retrofitting strategy for improving thermal comfort in an occupied residential building. Unlike controlled laboratory tests, this research examined PCM performance in real-world conditions, incorporating both experimental monitoring and building energy simulations.

FIGURE 13.26 Exterior view of the monitored duplex house in Melbourne, Australia, used to evaluate PCM retrofitting for improved thermal comfort in occupied residential conditions. (Jamil et al., 2016.)

The experiment took place in a modern duplex house in Melbourne, Australia, built in 2009 with a five-star energy rating under the Nationwide House Energy Rating Scheme (NatHERS). The house has four bedrooms and a total floor area of 326 m^2. PCM was installed in the ceiling of one bedroom (BED 2), positioned between the insulation and plasterboard, while an adjacent bedroom without PCM served as a reference space. The experimental house is shown in Figure 13.26.

The BioPCM™ Q25 used in the study is a fatty-acid-based organic PCM, with a melting temperature of 25°C and a freezing temperature of 23°C. A total of 58.8 kg of PCM mats were distributed across the ceiling, designed to store heat during the day and release it at night.

To assess PCM effectiveness, indoor and outdoor temperatures were continuously recorded, and a building simulation model was developed using EnergyPlus software. The model was validated against experimental data and used to analyze how different ventilation and door operation strategies influenced PCM performance.

13.2.12.2 Results

The study found that PCM reduced peak indoor temperatures by up to 1.1°C and lowered thermal discomfort hours by 34% compared to the reference room. Night ventilation significantly improved PCM performance, as cooling the material overnight allowed it to absorb more heat the following day. Conversely, leaving internal doors open reduced PCM effectiveness by allowing heat transfer between zones.

Simulations also showed that spreading PCM at the ceiling and walls further enhanced its cooling effect, leading to greater temperature reductions and improved indoor comfort compared to ceiling-only placement.

13.2.12.3 Conclusion

The study confirmed that PCM can be an effective retrofitting option for improving thermal comfort in residential buildings, particularly in moderate to warm climates where overheating is a common issue. However, the study also highlighted limitations, particularly the strong dependence of PCM effectiveness on nighttime cooling conditions. Without adequate ventilation, PCM may not fully transition back to its solid phase, reducing its ability to absorb heat the following day. User behavior, such as keeping windows open at night and internal doors closed, played a crucial role in optimizing PCM performance. While PCM integration proved beneficial, the study emphasized that future retrofitting efforts should consider optimal placement, ventilation strategies, and user interaction to maximize performance (Jamil et al., 2016).

Additionally, the results indicate that an important role in PCM performance plays its distribution, providing better thermal stability than ceiling-only installation. These findings suggest that PCM retrofitting should be carefully planned alongside ventilation strategies to achieve the best results in residential applications (Jamil et al., 2016).

13.2.13 PCM in Super-Insulated Buildings: Impact on Thermal Comfort

13.2.13.1 Methodology

The study conducted by Sage-Lauck and Sailor (2014) investigated the use of phase change materials (PCM) to mitigate overheating in a Passive House-certified duplex in Portland, Oregon, USA. The experiment was conducted in a newly constructed duplex, with one unit serving as a test space equipped with PCM and the other as a control unit without PCM (Figure 13.27). PCM was installed in the second-floor living area of the test unit, incorporating 130 kg of BioPCM™ mats with a melting temperature of 25°C. These mats were positioned behind the gypsum board in the living room party wall, ceiling, and partition walls to enhance thermal storage capacity.

A comprehensive monitoring system was installed, including thermocouples for air and surface temperature measurements, CO_2 sensors, and door/window switches to track occupant behavior. A weather station was mounted on the roof to record outdoor temperature, humidity, wind speed, and solar radiation. Data were logged at five-minute intervals, allowing for continuous observation of PCM performance.

Building energy simulations were also performed using EnergyPlus, incorporating observed occupancy schedules, internal heat gains, and window operations. The simulated scenarios compared different PCM melt temperatures and placements to evaluate optimal design configurations.

13.2.13.2 Results

The study found that PCM reduced peak indoor temperatures by an average of 0.8°C, with the most significant cooling effects observed during the hottest month, August, when PCM lowered temperatures by up to 1.1°C.

The modeled results confirmed that the PCM installation reduced overheated hours by approximately 50% in the monitored room. However, PCM performance was strongly influenced by occupant behavior, particularly window operation. The

FIGURE 13.27 Exterior view of the Passive House-certified duplex in Portland, Oregon, used to evaluate the performance of PCM retrofitting for overheating mitigation. (Sage-Lauck & Sailor, 2014.)

PCM-equipped unit remained cooler when nighttime ventilation was used, allowing the material to reset by releasing stored heat.

Alternative simulation scenarios showed that relocating PCM to the interior wall surfaces further improved thermal performance, providing a greater reduction in overheated hours compared to placement behind the drywall.

13.2.13.3 Conclusion

The study confirmed that PCM can effectively reduce overheating in Passive House buildings, lowering peak temperatures and improving indoor thermal comfort. The BioPCM™ mats reduced the number of overheated hours by approximately 50%, demonstrating the potential of PCM as a passive cooling strategy. However, PCM performance was highly dependent on nighttime cooling, as insufficient cooling prevented the material from fully resetting for the following day. The study also highlighted that placing PCM closer to interior surfaces enhanced its cooling effectiveness, as it improved thermal coupling with indoor air.

One important limitation of the study was that occupant behavior in the PCM and reference units differed significantly, particularly regarding window ventilation at night. Since the PCM unit had greater window use, the authors were uncertain whether the observed temperature reductions were entirely due to PCM or influenced by enhanced ventilation.

13.2.14 Effectiveness of PCM in Cold Climates: Experimental and Simulation-Based Evidence from Northeast China

13.2.14.1 Introduction

The application of phase-change materials (PCMs) in buildings has been widely studied in warm and temperate climates, yet their performance in cold regions remains poorly documented. Xu et al. (2025) addressed this gap by combining field experiments and building simulations to assess the effect of PCM-integrated envelopes on thermal comfort and energy performance in Harbin, a city located in one of the coldest regions of China. The study examined both summer overheating mitigation and annual energy consumption across current and future climate scenarios, offering valuable insight into PCM viability in extreme seasonal climates.

13.2.14.2 Methodology

Two small-scale test buildings were constructed using standardized materials and geometry in accordance with regional energy design codes. One building was equipped with PCM panels on the internal surfaces of walls and roof, while the other served as a control without PCM. Temperature sensors were installed at non-sunlit positions to ensure accurate measurements. Data were collected every five minutes from May to September 2023.

The design and construction process of both buildings, including insulation layering, internal framing, and PCM integration, is illustrated in Figure 13.28, which highlights the structural differences between the test and reference units.

The design and construction details of the experimental buildings, including the integration of PCM panels, are shown in Figure 13.7. Each structure consisted of insulated external walls using XPS panels and interior timber framing, with PCM boards selected for a phase-transition temperature of 25°C, appropriate for typical summer peaks in the region. The panels were made of paraffin-based material with encapsulation for stability and longevity.

Parallel to the experimental setup, a four-story residential building in Harbin was modeled and calibrated using real temperature data. Simulations were conducted in EnergyPlus, with and without PCM, under both natural ventilation and active cooling conditions. In addition, future projections for 2040 and 2060 were included using RCP 2.6, 4.5, and 8.5 emission scenarios to assess the long-term utility of PCMs under climate change.

13.2.14.3 Results

The field experiment revealed that the PCM-integrated building maintained more stable temperatures. While the average indoor air temperature in the PCM building was slightly higher (20.79°C vs 19.99°C), it exhibited significantly fewer extreme highs and lows. The PCM building avoided all temperature readings above 28°C or below 8°C, while the control structure exceeded both thresholds multiple times.

In the residential simulation, PCM reduced the number of static overheating hours (above 26°C) by 3.27–3.83%, and dynamic overheating hours (CIBSE TM52 criteria) by up to 4.88%. Cooling energy demand dropped by 11.56–15.04%, depending on

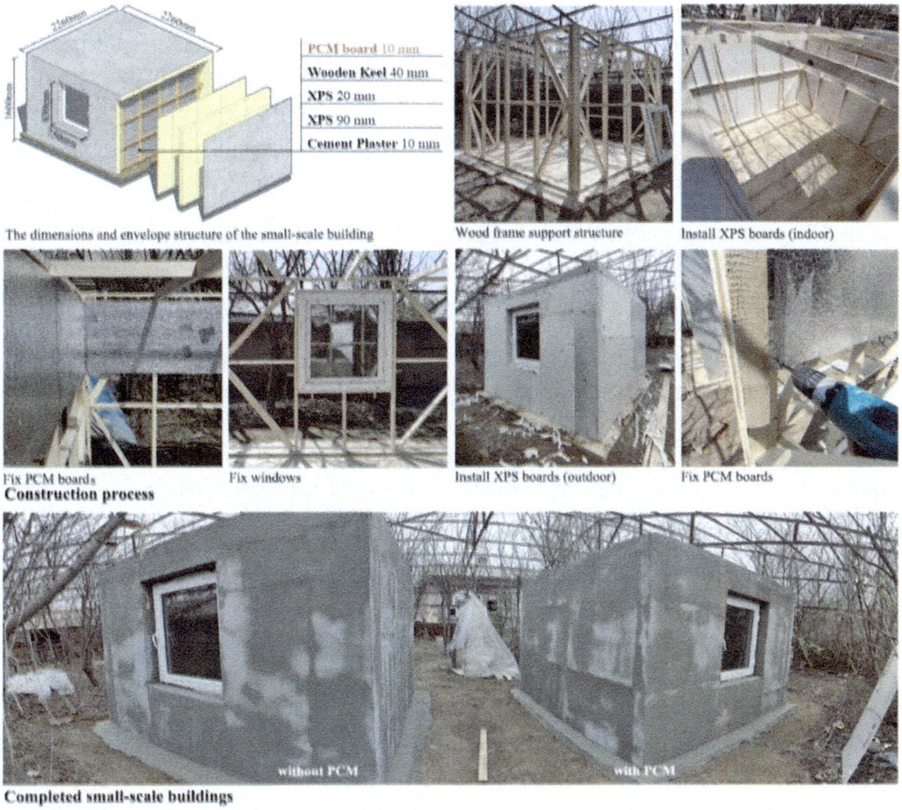

FIGURE 13.28 Design and construction of the small-scale experimental buildings with and without PCM boards in Harbin, China (2023).

room orientation, while heating demand decreased by up to 1.77%. These dual savings were achieved without compromising winter insulation.

Simulations under future climate scenarios showed that even in 2060, PCMs could reduce summer overheating duration by 3–11%, depending on room type and emission scenario. While heating energy savings were modest, they remained positive across all years and cases.

13.2.14.4 Conclusion

This study offers strong empirical and simulation-based evidence that PCM-integrated envelopes improve thermal performance in cold climates, not only by reducing cooling energy use but also by increasing temperature stability during summer. Unlike conventional solutions that compromise winter insulation, PCMs provide a dual benefit by offering passive cooling in summer and maintaining wall thermal inertia in winter.

In small-scale and full-scale buildings, PCMs consistently reduced overheating risk, peak indoor temperatures, and total annual energy use, particularly in cooling-dominated months. These results support the inclusion of PCMs in residential retrofits and new designs in cold regions facing intensified summer heat due to climate change. The findings also highlight the need for regulatory revisions in China's building codes to include summer overheating mitigation in zones traditionally focused on winter performance.

13.2.15 Performance of Phase Change Materials in Mitigating Indoor and Outdoor Overheating in Cold Regions

13.2.15.1 Introduction

While phase change materials (PCMs) have been widely studied in temperate and hot climates, their application in cold regions, particularly during summer, remains relatively unexplored. Zhao et al. (2025) addressed this gap by evaluating the potential of PCMs to mitigate both indoor and outdoor overheating in residential environments across three cities in China's severe cold regions: Yichun, Harbin, and Shenyang. Using a combination of field measurements and dynamic simulations, the study assessed how PCM-integrated walls could reduce thermal discomfort and lower ambient temperatures under summer conditions. Unlike previous studies focused solely on indoor benefits, this research also examined block-scale effects using ENVI-met modeling.

13.2.15.2 Methodology

The study combined in situ monitoring with dynamic simulations to assess PCM performance on two scales: individual residential buildings and urban blocks. Field measurements of indoor and outdoor air temperatures were carried out between May and September 2021. A total of 11 residential buildings across three cities were monitored using calibrated temperature and humidity sensors. For outdoor monitoring, five sensors were installed per residential block at 1.5 m above ground level, with locations including paved areas and green spaces, as shown in Figure 13.29.

Each building's envelope was simulated in EnergyPlus, both with and without PCM layers. The PCM used was a 30 mm-thick paraffin-based phase change mortar, installed on the inner surfaces of external walls. The phase transition temperatures were optimized for each city: 21–25°C in Yichun, 24–27°C in Harbin, and 27–29°C in Shenyang, aligning with observed summer indoor thermal profiles. Building use schedules, ventilation patterns, internal gains, and infiltration rates were adapted to local occupant behavior, including nighttime window opening in naturally ventilated spaces.

At the urban scale, ENVI-met models were created for representative blocks in each city. The block geometry, vegetation layout, and surface materials were replicated in detail. Simulations compared ambient conditions for scenarios with and without PCMs applied to the building facades.

13.2.15.3 Results

The monitoring results revealed extensive overheating in all three cities. In south-facing bedrooms, indoor operative temperatures exceeded 26°C for 26–56% of

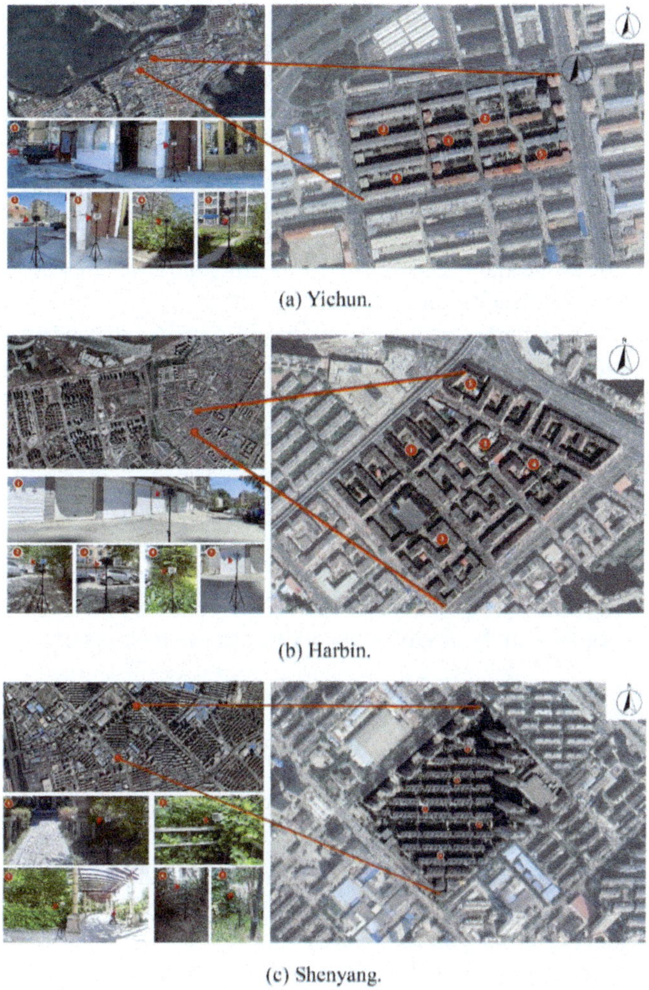

FIGURE 13.29 Locations of outdoor temperature monitoring points in three typical residential blocks in Yichun (a), Harbin (b), and Shenyang (c).

occupied nighttime hours, far surpassing CIBSE TM59 comfort thresholds. In Harbin, 40.4% of the monitored period fell into the overheating range, while Shenyang reached 56%.

Simulations showed that PCM application reduced overheating duration in bedrooms by 66% in Yichun, 27% in Harbin, and 39% in Shenyang. Similar reductions were observed in living rooms, where the cooling impact of PCMs led to decreased periods of discomfort by up to 43–44%. Peak indoor temperatures dropped by 2.0–2.8°C, depending on the city and room orientation.

At the urban block level, ENVI-met simulations demonstrated that PCMs can also reduce outdoor ambient temperatures. The average temperature at pedestrian level (1.5 m) declined by 0.21°C in Yichun, 0.27°C in Harbin, and 0.34°C in Shenyang.

Maximum temperatures dropped by up to 0.51°C during the hottest hours of the day. These effects were especially visible in paved courtyard areas, where solar exposure is high and building façades contribute significantly to heat accumulation.

13.2.15.4 Conclusion

The findings confirm that PCMs are not only effective in reducing indoor overheating in cold-climate buildings during summer but can also produce measurable reductions in outdoor urban temperatures. The integration of PCM panels into envelope structures led to substantial decreases in overheating duration, peak indoor temperatures, and block-level heat buildup.

This dual benefit is particularly relevant in cold-climate cities where summer heat is becoming more frequent due to climate change, but design regulations remain focused primarily on winter insulation. The study supports the adoption of PCM-based retrofit strategies in these regions and suggests revising national design standards to incorporate summer overheating risk in severe cold zones. The documented outdoor cooling effects also position PCMs as a complementary strategy within urban heat island mitigation efforts.

13.2.16 Phase Change Materials in South-Facing Top-Floor Rooms: Experimental Insights from a Decade of Research

Summary Box—Phase Change Materials in Top-Floor Rooms

- Three generations of PCM (2015–2023) tested in top-floor rooms with large south-facing glazing.
- Low- and mid-range PCMs (17.8°C–26°C) showed negligible cooling impact due to inadequate night regeneration.
- High-density PCM (26°C–29°C, 40 kg/m^2) achieved up to 1.2°C reduction in peak indoor temperatures under 10 ACH ventilation, but performance declined under prolonged heat waves.
- Heavy masonry reference consistently outperformed PCM-enhanced rooms, lowering peaks by 2.7°C and reducing cooling energy demand by over 12%.
- PCM effects proved conditional, dependent on night ventilation and diurnal temperature range, with measured energy savings <2%.

13.2.16.1 Methodology and Experimental Background

This study presents the results of a long-term experimental program (2015–2023) carried out in top-floor test rooms with large south-facing glazing, exposed to direct solar gains. The research took place in the full-scale facility of the Centre for Sustainable Building and Energy (CSBE) at the University of Zielona Góra (Figure 13.30), located in central-western Poland (Köppen–Geiger Cfb). The building contains eight identical upper-story rooms; three were selected for PCM testing and compared with a lightweight reference and a heavyweight benchmark (Figure 13.31). Their unshaded location and large glazed areas created overheating-prone conditions typical of attic-level dwellings.

FIGURE 13.30 Exterior view of the Centre for Sustainable Building and Energy (CSBE) located at the Science and Technology Park of the University of Zielona Góra in Nowy Kisielin.

The experimental program consisted of three phases. In the first (Staszczuk & Kuczyński & Staszczuk, 2021), a low-melting PCM (17.8°C–22.3°C, Dupont™ Energain®) was applied to all interior surfaces of a heavyweight concrete room. The PCM failed to activate, as indoor temperatures exceeded its solidification range for most of the summer, while the heavy structure alone reduced peaks by >2 K. The second phase (Kuczyński & Staszczuk, 2023) tested Rubitherm® RT25HC

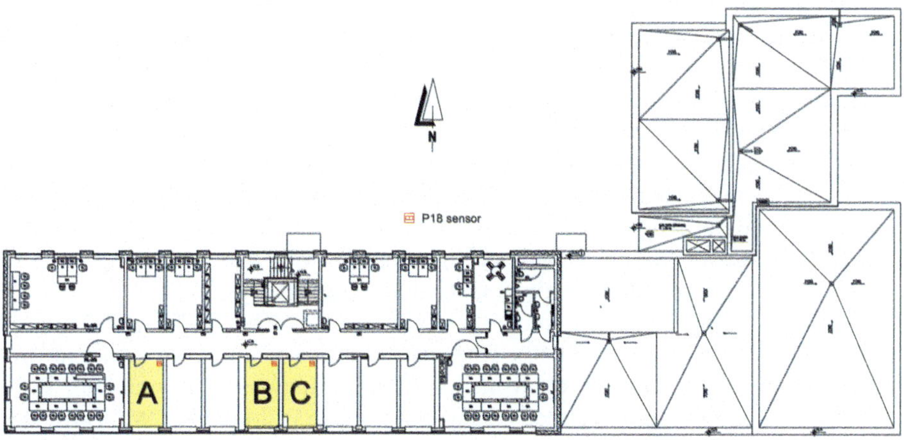

FIGURE 13.31 Top floor layout of the CSBE facility with indication of test rooms used in PCM experiments.

(22°C–26°C), applied on vertical walls of a lightweight room (200 kg, 11 kg/m²). Some activation occurred during cooler nights, but the effect was modest and disappeared during heat waves.

The final and most comprehensive stage (Kuczyński et al., 2025) employed Rubitherm® SP25E2 (26°C–29°C), a PCM with extended enthalpy up to 31°C and higher density (1.6 kg/l). It was installed on all walls and the ceiling at 40 kg/m² (≈720 kg per room) and combined with controlled night ventilation at 0.6, 3.0, and 10.0 ACH. This design aimed to maximize thermal interaction and enable consistent latent storage cycles.

All rooms shared identical geometry, envelope properties, and glazing. No internal heat gains or mechanical cooling were introduced. Night ventilation served as the only active element to regenerate the PCM. Thermal monitoring included indoor air and surface temperatures, PCM interface measurements, and outdoor meteorological data from an on-site rooftop weather station.

This progression, from low-range PCM in a heavy envelope, through wall-mounted PCM in a lightweight structure, to optimized full-surface application with active night ventilation, represents a systematic attempt to refine PCM performance under real summer conditions. The following section presents a comparative analysis of the results with emphasis on thermal buffering and temperature regulation.

13.2.16.2 Results and Comparative Analysis

The final phase of the experimental program confirmed that, even when the phase change material (Rubitherm® SP25E2) was carefully matched to indoor conditions and applied to all major internal surfaces, its contribution to summer performance remained limited. Effectiveness depended strongly on night ventilation and the persistence of elevated night temperatures.

At the lowest ventilation rate (0.6 ACH), the PCM rarely crystallized, remaining in a melted state and showing no measurable cooling effect compared to the lightweight reference room. With moderate ventilation (3.0 ACH), partial regeneration occurred, reducing peaks by about 0.6°C, though indoor temperatures still exceeded 28°C for much of the occupied period. Only at the highest ventilation (10.0 ACH) did the PCM achieve a reduction of up to 1.2°C, cutting hours above 28°C by about 40%. Even then, overheating persisted during heat waves, confirming the dependence of PCM on effective night cooling.

By contrast, the heavyweight reference room consistently maintained lower maximum and average indoor temperatures in all scenarios. During a two-week August heat wave, daily peaks were 2.7°C lower than in the lightweight room, compared to just 1.0°C for the PCM-enhanced case. Nighttime recovery was also stronger, ensuring stable comfort conditions without mechanical support.

Cooling energy data reinforced these differences. The PCM room showed only a marginal reduction of 1.6%, whereas the heavyweight room achieved over 12%. Earlier test campaigns confirmed similar limitations: a low-melting PCM (17.8°C–22.3°C) was fully ineffective, and a mid-range PCM (22°C–26°C) provided at most 1°C peak reduction. Even with increased mass and partial wall coverage, performance remained modest due to incomplete night regeneration and lack of ceiling application.

In summary, across all three phases of research, PCM effects proved conditional, modest, and time-limited, while heavyweight construction delivered robust and consistent buffering. Under typical unshaded top-floor conditions, PCMs should be considered as supplementary elements rather than alternatives to thermal mass.

13.2.16.3 Discussion

The results of this study confirm recurring discrepancies between modeled and observed PCM performance in real buildings. Despite optimized material selection and large installation coverage, effectiveness under prolonged heat stress remained limited. This outcome is consistent with Sage-Lauck and Sailor (2014), who reported reductions of about 1°C under favorable diurnal conditions, with rapid loss of impact during warm nights. Similar results were observed by Jamil et al. (2016), who showed that PCMs in lightweight retrofits only perform when supported by consistent night ventilation. Our findings mirror this: even with up to 40 kg/m^2 of PCM, the benefits diminished when nighttime regeneration was incomplete, suggesting diminishing returns with increased mass under suboptimal conditions.

Kuznik et al. (2011) likewise reported peak reductions of 1°C–2°C in in-situ tests, stressing that the timing of melting–solidification cycles is decisive. In our case, incomplete crystallization during warm nights, even with ventilation, confirmed this limitation. Laboratory studies, such as Kuznik and Virgone (2009), showed larger peak reductions under controlled conditions, but their transferability to real residential environments is limited. Comparisons of different materials also support earlier claims by De Gracia et al. (2010): extended enthalpy ranges may offer modest improvements, yet they did not match the consistent buffering observed in high-mass reference rooms. Similar conclusions were reached by Brambilla and Jusselme (2017) and Castell et al. (2010), who found that PCM reduces peaks by 1°C–1.5°C but only with effective night cooling.

Surface distribution also plays a role. Schossig et al. (2005) showed that fully integrated PCM plaster improved comfort significantly, but only under controlled ventilation of 4 ACH. Later studies (Jamil et al., 2016; Sage-Lauck & Sailor, 2014) and our own results confirm that multi-surface application enhances performance, though still below the level of heavy-mass envelopes. A recurring critique, highlighted by Vega et al. (2024), is the reliance on short-term laboratory studies. Our long-term monitoring across three summers supports this concern: measured reductions in cooling demand did not exceed 1.6%, contrasting sharply with simulation-based projections of 50% or more (Saffari et al., 2017; Thiele et al., 2015).

This discrepancy is particularly important in temperate continental climates, where the diurnal temperature swing is too small to fully regenerate PCM systems. Artmann et al. (2008) demonstrated that the passive cooling potential in Central Europe is marginal compared to Mediterranean or arid regions. By contrast, our heavy-mass reference room consistently provided reliable buffering, confirming earlier observations by Givoni (1998) and Tink et al. (2018) that traditional mass materials distribute and delay heat fluxes more effectively than latent storage.

In summary, while PCMs can be optimized in terms of melting range, coverage, and ventilation support, their performance remains inferior to heavyweight construction. These results reinforce the need to evaluate PCM use critically, avoiding overreliance on idealized simulation models. PCMs may still have a role in lightweight retrofits where mass cannot be added, but their success depends on climate, ventilation potential, and occupant behavior.

13.2.16.4 Implications for Design and Policy

The decade-long PCM experiments confirm that, despite methodological refinements, PCM performance remains limited compared to the robust buffering capacity of traditional heavyweight construction. While PCM systems may be suitable in lightweight retrofit scenarios where mass cannot be increased, their effectiveness is conditional upon sufficient night cooling, favorable diurnal cycles, and consistent user operation. For practice, this means PCMs should not be presented as a substitute for thermal mass, but rather as a supplementary element in specific contexts. Performance projections should be grounded in long-term, full-scale monitoring rather than simulation alone.

13.2.17 Summary

The evidence covers laboratory cells, full-scale prototypes, and occupied dwellings across cool temperate, oceanic, Mediterranean, semi-arid, and severe cold climates. Effectiveness is reported using indoor and surface temperatures, peak suppression and delay, decrement factors, overheating hours, energy use, and, in several cases, life-cycle outcomes.

Traditional high thermal mass. Heavy envelopes reliably damp daily swings, lower peaks, and delay heat transfer where solar or internal gains are significant. In New Zealand side-by-side houses, exterior-insulated concrete walls reduced summer peaks by up to 5°C and cut sunny-day winter heating by more than 50%, with further gains from exposing the concrete floor (Bellamy & Mackenzie, 2001). Office-scale rooms in Switzerland showed compressed earth bricks reduced peaks by about 3°C and narrowed daily amplitude, with night ventilation the most effective discharge strategy (Brambilla & Jusselme, 2017). In the UK, solid-wall dwellings, internal wall insulation raised summertime operative temperatures and overheated hours, but daytime shading plus night ventilation largely neutralized the penalty under current weather; future warming is likely to leave a residual risk (Tink et al., 2018). In an insulated concrete formwork dwelling, the concrete core stabilized the wall at about 16°C, lowered summer indoor air temperature by roughly 2°C compared with a low-mass frame, and reduced heat losses by 37% for similar U-values; standard methods tended to under-credit storage (Mantesi et al., 2019). Over 100 years, medium- to heavyweight construction delayed air-conditioning adoption, lowered heating demand, and reduced life-cycle carbon by up to 17%, offsetting higher embodied carbon within one to two decades (Hacker et al., 2008). Briefly, full-scale comparisons in Poland found that a masonry house maintained lower summer peaks and far fewer overheating hours than a matched lightweight house under a natural heat wave (Kuczyński & Staszczuk, 2020).

Ground coupling amplifies mass effects. Where detailing controls losses, enhancing conductive exchange with the subsoil further stabilizes indoor conditions. A Polish full-scale study reported season-long indoor temperatures below 26°C in a ground-coupled masonry house, while a matched lightweight insulated house frequently exceeded 30°C; perimeter insulation is recommended to curb lateral soil losses (Kuczyński & Staszczuk, 2023).

Phase Change Materials. PCM can lower peaks and flatten daily profiles, but effects are typically modest and depend on nighttime regeneration. Controlled cell tests with PCM wallboards reported up to 4.2°C peak reduction and a 40–60 minute peak delay when night ventilation enabled full crystallization (Kuznik & Virgone, 2009). In-situ offices showed up to 2.2°C peak reduction and improved mean radiant temperature with adequate night cooling (Kuznik et al., 2011). Brick cubicles in Spain recorded 15–17% cooling-energy savings at a 24°C set-point and about 1°C free-running peak reduction, with incomplete solidification on hot nights limiting performance (Castell et al., 2010). Super-insulated housing saw peak cuts of about 0.8°C–1.1°C and roughly 50% fewer overheated hours when occupants used night ventilation, with better results when PCM was placed on interior surfaces (Sage-Lauck & Sailor, 2014). In severe cold regions, experiments and simulations reported peak cuts of 2.0°C–2.8°C, 3%–11% fewer overheating hours, and 11–15% lower cooling energy, with small outdoor cooling at block scale (Xu et al., 2025; Zhao et al., 2025). Long-duration tests in Polish top-floor rooms confirmed this boundary-condition dependence and indicated a supplementary, not substitutive, role for PCM under constrained night cooling (Kuczyński et al., 2025).

Design guidance. Traditional mass provides robust 2°C–5°C peak suppression and reliable stabilization, especially with shading and night ventilation. Ground coupling can strongly amplify buffering where vertical perimeter insulation and moisture detailing control losses. Phase Change Materials are best treated as a complementary measure in lightweight or retrofit contexts where adding mass is impractical. Typical real-world benefits are peak reductions of about 0.8°C–2.2°C and modest energy savings, rising in controlled cubicles but constrained without nighttime regeneration. Selection of melt range, broad interior distribution, and integration with shading and purge ventilation are decisive. Across both strategies, dynamic comfort metrics and, where relevant, prospective life-cycle assessment provide a clearer basis for climate-ready design than static envelope properties alone.

13.3 VENTILATIVE COOLING STRATEGIES

Ventilative cooling is a key passive strategy for mitigating overheating by enhancing air exchange and removing excess indoor heat. This section examines natural ventilation techniques, focusing on their effectiveness across different building typologies and climatic conditions.

The discussion begins with an analysis of night ventilation, which utilizes cooler nighttime air to flush accumulated heat from indoor spaces. Studies explore the impact of air change rates, window operation strategies, and urban microclimate conditions on cooling efficiency. The role of cross-ventilation, which promotes airflow between openings, and stack effect ventilation, driven by temperature differences, is also assessed.

Further, the section examines ventilation in high-rise and low-rise buildings, emphasizing its role in reducing reliance on active cooling. Studies investigate the influence of ventilation intensity, seasonal variations, and urban airflow patterns on cooling effectiveness.

Findings indicate that ventilative cooling significantly reduces indoor temperatures and cooling energy demand, though its effectiveness depends on climate conditions and airflow dynamics. The section provides insights into optimizing natural airflow strategies, emphasizing their potential for improving thermal comfort in both traditional and modern buildings.

13.3.1 Summer Night Ventilation and Urban Building Performance: A Numerical Study

13.3.1.1 Methodology

The study conducted by Tsoka (2021) aimed to evaluate the impact of night ventilation on the thermal and energy performance of urban residential buildings using numerical simulations. The analysis was conducted in four different urban areas in Thessaloniki, Greece, each presenting distinct urban morphologies that could influence ventilation effectiveness.

The simulations were performed using the EnergyPlus dynamic building energy simulation model. The study focused on eight generic building units, located on the first and third floors, to assess differences in night ventilation performance between lower and intermediate-level apartments. The plan and 3D view of the generic building unit used in the simulations are shown in Figure 13.32.

The investigated buildings were constructed before 1979, meaning they lacked thermal insulation. The external walls consisted of brick masonry with a U-value of 1.64 W/m^{2}K, while reinforced concrete components had U-values ranging from 2.53

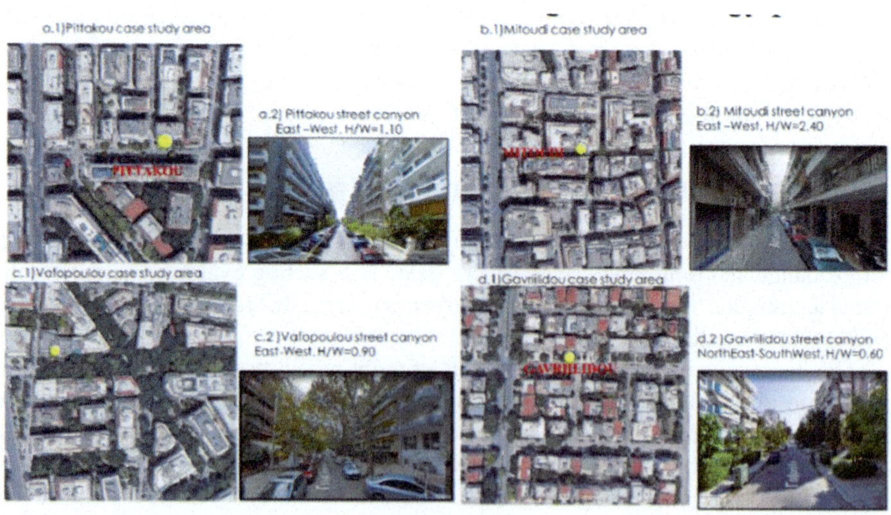

FIGURE 13.32 Case study areas in Thessaloniki, Greece, showing urban street canyons with different height-to-width ratios. (Tsoka, 2021.)

to 3.17 W/m^2K. Window characteristics were also considered, with synthetic frames (Uf = 2.8 W/m^2K) and double-glazed glass (Ug = 2.8 W/m^2K).

The study examined three different night ventilation rates (5 ACH, 10 ACH, and 15 ACH) in addition to a base case scenario with 3 ACH, which represented typical urban night ventilation conditions. Single-sided ventilation was considered instead of cross ventilation, simulating a less optimal but common building configuration. The night ventilation was activated only when the indoor temperature exceeded the outdoor temperature by at least 1°C, and it was applied between 00:00 and 08:00 AM. The simulation results assessed the annual cooling energy savings under different ventilation rates and analyzed how night ventilation affected the time delay of the next day's peak indoor temperatures.

13.3.1.2 Results

The study found that increasing the night ventilation rate significantly reduced cooling energy demand, with the highest energy savings observed in buildings located in less dense urban areas. For first-floor units, increasing night ventilation from 3 ACH to 5 ACH resulted in cooling energy savings ranging from 5.3% to 12.0%, depending on the urban morphology. At higher ventilation rates of 10 ACH and 15 ACH, the cooling energy savings increased to a maximum of 40.8% in the least dense area.

For third-floor apartments, which typically experience higher solar gains, the impact of night ventilation was even more pronounced. A moderate increase from 3 ACH to 5 ACH led to cooling energy savings of 5.9% to 8.0%, while 10 ACH and 15 ACH further reduced cooling energy demand by up to 20.2%, depending on the urban setting.

In addition to reducing energy demand, night ventilation delayed the occurrence of peak indoor temperatures the following day. In the most favorable cases, the peak temperature was delayed by up to 3 hours at 5 ACH, while at 10 ACH and 15 ACH, the peak temperature was significantly flattened, preventing overheating during the hottest part of the day.

However, the effectiveness of night ventilation varied depending on the surrounding urban morphology. In densely built areas, where airflow is restricted due to narrow street canyons and reduced wind speeds, the impact of night ventilation was less pronounced, with energy savings limited to 5–8% at 5 ACH. Conversely, in less dense areas with greater air movement, cooling energy savings exceeded 20% at 10 ACH, demonstrating that the effectiveness of night ventilation is strongly dependent on external climatic and urban conditions.

13.3.1.3 Conclusions

The study confirmed that night ventilation is an effective passive cooling strategy for reducing cooling energy demand and improving indoor thermal conditions in urban residential buildings. Higher ventilation rates led to greater energy savings and delayed peak indoor temperatures, mitigating overheating. However, the effectiveness of night ventilation strongly depends on the surrounding urban environment.

In less dense areas with greater airflow, night ventilation significantly reduced cooling loads, particularly at 10 ACH and 15 ACH, where energy savings reached 40.8%. However, in densely built urban areas, where airflow is restricted by high building density, the benefits were more limited, with maximum energy savings of around 8% at 5 ACH.

The findings suggest that night ventilation should be carefully adapted to local urban conditions. In highly urbanized settings, complementary strategies such as external shading or mechanical ventilation may be needed to enhance cooling performance. Additionally, the study highlighted the importance of optimizing window operation strategies, ensuring that ventilation is activated only when outdoor temperatures are lower than indoor temperatures to maximize cooling efficiency (Tsoka, 2021).

13.3.2 Night Ventilation in a Nearly Zero-Energy Office during Heat Waves in Brussels

13.3.2.1 Methodology

The study conducted by Amaripadath et al. (2023) evaluated the effectiveness of natural and mechanical night ventilation in mitigating indoor overheating and reducing energy demand in a nearly zero-energy office building (nZEB) located in Brussels, Belgium. The research analyzed night ventilation performance under both heat wave and non-heat wave conditions, considering urban and rural microclimates to assess external environmental influences.

A calibrated building simulation model was used to simulate thermal and energy performance across different ventilation strategies. The reference office building used in the study, including its thermal zones and ventilation configurations, is shown in Figure 13.33. The building, constructed to meet nZEB standards, had a highly insulated and airtight envelope, requiring passive strategies to prevent overheating.

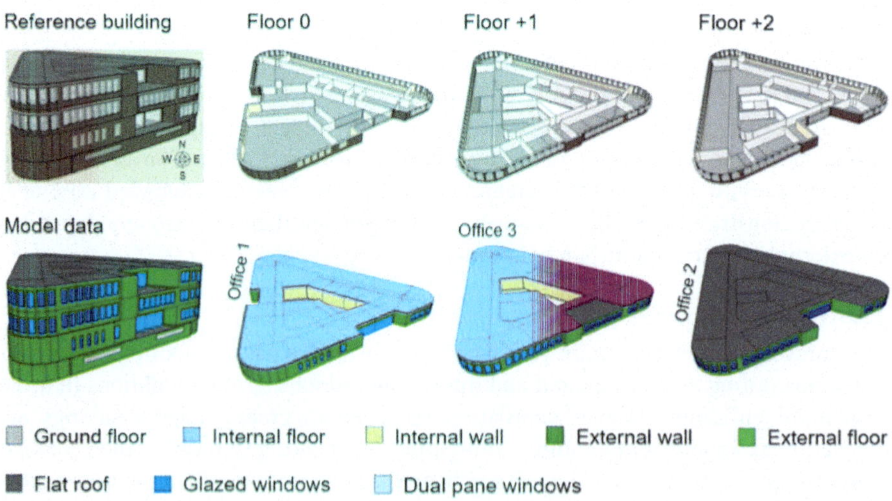

FIGURE 13.33 Reference nearly zero-energy office building in Brussels, Belgium, illustrating thermal zones and night ventilation strategies applied in the study.

Three HVAC configurations were tested:

- Baseline scenario—Active cooling without night ventilation.
- Strategy 1—Active cooling combined with natural night ventilation (window opening at 3 ACH).
- Strategy 2—Active cooling combined with mechanical night ventilation (ventilation system at 1.1 $l/s/m^2$).

Night ventilation was activated from 22:00 to 06:00, simulating practical building operation. The study assessed key performance indicators (KPIs) such as the Standard Effective Temperature (SET), peak cooling demand, exceedance hours, and energy consumption, providing a comprehensive evaluation of night ventilation strategies in extreme heat conditions.

13.3.2.2 Results

The study found that natural night ventilation was more effective in reducing indoor overheating than mechanical night ventilation, particularly in the rural microclimate. Strategy 1 (natural night ventilation with active cooling) reduced overheating by 0.39°C in the urban microclimate and 0.50°C in the rural microclimate compared to the baseline scenario.

However, when analyzing cooling energy consumption, natural night ventilation had no significant impact, while mechanical night ventilation increased energy use by 0.54 kWh/m^2 in the urban microclimate and 0.40 kWh/m^2 in the rural microclimate due to prolonged fan operation. This suggests that while mechanical ventilation aids in temperature control, its increased energy demand may offset its benefits.

The heat index analysis revealed that natural night ventilation contributed to safer indoor conditions during extreme heat events, reducing the time spent in high-risk temperature ranges. However, mechanical night ventilation provided only limited additional cooling compared to the baseline case.

In both urban and rural scenarios, night ventilation was more effective during heat waves than in normal summer conditions, confirming its role as a passive cooling strategy during extreme events. However, the effectiveness of night ventilation varied depending on building orientation, shading, and urban microclimate effects, with buildings in rural settings benefiting more from passive ventilation strategies due to lower ambient nighttime temperatures.

13.3.2.3 Conclusion

The study confirmed that natural night ventilation is an effective passive cooling strategy, particularly in rural settings, where it reduced indoor overheating and improved thermal comfort without increasing energy consumption. In contrast, mechanical night ventilation led to higher energy use, making it a less efficient option for cooling in low-energy buildings.

The findings suggest that night ventilation is most effective when combined with passive design elements such as shading and thermal mass, particularly in highly insulated buildings where overheating is a concern. While both urban and rural microclimates benefited from night ventilation, the cooling potential was greater in rural environments, where nighttime temperatures were lower.

The study highlights that future building designs should prioritize natural ventilation strategies before relying on mechanical systems to optimize both thermal comfort and energy efficiency. Additionally, further research should explore hybrid ventilation approaches that integrate natural and mechanical ventilation based on real-time climate conditions (Amaripadath et al., 2023).

13.3.3 Ventilative Cooling in Chinese Homes: Simulation of Modular Constructions Under Climate Change

13.3.3.1 Introduction

Hu et al. (2024) conducted a study on the effectiveness of ventilative cooling in a multi-family nearly zero-energy building (nZEB). The analysis was performed for the Holon Garden A-1, a modular building completed in 2021, which has won multiple international awards and was certified as compliant with the European Commission's Construction Products Regulation. The building, with a gross floor area of approximately 2,968 m^2 and a height of 33 meters, consists of 56 prefabricated modules, which were produced in 15 days and assembled on-site in just 29 hours. A north-facing second-floor apartment was selected for analysis. The study compared traditional heavyweight construction (HWC), with an exterior façade made of porous cement bricks with insulation, to a lightweight construction (LWC) with an envelope mass of approximately 41 kg/m^2.

13.3.3.2 Methodology

The study employed EnergyPlus simulations to assess energy performance under current and future climate scenarios (2050s and 2080s), based on the CMIP6 SSP5-8.5 projection. The comparison was conducted for five climate zones in China: severe cold (Dwb, Harbin), cold (Dwa, Beijing), hot summer cold winter (Cwa, Changsha), hot summer warm winter (Cfa, Guangzhou), and mild (Cwb, Kunming). Both construction types met the 2016 (S2016), 2021 (S2021), and nZEB energy-efficiency standards. The analysis included variations in thermal transmittance (U-value) and thermal inertia (D-value) to assess the impact of different envelope designs on annual heating and cooling loads.

The case study building, Holon Garden A-1, was chosen as a representative modular integrated construction (MIC) project, as shown in Figure 13.34a. The building's 3D SketchUp model and layout are presented in Figures 13.34b and 13.34c, illustrating the structural composition and spatial configuration. The study focused on a north-facing second-floor apartment (highlighted in the floor plan), allowing for controlled analysis of thermal performance and ventilation strategies while minimizing external variability from orientation effects.

The ventilation airflow rate was set at 2, 4, 6, 8, 10, and 12 ACH, and indoor thermal comfort was evaluated using the ASHRAE 55 adaptive comfort model, with mechanical cooling activated above 26°C and heating below 20°C.

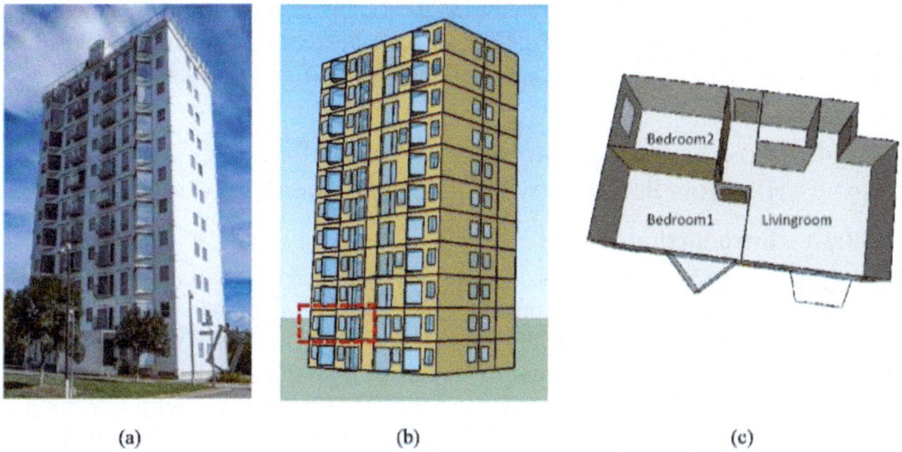

FIGURE 13.34 Holon Garden A-1 modular integrated construction (MIC) building and its simulation model, showing the real structure, 3D representation, and selected apartment layout used for energy performance and ventilation analysis. (Hu et al., 2024.)

13.3.3.3 Results

In the current climate scenario, LWC reduced energy consumption by 9% compared to HWC in the severe cold, cold, hot summer cold winter, and hot summer warm winter zones, while HWC reduced energy demand by 52% in mild climates due to its higher thermal mass. However, in future warming scenarios, the efficiency of HWC decreased, as rising outdoor temperatures limited the benefits of thermal mass.

Night ventilation was the most effective energy-saving strategy, although it was less efficient for LWC, reducing energy demand by 8% in severe cold climates, 7% in cold zones, 17% in hot summer cold winter regions, 20% in hot summer warm winter climates, and 50% in mild climates. The effectiveness of night ventilation in HWC also declined under future climate conditions, as increasing outdoor temperatures reduced the nighttime cooling potential.

For Harbin (Dwb), the annual heating and cooling demand of LWC was comparable to HWC across all weather scenarios, suggesting that LWC could be promoted in climates dominated by heating demand, as future climate change would not alter this outcome. The study also identified a turning point in airflow rate, beyond which HWC became more efficient than LWC. In hot summer climates, LWC performed better at lower ventilation rates, while in cold and mild climates, HWC retained its efficiency due to greater heat storage capacity.

13.3.3.4 Conclusion

The findings underscore the importance of adapting building strategies to local climate conditions and considering both thermal mass and ventilation efficiency for optimal energy performance. LWC proved to be more energy-efficient in heating-dominated climates, while HWC remained advantageous in moderate climates but lost efficiency in future warming scenarios. The study confirms that ventilative cooling is a key energy-saving strategy, but its effectiveness varies with construction type and future climate projections. Future nZEB standards should integrate thermal mass considerations to enhance energy efficiency in different climate zones.

13.3.4 Cross-Ventilation and Thermal Comfort in High-Rise Buildings in Hot-Humid Climates

13.3.4.1 Introduction

Stasi et al. (2024) investigated the cooling potential of cross-ventilation in high-rise buildings to enhance thermal comfort and reduce reliance on mechanical cooling systems in hot and humid climates. The study focused on Charles Correa's Kanchanjunga Apartments in Mumbai, India, a 28-story naturally ventilated high-rise building designed to integrate cross-ventilation through large open terraces and verandas. The research employed Computational Fluid Dynamics (CFD) simulations coupled with Building Energy Simulation (BES) to assess airflow patterns, indoor thermal conditions, and energy savings potential. The architectural features and ventilation strategy of Kanchanjunga Apartments are illustrated in Figure 13.35.

13.3.4.2 Methodology

The research used a multi-step computational approach to evaluate the thermal performance of cross-ventilation under varying climate conditions. The study modeled the building's airflow dynamics using Ansys Fluent for outdoor wind pressure distribution and validated indoor ventilation efficiency using CFD-based air velocity analysis. The simulation process incorporated site-specific meteorological data from

FIGURE 13.35 Kanchanjunga Apartments: (a) south-east façade; (b) south-east terrace view; (c) south-west façade; (d) north-west terrace; (e) double-height living room. (Stasi et al., 2024.)

Mumbai, where average annual temperatures exceed 26.7°C, with summer peaks reaching 38.5°C.

To analyze the effectiveness of cross-ventilation, the study compared four ventilation scenarios with different air change rates (ACH):

1. WD0—No natural ventilation (ACH = 0)
2. WD1—Minimum ventilation (ACH = 0.5, ASHRAE standard for air quality)
3. WD2—Natural ventilation with high air exchange (ACH = 12)
4. WD3—Cross-ventilation based on actual wind conditions (ACH = 67.2, determined via CFD analysis)

The CFD results were integrated into Building Energy Simulations (BES) conducted in DesignBuilder, assessing thermal performance and adaptive comfort levels under the Indian Model for Adaptive Comfort (IMAC). The study also examined the spatial configuration of indoor airflow, emphasizing the role of verandas, double-height terraces, and open staircases in enhancing ventilation performance.

13.3.4.3 Results

The study demonstrated that cross-ventilation played a crucial role in reducing indoor temperatures and minimizing discomfort hours, with the highest effectiveness observed under high air change rates and nighttime structural cooling. The cooling potential of cross-ventilation was particularly pronounced, as it reduced peak indoor operative temperatures by up to 5.3°C compared to non-ventilated conditions. Additionally, nighttime structural cooling further lowered temperatures by 6.4°C relative to outdoor temperature peaks, reinforcing the benefits of enhanced airflow. As a result, discomfort hours were reduced by up to 58% on the hottest days and were completely eliminated during monsoon and winter conditions.

The study also highlighted the role of thermal comfort and adaptive cooling. In the WD3 scenario, characterized by high cross-ventilation, indoor temperatures remained within the IMAC acceptability limits for most of the day, ensuring comfort without mechanical cooling. The effectiveness of nighttime ventilation increased as diurnal temperature variations grew, demonstrating the importance of controlled airflow strategies. However, while higher air change rates (ACH > 12) contributed to cooling, they provided diminishing returns, indicating the existence of an optimal range for ventilation efficiency.

The building design and optimization of cross-ventilation further enhanced cooling performance. The architectural configuration of Kanchanjunga Apartments facilitated airflow through double-height terraces and open stairwells, allowing for natural air movement between floors. Verandas located on the windward side acted as controlled inlets, regulating air inflow and preventing excessive turbulence. The inlet-to-outlet ratio (0.28) was optimized to maximize airflow efficiency, ensuring stable indoor conditions while minimizing excessive cooling losses. These findings emphasize the importance of integrating passive ventilation strategies into architectural design to enhance indoor thermal comfort in warm climates.

13.3.4.4 Conclusion

This study confirmed that cross-ventilation is a highly effective passive cooling strategy in high-rise buildings, particularly when airflow pathways and opening configurations are optimized. The findings suggest that nighttime structural cooling through cross-ventilation is critical for improving indoor thermal comfort in hot and humid climates.

The research provides key design recommendations for enhancing ventilation performance in high-rise buildings, including dual façade exposure, staggered floor layouts, and controlled inlet-to-outlet ratios. Future studies should explore hybrid ventilation strategies, real-time airflow control systems, and the integration of dynamic façade elements to further improve building resilience to climate change.

13.3.5 Cross-Ventilation as a Passive Cooling Strategy in a Historic Building in Écija, Spain

Palomo Amores et al. (2025) examined the effectiveness of cross-ventilation as a passive cooling technique in a historic nineteenth-century building located in Écija, Spain. The research aimed to validate how natural airflow between exterior openings and a central courtyard could reduce indoor temperatures and enhance occupant comfort during the region's hot summer months. The study was conducted in a two-story residence with 60 cm thick rammed earth walls, known for their thermal inertia, making them well-suited for evaluating long-term passive cooling effects.

Figure 13.36 provides a plan and section of the building, illustrating the layout of the rooms, openings, and airflow pathways used to facilitate cross-ventilation. The structure features a symmetrical distribution of rooms surrounding a courtyard, with windows and doors aligned to encourage air movement from the street to the inner spaces. The design allows for the introduction of cooler night air, while the building's high thermal mass helps regulate indoor temperatures by delaying heat transfer throughout the day.

FIGURE 13.36 Architectural plan and section of the historic house in Écija, Spain, illustrating the layout of openings and airflow pathways for cross-ventilation. (Palomo Amores et al., 2025.)

13.3.5.1 Methodology and Experimental Setup

A hybrid research approach was employed, integrating field monitoring, computational fluid dynamics (CFD) simulations, and thermal energy modeling to assess ventilation effectiveness and cooling performance.

- Field Monitoring: Temperature and relative humidity sensors were placed in multiple rooms to capture indoor thermal variations under natural conditions and controlled cross-ventilation experiments. Data was collected over a three-month summer period (July–September 2023), with specific focus on 20 experimental nights where ventilation was activated between 21:00 and 10:00.
- CFD Simulations: Airflow distribution was modeled using computational fluid dynamics (CFD) to evaluate pressure differences between street and courtyard-facing openings, providing insight into air movement dynamics in ventilated and non-ventilated conditions.
- Thermal Modeling (HULC Software): A dynamic energy model of the building was calibrated using the measured data, achieving an accuracy of ±5%. The model simulated indoor temperature variations, assessing cooling effects under different ventilation intensities.

13.3.5.2 Results

The study confirmed that cross-ventilation effectively reduced peak indoor temperatures, achieving an average cooling effect of 3.8°C. The most pronounced temperature reductions were observed in first-floor rooms, where airflow pathways were unobstructed, reaching a peak decrease of 4.5°C. The cooling effect extended beyond the ventilation period, demonstrating the thermal inertia properties of rammed earth walls, which delayed heat buildup and prolonged the cooling impact for several hours.

A cumulative cooling effect was observed over multiple nights, preventing heat accumulation and maintaining lower temperatures for up to 10 days, even after ventilation ceased. When ventilation was completely discontinued, indoor temperatures gradually returned to their pre-ventilation levels within 10–20 days, confirming that strategic nighttime cooling can sustain comfortable conditions for extended periods.

From a thermal comfort perspective, the number of hours within the optimal indoor temperature range (25°C–27°C) increased by 30% when cross-ventilation was systematically applied. Adaptive comfort modeling (Fanger's model) indicated a 14% increase in comfort hours and a 28% reduction in exposure to extreme heat conditions, confirming that cross-ventilation significantly enhances occupant well-being in hot climates.

13.3.5.3 Conclusion

The findings demonstrate that cross-ventilation, in combination with the thermal mass of rammed earth walls, can effectively mitigate overheating risks in historic buildings. The study highlights the long-term benefits of strategic nighttime ventilation, particularly in regions with hot summers and significant diurnal temperature variations. The research underscores the potential for reintegrating traditional passive cooling strategies into heritage conservation efforts, demonstrating that historic buildings can maintain comfort without mechanical cooling systems.

13.3.6 Hybrid Ventilation in Arid Climates: A Systematic Review and Case Study in Dubai

13.3.6.1 Introduction

Hybrid ventilation systems, combining natural and mechanical airflow, are increasingly promoted as energy-efficient alternatives to conventional air conditioning, particularly in hot climates where cooling loads are extreme. In this context, Al Niyadi and Elnabawi (2025) conducted a two-part investigation to assess the viability of hybrid systems in Dubai's hot, arid conditions. The first phase consisted of a systematic review of international literature on hybrid ventilation strategies, focusing specifically on research relevant to arid climates. The second phase involved a detailed case study of a nine-story commercial office building in Dubai, simulated using DesignBuilder and validated against real energy consumption data. The research aimed to quantify energy and carbon savings, understand operational dynamics, and evaluate thermal comfort under mixed-mode ventilation.

13.3.6.2 Methodology

The systematic review encompassed 84 articles published between 2010 and early 2024, narrowing the final selection to 21 studies that met predefined criteria. These included a focus on mixed-mode ventilation in hot, arid climates, use of validated simulations or real-world monitoring, and assessment of thermal comfort using recognized standards.

Building on the literature synthesis, the authors developed a detailed energy model of a Dubai office building. The structure featured a reinforced concrete frame with hollow-core slabs, no heating system, and extensive glazing (28.9% window-to-wall ratio). HVAC was supplied by roof-mounted chillers and air handling units operating at a constant 24°C setpoint. Internal loads included lighting (10 W/m^2), computing (7 W/m^2), and assumed occupancy of 1 person per 10 m^2. The hybrid ventilation model incorporated a changeover control logic, switching to natural ventilation when outdoor temperatures were favorable, and openable glazing set to 15% free area.

The simulation was calibrated using actual electricity data, achieving a root mean square error (RMSE) of 5.3% and a Pearson correlation coefficient of 0.78. Cooling energy represented 69% of the building's total energy use, confirming the high impact of thermal loads in arid climates.

13.3.6.3 Results

The hybrid ventilation scenario achieved an overall annual energy saving of 20.5% compared to the base case. Seasonal variation was significant: energy savings peaked at 42% in January and 35% in December, but dropped to just 11% during the hottest summer months (July–August). The reduction in mechanical cooling demand alone reached 29%, underscoring the system's effectiveness during moderate seasons.

Monthly CO_2 emissions were also reduced by 20% annually, with the greatest reductions occurring in the cooler months when natural ventilation replaced air conditioning for extended periods.

An in-depth seasonal analysis revealed that natural ventilation was feasible for several hours daily between November and March but nearly negligible during summer. Design weeks for both winter and summer showed significantly different air exchange profiles, with infiltration and mechanical cooling dominating during hot months.

Thermal comfort was evaluated using the Givoni adaptive model overlaid on psychrometric charts for both seasons. The results confirmed that the indoor environment remained within the comfort envelope when setpoint temperatures ranged from 25°C to 26.5°C during summer and 20°C to 24°C in winter. Occupants in hot climates demonstrated higher thermal tolerance, supporting the potential to increase setpoints and further reduce cooling demand. The adaptive range extended beyond standard ASHRAE thresholds, indicating that context-specific calibration of comfort models is crucial for hybrid system optimization.

13.3.6.4 Conclusion

This study confirms the potential of hybrid ventilation systems to reduce energy use and CO_2 emissions in arid climates without compromising indoor comfort, provided they are carefully calibrated to seasonal and climatic conditions. While energy savings were less pronounced during peak summer heat, the system proved highly effective in transitional seasons, offering up to 42% cooling load reduction in winter.

The findings also emphasize the importance of flexible control strategies, window operability, and adaptive comfort standards. Rigid reliance on conventional cooling setpoints limits the effectiveness of passive systems in hot regions. Instead, thermal comfort should be defined based on local acclimatization patterns and expanded temperature tolerance.

Although the current application focused on a commercial building, the methodological framework and insights can be extended to residential and institutional buildings in similar climatic zones. The study supports broader adoption of mixed-mode systems and reinforces the value of integrating simulation with empirical validation to inform regionally appropriate design.

13.3.7 Thermal Comfort and Air Movement in Naturally Ventilated Dwellings in a Tropical Climate

13.3.7.1 Introduction

Aqilah et al. (2025) conducted a field-based thermal comfort study in naturally ventilated residential dwellings in Kota Bharu, Kelantan, located in the northeast of Peninsular Malaysia. The research investigated the indoor thermal environment and examined how air movement, particularly through the use of fans, influences perceived comfort temperatures. Unlike conventional studies focused on mechanically cooled or air-conditioned buildings, this work targeted low-energy dwellings dependent exclusively on natural ventilation. The study is set within a tropical climate, where high humidity and persistently elevated temperatures present significant challenges to passive comfort strategies. Its findings provide new evidence for adaptive comfort approaches in warm-humid housing and contribute to the development of region-specific comfort benchmarks.

13.3.7.2 Methodology

The study combined continuous and spot thermal monitoring with structured occupant surveys. In total, 53 dwellings were included in the spot monitoring campaign, while 4 houses were selected for continuous measurements. All buildings were located in a semi-rural region of Kelantan, where detached single-family homes with relatively open surroundings are common. These spatial configurations allow more effective cross-ventilation and reduce urban heat gain, making them ideal for assessing natural ventilation performance.

The description and visual documentation of the study setting and monitored buildings are presented in Figure 13.37, which shows both the regional location and photographs of typical dwellings included in the study. The climate is classified as tropical rainforest, with high year-round humidity and minimal seasonal variation. All monitoring was conducted during the month of August, coinciding with the hot and humid monsoon period.

Thermal measurements included globe temperature, air temperature, relative humidity, and air velocity, with air movement data collected using KANOMAX sensors during spot campaigns. Surveys were conducted simultaneously, capturing thermal sensation, air movement perception, and clothing levels. A seven-point scale adapted from the Bedford scale was used to assess thermal sensation (TSV) and air movement votes (AMV), ranging from "very cold" (1) to "very hot" (7), and "very low" to "very high," respectively.

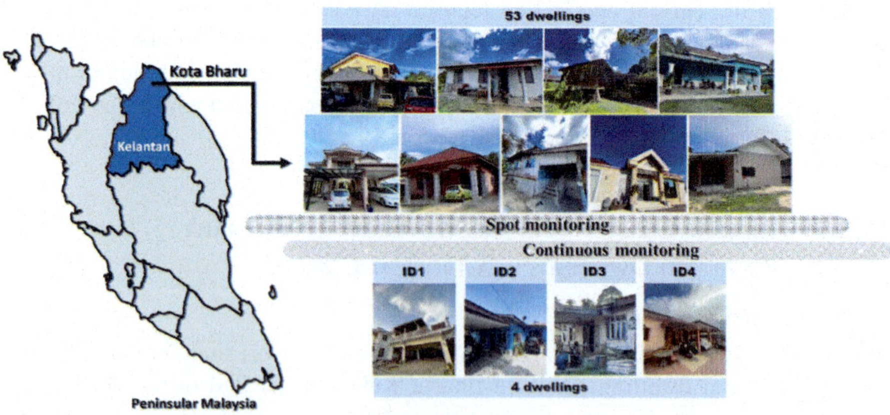

FIGURE 13.37 Location and documentation of naturally ventilated dwellings in Kota Bharu, Malaysia, showing the regional map (left) and photographs of representative houses used for spot and continuous monitoring (right). (Aqilah et al., 2025.)

13.3.7.3 Results

The results revealed that indoor temperatures in naturally ventilated homes were consistently higher than outdoor temperatures, with globe temperatures ranging from 26°C to 32°C when the outdoor air temperature was 25°C. Wooden dwellings tended to exhibit smaller temperature differences compared to concrete homes, likely due to their lower thermal mass and greater permeability to outdoor airflow.

The calculated mean comfort temperature was 30.5°C with a standard deviation of 1.7°C, notably higher than values reported in comparable studies from temperate or air-conditioned contexts. The Griffiths method was used to derive this comfort temperature from occupant responses. The study also quantified the impact of air movement on thermal perception: with fan use, comfort temperature increased by approximately 1°C, supporting the hypothesis that elevated airflow enhances evaporative and convective cooling.

The analysis of thermal sensation and air movement votes showed that 80% of respondents reported comfort at temperatures between 26°C and 33°C, while 71% rated the perceived air movement as "neutral" to "slightly high." Probit analysis of the TSV data revealed that thermal neutrality was achieved at approximately 30°C, and comfort acceptability remained high within a broader range due to adaptive behavior and cultural acclimatization.

Fan usage patterns confirmed this adaptive behavior: 56% of residents reported using ceiling fans during the survey period, while 46% used standing fans. A subset (12%) used both simultaneously, especially during peak afternoon hours. This widespread reliance on mechanical airflow, despite the absence of air conditioning, underscores the importance of integrating air movement into thermal comfort standards for tropical housing.

13.3.7.4 Conclusion

This study offers robust empirical evidence that residents of naturally ventilated homes in hot-humid climates adapt to higher indoor temperatures through both physiological and behavioral strategies. The high mean comfort temperature (30.5°C) and the strong preference for air movement indicate that tropical dwellings can provide acceptable thermal conditions without mechanical cooling, provided that airflow is sufficient and adaptive comfort standards are applied.

Importantly, the findings demonstrate that fan use significantly increases perceived comfort thresholds, supporting design strategies that prioritize ceiling fan infrastructure, cross-ventilation, and openable windows. The wide comfort range observed in this study also calls into question the applicability of global comfort standards developed for air-conditioned buildings, reinforcing the need for region-specific thermal benchmarks.

Future comfort models for tropical housing should explicitly account for the modifying effect of air movement and the cultural acceptability of warmer environments. Moreover, continued field studies across a broader range of building typologies will be essential for establishing reliable, low-energy thermal comfort guidelines in the context of climate adaptation.

13.3.7.5 Summary

The evidence spans simulation studies, full-scale case analyses, monitored dwellings, and systematic reviews across Mediterranean, temperate, humid tropical, and arid climates. Effectiveness is reported using cooling energy savings, peak suppression and delay, discomfort or exceedance hours, adaptive comfort models, and, in some cases, life-cycle carbon outcomes.

Night ventilation. Dynamic simulations in Greece confirmed that increasing nighttime air change rates reduced cooling loads by 5–40% and delayed the following day's peak by up to three hours, though benefits were sharply curtailed in dense urban canyons (Tsoka, 2021). In a Brussels nZEB office, natural window ventilation reduced heat stress more effectively than mechanical purge, improving comfort without raising energy demand, while fan-driven systems increased electricity use despite marginal temperature gains (Amaripadath et al., 2023). In modular housing in China, ventilative cooling reduced annual loads by 7–50% depending on climate, with heavy walls favored in mild regions and lightweight envelopes in heating-dominated zones; under future warming, thermal mass benefits diminished and ventilation emerged as the most consistent adaptive strategy (Hu et al., 2024).

Cross-ventilation. High-rise modelling in Mumbai demonstrated that optimized openings and terraces enabled peak reductions of 5°C–6°C and eliminated discomfort hours in monsoon and winter, with diminishing returns beyond 12 ACH but strong gains from nighttime purge (Stasi et al., 2024). Field and CFD studies in a historic Andalusian courtyard house reported mean reductions of 3.8°C, cumulative cooling effects persisting over several nights, and up to 30% more hours within the adaptive comfort band, confirming the combined role of cross-breezes and high thermal inertia (Palomo Amores et al., 2025).

Hybrid systems. A Dubai case study and systematic review found mixed-mode strategies saved about 20% annually, with peak winter savings above 40% but minimal summer gains when outdoor air was too hot. Comfort analysis showed wider adaptive tolerance in occupants than fixed setpoints assume, underscoring the need for flexible thresholds and seasonal control logic (Al Niyadi & Elnabawi, 2025).

Air movement and adaptive comfort. In Malaysian naturally ventilated homes, residents accepted comfort temperatures around 30.5°C, 3–5 degrees higher than conventional standards, and fan use raised thresholds by about 1°C. More than 70% of occupants reported comfort at 26–33°C, confirming the central role of airflow perception and behavioral adaptation in tropical housing (Aqilah et al., 2025).

Design guidance. Ventilative cooling offers consistent though climate-dependent benefits: 3–6°C peak suppression and 5–40% cooling-load reduction are achievable with well-timed purge and cross-flows. Effectiveness is highest in less dense settings, in buildings with dual façades or courtyards, and when coupled with mass and shading. Hybrid ventilation is most valuable in transitional seasons, with adaptive setpoints and occupant tolerance widening the usable range. In tropical dwellings, integrating ceiling fans and ensuring cross-ventilation pathways is decisive. Overall, climate-responsive control of natural airflow should be prioritized before mechanical means, with hybrid or assisted ventilation reserved for periods when outdoor conditions limit passive potential.

13.4 INTEGRATED PASSIVE STRATEGIES

This chapter will examine how passive measures can be integrated to provide reliable summer comfort and reduce energy demand in residential and urban contexts. The discussion will cover studies of detached houses, apartments, high-rise buildings, social housing, test cells, and dense urban districts across temperate, Mediterranean, humid, and arid climates. The assessment will consider indoor air and operative temperatures, peak suppression and delay, overheating or exceedance hours, seasonal heating and cooling demand, and, in selected cases, life-cycle and district-scale implications.

The chapter will review how passive solar and energy efficiency strategies, combined with higher thermal mass, are expected to lower heating and cooling loads in housing. Roof-integrated measures such as hybrid radiant–evaporative green roofs or ventilated reflective roof systems will be analyzed as potential solutions to mitigate overheating in hot climates. Bundled approaches, including natural ventilation, external shading, and reflective cool roofs, will be presented as strategies for stabilizing indoor conditions. At the urban scale, high-albedo materials and vegetation will be assessed in terms of their role in reducing air temperature and improving local comfort, with attention to placement and density.

Overall, the chapter will argue that single measures rarely provide sufficient resilience, while integrated strategies that combine solar protection, purge ventilation, and thermal storage offer the most robust and adaptable pathway for climate-ready building design.

13.4.1 Energy Conservation in Residential Buildings: Passive Solar, Efficiency, and Thermal Mass

13.4.1.1 Introduction

Albayyaa et al. (2019) conducted a study to evaluate the impact of Passive Solar and Energy Efficiency Design Strategies (PSEEDS) and increased thermal mass on energy consumption and thermal comfort in residential buildings. The experiment analyzed the performance of two common detached house types in Sydney, Australia—a fibro house with lightweight construction and a brick veneer house with higher thermal mass. The study aimed to determine how different external walls and flooring systems, combined with PSEEDS, influenced cooling and heating energy demand in residential buildings.

13.4.1.2 Methodology

The study was conducted using IDA Indoor Climate and Energy (IDA-ICE) simulation software to assess the thermal performance of two house types, each covering 400 m^2 over two floors. The layout of the test houses, including their orientation and construction details, is shown in Figure 13.38.

The simulations included two construction types:

- Fibro house—Lightweight construction with raised timber flooring and fibro walls (R-value = 1.2 m^2K/W).

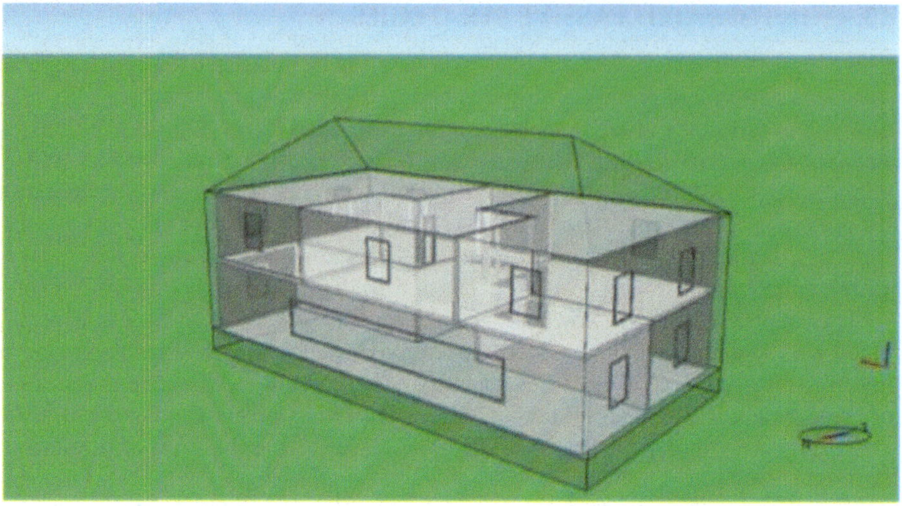

FIGURE 13.38 Schematic layout of the test houses used in the study, illustrating building orientation, construction details, and monitored zones for energy performance assessment. (Albayyaa et al., 2019.)

- Brick veneer house—Higher thermal mass construction with concrete slab flooring and brick veneer walls (R-value = 3.2 m^2K/W).

Additionally, the study examined the impact of PSEEDS, implementing four scenarios:

1. Standard fibro house (no PSEEDS).
2. Improved fibro house (with PSEEDS).
3. Standard brick veneer house (no PSEEDS).
4. Improved brick veneer house (with PSEEDS).

The PSEEDS measures included:

- Double-glazed windows (U-value = 2.9 W/m^2K).
- Internal shading devices.
- Improved ventilation and infiltration rates (0.5 ACH instead of 2.0 ACH in standard houses).

To validate the simulation results, the study compared simulated energy consumption with real-world electricity usage from two actual houses in Sydney, a fibro house and a brick veneer house. The site plan for the actual houses used for validation is shown in Figure 13.39.

13.4.1.3 Results

The results indicated that applying PSEEDS significantly reduced energy demand. In the fibro house, incorporating PSEEDS led to a 37% reduction in annual heating

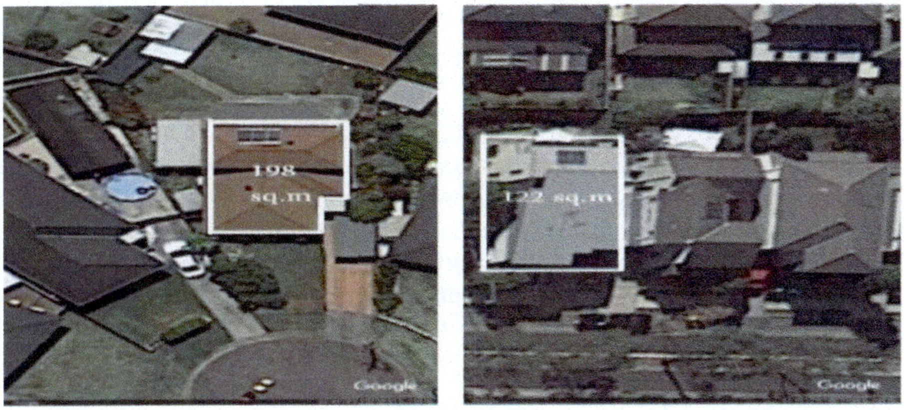

FIGURE 13.39 Site plan of the actual residential houses used for energy validation, showing their locations in Sydney, Australia, and orientation relative to solar exposure. (Albayyaa et al., 2019.)

and cooling energy consumption, while in the brick veneer house, energy demand decreased by 36%.

Thermal mass played a crucial role in reducing cooling and heating loads. Compared to the fibro house, the brick veneer house reduced total energy consumption by 35%, demonstrating the benefits of higher thermal inertia. The simulation results showed that the brick veneer house maintained a more stable indoor temperature, while the fibro house exhibited greater temperature fluctuations, closely following outdoor conditions.

Additionally, the validation results showed that the simulated energy consumption was within a 16% error margin compared to actual measured household energy use, confirming the reliability of the modeling approach.

13.4.1.4 Conclusion

The study confirmed that incorporating Passive Solar and Energy Efficiency Design Strategies (PSEEDS) and increasing thermal mass significantly improves energy efficiency in residential buildings. The results demonstrated that upgrading a fibro house with PSEEDS can reduce energy demand by 37%, while a brick veneer house with PSEEDS achieved up to 58% energy savings compared to a standard fibro house.

The findings emphasize the importance of integrating passive solar strategies during the design phase, including double glazing, shading devices, and optimized ventilation rates, to enhance energy efficiency and indoor thermal comfort. Additionally, using high thermal mass materials such as brick veneer and concrete slab flooring effectively reduces temperature fluctuations and improves heat retention in winter and cooling in summer.

This study provides valuable insights into cost-effective strategies for reducing energy consumption in residential buildings, supporting the adoption of passive solar design principles and thermal mass optimization in future housing developments (Albayyaa et al., 2019).

13.4.2 Passive Cooling with a Hybrid Green Roof in Extreme Climates

13.4.2.1 Introduction

The study conducted by La Roche et al. (2020) investigated the effectiveness of a hybrid green roof system integrating radiant and evaporative cooling for improving thermal comfort in hot and dry climates. The research aimed to assess how a combination of vegetative shading, evaporative cooling, and radiant heat transfer could enhance passive cooling performance while minimizing energy consumption. Additionally, the study explored real-world applications of this cooling approach, particularly in outdoor urban environments. The research examined design modifications of the hybrid green roof system, integrating elements such as wind capture areas, misting systems, and radiant cooling panels.

13.4.2.2 Methodology

The study was conducted at the Lyle Center for Regenerative Studies at California State Polytechnic University, Pomona, USA. The experimental setup consisted of outdoor test cells, each measuring 1.35 m × 1.35 m × 1.35 m, designed to analyze different green roof configurations under real climatic conditions. The test cells were exposed to external weather conditions, simulating real-world heat loads while varying roof insulation, vegetation coverage, and ventilation rates. The test setup and monitored parameters are shown in Figure 13.40.

The hybrid green roof system incorporated a radiant cooling mechanism with a closed-loop water pipe embedded in the soil layer, allowing heat transfer from the

FIGURE 13.40 Experimental setup for testing a hybrid green roof system, with test cells evaluating various passive cooling configurations. (La Roche et al., 2020.)

indoor space to the roof. Additionally, an evaporative sprinkler system was integrated to enhance cooling efficiency by reducing soil and air temperatures through evaporative processes. The water flow rate and sprinkler operation were optimized based on external humidity and diurnal temperature cycles to maximize cooling effects. To further enhance cooling efficiency, night ventilation was applied using an exhaust fan, promoting heat dissipation from the indoor space during nighttime hours. The cooling effect of different configurations was compared by monitoring indoor air temperatures, roof surface temperatures, and relative humidity levels over an extended period. Additionally, the study explored urban-scale adaptations of the hybrid green roof system in public spaces and courtyards, optimizing design modifications to improve thermal comfort in outdoor settings. The design integration of passive cooling strategies within an urban context is illustrated in Figure 13.41.

13.4.2.3 Results

The study demonstrated that the hybrid green roof significantly reduced peak indoor temperatures compared to conventional roofs. The maximum indoor temperature in the test cell with the hybrid green roof was 9°C–13°C lower than outdoor temperatures, confirming the system's ability to mitigate extreme heat loads.

Among the tested configurations, the radiant-evaporative green roof system exhibited the highest cooling efficiency, maintaining an average indoor temperature of 27.6°C, compared to 29.7°C in the control insulated roof, 30.3°C in the uninsulated green roof, and 36.4°C outside.

The study also found that combining the green roof with night ventilation further improved cooling efficiency, particularly on hot and dry days with significant

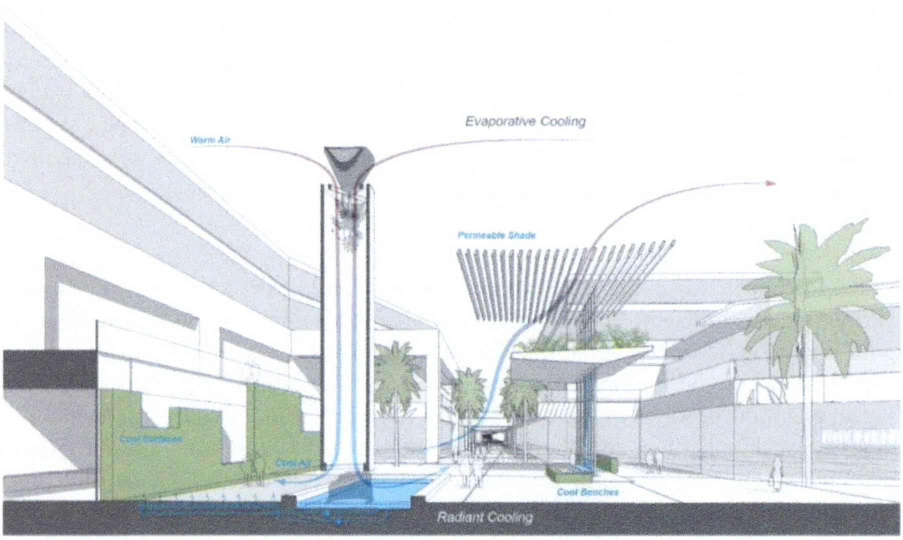

FIGURE 13.41 Concept design for integrating passive cooling strategies into urban public spaces, showing evaporative cooling towers, permeable shading, radiant-cooled seating, and water features to improve thermal comfort. (La Roche et al., 2020.)

diurnal temperature swings. The greatest cooling effects were observed when temperature swings exceeded 12°C, indicating that passive cooling strategies are most effective under conditions with high daytime heat and significant nighttime cooling potential.

13.4.2.4 Conclusion

The study confirmed that hybrid green roofs integrating radiant and evaporative cooling can effectively enhance thermal comfort in extreme climates, significantly reducing indoor temperatures without mechanical cooling systems. The combination of green vegetation, water-based radiant cooling, and evaporative effects provided substantial cooling benefits, particularly when paired with night ventilation strategies.

Additionally, the study highlighted the potential for adapting hybrid green roofs beyond building applications, incorporating them into urban cooling infrastructure. The design variations tested for urban applications (Figure 13.41) demonstrated the potential for integrating misting systems, shading structures, and wind-assisted cooling features to improve thermal comfort in outdoor spaces.

The findings suggest that future urban developments in hot climates should consider integrating hybrid passive cooling solutions, utilizing radiant-evaporative green roofs in combination with shading, night ventilation, and water-based cooling technologies (La Roche et al., 2020).

13.4.3 Natural Ventilation, Shading, Cool Roof, and Thermal Mass: Field and Simulation Study

13.4.3.1 Introduction

Li et al. (2024) investigated the impact of natural ventilation, external shading, cool roofs, and thermal mass on indoor thermal conditions in a cooling-dominated climate. The study was conducted in Tongling, Anhui Province, China, which is classified as Cfa (warm temperate, fully humid, hot summer) in the Köppen system. Both short- and long-term effects of these passive cooling measures were assessed through a combination of field monitoring and numerical simulations.

13.4.3.2 Methodology

The experiment was conducted on two identical single-story chambers, each measuring 3 × 3 × 3 meters. Both chambers had the same construction and orientation, with a window-to-wall ratio of 0.2. The test chamber incorporated passive cooling strategies, while the reference chamber remained unmodified for comparison. The layout and modifications of the test chamber are illustrated in Figure 13.42a.

The test chamber was equipped with a natural ventilation system, a cool roof, and external shading. The natural ventilation system used an earth-air heat exchanger (EAHE) and a solar chimney to improve airflow and maintain continuous ventilation. The cool roof was coated with a reflective material with an albedo of 0.84 and thermal emittance of 0.92, while the external shading device consisted of adjustable louvers positioned at 90° to block direct solar radiation. The configurations of these passive cooling strategies are shown in Figures 13.42b, 13.42c, and 13.42d.

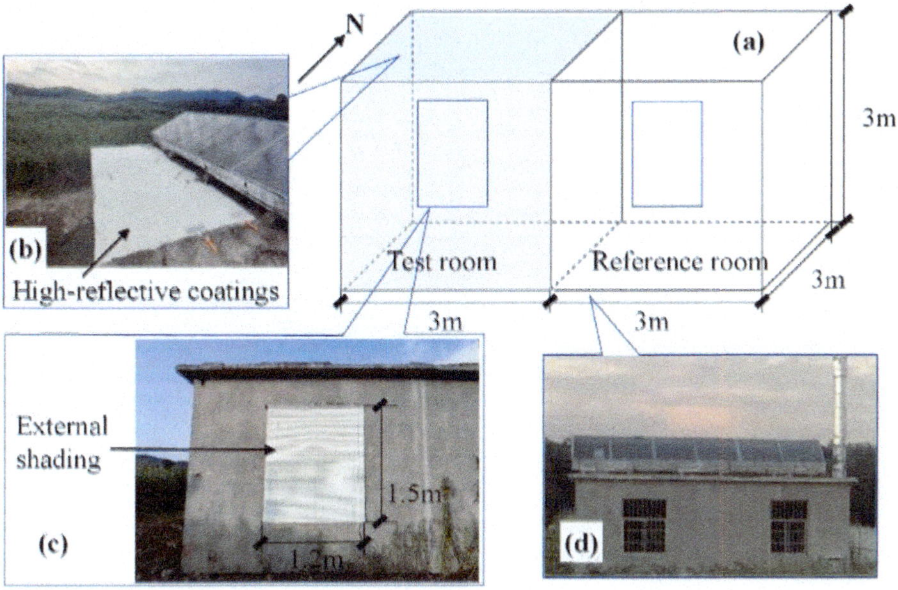

FIGURE 13.42 Experimental setup of the test and reference chambers. (Li et al., 2024.)

Figure 13.43 illustrates the four experimental scenarios tested. In the natural ventilation scenario, a vertical chimney with a solar collector induced airflow through the chamber, while ambient air entered via an underground earth-air heat exchanger. In the external shading scenario, shading panels blocked direct solar radiation on the south-facing wall. The cool roof scenario used a reflective roof surface to reduce heat gain. The combined system integrated all three measures: natural ventilation via the solar chimney and EAHE, external wall shading, and a reflective cool roof. In the diagrams, arrows indicate airflow direction, sun icons represent solar exposure, and blue and red lines show the movement of cool and warm air.

The study was conducted under these four scenarios between August 10 and August 28, 2021, with each configuration tested for 48 hours. Daily outdoor temperatures ranged from 19.2°C to 38.6°C, with peak solar radiation reaching 1088 W/m^2. Temperature and airflow data were recorded continuously to evaluate the effectiveness of each passive cooling strategy.

13.4.3.3 Results

The natural ventilation system reduced the mean indoor air temperature by 1.3°C, with a maximum reduction of 3.0°C compared to the reference chamber. The airflow rate varied between 50 and 245 m^3/h, enhancing heat dissipation. Internal wall temperatures in the test chamber ranged from 27.2°C to 29.1°C, while in the reference chamber, they varied from 27.8°C to 30.6°C, resulting in an average reduction of 1.0°C. The window glass temperature in the test chamber was lowered by an average of 1.1°C, with a peak reduction of 4.9°C due to the airflow effect.

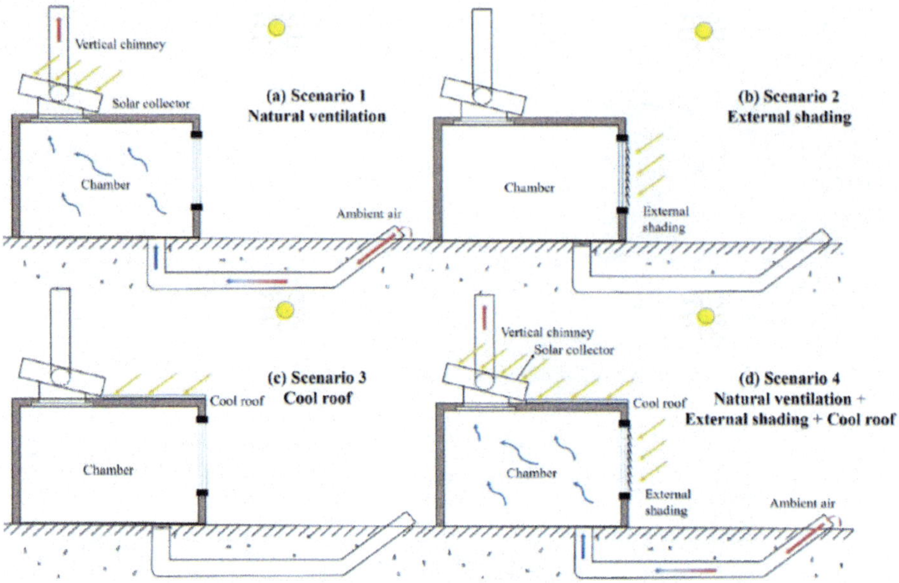

FIGURE 13.43 A conceptual diagram displaying the four experimental conditions: Natural ventilation only, external shading only, cool roof only, and a combined system using all three passive strategies. Li et al. (2024) investigated the impact of natural ventilation, external shading, cool roofs, and thermal mass on indoor thermal conditions in a cooling-dominated climate. The study was conducted in Tongling, Anhui Province, China, which is classified as Cfa (warm temperate, fully humid, hot summer) in the Köppen system. Both short- and long-term effects of these passive cooling measures were assessed through a combination of field monitoring and numerical simulations.

The external shading system demonstrated a stronger cooling effect, reducing the mean indoor air temperature by 2.0°C, with a maximum reduction of 3.1°C. Internal wall temperatures in the test chamber ranged from 27.1°C to 28.5°C, compared to 28.4°C to 30.0°C in the reference chamber, showing a temperature drop of 1.3°C. The window glass temperature was significantly reduced by an average of 2.6°C, with a peak reduction of 14.6°C, highlighting the effectiveness of external shading in preventing heat gain.

The cool roof system reduced external surface temperatures from 25.2°C to 43.2°C (reference chamber) to 23.1°C to 37.1°C (test chamber). The internal roof surface temperature in the test chamber was on average 1.3°C lower than in the reference chamber, leading to a mean indoor air temperature reduction of 1.1°C. The maximum temperature reduction was 1.3°C, with the cooling effect being more noticeable during nighttime, as the cool roof enhanced heat dissipation.

The combined strategy of natural ventilation, external shading, and cool roof produced the greatest cooling effect, reducing the mean indoor air temperature by 2.8°C, with a maximum reduction of 4.3°C. The test chamber maintained indoor temperatures between 26.2°C and 28.1°C, while the reference chamber ranged from 28.2°C to 32.3°C. Internal wall temperatures were lowered by 1.5°C on average, and roof temperatures decreased by 1.4°C. The window glass temperature saw an average reduction of 2.6°C, with a maximum drop of 13.4°C, confirming that combining multiple passive cooling measures enhances overall efficiency.

13.4.3.4 Conclusion

The study confirmed that passive cooling strategies significantly enhance indoor thermal comfort and reduce heat gain in buildings situated in hot and humid climates. Among the individual measures, external shading proved to be the most effective, offering the greatest reduction in indoor temperature. Natural ventilation and cool roof strategies also contributed to cooling, although their impact was more dependent on airflow conditions and roof exposure.

The combined use of all three passive strategies provided the highest overall cooling effect, demonstrating that an integrated approach is the most effective way to regulate indoor temperatures. This strategy not only reduced peak temperatures but also stabilized daily fluctuations, preventing excessive heat accumulation. The findings indicate that layering multiple passive cooling techniques optimizes heat dissipation and minimizes indoor overheating, especially during peak temperature hours.

The study also highlighted the role of thermal mass in moderating temperature variations. While increasing specific heat capacity can contribute to indoor stability, excessive thermal mass may lead to higher heat retention in certain conditions. Similarly, reducing heat transfer through building materials was shown to significantly lower indoor temperatures, reinforcing the importance of material selection in energy-efficient building design.

13.4.4 Predicted Effectiveness of Climate Adaptation Measures for Residential Buildings

13.4.4.1 Introduction

Van Hooff et al. (2015) evaluated the effectiveness of passive climate adaptation measures in residential buildings to mitigate overheating risks due to climate change. The focus was on six adaptation strategies: increasing thermal resistance, modifying thermal mass, enhancing short-wave reflectivity, implementing vegetated roofs, applying solar shading, and utilizing additional natural ventilation. The research analyzed three types of dwellings—detached houses, terraced houses, and apartments—built according to 2012 regulations and 1970s construction standards. The study used dynamic thermal simulations in EnergyPlus to assess how these measures impact indoor thermal comfort in the context of projected future climate conditions in the Netherlands.

13.4.4.2 Methodology

The simulations were conducted for three generic residential buildings: a detached house (451 m^3), a terraced house (348 m^3), and an apartment (210 m^3). Each building was modeled in two versions, one based on 1970s construction with lower insulation values and another following 2012 Dutch building regulations with higher thermal resistance. The apartment unit was located on the top floor of a multi-story building, exposing its roof and two façades to external conditions. The terraced house had only two exposed sides, while the detached house was fully exposed.

The six passive climate adaptation measures were tested under four different orientations (0°, 90°, 180°, and 270°) to assess the impact of solar exposure. The primary performance indicator was the number of overheating hours, defined as hours exceeding adaptive comfort temperature limits based on Peeters et al. (2009). The weather data used was from De Bilt, the Netherlands, during the summer of 2006, which experienced multiple heat waves and served as a proxy for future climate conditions.

The six adaptation measures included:

- Increased thermal resistance (RC50, RC65): Raising insulation levels to RC = 5.0 m^2K/W and 6.5 m^2K/W.
- Modified thermal mass (TM_low): Replacing concrete walls and ceilings with wooden elements, reducing the building's heat storage capacity.
- Increased short-wave reflectivity (SWR06, SWR08): Raising external albedo from 0.3 to 0.6 and 0.8, respectively.
- Vegetated roofs (VR): Installing extensive green roofs with 0.1 m high sedum plants and a Leaf Area Index (LAI) of 5.
- Solar shading (SH): Installing automated external blinds that deploy when solar radiation exceeds 150 W/m^2.
- Natural ventilation (NV_all, NV_day): Enabling window opening based on temperature thresholds (24°C) and air quality, either 24/7 (NV_all) or only during the day (08:00–20:00, NV_day).

The dynamic thermal simulations were performed with six time steps per hour in EnergyPlus, incorporating airflow network models to capture ventilation effects. The operative temperature was used to assess indoor comfort, calculated as the average of mean radiant and air temperatures. The upper comfort threshold varied depending on the running mean outdoor temperature. Overheating was quantified as both total hours exceeding the threshold and degree-hours, which account for temperature deviations above the threshold.

13.4.4.3 Results

13.4.4.3.1 Detached House

For the 1970s detached house, increasing thermal resistance (RC50, RC65) led to higher indoor temperatures, trapping solar heat gains from windows. The lowering of thermal mass (TM_low) resulted in 53% more overheating hours on the ground floor, while reducing them by 45% on the first floor, due to faster nighttime cooling. Increasing albedo (SWR06, SWR08) reduced overheating by up to 89%, while vegetated roofs had a limited impact, lowering overheating by only 17%. Natural ventilation (NV_all) and solar shading (SH) were the most effective, nearly eliminating overheating hours.

For the 2012 detached house, overheating was significantly worse than in the 1970s version, emphasizing the unintended effects of modern insulation. Albedo effects were less pronounced, reducing overheating by only 27% compared to 89%

in the older model. Natural ventilation and solar shading remained the most effective solutions, again eliminating nearly all overheating issues.

13.4.4.3.2 Terraced House

The terraced house exhibited strong orientation effects, with west- and east-facing homes experiencing nearly twice as many overheating hours as north- and south-facing ones. Similar to the detached house, increasing thermal resistance raised overheating risks. Reducing thermal mass (TM_low) increased overheating on the ground floor (40%) but decreased it on the first floor (18%). Albedo improvements reduced overheating but were less effective than in detached homes, due to fewer exposed surfaces. Solar shading (SH) and full-day natural ventilation (NV_all) were the most effective strategies, significantly cutting overheating risks.

13.4.4.3.3 Apartment

The apartment suffered the highest overheating levels, due to its exposure to direct solar radiation on the roof and façades. Overheating was significantly worse than in the other building types, with degree-hours exceeding 8000 in the living room for the 2012 version. While natural ventilation and solar shading remained effective, they could not eliminate overheating entirely. Increasing thermal mass (TM_low) helped reduce nighttime temperatures, but only when ventilation was available. Albedo adjustments had less impact due to limited exposed surface area.

13.4.4.3.4 Conclusion

The study highlighted that passive climate adaptation measures can significantly impact indoor thermal comfort, but their effectiveness depends on building type, insulation levels, and ventilation strategies. Modern, well-insulated homes (2012 version) experienced greater overheating risks than older, less insulated homes, demonstrating that insulation improvements must be paired with effective cooling strategies.

Among the tested measures, solar shading and natural ventilation were the most effective, particularly for detached and terraced houses. Increasing albedo had a strong effect on older buildings but was less effective in modern constructions. Lowering thermal mass helped with nighttime cooling, but only when combined with ventilation. Vegetated roofs provided limited benefits, as their insulative properties could sometimes trap heat rather than dissipate it.

The findings emphasize the importance of integrated climate adaptation strategies, particularly for future-proofing highly insulated buildings. New construction and renovation projects should prioritize shading, ventilation, and reflective materials to mitigate overheating risks. Further research should explore adaptive façade technologies, hybrid passive-active cooling systems, and human behavior modeling in ventilation control.

13.4.5 Passive Design Strategies for Future Climate in Hot-Humid Hong Kong

13.4.5.1 Introduction

Liu et al. (2020) conducted a numerical study assessing the effectiveness of passive design strategies in mitigating the impact of future climate change on residential

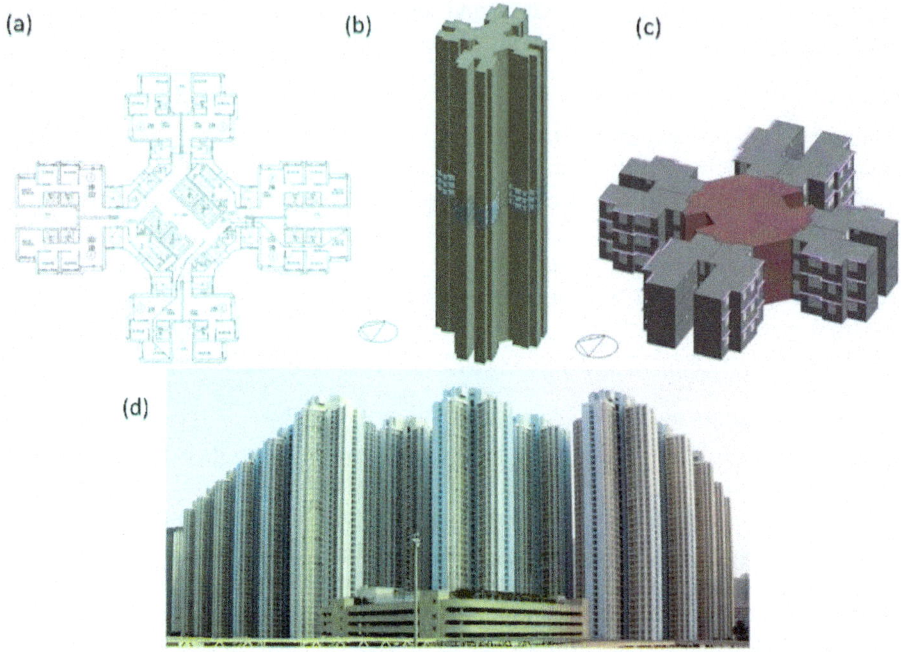

FIGURE 13.44 Typical high-rise public rental housing (PRH) buildings in Hong Kong: (a) typical floor plan; (b) whole-building simulation model; (c) mid-floor residential units; (d) photograph of the building complex. (Liu et al., 2020.)

high-rise buildings in Hong Kong. Using building energy simulations, the study examined how airtightness, natural ventilation, solar protection, and insulation perform under projected climate conditions up to 2090. The selected case study building was a 30–40-story public rental housing (PRH) development, representative of Hong Kong's dominant residential typology. The building layout and simulation approach are illustrated in Figure 13.44.

13.4.5.2 Methodology

The study employed EnergyPlus simulations to analyze a typical high-rise PRH building, focusing on three climate scenarios: the current Typical Meteorological Year (TMY), the moderate climate projection (RCP4.5), and the extreme climate projection (RCP8.5) for 2035, 2065, and 2090. The model included habitable units from mid-level floors, while the remaining structure was modeled as adiabatic blocks to simulate shading effects. Simulations incorporated occupancy, lighting, and equipment loads based on Hong Kong residential standards and evaluated thermal comfort using the Adaptive Comfort Standard (ACS) model.

The study examined passive strategies such as solar protection, airtightness improvements, enhanced insulation, and window configurations. The Building Bio-Climatic Chart (BBCC) and sensitivity analysis (SA) were used to evaluate the time-dependent effectiveness of these strategies under changing climatic conditions.

13.4.5.3 Results

Future climate change will significantly increase the number of discomfort hours requiring cooling, even with adaptive thermal models. In current conditions (TMY), 2058 discomfort hours per year were recorded, but this figure doubled to 4106 under the RCP8.5-2090 scenario, highlighting the growing demand for cooling.

Implementing passive strategies today reduces annual cooling energy demand by 55.1% and peak demand by 38.1%. In the worst-case climate scenario (RCP8.5-2090), these savings increase to 56.7% and 64.5%, respectively. Solar protection strategies, including overhang shading and reflective glazing, remain effective but gradually lose efficiency in extreme heat. Airtightness improvements gain importance, with their effectiveness increasing by 157% in RCP4.5-2090 and 329% in RCP8.5-2090 due to rising temperatures amplifying air leakage-related energy losses.

Natural ventilation, currently an effective cooling method, loses efficiency as outdoor temperatures rise, reducing its viability as a primary cooling strategy. Wall insulation and thermal mass become more critical, helping stabilize indoor conditions despite climate fluctuations. Passive retrofitting measures, including solar protection, airtightness improvements, and insulation, resulted in peak indoor temperature reductions of 3.6°C (TMY) and 3.1°C (RCP8.5-2090), demonstrating their potential for long-term energy savings.

13.4.5.4 Conclusion

The study confirms that climate change will drastically increase cooling demand in residential buildings, making passive design strategies essential for long-term energy efficiency. While natural ventilation remains beneficial today, it will become ineffective in future climates, shifting the focus to airtightness and insulation as the most important factors. Solar protection will continue to play a role, but will gradually lose dominance in favor of building envelope improvements.

By 2090, insulation and airtightness will be key in reducing energy consumption, while natural ventilation will no longer suffice as a cooling method. Future research should explore hybrid strategies integrating passive and mechanical cooling approaches to enhance building performance in a warming climate.

13.4.6 Improving Social Housing Habitability through Passive Cooling: A Case Study in Mengíbar, Spain

13.4.6.1 Introduction

Delgado et al. (2022) conducted a study on the renovation of social housing in Mengíbar, Spain, focusing on passive cooling strategies to combat energy poverty and overheating. The study aimed to enhance thermal comfort and energy efficiency in 14 three-story residential buildings constructed in 1984, which suffer from poor insulation and high summer temperatures. Many residents, due to financial constraints, lack air conditioning, leading to indoor temperatures frequently exceeding 30°C in summer. The research sought to develop cost-effective passive cooling technologies that would be both economically viable and highly effective in reducing indoor overheating and energy consumption.

13.4.6.2 Methodology

The study was conducted in Mengíbar, Jaén, Spain, in a social housing complex comprising 14 buildings. The district layout and architectural characteristics are presented in Figure 13.45.

The buildings, constructed with limited insulation and outdated thermal performance, had measured U-values of 1.4 W/m^{2}K for walls and 5.6 W/m^{2}K for windows, contributing to high thermal discomfort levels.

A monitoring campaign from April to September 2019 was conducted, recording indoor temperature and humidity levels in six representative apartments. Sensors were installed in living rooms and bedrooms, collecting data at 10-minute intervals. Temperature readings confirmed severe overheating, particularly in top-floor apartments, where peak indoor temperatures were consistently higher than in lower floors. The temperature differences ranged up to 7°C due to excessive heat gains through uninsulated roofs.

To mitigate overheating, the study implemented a ventilated double-layer roof system, designed to reduce solar gains and enhance passive cooling. The roof structure,

FIGURE 13.45 Residential housing district in Mengíbar before renovation. (Delgado et al., 2022.)

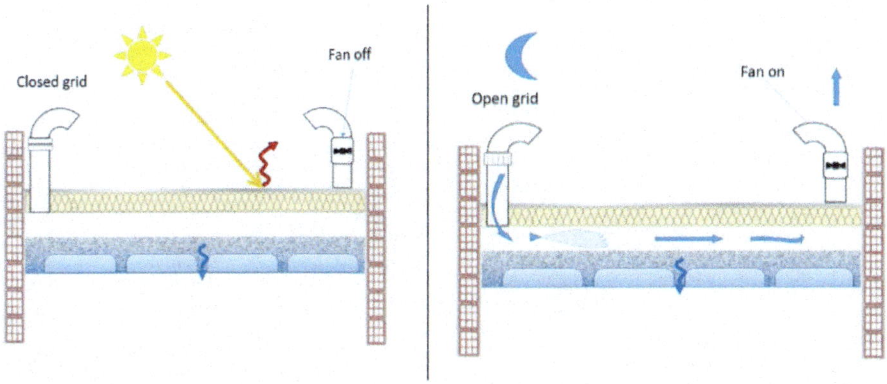

FIGURE 13.46 Schematic of the ventilated roof system. (Delgado et al., 2022.)

shown in Figure 13.46, consists of an external EPS panel (8 cm thick) with high solar reflectance and an internal concrete layer that acts as thermal mass, stabilizing indoor temperatures. A 5 cm ventilated air cavity was introduced between the layers to allow for controlled airflow. The ventilation system activated at night when outdoor temperatures dropped below 22°C, allowing cool air to circulate and dissipate accumulated heat. In extreme heat conditions, evaporative cooling was integrated to further enhance thermal regulation.

To ensure optimal airflow distribution, Computational Fluid Dynamics (CFD) simulations were performed, analyzing air movement across different roof module configurations. Three roof module types were designed, adapted to specific building layouts and orientations. In total, 2500 m^2 of ventilated roofing was installed across the complex. The final renovated district, incorporating the new roofing system, is shown in Figure 13.47.

13.4.6.3 Results

The ventilated roof system significantly improved thermal comfort and energy efficiency. Cooling energy demand decreased by 65% when night ventilation and evaporative cooling were used together, while heating demand dropped by 50% due to improved insulation. The number of hours of thermal discomfort was reduced by 80%, with peak summer temperatures in top-floor apartments lowering by 4°C.

When the ventilated roof operated in passive mode, cooling demand decreased by 28%, while active night ventilation reduced demand by 39%. The greatest cooling efficiency (65% reduction) was achieved when evaporative cooling was used throughout the day. The ventilated roof successfully stabilized indoor temperatures, preventing excessive heat accumulation and reducing overreliance on air conditioning systems.

Residents reported high satisfaction levels, noting improved nighttime comfort and a reduction in summer heat stress. The integration of cost-effective passive cooling solutions proved to be a viable strategy for addressing overheating and energy poverty in social housing.

FIGURE 13.47 Final condition of the district after renovation. (Delgado et al., 2022.)

13.4.6.4 Conclusion

The study highlights the effectiveness of ventilated double-layer roofs as a passive cooling strategy for hot, arid climates. By combining solar-reflective insulation, thermal mass, and controlled ventilation, the system significantly reduced cooling demand while maintaining thermal stability.

The results demonstrate that passive cooling technologies can be highly effective even in extreme summer climates, providing a low-cost alternative to energy-intensive air conditioning systems. The ventilated roof system reduced the need for mechanical cooling while ensuring stable indoor temperatures, making it a sustainable and affordable solution for low-income housing.

Another significant outcome is the importance of integrating advanced simulation techniques, such as CFD modeling, to optimize airflow patterns and cooling efficiency. The study also emphasizes the scalability of the ventilated roof system, making it replicable for similar housing contexts in other regions facing overheating and energy poverty challenges.

In a broader perspective, this research provides strong evidence that passive cooling strategies can improve urban resilience, particularly in regions where low-income populations are most vulnerable to heat stress. By reducing cooling energy demand and improving indoor living conditions, such interventions can play a crucial role in sustainable urban planning and building retrofitting programs.

13.4.7 Cooling and Energy Savings from Vegetation and Albedo: A Case Study in Gyeonggi-Do, South Korea

13.4.7.1 Introduction

Wang et al. (2025) conducted a study to evaluate the impact of vegetation and reflective surfaces (albedo) on urban cooling and building energy consumption in high-density, low-rise residential areas of Gyeonggi-Do, South Korea. The research aimed to assess the effectiveness of these strategies in mitigating the urban heat island effect and reducing cooling energy demand, with a specific focus on different vegetation configurations and material reflectivity. The study employed numerical simulations using ENVI-met and DesignBuilder to quantify the impact of these strategies under various urban scenarios.

13.4.7.2 Methodology

The research was conducted in six low-rise, high-density residential areas within Gwacheon-si and Uiwang-si, Gyeonggi Province, chosen based on their compact urban form, limited vegetation cover, and high solar heat exposure. The site locations and study areas are illustrated in Figure 13.48, showing selected neighborhoods and on-site meteorological measurements used for model validation.

The study analyzed five simulation scenarios: (1) existing conditions, (2) high-albedo cooling materials, (3) increased vegetation, (4) removal of vegetation, and (5) a combined strategy of cooling materials with vegetation. Microclimate effects were simulated using ENVI-met (V5.6), a three-dimensional urban climate modeling tool, while building energy consumption was evaluated using DesignBuilder software, integrating modified EPW weather files based on ENVI-met output.

To enhance model accuracy, the study included on-site meteorological validation, comparing simulated air temperature and humidity data with ground measurements from urban sensors. The urban layout, block morphology, and land cover properties were classified based on satellite imagery, LiDAR urban data, and GIS mapping. Cooling load simulations accounted for varied street orientations, vegetation types, and material reflectivity levels, ensuring a comprehensive analysis of energy efficiency improvements.

13.4.7.3 Results

The study demonstrated that cooling materials and vegetation exerted distinct yet complementary effects in reducing urban temperatures and building energy consumption. High-albedo materials provided the most substantial temperature reduction, whereas the effectiveness of vegetation varied depending on placement and density.

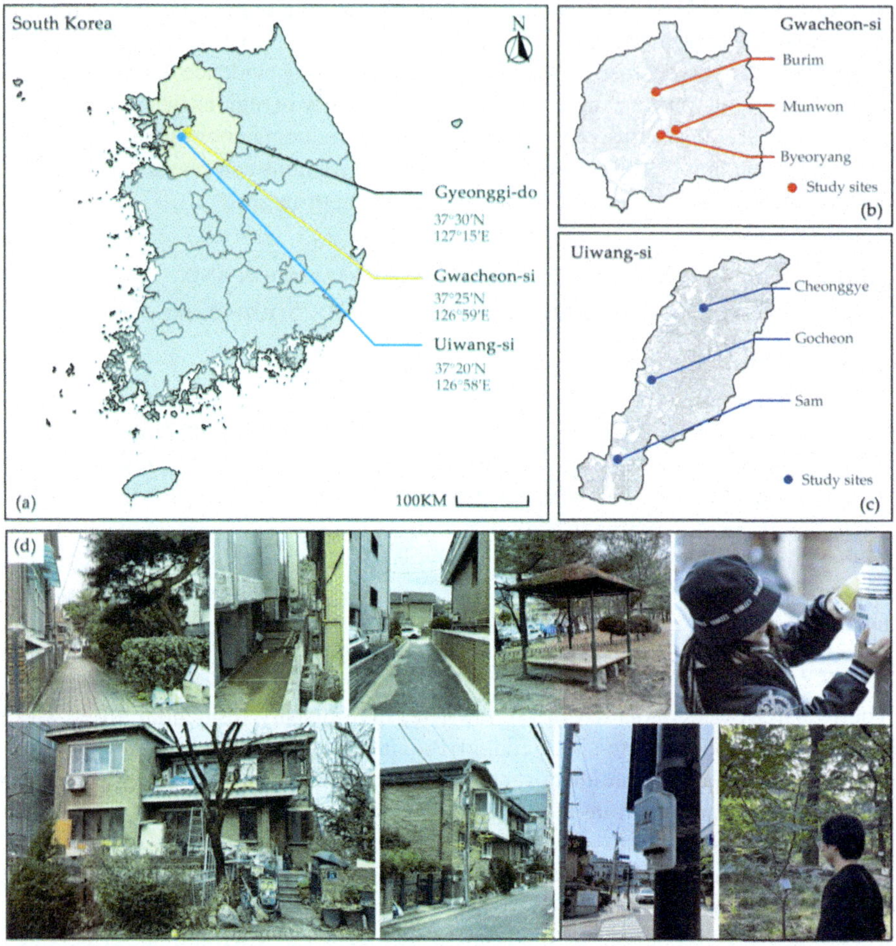

FIGURE 13.48 Site location and on-site measurement: (a) the location of Gyeonggi-do, Gwacheon-si, and Uiwang-si; (b,c) study sites in Gwacheon-si and Uiwang-si; (d) site condition and on-site meteorological measurement. (Wang et al., 2025.)

13.4.7.3.1 Impact on Urban Air Temperature

Cooling materials reduced air temperatures by 1.9°C, with the greatest effect occurring at peak solar hours. Vegetation contributed a slight reduction in daytime temperatures, but its heat retention properties limited nighttime cooling efficiency. The highest daily urban heat island (UHI) mitigation was achieved with cooling materials (2.4°C), followed by vegetation (1.6°C), highlighting the different thermal behaviors of these strategies.

13.4.7.3.2 Reduction in Building Energy Consumption

The use of cooling materials led to a 10.37% reduction in summer cooling energy demand, making it the most effective passive strategy for improving building energy efficiency. In contrast, vegetation alone resulted in moderate energy savings (4–6%), with shrubs and grass offering better cooling effects than tall trees due to improved airflow and reduced shading effects on facades. The combination of high-albedo surfaces and vegetation optimized the microclimate, leading to a total energy demand reduction of 12.5%, underscoring the potential of integrated approaches in urban cooling strategies.

13.4.7.3.3 Effectiveness of Cooling Strategies in Dense Urban Areas

The impact of cooling measures was influenced by urban density. In compact urban environments, excessive vegetation hindered ventilation, causing higher nighttime temperatures and increased cooling loads. The removal of vegetation resulted in a temperature rise of 0.29°C and an increase in energy demand by 4.3 kWh/day, further emphasizing the importance of balanced greenery distribution. Notably, cooling materials consistently lowered temperatures and reduced building cooling demand, regardless of vegetation cover, reaffirming their reliability in diverse urban settings.

These findings highlight the need for context-specific urban cooling strategies, where material reflectivity and vegetation placement are carefully optimized to enhance thermal comfort and energy efficiency in built environments.

13.4.7.4 Conclusion

The study highlighted that high-albedo materials are the most effective strategy for reducing air temperature and cooling energy consumption in dense urban environments. Vegetation can enhance cooling efficiency when carefully placed, but may reduce nighttime heat dissipation in compact areas. The combination of reflective materials with optimized vegetation cover provides the best balance between urban cooling and energy efficiency.

These findings suggest that future urban planning should prioritize high-albedo surfaces in dense areas while strategically integrating greenery to enhance cooling without impeding ventilation. Future studies should explore adaptive shading systems, hybrid cooling strategies, and seasonal variations in vegetation performance to further improve urban thermal comfort and energy sustainability.

13.4.8 Vegetation Effects on Indoor Temperature, Airflow, and Comfort in Courtyard Buildings

Sun et al. (2021) investigated how courtyard vegetation influences indoor thermal conditions, natural ventilation, and thermal comfort in hot climates. The study aimed to determine the cooling potential of different vegetation configurations and their effects on airflow patterns and occupant comfort in surrounding rooms.

Figure 13.49 presents the study layout, illustrating the spatial organization of the courtyard, monitored rooms, and ventilation pathways used in the analysis. The research focused on a courtyard building where natural airflow and shading were expected to influence indoor temperature and comfort conditions.

The study compared three configurations of the courtyard. The first scenario was a baseline case, where the courtyard was left open without vegetation, relying solely on window openings for passive cooling. In the second scenario, small trees were planted in each corner of the courtyard, introducing shading and evapotranspiration effects. The third scenario included larger trees, increasing canopy coverage and modifying airflow patterns.

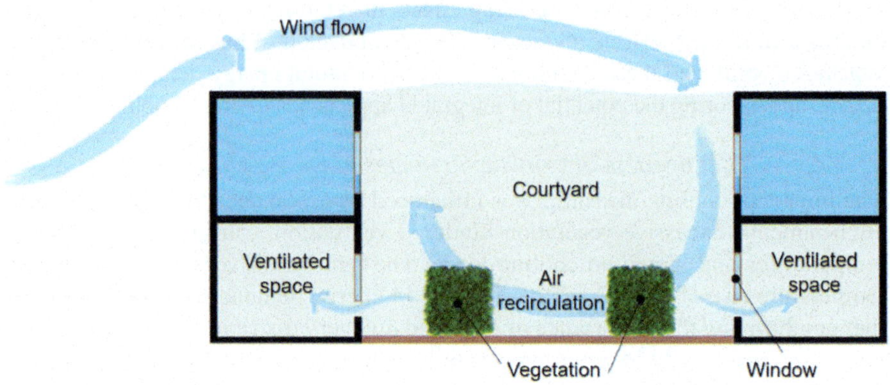

FIGURE 13.49 Study layout of the courtyard building, showing monitored rooms, vegetation placement, and airflow pathways for assessing the cooling impact of courtyard greenery. (Sun et al., 2021.)

13.4.8.1 Methodology and Experimental Setup

A Computational Fluid Dynamics (CFD) model was developed to simulate airflow dynamics, temperature distribution, and heat transfer interactions between the courtyard, vegetation, and indoor spaces. The ANSYS Fluent v18.1 software was used for simulations, validated through previous experimental studies to ensure model accuracy.

The vegetation was modeled as a porous medium, accounting for airflow resistance and evapotranspiration cooling effects. A cooling power of 350 W/m³ per unit of leaf area density (LAD) was applied, based on existing empirical data. The domain was discretized using an unstructured mesh, with finer elements near vegetation and window openings to capture detailed convective heat exchange.

Boundary conditions were defined based on climate conditions representative of hot-arid environments, with an inlet air temperature of 35.85°C. The influence of vegetation on airflow velocities, cooling rates, and thermal comfort was analyzed using three key performance indicators:

- Mean indoor air temperature reduction (°C)
- Wind speed magnitude inside rooms (m/s)
- Thermal comfort metrics (PMV and PPD) based on ASHRAE 55 standards (%)

13.4.8.2 Results

13.4.8.2.1 Indoor Temperature Distribution

The introduction of vegetation significantly reduced indoor temperatures compared to the baseline case, where all rooms maintained a constant temperature of 35.85°C. Small trees reduced average indoor temperatures by 7.4% (ground floor) and 9.3% (first floor). The greatest cooling effect was observed in north- and south-facing rooms (RSF1 and RNF1), where temperatures decreased by 5.31°C and 5.33°C, respectively. In the case of larger trees, indoor temperatures dropped by 6.58°C in RSF1 and 6.46°C in RNF1. The lowest temperature recorded was 26.0°C in rooms closest to the courtyard, where shading and evapotranspiration effects were strongest. Despite these reductions, east-facing rooms (REG1 and REG2) remained the warmest, with temperatures only decreasing by 1.2°C due to limited exposure to direct airflow.

13.4.8.2.2 Airflow and Ventilation Performance

The study revealed that vegetation altered airflow patterns inside the courtyard, influencing wind speed within surrounding rooms, and did not uniformly improve air circulation. While some rooms experienced increased ventilation, others saw a decline due to airflow obstruction by trees.

13.4.8.2.3 Thermal Comfort Assessment (PMV and PPD Analysis)

Thermal comfort was evaluated using the Predicted Mean Vote (PMV) and Predicted Percentage of Dissatisfied (PPD) indices, based on ASHRAE 55 comfort standards. In the baseline case, all rooms maintained a PMV of 4.10 and a PPD of 100%, indicating extreme discomfort due to high temperatures and low air movement. With small trees, PMV dropped to 2.38 (91% PPD) in RSG1 and 2.63 (96% PPD) in RNG1, while east-facing rooms (REG1 and REG2) remained above a PMV of 3.6, with a PPD of 100%, indicating insufficient cooling improvements. The introduction of larger trees provided the most significant improvements, reducing PMV to 1.55 (54% PPD) in RSF1 and 1.60 (56% PPD) in RNF1, confirming that larger vegetation coverage enhances thermal comfort more effectively than smaller plants. However, no scenario fully met optimal comfort thresholds, highlighting the need for additional passive cooling measures, such as increased natural ventilation openings or reflective surfaces.

13.4.8.3 Conclusion

The study demonstrated that vegetation plays a critical role in regulating indoor temperature and airflow within courtyard buildings in hot climates. The introduction of trees lowered indoor temperatures by up to 6.58°C, while also influencing air movement patterns. Larger trees provided greater cooling benefits, but their impact on airflow distribution was mixed, leading to improvements in some areas while reducing ventilation in others.

Thermal comfort analysis confirmed that vegetation reduces overheating risks, but does not fully eliminate thermal discomfort, suggesting that additional strategies, such as increased ventilation rates or phase change materials, may be required for optimal performance.

The authors recommend that future studies explore the effects of different plant species, tree densities, and courtyard geometries to optimize passive cooling performance in urban environments.

13.4.9 PCM Wallboard and Night Ventilation: Energy Efficiency and Overheating Risk in Offices

13.4.9.1 Introduction

Tamer et al. (2023) conducted a simulation-based study to evaluate the combined effect of phase change materials (PCM) and night ventilation (NV) on heating and

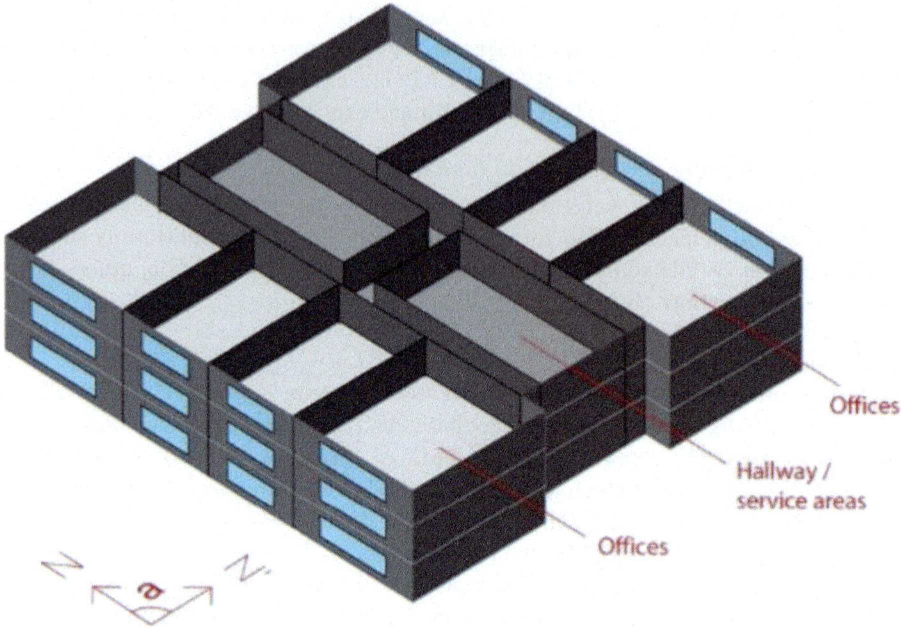

FIGURE 13.50 Floor plan of the simulated three-story office building in Ankara, showing the arrangement of office rooms, central hallways, and service areas used in the PCM and night ventilation performance study. (Tamer et al., 2023.)

cooling energy demand in office buildings, considering both current and projected 2050 climate conditions. The research focused on a three-story office building located in Ankara, Turkey, with a total floor area of 6,330 m^2. Each floor comprised eight office rooms (157–217 m^2 each), connected by a central corridor and service areas totaling 625 m^2, as illustrated in Figure 13.50.

The study aimed to assess the performance of two PCM types—PCM with a melting temperature of 19°C (PCM19), optimized for heating, and PCM with a melting temperature of 25°C (PCM25), optimized for cooling. The study analyzed their energy-saving potential when integrated with night ventilation, under both present and future climate scenarios.

13.4.9.2 Methodology

The study utilized EnergyPlus simulations to model building energy consumption, considering thermal storage characteristics of PCM and the impact of night ventilation. Climate data for Ankara were obtained from Climate.OneBuilding.org (2020 scenario), and future weather files for 2050 were generated using CCWorldWeatherGen, based on the HadCM3 A2 high-emission scenario.

The office building model featured external walls with insulation (U = 0.33 W/m^{2}K), windows with a solar heat gain coefficient (SHGC) of 0.85, and a window-to-wall ratio of 0.3 in all directions. PCM wallboards were installed on the internal surfaces of walls, ceilings, and floors, replacing gypsum plasterboard. The building's ventilation and HVAC settings included:

- Four scenarios:
  1. No night ventilation, no air conditioning (NNV-NAC)
  2. Night ventilation, no air conditioning (NV-NAC)
  3. No night ventilation, with air conditioning (NNV-AC)
  4. Night ventilation, with air conditioning (NV-AC)
- Heating: October–April, setpoint 21.1°C (7 AM–6 PM) with setback 15.6°C (6 PM–7 AM)
- Cooling: May–September, setpoint 25.5°C with mechanical cooling or passive ventilation

Thermal comfort was assessed using operative temperature (TO) and indoor overheating degrees (IOD). PCM enthalpy-temperature (h-T) curves were generated to simulate their phase transitions, accounting for latent heat storage (110 J/g) and melting range width (3°C).

13.4.9.3 Results

The study demonstrated that phase change materials (PCMs) played a crucial role in reducing heating and cooling energy demand, with PCM19 effectively minimizing heating loads, while PCM25 contributed to cooling energy savings, particularly when integrated with night ventilation (NV).

13.4.9.4 Heating Energy Savings

PCM19 provided the greatest reduction in heating energy demand, achieving 20–30% savings in November, both under 2020 and projected 2050 climate conditions. In contrast, PCM25 had minimal impact on heating demand, as its melting temperature exceeded the heating setpoint, preventing effective thermal storage during colder months.

13.4.9.5 Cooling Energy Savings

The combination of PCM25 and NV proved highly effective in reducing cooling energy demand, leading to 10–20% savings in June (2020) and 5–10% savings in June (2050). However, in August, the cooling reductions were approximately half of those in June, as higher external temperatures limited the cooling effect of NV. Meanwhile, PCM19 showed no significant reduction in cooling energy demand, reinforcing its primary role in winter energy efficiency rather than summer cooling.

13.4.9.6 Impact of Night Ventilation (NV)

NV alone contributed to a 23–26% reduction in cooling demand in 2020 and 14–16% in 2050, highlighting its effectiveness in current climatic conditions. When PCM25 was integrated with NV, cooling energy savings increased to 29–31% in 2020 and 17–18% in 2050, demonstrating the synergistic effect of thermal storage and night-time cooling strategies. However, the overall effectiveness of NV declined in future climate scenarios, as higher night temperatures reduced its cooling potential.

13.4.9.7 Thermal Comfort and Overheating Risk

The study also assessed the impact of PCMs on thermal comfort and overheating risk. PCM19 helped stabilize winter temperatures, lowering operative temperature (TO) fluctuations. In summer, PCM25 improved TO stability, but its effectiveness was limited without NV. The Indoor Overheating Degree (IOD) values more than doubled between 2020 and 2050, reflecting the growing overheating risk due to climate change. While NV significantly reduced IOD values by up to 58% in 2020 and 37% in 2050, PCMs alone had a limited effect on overheating mitigation, emphasizing the importance of ventilation strategies in future climate adaptation.

13.4.9.8 Conclusion

The study confirmed that PCM19 is effective for heating energy savings, while PCM25 is beneficial for cooling energy savings when integrated with night ventilation. NV alone reduced cooling energy demand, but its efficiency declined in future scenarios due to rising night temperatures. PCM25 coupled with NV provided the greatest cooling energy savings, mitigating peak cooling loads and reducing HVAC dependency. However, PCM alone was ineffective in reducing overheating risk in naturally ventilated buildings, as high indoor temperatures often exceeded the PCM activation threshold.

Future studies should explore long-term PCM degradation, optimization of PCM placement, and integration with dynamic HVAC control strategies. The results highlight the importance of climate-adaptive design, combining PCM with ventilation strategies to enhance energy efficiency and occupant thermal comfort in a warming climate.

13.4.10 Analytical Model for Nighttime Cooling of Internal Thermal Mass

13.4.10.1 Introduction

Wu et al. (2023) developed an analytical model to evaluate the effectiveness of nighttime cooling applied to internal thermal mass as a passive strategy for reducing cooling demand in buildings. The research aimed to quantify free cooling energy storage and assess the potential energy savings across different climates. While previous studies have focused on external thermal mass, this research addressed the cooling capacity of internal thermal mass, particularly floor slabs, when flushed with cool nighttime air. The study introduced a simplified mathematical model that integrates mass dimension and real-world climate data to predict cooling performance and energy demand reduction.

13.4.10.2 Methodology

The study analyzed nighttime cooling of internal thermal mass, specifically concrete floor slabs, where unconditioned outdoor air is used to flush the thermal mass at night. During the following day, the cooled slab acts as a heat sink, absorbing indoor heat gains through convection and radiation. The conceptual setup and operation of this system are illustrated in Figure 13.51.

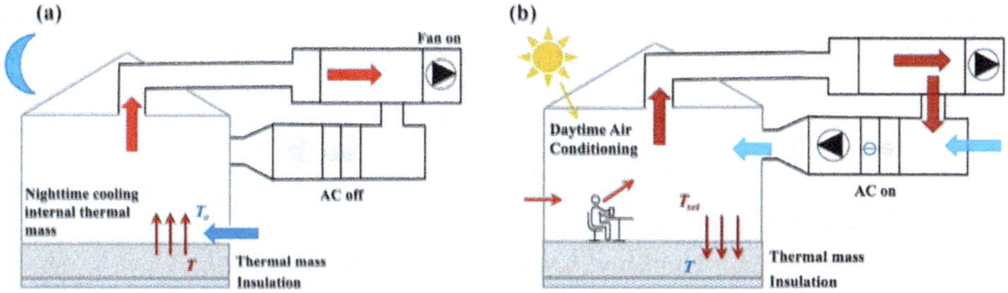

FIGURE 13.51 Nighttime cooling of internal thermal mass in concrete floor slabs, showing (a) the night flushing phase with unconditioned outdoor air to cool the slab, and (b) the daytime phase where the cooled slab absorbs indoor heat gains. (Wu et al., 2023.)

In the nighttime cooling phase (Figure 13.51a), cool outdoor air is drawn into the building and circulated across the exposed concrete slab, which serves as internal thermal mass. The slab releases stored heat to the incoming air, which is then expelled through an exhaust fan, lowering the slab temperature and enabling energy storage for the following day. The air conditioning system remains off during this process.

In the daytime operation phase (Figure 13.51b), the cooled slab absorbs heat from indoor air and radiant sources, delaying the rise of indoor temperatures. This moderates the cooling load while the air conditioning system operates to maintain comfort. Red arrows in the diagram indicate heat gains, blue arrows represent cooling flows, and sun and moon icons mark the diurnal cycle of operation.

A one-dimensional heat transfer model was developed, incorporating hourly outdoor temperature variations and a constant air exchange rate. The model was verified using numerical simulations, the conduction transfer function (CTF) method, and ANSI/ASHRAE Standard 140-Case 600 validation tests. The research assessed 48 U.S. cities across different climate zones, as well as 16 climate zones in California, using Typical Meteorological Year (TMY3) datasets. Cities were selected based on nighttime temperature profiles, ensuring effective free cooling conditions where nighttime temperatures remained below 18°C with significant diurnal temperature variation.

The study also incorporated Computational Fluid Dynamics (CFD) simulations to analyze temperature distribution and airflow characteristics within the thermal mass. The heat balance equations accounted for convective and radiative interactions between the slab and room air, predicting cooling load reductions under different ventilation schedules and climate.

13.4.10.3 Results

The study confirmed that nighttime cooling of internal thermal mass is an effective strategy for reducing peak cooling demand. The total free cooling energy storage varied significantly across different climate zones, with the highest recorded in Santa Fe, NM (19.1 kWh/m^2 per year) and the lowest in humid coastal cities. In California, the maximum total free cooling energy storage was 27.5 kWh/m^2 per year, while net free cooling energy (after deducting fan energy consumption) peaked at 6.11 kWh/m^2 per year.

The research also highlighted regional differences in the effectiveness of this strategy. Coastal regions with high humidity, such as Seattle and Honolulu, exhibited limited free cooling potential due to higher nighttime temperatures, while hot-arid inland regions, including parts of New Mexico and Arizona, showed the highest energy savings. In Las Vegas, NV, and Phoenix, AZ, nighttime cooling was effective for over 100 days per year, but net cooling energy savings were reduced due to high fan energy consumption.

The effectiveness of nighttime cooling depended on key climate parameters, particularly the diurnal temperature range and nighttime temperature levels. The model demonstrated that when nighttime temperatures dropped significantly below indoor setpoints, cooling energy savings were maximized, with reductions of up to 26% in total annual cooling demand.

The study also validated the impact of nighttime cooling on shifting peak cooling loads. Peak cooling demand reductions of up to 30% were observed in the hottest months, with energy shifting effects extending into the following day. Internal thermal mass delayed the onset of peak indoor temperatures by up to 4 hours, further improving thermal stability and occupant comfort.

13.4.10.4 Conclusion

This study confirms that nighttime cooling of internal thermal mass is a viable strategy for reducing energy consumption and improving energy flexibility in buildings. The free cooling potential is highest in hot, arid climates with significant diurnal temperature variations, while humid and coastal climates exhibit lower efficiency due to reduced nighttime temperature drops.

The analytical model developed in this research provides a simplified yet accurate approach for predicting nighttime cooling effectiveness and can be integrated into whole-building energy simulation tools such as EnergyPlus. Future research should focus on dynamic ventilation strategies, hybrid cooling approaches, and real-world implementation studies to further optimize thermal mass-based cooling solutions.

13.4.11 Thermal Mass and Night Ventilation: Overheating Mitigation across Chinese Climates

13.4.11.1 Introduction

Sun et al. (2022) conducted a study to evaluate the effectiveness of thermal mass and night ventilation in mitigating overheating in buildings under different climate conditions in China. The research analyzed the combined effect of thermal mass and night ventilation on heat retention and dissipation, particularly in response to projected climate scenarios for 2050 and 2080. The study focused on five representative cities-Harbin, Beijing, Shanghai, Kunming, and Guangzhou-spanning distinct climate zones.

13.4.11.2 Methodology

The study utilized EnergyPlus dynamic simulations to assess the performance of thermal mass and night ventilation strategies across different climate zones. Two base models were developed:

1. Lightweight buildings (low thermal mass), which store less heat but cool down quickly at night.

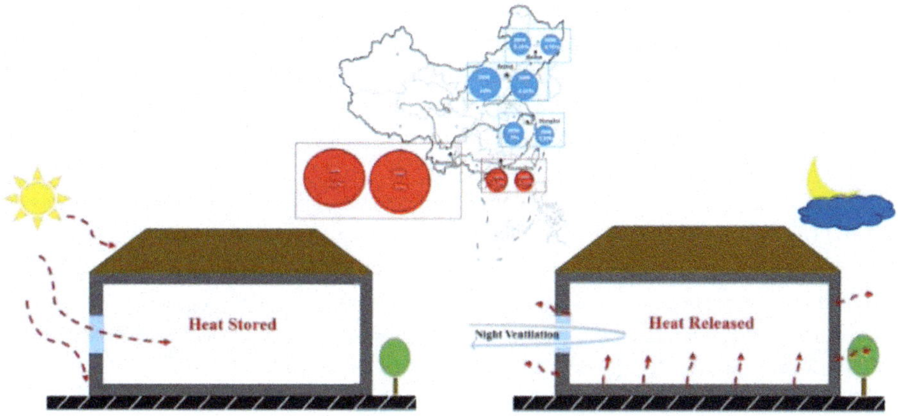

FIGURE 13.52 Conceptual representation of thermal mass storage and heat dissipation through night ventilation in buildings across different climate zones. The diagram shows daytime heat accumulation in the thermal mass and passive cooling through night ventilation.

2. Heavyweight buildings (high thermal mass), which store more heat and release it gradually over time.

A key focus was on the effectiveness of night ventilation in releasing stored heat, particularly in heavyweight buildings where heat buildup is significant. Night ventilation was applied between 18:00 and 07:00, simulating a practical approach to cooling buildings without increasing energy demand.

The conceptual diagram in Figure 13.52 illustrates this process. During the day (left panel), solar radiation and warm outdoor air cause heat to accumulate within the building's thermal mass, which stores the energy in walls, floors, or other massive elements. At night (right panel), cooler outdoor air is drawn into the building through ventilation openings, while warmer indoor air is expelled. This airflow flushes the thermal mass, releasing stored heat and lowering indoor temperatures for the next day. The background map and climate zone indicators highlight that the effectiveness of this cycle depends on local climate conditions, particularly diurnal temperature variations.

The study also incorporated future climate projections for 2050 and 2080 based on Intergovernmental Panel on Climate Change (IPCC) greenhouse gas emission scenarios. The analysis examined how thermal mass strategies could be optimized for different climate zones and future weather conditions.

13.4.11.3 Results

The study found that thermal mass alone effectively reduced overheating in colder climates, but in warmer climates, it increased overheating unless paired with night ventilation. In Harbin, Beijing, and Shanghai, high thermal mass reduced overheating hours by 5.25%, 10%, and 5%, respectively, by delaying heat transfer into the indoor space. However, in Kunming and Guangzhou, heavyweight materials extended overheating periods by 17.5% and 4.5%, as stored heat was not fully dissipated overnight.

Night ventilation played a critical role in cooling heavy thermal mass buildings. In Kunming, overheating hours decreased by nearly 60% when night ventilation was applied, highlighting its importance in warm and humid climates. In contrast, in Harbin, Beijing, Shanghai, and Guangzhou, night ventilation reduced overheating by 21.75%, 23%, 30%, and 22.5%, respectively, showing significant benefits across different climates.

13.4.11.4 Conclusion

The study confirmed that thermal mass and night ventilation are essential strategies for reducing overheating in buildings, but their effectiveness depends on climate conditions. In colder regions, high thermal mass alone effectively stabilizes indoor temperatures, while in warmer climates, it must be combined with night ventilation to prevent overheating.

The findings suggest that urban planners and architects should tailor thermal mass strategies to regional climates, ensuring that buildings in warm regions incorporate sufficient ventilation to facilitate heat dissipation. Future research should explore hybrid cooling strategies that integrate phase change materials (PCM) and adaptive ventilation systems to enhance.

13.4.12 Combined Impact of Thermal Mass, Night Ventilation, and Window Shading in a Temperate Climate

13.4.12.1 Summary Box—Combined Passive Strategies

- Lightweight buildings without adaptation (Strategy A) exceeded 90% of TM59 comfort thresholds.
- Heavyweight walls alone (Strategy B) reduced peak indoor temperatures by about 2 K but did not prevent cumulative overheating.
- Lightweight structures with night ventilation (Strategy C) improved night-time relief but offered limited daytime protection and produced large diurnal temperature swings.
- Heavyweight with night ventilation (Strategy D) reduced overheating to less than 20% of occupied hours.
- Full integration of thermal mass, night ventilation, and external shading (Strategy E) maintained indoor conditions below 28°C throughout the monitoring period, fully meeting TM59 criteria.

13.4.12.2 Introduction and Experimental Design

Thermal mass, night ventilation, and external shading are widely recognized as the most effective passive strategies for mitigating summer overheating, yet relatively few studies have quantified their combined performance under real-world conditions. To address this gap, Kuczyński and Staszczuk (2021) conducted a full-scale experiment in the Centre for Sustainable Building and Energy (CSBE) facility in Zielona Góra, Poland. The experiment took place during the exceptionally hot summer of 2019, which included three consecutive heat waves with outdoor maxima above 32°C and sustained night temperatures above 20°C, creating a demanding test environment.

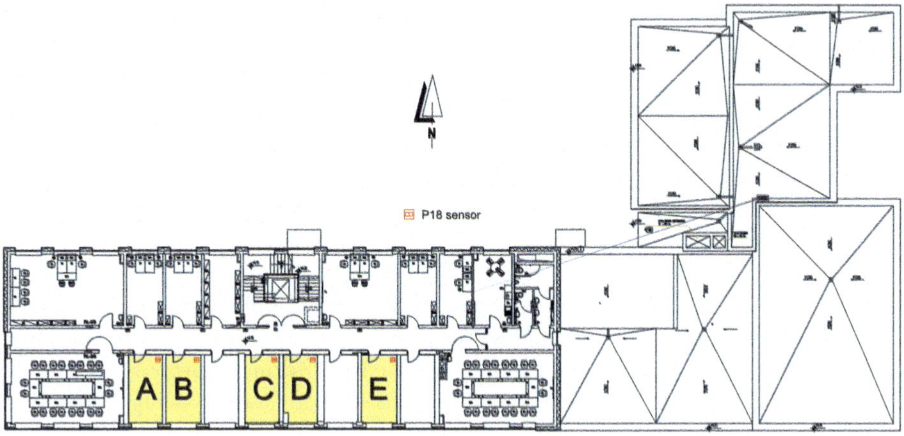

FIGURE 13.53 Top floor plan of the CSBE building with eight experimental rooms (A–E) and the location of thermo-hygrometers marked in red.

The research focused on five rooms located on the upper floor of the CSBE building. As shown in Figure 13.53, the top floor contained eight identical test chambers (A–H) aligned along a glazed south-facing façade. Five of these rooms (A–E) were selected for this study, each measuring 6.0 × 3.0 × 3.2 m. Their geometry, orientation, roof assemblies, and glazing were identical, ensuring uniform boundary conditions. Differences concerned only wall construction and the application of adaptive measures, enabling controlled comparison of passive strategies.

The five configurations were defined as follows:

- **A:** Lightweight timber-frame envelope without shading or night ventilation (reference)
- **B:** Heavyweight masonry envelope without adaptive measures
- **C:** Lightweight structure with night ventilation (10 ACH between 22:00 and 07:00)
- **D:** Heavyweight structure with night ventilation (10 ACH)
- **E:** Heavyweight structure with night ventilation (10 ACH) and automated external shading

All windows were identical south-facing double-glazed units of 4.2 m^2. Night ventilation was applied using a peak air exchange of 10 ACH, consistent with earlier recommendations by Orme and Palmer (2003) and Finn et al. (2007), who identified this as a realistic upper bound for residential cooling. No internal heat gains were introduced. Sensors continuously recorded indoor air temperature and humidity at consistent points in each room, alongside outdoor meteorological data.

13.4.12.3 Results and Discussion

The lightweight reference building (Strategy A) performed the worst. Daily peaks frequently exceeded 31°C, and during the hottest week in July, average daytime indoor temperature reached 30.4°C, with nighttime lows of 27.8°C. This confirmed the absence of thermal damping and poor night recovery.

Strategy B, heavyweight without ventilation, showed partial improvement. The masonry walls delayed the rise of indoor temperatures and reduced daily peaks by more than 2 K relative to Strategy A. Average maxima were 28.2°C, while minima reached 25.1°C. The decrement factor was 0.08 compared to 0.12 in Strategy A, indicating stronger buffering. However, in the absence of ventilation, cumulative overheating remained significant.

Strategy C, lightweight with night ventilation, improved nocturnal relief but not daytime conditions. Average peaks were 29.3°C, with nighttime lows of 24.4°C. Although night flushing helped remove heat, the low thermal inertia of the envelope limited daytime absorption, and overheating persisted during afternoons. The result was pronounced diurnal variability, with large daily temperature swings that undermined overall comfort stability.

Strategy D, heavyweight with night ventilation, produced much more stable conditions. Average daily peaks were 26.9°C, and minima fell below 24°C. Indoor air stayed below 28°C for 93% of the monitoring period. The synergy of thermal inertia and active night cooling enabled effective storage and release cycles, ensuring sustained comfort across consecutive hot days.

Strategy E, which added automated external shading to the setup of Strategy D, achieved the best results. Indoor peaks averaged 25.4°C during heat waves, with minima around 23°C. Compared to the reference lightweight room, Strategy E reduced peak temperatures by more than 5 K and mean 24-hour values by 3.8 K. Indoor air never exceeded 28°C, even during days when outdoor maxima reached 35–36°C. The decrement factor was 0.05, the lowest of all configurations, reflecting strong damping of daily swings.

13.4.12.4 Thermal Comfort Assessment

The CIBSE TM59 framework was used to assess comfort, applying thresholds of 26°C for bedrooms (22:00–07:00) and 28°C for living rooms (07:00–22:00). In Strategy A, thresholds were exceeded during over 90% of occupied hours: 91.5% in bedrooms and 88.6% in the living room. In Strategy B, exceedances dropped by about one third, yet still reached 60.2% of bedroom hours and 52.4% in the living room. Strategy C reduced exceedances in bedrooms to 58.3% but living rooms still overheated during 70.8% of hours.

Strategy D achieved a substantial improvement, with overheating reduced to 19.6% of occupied hours in bedrooms and 13.8% in the living room. Strategy E provided the most robust performance: only 23 hours of exceedance in bedrooms across the entire summer, while living rooms never surpassed 28°C. Both TM59 criteria were met in full, indicating that the integration of mass, night ventilation, and shading can ensure compliance even under record-breaking summers.

13.4.12.5 Comparative Context

These results reinforce established principles in passive design. Givoni (1998) and Artmann et al. (2007) emphasized the synergy of mass, ventilation, and shading. Finn et al. (2007) confirmed that ventilation benefits plateau around 10 ACH. Porritt et al. (2011) and O'Donovan et al. (2021) showed that integrated measures reduce overheating hours by 60–85%, a range consistent with our findings for Strategy E.

Experimental studies also align with these conclusions. Tink et al. (2018) observed that thermal mass combined with ventilation consistently outperformed either measure alone. Mantesi et al. (2019) noted that the benefits of mass plateau when used in isolation, a limitation visible in Strategy B. Kuczyński and Staszczuk (2020, 2021) demonstrated that integrating mass, shading, and ventilation reduced peak indoor temperatures by over 7°C and eliminated overheating in Polish conditions. Our results confirm these outcomes with full-scale, long-term monitoring.

The comparison between Strategies D and E highlights the decisive role of shading. Without blinds, the masonry ventilated building still overheated during the hottest days, while with shading it remained fully within comfort limits. This finding underscores that shading is not secondary but essential, particularly in mid-latitude temperate climates where afternoon solar loads coincide with peak occupancy.

13.4.12.6 Implications for Design and Policy

This study confirms that no single passive measure is sufficient under extreme climate stress. Thermal mass delays overheating but requires effective night discharge. Night ventilation improves comfort but is limited by external diurnal ranges. Shading is necessary to reduce solar gains and protect other strategies from being overwhelmed.

The integration of thermal mass, night ventilation, and shading provided full compliance with TM59 during the hottest summer on record in Poland. For design practice, this indicates that passive strategies should be applied in combination, not in isolation, and evaluated with dynamic performance indicators rather than static insulation metrics. For policy, the results highlight the need to include overheating mitigation criteria in building regulations, as advocated by De Masi et al. (2021) and Norouzi et al. (2023).

In practical terms, Strategy E demonstrates a replicable pathway for resilient housing. Automated external blinds, when paired with masonry and ventilative cooling, delivered robust comfort without mechanical systems. Such integrated passive measures represent a cost-effective approach to lowering energy demand, improving resilience, and preparing European housing for the more frequent and severe heat waves projected under climate change.

13.4.13 Summary

The reviewed evidence shows that combining passive measures yields larger, more reliable gains than any single technique used in isolation. Studies span laboratory cells, full-scale test chambers, and occupied buildings, and cover cool temperate, Mediterranean, hot–humid, hot–arid and semi-arid contexts. Effectiveness is reported through indoor air and operative temperatures, peak suppression and delay, overheating or exceedance hours, cooling and heating energy, and in several cases, life-cycle or district-scale outcomes. The most informative combinations pair solar control with nighttime purge and either sensible storage in thermal mass or latent storage in Phase Change Materials (PCM), often supported by high-albedo surfaces or vegetated elements.

13.4.13.1 Passive Solar, Envelope Efficiency and Thermal Mass in Housing

In Sydney simulations, Passive Solar and Energy Efficiency Design Strategies (double glazing, internal shading, lower infiltration) coupled with higher thermal mass reduced annual heating and cooling energy by about 36–37% for typical detached houses; a brick veneer house with Passive Solar and Energy Efficiency Design Strategies achieved up to 58% savings relative to a standard lightweight control, with validation against measured household electricity within 16% (Albayyaa et al., 2019). These results confirm that basic solar control and airtightness amplify the stabilizing effect of mass in both seasons.

13.4.13.2 Roof-Integrated Combinations

In hot–arid California test cells, a hybrid green roof that combined vegetative shading with radiant and evaporative cooling maintained indoor temperatures 9–13°C below outdoors. The radiant–evaporative variant averaged 27.6°C vs 29.7°C for an insulated control roof and 36.4°C outdoors; coupling with night ventilation strengthened the effect when diurnal swings exceeded about 12°C (La Roche et al., 2020). In Andalusian social housing, a ventilated double-layer roof with reflective outer skin, ventilated air cavity and concrete thermal mass cut cooling demand by up to 65% when night ventilation and evaporative assist were used, reduced discomfort hours by 80% and lowered top-floor summer peaks by about 4°C; heating demand also fell by about 50% due to improved insulation (Delgado et al., 2022).

13.4.13.3 Bundle of Natural Ventilation, Shading, Cool Roof, and Mass

Field and simulation work in Anhui showed that, used alone, external shading reduced mean indoor air temperature by about 2.0°C, natural ventilation by about 1.3°C, and a cool roof by about 1.1°C. The combined system lowered mean indoor temperature by about 2.8°C, with peaks reduced by up to 4.3°C, and consistently decreased wall, roof, and glazing temperatures, confirming the synergy of stacked measures (Li et al., 2024). In the Netherlands, dynamic simulations across detached, terraced, and top-floor apartment archetypes found that natural ventilation and external blinds almost eliminated overheating under proxy future summer weather, while increased albedo had strong benefits in older, low-insulation stock but weaker effects in highly insulated 2012-era buildings. Raising thermal resistance without shading and purge increased overheating, and lowering thermal mass helped only where night ventilation could consistently discharge stored heat (Van Hooff et al., 2015).

13.4.13.4 Future Climate and High-Rise Contexts

In hot–humid Hong Kong, simulations of a 30–40 story residential block to 2090 showed that passive measures implemented today reduce annual cooling demand by about 55% and peak demand by about 38%; under the most extreme 2090 scenario, these savings rise further in percentage terms as loads grow. The relative importance of measures shifts with warming: solar protection remains useful but gradually gives way to airtightness and insulation as key levers, while natural ventilation loses efficacy as nighttime temperatures rise (Liu et al., 2020).

13.4.13.5 Urban Fabric, Albedo, and Vegetation

Microclimate–energy co-simulation in dense low-rise neighborhoods in Gyeonggi-Do found that high-albedo materials lowered air temperature by about 1.9°C and reduced building cooling energy by about 10%; vegetation alone yielded 4 to 6% energy savings, with shrubs and grass outperforming tall trees for façade cooling and airflow. A combined albedo–vegetation strategy reduced energy by about 12.5%, although excessive greenery in compact canyons impeded night cooling and raised demand, highlighting the need for balanced planting to preserve ventilation (Wang et al., 2025). In courtyard buildings, strategically placed trees reduced indoor temperatures by up to about 6.6°C and improved Predicted Mean Vote and Predicted Percentage of Dissatisfied, but airflow effects were mixed across rooms, indicating that vegetation should be coordinated with ventilation openings to avoid local stagnation (Sun et al., 2021).

13.4.13.6 Thermal Storage Aligned with Night Purge

Coupling latent storage with purge can extend the benefit window. In Ankara offices, PCM tuned to 25°C combined with night ventilation reduced June cooling energy by 10–20% in the present climate and 5–10% in 2050, while night ventilation alone cut cooling by 23–26% today and 14–16% in 2050. Pairing PCM and night ventilation produced 29–31% savings today and 17–18% in 2050, although PCM on its own had little effect on overheating in naturally ventilated mode (Tamer et al., 2023). An analytical study of nighttime cooling of internal thermal mass predicted annual free-cooling storage up to 19.1 kWh/m^2 in hot–arid U.S. cities and peak-load reductions up to 30%, with the strongest gains where nights are cool and diurnal ranges are large (Wu et al., 2023). Across five Chinese climate zones, thermal mass alone reduced overheating in colder cities but increased it in warm zones unless paired with night ventilation; with purge, overheating fell by about 22–30% in Beijing, Shanghai, and Guangzhou and by nearly 60% in Kunming (Sun et al., 2022).

13.4.13.7 Full-Scale Integrated Evidence

In temperate Europe, full-scale experiments with five matched residential test rooms showed a clear gradient of performance from a lightweight reference to heavyweight with night ventilation, and then to heavyweight with night ventilation plus automated external blinds. The last configuration kept indoor temperatures below 26°C for 97% of hours through successive heat waves, lowered peak temperatures by more than 5°C relative to the lightweight control, reduced daily amplitude to below 2 K, and met both CIBSE TM59 criteria; thermal mass or night ventilation alone did not fully prevent heat build-up (Kuczyński & Staszczuk, 2021).

13.4.13.8 Synthesis and Design Guidance

1. Combine measures that address the three drivers of summer load: gains, storage, and purge. External shading or high-albedo surfaces limit gains, thermal mass or calibrated PCM stabilize and shift loads, and night ventilation or change-over hybrid control removes stored heat.

2. Target the package to climate and typology. In hot–arid and warm–temperate zones with cool nights, night ventilation paired with mass delivers large benefits; in hot–humid zones and future climates with warm nights, rely more on solar control, airtightness, and insulation, and consider hybrid ventilation with elevated set-points under adaptive comfort models.
3. Use microclimate-aware modeling. Coupled urban and building simulation avoids underestimating benefits by ignoring street-canyon aerodynamics or reflective surfaces.
4. Detail for airflow and maintenance. Cross-ventilation pathways, stack drivers, operable blinds, and durable high-reflectance finishes are essential for sustained performance.
5. Treat PCM as a complement, not a substitute, for mass, and always pair it with a dependable night purge.
6. Evaluate with dynamic, time-resolved metrics and, where relevant, prospective life-cycle assessment to capture the long-term value of suppressing peaks and reducing cooling growth.

Together, these studies show that integrated passive strategies can reduce cooling energy by roughly 20%–60% in housing and offices, keep indoor temperatures within adaptive comfort limits through heat waves, and cut peak loads substantially when designs are tuned to climate, microclimate, and building.

REFERENCES

Al Niyadi, S., & Elnabawi, M. H. (2025). Optimizing hybrid ventilation in arid climates: A systematic review and Dubai case study. *City and Environment Interactions*, *26*, 100191. https://doi.org/10.1016/j.cacint.2025.100191

Albayyaa, H., Hagare, D., & Saha, S. (2019). Energy conservation in residential buildings by incorporating passive solar and energy efficiency design strategies and higher thermal mass. *Energy and Buildings*, *182*, 205–213. https://doi.org/10.1016/j.enbuild.2018.09.036

Amaripadath, D., Paolini, R., Sailor, D. J., & Attia, S. (2023). Comparative assessment of night ventilation performance in a nearly zero-energy office building during heat waves in Brussels. *Journal of Building Engineering*, *78*, 107611. https://doi.org/10.1016/j.jobe.2023.107611

Aqilah, N., Rijal, H. B., & Zaki, S. A. (2025). Indoor thermal environment and effect of air movement on comfort temperature in Malaysian naturally ventilated dwellings. *Journal of Building Engineering*, *104*, 112151. https://doi.org/10.1016/j.jobe.2025.112151

Artmann, N., Gyalistras, D., Manz, H., & Heiselberg, P. (2008). Impact of climate warming on passive night cooling potential. *Building Research & Information*, *36*, 111–128. https://doi.org/10.1080/09613210701621919

Artmann, N., Manz, H., & Heiselberg, P. (2007). Climatic potential for passive cooling of buildings by night-time ventilation in *Europe*. *Applied Energy*, *84*(2), 187–201. https://doi.org/10.1016/j.apenergy.2006.04.001

Bellamy, L. A., & Mackenzie, D. W. (2001). *Thermal performance of buildings with heavy walls* (Study Report No. 108). Building Research Association of New Zealand. https://www.branz.co.nz/pubs/research-reports/sr108/ (accessed February 5, 2025)

Brambilla, A., & Jusselme, T. (2017). Preventing overheating in offices through thermal inertial properties of compressed earth bricks: A study on a real-scale prototype. *Energy and Buildings*, *156*, 281–292. https://doi.org/10.1016/j.enbuild.2017.09.070

Cardinale, N., Rospi, G., & Stefanizzi, P. (2013). Energy and microclimatic performance of Mediterranean vernacular buildings: The Sassi district of Matera and the Trulli district of Alberobello. *Building and Environment*, *59*, 590–598. https://doi.org/10.1016/j.buildenv.2012.10.006

Castell, A., Martorell, I., Medrano, M., Pérez, G., & Cabeza, L. F. (2010). Experimental study of using PCM in brick constructive solutions for passive cooling. *Energy and Buildings*, *42*(4), 534–540. https://doi.org/10.1016/j.enbuild.2009.10.022

Chandrasekaran, K. S., & Madhumathi, A. (2023). Energy-efficient retrofitting with exterior shading device in hot and humid climate—case studies from fully glazed multi-storied office buildings in Chennai, India. *Journal of Asian Architecture and Building Engineering*, *22*(4), 2209–2223. https://doi.org/10.1080/13467581.2022.2145208

Coma, J., Pérez, G., de Gracia, A., Burés, S., Urrestarazu, M., & Cabeza, L. F. (2017). Vertical greenery systems for energy savings in buildings: A comparative study between green walls and green facades. *Building and Environment*, *111*, 228–237. https://doi.org/10.1016/j.buildenv.2016.11.014

De Gracia, A., Rincón, L., Castell, A., Jiménez, M., Boer, D., Medrano, M., & Cabeza, L. F. (2010). Life cycle assessment of the inclusion of phase change materials (PCM) in experimental buildings. *Energy and Buildings*, *42*, 1517–1523. https://doi.org/10.1016/j.enbuild.2010.03.001

De Masi, R. F., Gigante, A., & Vanoli, G. P. (2021). Are nZEB design solutions environmentally sustainable? Sensitivity analysis for building envelope configurations and photovoltaic integration in different climates. *Journal of Building Engineering*, *39*, 102292. https://doi.org/10.1016/j.jobe.2021.102292

Delgado, M. G., Sánchez Ramos, J., Palomo Amores, T. R., Castro Medina, D., & Álvarez Domínguez, S. (2022). Improving habitability in social housing through passive cooling: A case study in Mengíbar (Jaén, Spain). *Sustainable Cities and Society*, *78*, 103642. https://doi.org/10.1016/j.scs.2021.103642

Detsi, M., Atsonios, I., Mandilaras, I., & Founti, M. (2024). Effect of smart glazing with electrochromic and thermochromic layers on building's energy efficiency and cost. *Energy and Buildings*, *319*, 114553. https://doi.org/10.1016/j.enbuild.2024.114553

Es-sakali, N., Kaitouni, S. I., Ait Laasri, I., Mghazli, M. O., Cherkaoui, M., & Pfafferott, J. (2024). Static and dynamic glazing integration for enhanced building efficiency and indoor comfort with thermochromic and electrochromic windows. *Thermal Science and Engineering Progress*, *52*, 102681. https://doi.org/10.1016/j.tsep.2024.102681

Fernandes, J., Pimenta, C., Mateus, R., Silva, S. M., & Bragança, L. (2015). Contribution of Portuguese vernacular building strategies to indoor thermal comfort and occupants' perception. *Buildings*, *5*(4), 1242–1264. https://doi.org/10.3390/buildings5041242

Finn, D. P., Connolly, D., & Kenny, P. (2007). Sensitivity analysis of a maritime located night ventilated library building. *Solar Energy*, *81*(6), 697–710. https://doi.org/10.1016/j.solener.2006.10.008

Givoni, B. (1998). Effectiveness of mass and night ventilation in lowering the indoor daytime temperature. Part I: 1993 experimental periods. *Energy and Buildings*, *28*(1), 25–32. https://doi.org/10.1016/S0378-7788(97)00056-X

Hacker, J. N., De Saulles, T. P., Minson, A. J., & Holmes, M. J. (2008). Embodied and operational carbon dioxide emissions from housing: A case study on the effects of thermal mass and climate change. *Energy and Buildings*, *40*(3), 375–384. https://doi.org/10.1016/j.enbuild.2007.03.005

Hao, X., Xing, Q., Long, P., Lin, Y., Hu, J., & Tan, H. (2020). Influence of vertical greenery systems and green roofs on the indoor operative temperature of air-conditioned rooms. *Journal of Building Engineering*, *31*, 101373. https://doi.org/10.1016/j.jobe.2020.101373

Hu, Y., Ai, Z., Wargocki, P., Chang, Y., & Zhang, G. (2024). Ventilative cooling of residential buildings in China: A simulation-based evaluation of lightweight modular integrated constructions considering climate change. *Energy and Buildings*, *317*, 114382. https://doi.org/10.1016/j.enbuild.2024.114382

Jamil, H., Alam, M., Sanjayan, J., & Wilson, J. (2016). Investigation of PCM as retrofitting option to enhance occupant thermal comfort in a modern residential building. *Energy and Buildings*, *133*, 217–229. https://doi.org/10.1016/j.enbuild.2016.09.064

Kuczyński, T., & Staszczuk, A. (2020). Experimental study of the influence of thermal mass on thermal comfort and cooling energy demand in residential buildings. *Energy*, *195*, 116984. https://doi.org/10.1016/j.energy.2020.116984

Kuczyński, T., & Staszczuk, A. (2021). The impact of wall and roof material on the summer thermal performance of a building in a temperate climate. *Energy*, *228*, 120482. https://doi.org/10.1016/j.energy.2021.120482

Kuczyński, T., & Staszczuk, A. (2023). Impact of uninsulated slab-on-grade and masonry walls on residential building overheating. *Energies*, *16*(22), 7558. https://doi.org/10.3390/en16227558

Kuczyński, T., Staszczuk, A., Gortych, M., & Stryjski, R. (2025). Phase change materials in south-facing top-floor rooms: Experimental insights from a decade of research. *Energy*, *325*, 136213. https://doi.org/10.1016/j.energy.2025.136213

Kuznik, F., & Virgone, J. (2009). Experimental assessment of a phase change material for wall building use. *Applied Energy*, *86*(10), 2038–2046. https://doi.org/10.1016/j.apenergy.2009.01.004

Kuznik, F., Virgone, J., & Johannes, K. (2011). In-situ study of thermal comfort enhancement in a renovated building equipped with phase change material wallboard. *Renewable Energy*, *36*, 1458–1462. https://doi.org/10.1016/j.renene.2010.11.008

La Roche, P., Yeom, D. J., & Ponce, A. (2020). Passive cooling with a hybrid green roof for extreme climates. *Energy and Buildings*, *224*, 110243. https://doi.org/10.1016/j.enbuild.2020.110243

Li, Y., Shah, K. W., Li, W., & Xiong, T. (2023). Experimental investigation on indoor thermal environment improvement and energy-saving of electrochromic window under Singapore's tropical climate. *Journal of Building Engineering*, *73*, 106779. https://doi.org/10.1016/j.jobe.2023.106779

Li, Y., Tao, X., Zhang, Y., & Li, W. (2024). Combining use of natural ventilation, external shading, cool roof, and thermal mass to improve indoor thermal environment: Field measurements and simulation study. *Journal of Building Engineering*, *86*, 108904. https://doi.org/10.1016/j.jobe.2024.108904

Liu, S., Kwok, Y. T., Lau, K. K.-L., Ouyang, W., & Ng, E. (2020). Effectiveness of passive design strategies in responding to future climate change for residential buildings in hot and humid Hong Kong. *Energy and Buildings*, *228*, 110469. https://doi.org/10.1016/j.enbuild.2020.110469

Mantesi, E., Hopfe, C. J., Mourkos, K., Glass, J., & Cook, M. (2019). Empirical and computational evidence for thermal mass assessment: The example of insulating concrete formwork. *Energy and Buildings*, *188–189*, 314–332. https://doi.org/10.1016/j.enbuild.2019.02.021

Norouzi, M., Haddad, A. N., Jiménez, L., Hoseinzadeh, S., & Boer, D. (2023). Carbon footprint of low-energy buildings in the United Kingdom: Effects of mitigating technological pathways and decarbonization strategies. *Science of the Total Environment*, *882*, 163490. https://doi.org/10.1016/j.scitotenv.2023.163490

O'Donovan, A., Murphy, M. D., & O'Sullivan, P. D. (2021). Passive control strategies for cooling a non-residential nearly zero energy office: Simulated comfort resilience now and in the future. *Energy and Buildings*, *231*, 110607. https://doi.org/10.1016/j.enbuild.2020.110607

Orme, M., & Palmer, J. 2003. Control of Overheating in Future Housing: Design Guidance for Low Energy Strategies. Faber Maunsell.

Palma, R. M., Medina, D.C., Delgado, M. C. G., Ramos, J.S., Montero-Gutiérrez, P., & Domínguez, S.A. (2024). Enhancing the building resilience in a changing climate through a passive cooling roof: A case study in Camas (Seville, Spain). *Energy and Buildings*, *321*, 114680. https://doi.org/10.1016/j.enbuild.2024.114680

Palomo Amores, T., Ruda Sarria, F., Medina, D. C., Valera, T. C., Sánchez Ramos, J., & Álvarez Domínguez, S. (2025). Experimental validation of the potential of cross-ventilation strategy as a natural cooling technique integrated in a real historic building. *Applied Sciences*, *15*(4), 2174. https://doi.org/10.3390/app15042174

Peeters, L., de Dear, R., Hensen, J., D'haeseleer, W. (2009). Thermal comfort in residential buildings: Comfort values and scales for building energy simulation. *Applied Energy*, *86*(5), 772–780. https://doi.org/10.1016/j.apenergy.2008.07.011

Porritt, S. M., Shao, L., Cropper, P., & Goodier, C. I. (2011). Adapting dwellings for heatwaves. *Sustainable Cities and Society*, *1*(2), 89–90. https://doi.org/10.1016/j.scs.2011.02.002

Roque, E., Vicente, R., & Almeida, R. M. S. F. (2021). Opportunities of light steel framing towards thermal comfort in southern European climates: Long-term monitoring and comparison with the heavyweight construction. *Building and Environment*, *200*, 107937. https://doi.org/10.1016/j.buildenv.2021.107937

Saffari, M., de Gracia, A., Fernández, C., & Cabeza, L. F. (2017). Simulation-based optimization of PCM melting temperature to improve the energy performance in buildings. *Applied Energy*, *202*, 420–434. https://doi.org/10.1016/j.apenergy.2017.05.107

Sage-Lauck, J. S., & Sailor, D. J. (2014). Evaluation of phase change materials for improving thermal comfort in a super-insulated residential building. *Energy and Buildings*, *79*, 32–40. https://doi.org/10.1016/j.enbuild.2014.04.028

Schossig, P., Henning, H.-M., Gschwander, S., & Haussmann, T. (2005). Micro-encapsulated phase-change materials integrated into construction materials. *Solar Energy Materials and Solar Cells*, *89*, 297–306. https://doi.org/10.1016/j.solmat.2005.01.017

Silva, R., Eggimann, S., Fierz, L., Fiorentini, M., Orehounig, K., & Baldini, L. (2022). Opportunities for passive cooling to mitigate the impact of climate change in Switzerland. *Building and Environment*, *208*, 108574. https://doi.org/10.1016/j.buildenv.2021.108574

Stasi, R., Ruggiero, F., & Berardi, U. (2024). Influence of cross-ventilation cooling potential on thermal comfort in high-rise buildings in a hot and humid climate. *Building and Environment*, *248*, 111096. https://doi.org/10.1016/j.buildenv.2023.111096

Sun, H., Calautit, J. K., & Jimenez-Bescos, C. (2022). Examining the regulating impact of thermal mass on overheating, and the role of night ventilation, within different climates and future scenarios across China. *Cleaner Engineering and Technology*, *9*, 100534. https://doi.org/10.1016/j.clet.2022.100534

Sun, H., Jimenez-Bescos, C., Mohammadi, M., Zhong, F., & Calautit, J. K. (2021). Numerical investigation of the influence of vegetation on the aero-thermal performance of buildings with courtyards in hot climates. *Energies*, *14*(17), 5388. https://doi.org/10.3390/en14175388

Tamer, T., Gürsel Dino, I., Baker, D. K., & Akgül, C. M. (2023). Coupling PCM wallboard utilization with night ventilation: Energy efficiency and overheating risk in office buildings under climate change impact. *Energy and Buildings*, *298*, 113482. https://doi.org/10.1016/j.enbuild.2023.113482

Tan, H., Kotamarthi, R., Wang, J., Qian, Y., & Chakraborty, T. C. (2023). Impact of different roofing mitigation strategies on near-surface temperature and energy consumption over the Chicago metropolitan area during a heatwave event. *Science of the Total Environment*, *860*, 160508. https://doi.org/10.1016/j.scitotenv.2022.160508

Thiele, A. M., Jamet, A., Sant, G., & Pilon, L. (2015). Annual energy analysis of concrete containing phase change materials for building envelopes. *Energy Conversion and Management, 103*, 374–386. https://doi.org/10.1016/j.enconman.2015.06.068

Tink, V., Porritt, S., Allinson, D., & Loveday, D. (2018). Measuring and mitigating overheating risk in solid wall dwellings retrofitted with internal wall insulation. *Building and Environment, 141*, 247–261. https://doi.org/10.1016/j.buildenv.2018.05.062

Tsoka, S. (2021). Investigating the effect of summer night ventilation on the urban buildings' thermal and energy performance using numerical approaches. *IOP Conference Series: Earth and Environmental Science, 899*(1), 012013. https://doi.org/10.1088/1755-1315/899/1/012013

Van Hooff, T., Blocken, B., Hensen, J. L. M., & Timmermans, H. J. P. (2015). Reprint of: On the predicted effectiveness of climate adaptation measures for residential buildings. *Building and Environment, 83*, 142–158. https://doi.org/10.1016/j.buildenv.2014.10.006

Vega, M., Marín, P. E., Ushak, S., & Shire, S. (2024). Research trends and gaps in experimental applications of phase change materials integrated in buildings. *Journal of Energy Storage, 75*, 109746. https://doi.org/10.1016/j.est.2023.109746

Wang, Y., Zhang, Y., Song, Y., Lee, J., Li, G., Song, Z., Zhou, Z., Zhang, J., Xu, J., Li, J., & Li, Y. (2025). Evaluating the cooling effects and building energy-saving potential of vegetation and albedo: a case study of Gyeonggi-Do, South Korea. *Buildings, 15*(4), 597. https://doi.org/10.3390/buildings15040597

Wu, W., Benner, J., & Luo, Z. (2023). Developing analytical model for nighttime cooling of internal thermal mass. *Applied Thermal Engineering, 220*, 119798. https://doi.org/10.1016/j.applthermaleng.2022.119798

Xu, Y., Zhao, B., Yu, J., Yin, X., Chang, W.-S., & Guo, H. (2025). Integrating phase-change materials to reduce overheating risk in residential buildings in cold regions of China. *Construction and Building Materials, 475*, 141153. https://doi.org/10.1016/j.conbuildmat.2025.141153

Yang, F., Yuan, F., Qian, F., Zhuang, Z., & Yao, J. (2018). Summertime thermal and energy performance of a double-skin green facade: A case study in Shanghai. *Sustainable Cities and Society, 39*, 43–51. https://doi.org/10.1016/j.scs.2018.01.049

Yannas, S., & Rodríguez-Álvarez, J. (2020). Domestic overheating in a temperate climate: Feedback from London residential schemes. *Sustainable Cities and Society, 59*, 102189. https://doi.org/10.1016/j.scs.2020.102189

Yu, H., Zhang, H., Qiu, L., & Lai, D. (2023). Evaluation of renovation strategies for existing residential districts in Shanghai based on coupling calculation of microclimate and energy consumption. *Energy and Buildings, 298*, 113569. https://doi.org/10.1016/j.enbuild.2023.113569

Zhao, Y., Yu, J., Zhao, B., Xu, Y., Chang, W.-S., & Guo, H. (2025). Potential of phase change materials to mitigate overheating in residential blocks in severe cold regions of China. *Sustainable Cities and Society, 119*, 106114. https://doi.org/10.1016/j.scs.2024.106114

14 Research on Building Thermal Resilience

14.1 INTRODUCTION

In recent years, increasing attention has been given to the ability of buildings to maintain indoor air temperatures at acceptable levels during extreme weather events, both cold and hot, particularly in the context of potential power outages. This resilience has become even more critical as the global transition away from fossil fuels gains momentum. Buildings with enhanced thermal resilience not only improve occupant comfort and safety during extreme events but also contribute to energy demand flexibility, allowing heating and cooling loads to shift in response to the availability of renewable energy.

Given these benefits and the current priorities in sustainable building design, integrating resilience to extreme temperature fluctuations is becoming a necessity. Residential spaces must remain habitable even in scenarios where heating systems fail during freezing conditions or cooling systems are rendered inoperative during heat waves due to power outages. Achieving this requires a fundamental reconsideration of building envelope design and retrofitting strategies, extending beyond the traditional focus on minimizing energy consumption to explicitly incorporate resilience to thermal extremes.

Despite its growing importance, scientific knowledge on how to integrate energy efficiency and thermal resilience into a cohesive design framework remains insufficient. This is particularly relevant because these two objectives—while theoretically complementary—often present practical conflicts. Traditional approaches to designing heating, ventilation, and air conditioning (HVAC) systems, typically based on typical meteorological year (TMY) data, do not sufficiently account for extreme weather events. Consequently, many buildings fail to provide adequate thermal comfort during power outages occurring in extreme winter cold or summer heat.

Building resilience in the face of global climate change and the increasing frequency of extreme weather events has become one of the most pressing challenges in modern architecture and urban planning. The concept refers to a building's capacity to maintain livable conditions under extreme stress while ensuring occupant well-being. Peri et al. (2024) argue that resilience analysis should adopt a bottom-up approach, incorporating user needs, adaptive behaviors, and supporting technologies. This perspective conceptualizes buildings as components of an integrated "building–systems–users" framework, in which each element plays a critical role in managing indoor thermal conditions.

User behavior is particularly crucial when addressing challenges posed by extreme temperatures and power outages. Research indicates that elevated nighttime

DOI: 10.1201/9781003645573-14

temperatures not only impact physical and mental health but may also contribute to social problems, such as increased domestic violence rates. Therefore, building design solutions must integrate technology with the behavioral patterns and constraints of occupants. For instance, educating residents on adaptive actions, such as window operation, shading control, and thermostat settings, is essential for optimizing thermal comfort and safety. Future buildings should be designed as dynamic systems that not only respond to changing external conditions but also adapt to the evolving needs of their occupants over time (Peri et al., 2024).

To advance the standardization of simulation practices in assessing thermal resilience, it is necessary to establish common simulation parameters and evaluation metrics. A unified approach would enable more consistent and comparable assessments across different building typologies and climatic conditions. One significant step in this direction is the development of guidelines for defining thermal stress thresholds, resilience metrics, and critical exposure durations, accounting for demographic variability and occupant vulnerabilities. Existing standards, such as RELi and passive survivability criteria in the LEED rating system, provide valuable foundations for resilience assessment. However, further validation across diverse climate contexts is required to ensure their broader applicability (Siu et al., 2023).

An additional challenge in standardizing resilience assessments lies in generating accurate and representative weather data for simulations. Such datasets must incorporate both long-term climate change trends and localized urban heat island effects to ensure a more realistic representation of future extreme temperature scenarios. Integrating these factors into thermal resilience modeling will enhance predictive capabilities and support mitigation strategies in urban environments (Peri et al., 2024).

The design of resilient buildings must also account for the demographic diversity of occupants, their thermal tolerance, health conditions, and adaptive capacities. A bottom-up approach recognizes not only the role of technological interventions but also the behavioral patterns and constraints of building users, which are fundamental to achieving comprehensive climate resilience. Only through such an integrated approach can the construction industry effectively address the growing demands of climate adaptation and ensure that buildings remain functional and habitable during extreme temperature events (Siu et al., 2023).

Moreover, as the complexity of simulation-based resilience assessments continues to grow, developing advanced analytical tools is essential to streamline the evaluation process. These tools should be designed in collaboration with architects and researchers to ensure alignment with best practices. Beyond enabling efficient data processing and result interpretation, such tools must also present findings transparently and intuitively, with clearly defined limitations. This will help integrate resilience considerations into mainstream architectural and engineering workflows, making building designs more responsive to the evolving challenges of climate change (Siu et al., 2023).

In this chapter, the case studies of buildings that have undergone resilience assessments, applying various indicators to evaluate their performance under extreme climate scenarios, will be presented.

14.2 ENHANCING THERMAL RESILIENCE OF US RESIDENTIAL HOMES IN HOT HUMID CLIMATES DURING EXTREME TEMPERATURE EVENTS

Wijesuriya et al. (2024) conducted a study on improving the thermal resilience of single-family homes in the United States to extreme weather events, such as heat waves and winter storms, particularly in the context of power outages. The research focused on implementing passive solutions-including additional insulation, airtight sealing, and phase change materials (PCMs)-to extend the duration for which buildings could maintain safe indoor conditions during extreme events. This duration, termed "hours of safety" (HST), is defined as the period during which indoor temperatures remained within safety thresholds: above 15°C during winter storms and below a heat index (HI) of 33°C during heat waves.

For the analysis, the researchers selected a single-family home model from the ResStock database, representing typical residential buildings in the United States. The case study was based on a 184 m^2 single-family home selected from the ResStock database, located in Houston, Texas. The house was divided into two thermal zones: a conditioned living space and an unconditioned garage. Figure 14.1 illustrates the floor plan and isometric view of the modeled dwelling.

The simulations were conducted using EnergyPlus, a building energy modeling tool that enabled an assessment of indoor conditions during power outages. Meteorological data for the simulations were based on actual records of extreme temperatures, including a winter storm from February 11, 1981 (with temperatures ranging from −6.1°C to 3.9°C), and a heat wave from September 4, 2000 (with temperatures between 30.7°C and 42.4°C).

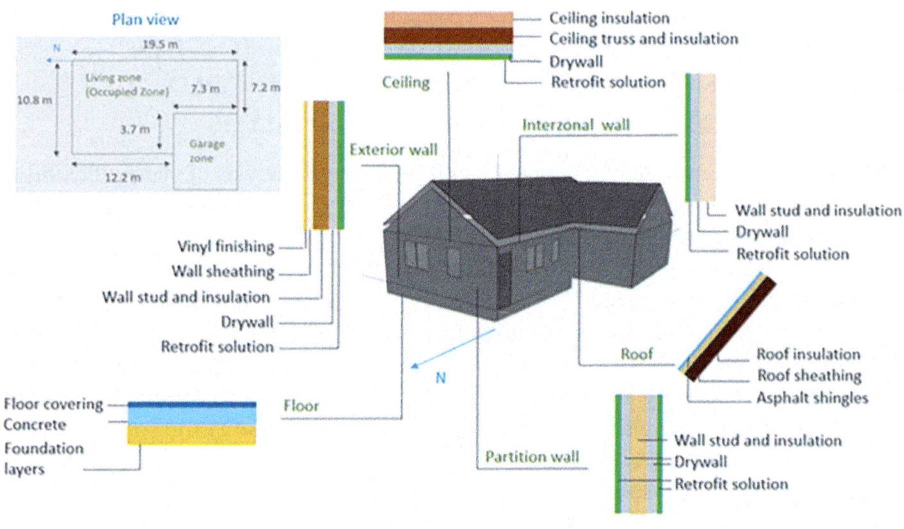

FIGURE 14.1 Floor plan and isometric view of the modeled single-family house with sectional details of walls, roof, ceiling, and floor showing passive retrofit layers (Wijesuriya et al., 2024).

The study first analyzed the baseline condition of the building, without any adaptive measures. The results showed that during the winter storm, indoor temperatures in the living space dropped below 15°C within two hours of the power outage. During the heat wave, the heat index exceeded the safety threshold after 12 hours. These findings demonstrated that without any adaptive measures, the building rapidly failed to provide safe conditions, posing potential health risks to occupants.

To enhance thermal resilience, the study tested three strategies: additional insulation, airtight sealing, and the incorporation of PCM. Insulation alone was effective only when combined with reduced air infiltration. With complete airtight sealing (ACH50 = 0), the hours of safety increased to 31 during the winter storm and 38 during the heat wave. However, PCM, applied to interior surfaces of exterior walls and ceilings, provided greater benefits. When tuned to appropriate melting ranges (17°C–21°C for winter and 25°C–29°C for summer), PCM extended the time before crossing the safety threshold to 44 hours during the winter storm and 37 hours during the heat wave. When combined with full airtight sealing, the hours of safety increased dramatically to 100 hours in winter and 185 hours in summer.

Preconditioning the building (heating or cooling the interior before an event) also played a significant role in improving thermal resilience. Raising or lowering indoor temperatures by 5°C before a power outage extended the hours of safety by an additional 10 to 25 hours, particularly when combined with PCM and airtight sealing.

The findings of Wijesuriya et al. (2024) highlight the importance of passive design strategies in enhancing the resilience of residential buildings to extreme weather events. The integration of airtight envelopes, thermal mass, and PCM can significantly delay indoor temperature extremes during power outages, reducing reliance on mechanical systems and improving occupant safety.

14.3 ENERGY FLEXIBILITY AND TOWARD RESILIENCE IN NEW AND OLD RESIDENTIAL HOUSES IN COLD CLIMATES: A TECHNO-ECONOMIC ANALYSIS

Rehman and Hasan (2023) conducted a comprehensive analysis of energy flexibility and thermal resilience in the context of Finnish residential buildings, focusing on the comparative performance of older and recently constructed single-family homes. The study sought to quantify energy flexibility through a dedicated coefficient that captures a building's capacity to temporally shift heating demand in response to dynamic electricity tariffs or variable climatic conditions. The spatial and infrastructural characteristics of the two case study buildings—representing pre- and post-regulation housing stock—are schematically illustrated in Figure 14.2, which also depicts the respective flows of electricity and thermal energy through the system infrastructure.

Energy flexibility was defined as a building's ability to shift heating demand over time in response to variable energy pricing or climatic conditions. This coefficient is calculated as the ratio of the difference between energy consumption during low and high tariff hours to the sum of these values, where higher percentages indicate greater adaptability for optimizing heating costs.

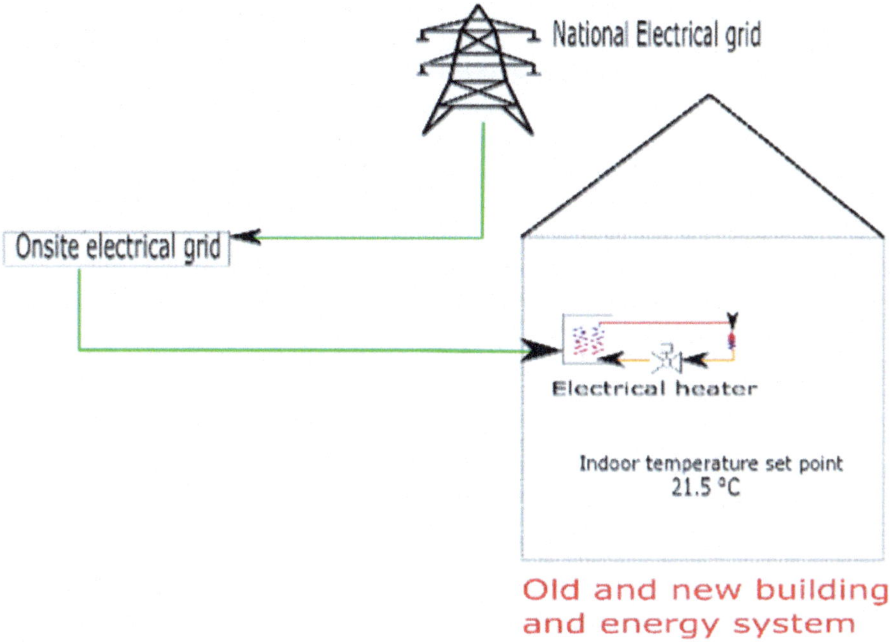

FIGURE 14.2 Floor plan and isometric view of the modeled single-family house, with envelope section details showing materials and passive retrofit configurations (Rehman and Hasan, 2023).

The findings revealed that older buildings exhibited significantly higher energy flexibility, with values ranging from 24% to 52%, depending on the activation strategy and its duration. This was attributed to the greater thermal inertia of older buildings, allowing them to store heat for longer periods and gradually release it during high-cost tariff hours. In contrast, modern buildings had much lower energy flexibility, ranging from 7% to 14%, due to their superior insulation and lower overall heating demand, which limited their ability to shift thermal loads.

An economic analysis evaluating the potential cost savings through dynamic energy pricing indicated that older buildings could achieve annual savings of approximately 400 euros, whereas modern buildings saved only about 70 euros. The savings were achieved by preheating the interior by 1.5°C during low-cost tariff periods, allowing the heat to be stored within the building's thermal mass and gradually released when energy prices were higher. In modern buildings, despite their lower overall energy consumption, the reduced thermal storage capacity meant that this strategy was less effective, resulting in significantly lower financial benefits.

Beyond energy flexibility, the study also assessed the buildings' resilience to power outages, defined as their ability to maintain indoor thermal comfort during heating system failures. The results showed that modern buildings were considerably more resistant to heat loss, maintaining indoor temperatures above 18°C for 17 hours after the heating system was turned off, whereas older buildings could only sustain comfortable conditions for three hours. The superior insulation and airtightness of modern buildings substantially increased their ability to retain heat, making them more resilient during power outages.

Rehman and Hasan (2023) concluded that effective energy management strategies must consider both energy flexibility and building resilience to power disruptions. They highlighted that these two factors compete for the same thermal storage capacity within a building. In older buildings, optimizing energy use through flexible heating strategies can yield significant financial savings, leveraging their higher thermal inertia. In contrast, for modern buildings, where resilience is already high, precise temperature control and minimizing energy losses should take priority through efficient heating system management.

14.4 QUANTIFYING AND RATING THE ENERGY RESILIENCE PERFORMANCE OF BUILDINGS INTEGRATED WITH RENEWABLES IN THE NORDICS UNDER TYPICAL AND EXTREME CLIMATIC CONDITIONS

Expanding on this research, Rehman et al. (2024) further investigated the thermal resilience of single-family homes in Finland, evaluating their ability to maintain comfort during power outages under extreme climatic conditions. The study focused on two contrasting climate scenarios—extreme cold (ECY) and extreme heat (EWY)—and compared houses built in the 1970s with modern constructions from the twenty-first century. As shown in Figure 14.3, the analysis considered both building typologies and their integrated energy systems.

The study introduced the Degree of Disruption (DoD) index, which quantifies the extent of thermal discomfort during power outages. Lower DoD values indicate better thermal resilience.

Under extreme cold conditions, older buildings exhibited significantly poorer performance, with DoD values reaching 0.47 due to their higher heat loss rates. Within just six hours of a power failure, indoor temperatures in these buildings dropped below the 18°C thermal comfort threshold. Conversely, modern buildings demonstrated much greater resilience, with a DoD of 0.26, maintaining acceptable indoor temperatures for up to 20 hours without heating.

In extreme heat conditions, older homes also struggled to maintain thermal comfort, with a DoD of 0.3. Their poor insulation led to rapid indoor temperature increases, particularly in the afternoon hours. In contrast, modern homes achieved better results, with a DoD of 0.22, benefiting from advanced insulation materials and improved ventilation systems. These newer buildings were able to sustain comfortable indoor temperatures for longer periods, even during prolonged heat waves.

The findings suggest that modern buildings are far better equipped to handle extreme climatic conditions, both in winter and summer. However, older homes require urgent retrofitting to improve their ability to withstand climate extremes. The study underscores the importance of integrating climate resilience considerations into future building regulations to ensure occupant comfort and safety under increasingly variable weather conditions.

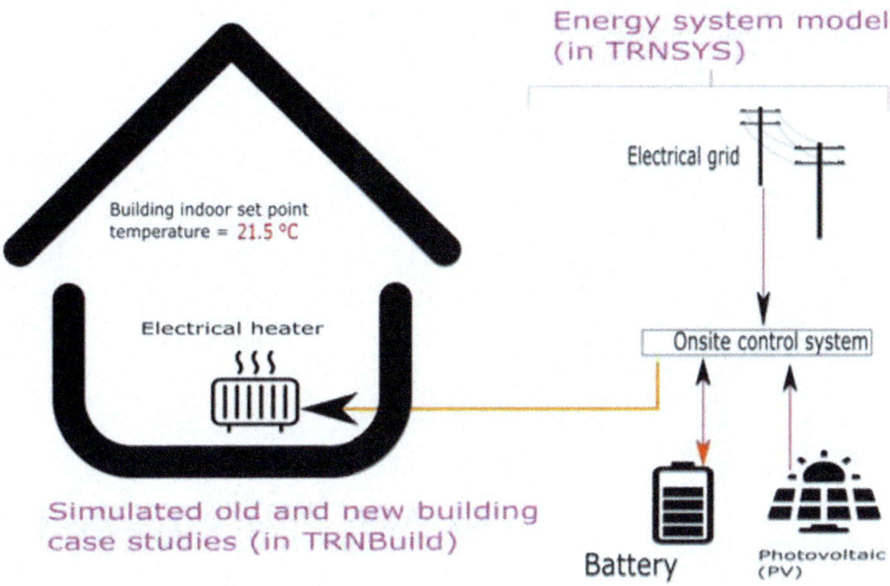

FIGURE 14.3 Case study buildings in Finland: Comparison of old and modern single-family (Rehman et al., 2024).

The research by Rehman et al. (2024) reinforces the need for a dual approach to energy management in buildings—balancing energy flexibility with thermal resilience. While modern constructions provide superior resistance to temperature fluctuations, older buildings can be optimized for cost savings through adaptive heating strategies. The findings also highlight the necessity of policy interventions to incorporate extreme climate scenarios into building standards, ensuring that future housing developments are both energy-efficient and capable of maintaining comfort during adverse weather events.

14.5 HEAT EXPOSURE MITIGATION IN RENOVATED NEARLY ZERO-ENERGY DWELLINGS DURING CONCURRENT HEAT WAVES AND POWER OUTAGES

Amaripadath et al. (2024) conducted an in-depth analysis of the risk of indoor overheating in nearly zero-energy buildings (nZEBs) in Brussels, Belgium. The study focused on assessing the impact of extreme weather conditions, particularly heat waves combined with power outages, on the indoor environment and thermal comfort of building occupants. Special attention was given to the interaction between fluctuating temperatures and high humidity—characteristic of the humid mixed climate—and the thermal performance of high-efficiency buildings. As illustrated in Figure 14.4, the research relied on in situ measurements from a local weather station and data from a representative renovated terraced dwelling in Brussels.

FIGURE 14.4 Study setup: (A) weather monitoring station installed on-site in Brussels; (B) reference renovated nearly zero-energy terraced house used in simulations (Amaripadath et al., 2024).

The research was based on a representative terraced house model, calibrated using real-world measurement data and energy consumption records to accurately reflect the behavior of buildings under actual conditions.

The simulations utilized meteorological data from the summer of 2022, which was characterized by record-breaking temperatures and intense heat waves. Additionally, scenarios including power outages were modeled to evaluate the consequences of losing access to mechanical cooling systems.

The findings indicated that despite their high energy efficiency, nZEBs are vulnerable to overheating during extreme heat events. The analysis showed that in highly exposed areas such as bedrooms, indoor temperatures frequently exceeded acceptable thermal comfort thresholds, potentially compromising occupant health and well-being. Such issues are particularly concerning in light of the increasing frequency of extreme weather events driven by climate change.

In light of the challenges associated with designing buildings resilient to changing climate conditions, this study provides valuable data to support decision-making among architects, engineers, and policymakers. The authors emphasize the necessity of incorporating adaptation strategies into future building designs, including additional passive and active cooling measures to mitigate overheating risks under extreme conditions. The results of this research could contribute to the development of more comprehensive building codes and renovation strategies aimed at effectively counteracting the effects of global climate change.

14.6 BUILDING AND SYSTEM DESIGN'S IMPACT ON THERMAL RESILIENCE TO OVERHEATING DURING HEAT WAVES: AN UNCERTAINTY AND SENSITIVITY ANALYSIS

Sengupta et al. (2024) carried out a detailed analysis of the thermal resilience of apartments to heat waves, focusing on a two-bedroom reference apartment measuring 85.2 m^2, located on a middle floor of a multi-family residential building in Ghent, Belgium. Their research was based on the concept of thermal shock, defined as a sudden and significant deviation from normal climatic conditions, leading to substantial changes in indoor temperatures.

The study examined extreme heat waves across three climate periods—historical (2001–2020), mid-century (2041–2060), and late-century (2081–2100)—identifying six of the most intense events, where maximum temperatures reached 43.8°C, and heat wave durations extended up to 45 days. The severity and duration of these events allowed for their classification into three categories: low, medium, and high thermal shock intensity.

The study employed a multi-stage methodology, encompassing the assessment of factors influencing building overheating resilience and computational simulations conducted in Modelica. The first phase involved defining thermal shock characteristics and determining key resilience indicators, such as overheating duration, heat absorption time, and thermal comfort recovery time. Subsequently, a detailed apartment model was developed, incorporating construction and system parameters. The final stages involved uncertainty and sensitivity analysis, identifying the most significant design variables affecting thermal resilience. The full methodological sequence—including shock characterization, model development, and uncertainty analysis—is outlined in Figure 14.5. A detailed reference apartment model, developed for simulation purposes, is shown in Figure 14.6, including thermal zoning and key propagated parameters.

The evaluation of thermal resilience to heat waves was conducted using three indicators that assessed a building's ability to dissipate heat and return to comfortable conditions once extreme external conditions subsided. The results indicated that thermal resilience is influenced by three key factors: the intensity of the thermal shock, the building's design parameters, and the cooling strategies implemented.

The analysis revealed that for short-term heat waves, most apartments remained resistant to overheating. However, under prolonged and intense heat events, the ability of buildings to maintain comfortable conditions deteriorated significantly. A critical factor influencing heat absorption and dissipation was the timing of peak temperatures. When the highest temperatures occurred early in the heat wave, heat absorption time was shorter, whereas in prolonged heat waves, cumulative heat buildup prolonged both overheating duration and thermal recovery time.

Among the most influential design parameters, three categories had the greatest impact: window-to-wall ratio (WWR), cooling system capacity, and passive cooling strategies. WWR was identified as a critical factor affecting heat absorption and retention. Larger window areas led to more intense indoor heating, significantly increasing the SET-Dh metric, which quantifies the total duration of excessive effective temperatures. Cooling system capacity was the most important factor influencing

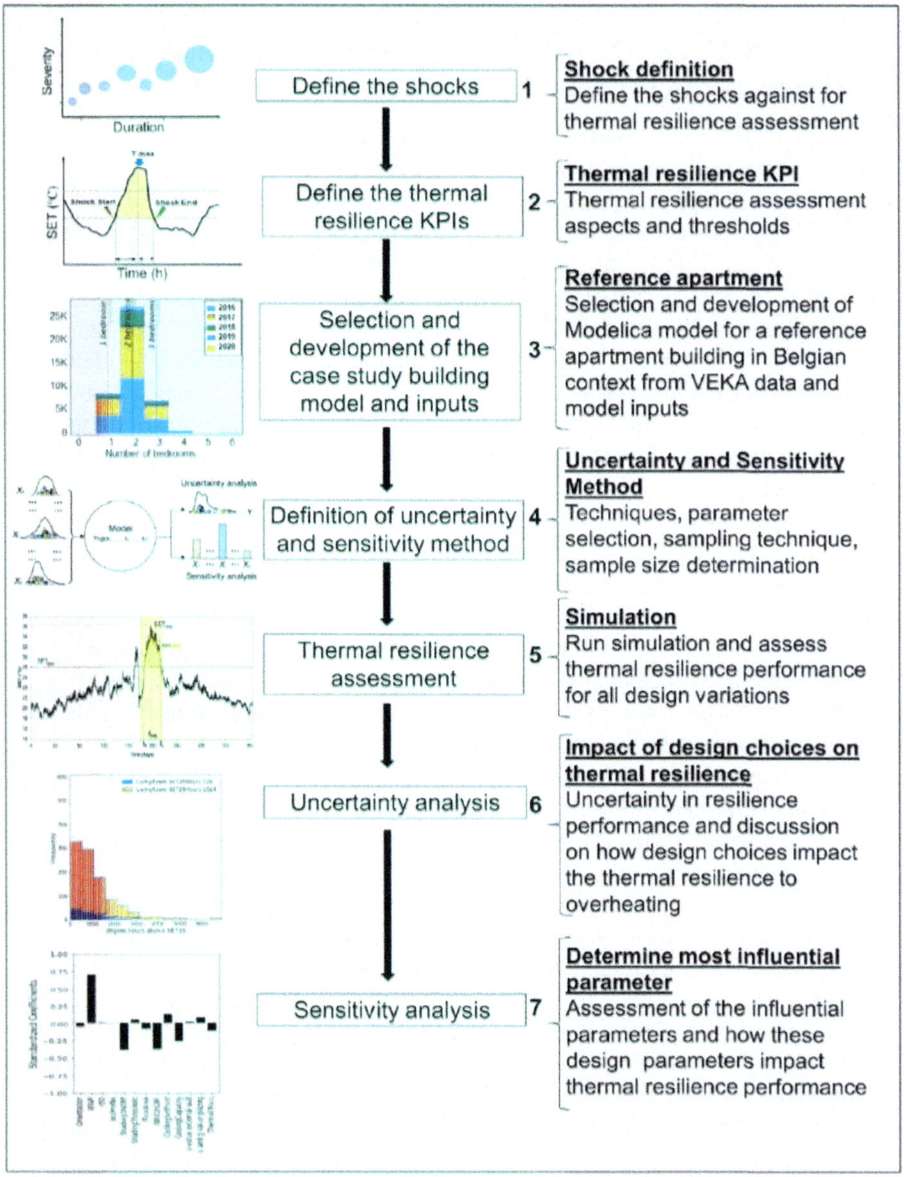

FIGURE 14.5 Flowchart of the thermal resilience assessment methodology, including model development, uncertainty and sensitivity analysis, and resilience simulation stages (Sengupta et al., 2024).

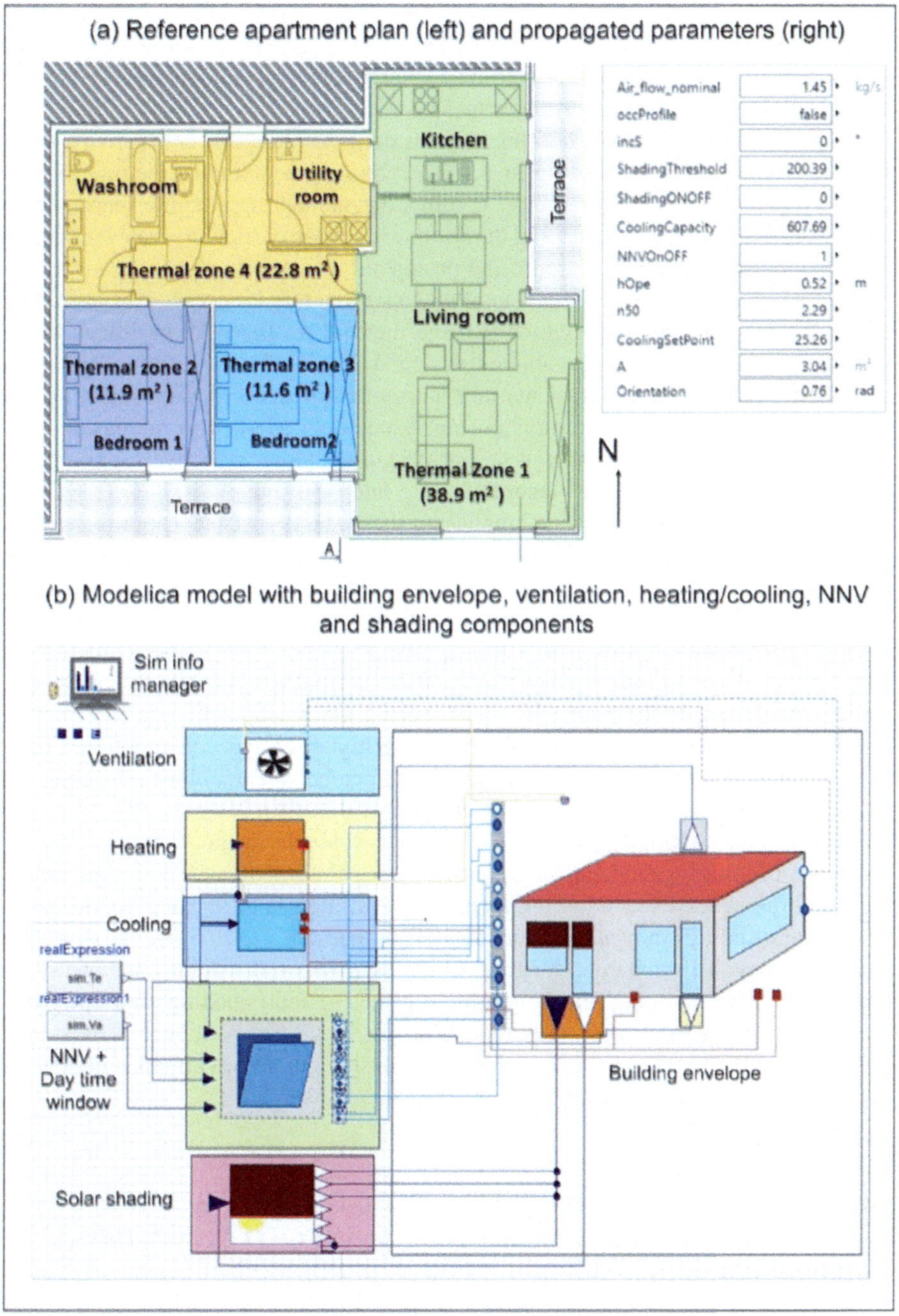

FIGURE 14.6 (a) Floor plan of the reference two-bedroom apartment in Ghent and simulation input parameters; (b) schematic Modelica model with system and envelope components (Sengupta et al., 2024).

thermal recovery time, as higher capacities allowed for a faster reduction in indoor temperatures once a heat wave subsided. Passive strategies, such as nighttime natural ventilation and solar shading, proved particularly effective in reducing indoor heat gains and extending heat absorption times.

A comparison of different cooling systems demonstrated that convective cooling was more effective in the short term, rapidly removing heat and shortening thermal recovery times. In contrast, radiative cooling, despite having a higher thermal storage capacity, prolonged overheating duration, which was particularly relevant during long-lasting heat waves. In extreme and prolonged thermal shock conditions, buildings with radiative cooling systems exhibited significantly longer recovery times.

Uncertainty analysis highlighted that the thermal resilience of buildings varied significantly depending on the applied design strategies. Under extreme heat wave conditions, only 21% of apartments met thermal resilience criteria, indicating that many current architectural designs are not adequately adapted to increasing overheating risks. These findings suggest that design strategies effective for short heat waves may be insufficient for longer and more intense events.

The study's recommendations emphasize the necessity of incorporating several key factors into future building designs. In addition to limiting excessive glazing and implementing effective cooling systems, buildings must be adapted to shifting climatic conditions. The role of occupant behavior and regional climate variations must also be considered in efforts to improve resilience to extreme temperatures. The authors also highlight the need to balance thermal resilience with energy efficiency. Increasing cooling capacity and lowering thermostat setpoints can improve thermal comfort, but such measures also lead to higher energy consumption. Future research should aim to identify an optimal balance between cooling efficiency and minimizing operational costs and CO_2 emissions.

In summary, Sengupta et al. (2024) provide valuable insights into the thermal resilience of buildings to heat waves, emphasizing the impact of key design parameters and the necessity of adapting architectural strategies to future climatic conditions. Their findings indicate that current design methods are effective for short heat waves but may fall short in prolonged and intense heat events. Additional measures, such as improved natural ventilation, more effective solar shading, and optimized construction parameters, are required to ensure long-term resilience. The study's conclusions could serve as the basis for future design guidelines aimed at enhancing thermal resilience in the face of ongoing climate change.

14.7 THERMAL RESILIENCY OF SINGLE-FAMILY HOUSING STOCK UNDER EXTREME HOT AND COLD CONDITIONS

Liyanage et al. (2024) conducted a comprehensive study on the thermal resilience of single-family homes under extreme weather conditions, analyzing both heat waves and extreme cold events. The research focused on assessing buildings' ability to passively maintain livable indoor conditions during power outages or HVAC system failures, as well as evaluating the effectiveness of passive adaptive strategies across different climatic scenarios. Special emphasis was placed on cold climates, where the analysis revealed that conventional building designs do not provide adequate protection against extreme weather conditions.

The study followed a three-stage methodology, incorporating climate projections, building energy modeling, and thermal resilience assessment. This structured approach is illustrated in Figure 14.7, which presents the overall analytical framework used to evaluate the performance of single-family homes under climate extremes.

The first stage involved extracting, statistically adjusting, and validating climate data. Climate projections were divided into three periods: recent past (1991–2020), near future (2021–2050), and distant future (2051–2080). These datasets were adapted to specific construction conditions in Canada, enabling realistic scenario development for extreme weather events.

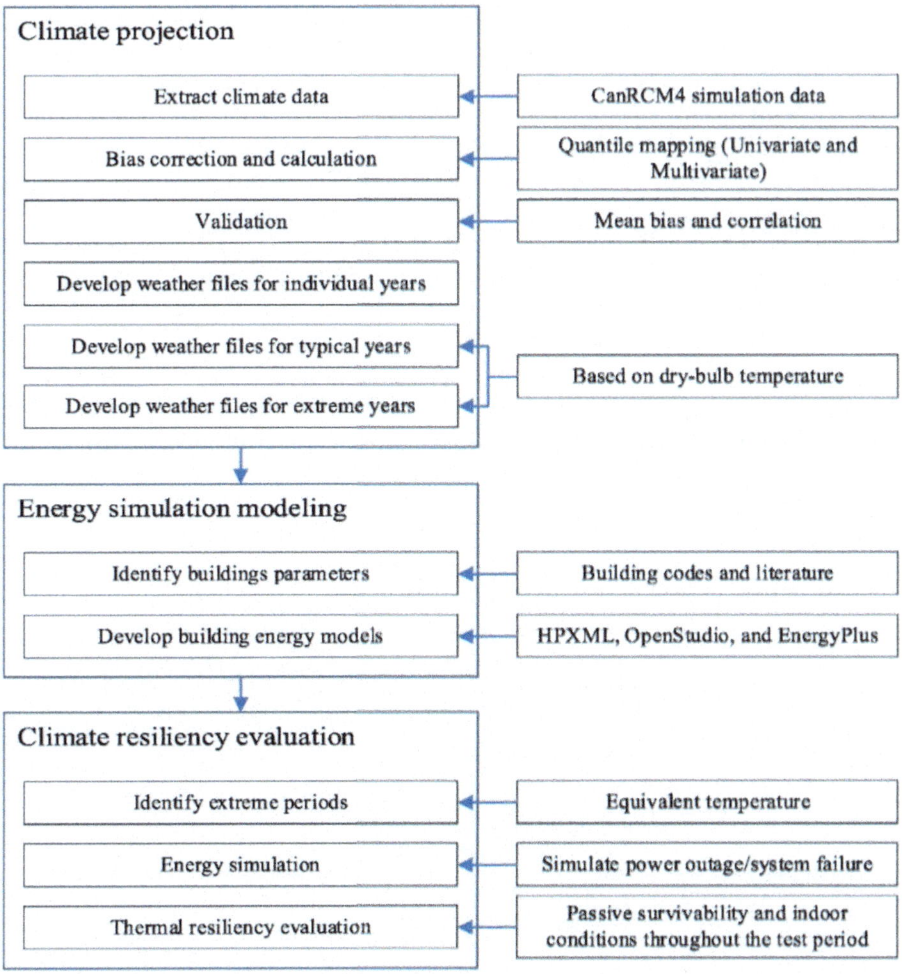

FIGURE 14.7 Overall methodological framework used for thermal resilience assessment (Liyanage et al., 2024).

In the second stage, energy models of single-family homes were developed according to Canadian building standards. Three climatic locations were selected for analysis:

- Vancouver (moderate maritime climate), characterized by mild winters and moderate summers,
- Toronto (cold and humid climate), known for significant seasonal temperature variations,
- Calgary (very cold climate), where extreme winter temperatures pose a major challenge to building energy efficiency.

Comprehensive data were collected on building geometry, thermal insulation values, HVAC system parameters, and occupant schedules, including presence patterns and appliance energy use. These inputs were used to create initial building models with HPXML (Home Performance XML) and EnergyPlus, which were then calibrated using historical weather data.

The developed models also incorporated the possibility of simulating power outages. HPXML enabled automatic deactivation of heating, ventilation, and air conditioning systems, as well as lighting and household appliances, allowing for a realistic representation of blackout conditions. These models were then used to evaluate the thermal resilience of buildings under extreme weather events.

These models were then used to evaluate the thermal resilience of buildings under extreme weather events. The building modeling workflow—from data acquisition to simulation under projected climate scenarios—is detailed in Figure 14.8.

Climate data analysis revealed that as climate change progresses, extreme weather events will become more frequent and intense. The most noticeable trend was the increasing frequency of heat waves, which are expected to occur significantly more often in the future. Projections indicate that heat wave frequency will

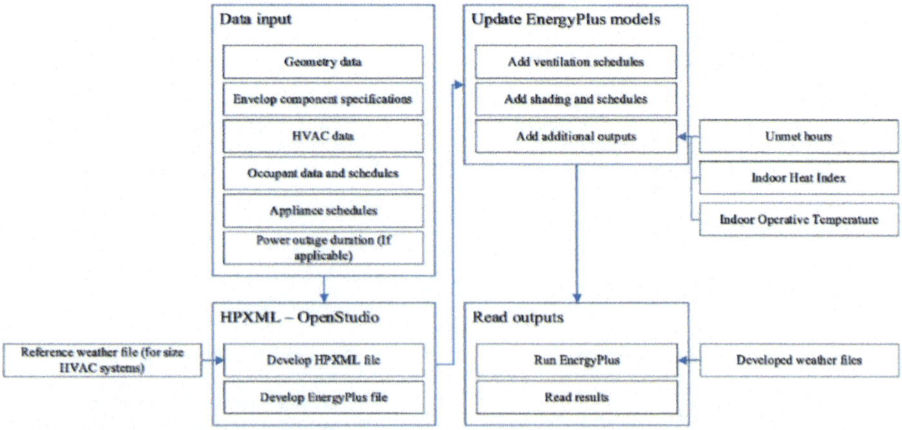

FIGURE 14.8 Simulation workflow showing climate data integration, development of EnergyPlus models using HPXML, and implementation of blackout conditions for resilience testing (Liyanage et al., 2024).

increase by 77–134% in Vancouver, 140–296% in Toronto, and 74–149% in Calgary. Simultaneously, the number of days with extreme cold temperatures is expected to decrease by 50–100%, depending on the region. This suggests that buildings in the future will be more exposed to overheating than to extreme cold, posing a challenge for architects and urban planners.

Further climate analysis showed that cooling degree days-a metric indicating cooling demand-will increase significantly across all study locations, while heating degree days, which indicate heating demand, will gradually decline. In Vancouver, cooling degree days are expected to rise by 286% by 2080, while heating demand will decrease by 30%. In Toronto, heating degree days are projected to drop by 28%, while cooling demand will more than double. In Calgary, cooling degree days are expected to increase by 207%, with heating demand decreasing by 21%. These trends indicate a gradual decline in heating demand in Canada, but at the same time, a rising energy demand for cooling. This necessitates a rethinking of current building design strategies to account for shifting climate conditions.

Climate projections also indicate a significant rise in average annual temperatures. Under the RCP 8.5 scenario, by 2080, the temperature in Vancouver is expected to rise by 2.9°C, in Toronto by 4.1°C, and in Calgary by 3.4°C. Despite these warming trends, buildings in colder regions will still face extreme temperature drops, particularly during prolonged power outages. This means that while global warming may reduce the number of extremely cold days, occasional cold spells will still pose serious risks.

The study found that buildings are generally better equipped to withstand heat waves than extreme cold, particularly when HVAC systems fail. During heat waves, the lack of active cooling could lead to critical indoor temperatures above 32.2°C within a few hours, but passive strategies, such as natural ventilation and solar shading, proved effective in mitigating overheating effects. In Vancouver, without proper ventilation, indoor operative temperatures exceeded 32.2°C for 69–84 hours, and projections indicate that this duration could increase to 47 hours in the danger zone, where temperatures surpass 39.4°C. In Toronto, extreme climatic conditions increased the number of hours above 32.2°C from 64 to 74. In Calgary, where the climate is generally cooler, temperatures rarely reached critical levels, but future warming trends may lead to overheating issues, especially if buildings lack effective ventilation.

Conversely, during extreme cold events, buildings were found to be highly vulnerable to heat loss, posing a severe risk to occupants. In prolonged heating system failures, indoor temperatures in Toronto and Calgary could drop to hypothermic levels within 32–43 hours. In Toronto, without additional insulation, indoor temperatures fell to 4.4°C within four days, making the building uninhabitable. In Calgary, the conditions were even more severe, with indoor temperatures dropping to –1°C, presenting a serious threat to health and safety. In such conditions, without access to electricity, heating, or adequate insulation, buildings rapidly lost their ability to maintain comfortable temperatures. This highlights the need for additional thermal protection measures, such as mobile insulation solutions or optimized passive heat retention strategies.

In conclusion, climate change will necessitate new approaches to building design. The growing threat of heat waves calls for an increased focus on passive cooling strategies, while cold regions will still require enhanced measures to prevent heat loss. Future building designs must integrate both passive cooling and advanced heating systems, along with improved insulation, to enhance resilience to shifting climatic conditions. These findings underscore the importance of climate-adaptive construction practices, which will be essential in ensuring occupant safety and thermal comfort in the decades to come.

14.8 HEALTH RISK ASSESSMENT OF RESIDENTIAL OVERHEATING UNDER THE HEAT WAVES IN GUANGZHOU

Chen et al. (2024) conducted a study on the risk of overheating in residential buildings during power outages amid heat waves in Guangzhou, China. The authors introduced a new index, HIHHCaution (Heat Index Hazard Hours at Caution Level), designed to more accurately quantify indoor overheating risk. This metric measures the duration of exposure to heat index values exceeding the caution threshold (27°C–32°C). The introduction of this index was justified by the need to adapt overheating assessment tools to the specific conditions of China's subtropical climate, where high humidity and temperature create unique challenges. Combined with the maximum heat index (HImax), HIHHCaution allows for a dual assessment of both the intensity and duration of health hazards. HImax considers the combined effect of air temperature and relative humidity on the human body, with values approaching 55°C indicating an extreme health risk. The study established that HIHHCaution signifies severe health risks when exceeding 1174°C·h for young adults and 850°C·h for elderly individuals.

The analysis focused on three main types of residential buildings: urban village housing (UVH), old housing (OH), and commodity housing (CH), each differing in thermal insulation quality and resilience to extreme climatic conditions.

To assess the physiological impact of overheating on occupants, the study utilized the JOS-3 thermal regulation model, which allowed for the determination of health risk thresholds based on indoor thermal conditions. The overall methodological approach—from building classification and climate data analysis to indoor simulation and thermophysiological modeling—is summarized in Figure 14.9.

As illustrated in Figure 14.9, the study followed several key stages: heat wave characterization, residential building modeling, indoor condition simulation during power outages, and the application of JOS-3 for physiological impact evaluation.

The initial phase involved an analysis of six years of meteorological data, revealing an increase in both the frequency and intensity of heat waves, which were starting earlier and lasting longer. In 2023, extreme temperatures were recorded as early as May, surpassing 40°C.

The next phase focused on analyzing the thermal resilience of different housing types. The study examined three predominant residential typologies in Guangzhou—urban village housing (UVH), old housing (OH), and commodity housing (CH)—each characterized by distinct thermal envelopes and spatial layouts. As shown in Figure 14.10, the spatial configurations of each building type were modeled in 3D and 2D to reflect their thermal zoning and internal structure.

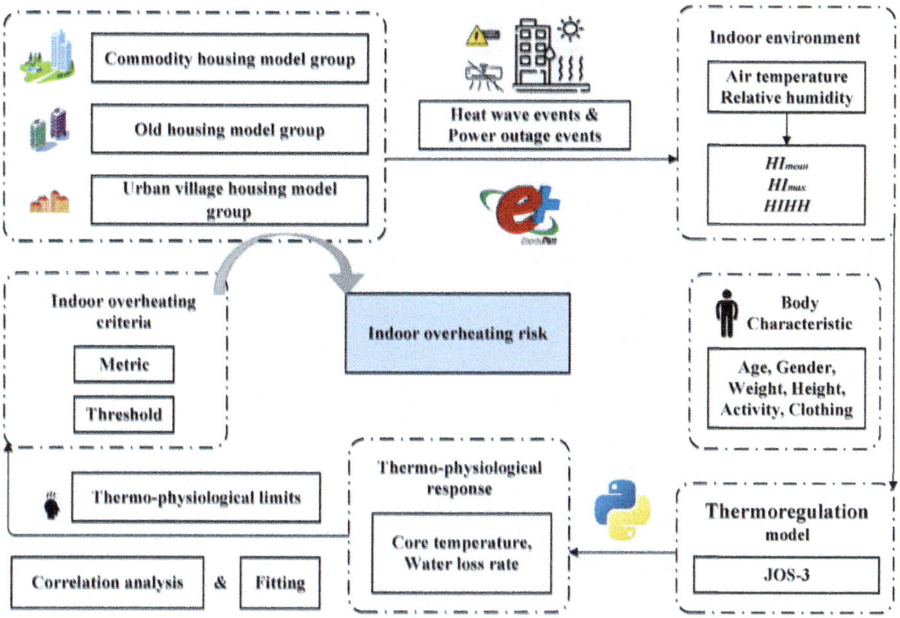

FIGURE 14.9 Workflow of the overheating risk assessment method combining housing typologies, heat wave and blackout scenarios, thermal simulation, and physiological modeling (Chen et al., 2024).

The study included simulations of indoor conditions during heat waves and assessed the impact of power outages on temperature and humidity levels inside dwellings. Using EnergyPlus, the researchers conducted simulations across various timeframes, ranging from several-hour outages to multi-day power failures. The JOS-3 model was then employed to assess physiological responses, tracking core body temperature fluctuations and dehydration levels to determine overheating thresholds.

The results indicated a strong correlation between overheating susceptibility and building thermal characteristics. Urban village housing, due to its poor insulation, exhibited overheating risks within a single day of a power outage. By the second day, the risk increased to 60%, reaching 100% after three days, meaning complete indoor overheating and critically unsafe conditions for residents. Old housing, with better insulation, remained more resistant. Overheating was not observed during the first two days of a power outage, but by the third day, the risk rose to 15%, escalating to 70% after four days and 78% after five days. Commodity housing, representing modern residential buildings compliant with the latest energy efficiency standards, provided safe conditions for three days. However, by the third day, overheating risk increased to 44%, by the fourth day to 58%, and by the fifth day reached 100%.

The study concluded that urban village housing is the most vulnerable to overheating, emphasizing the need for adaptive strategies, such as improved thermal insulation and enhanced ventilation systems. While commodity housing exhibited

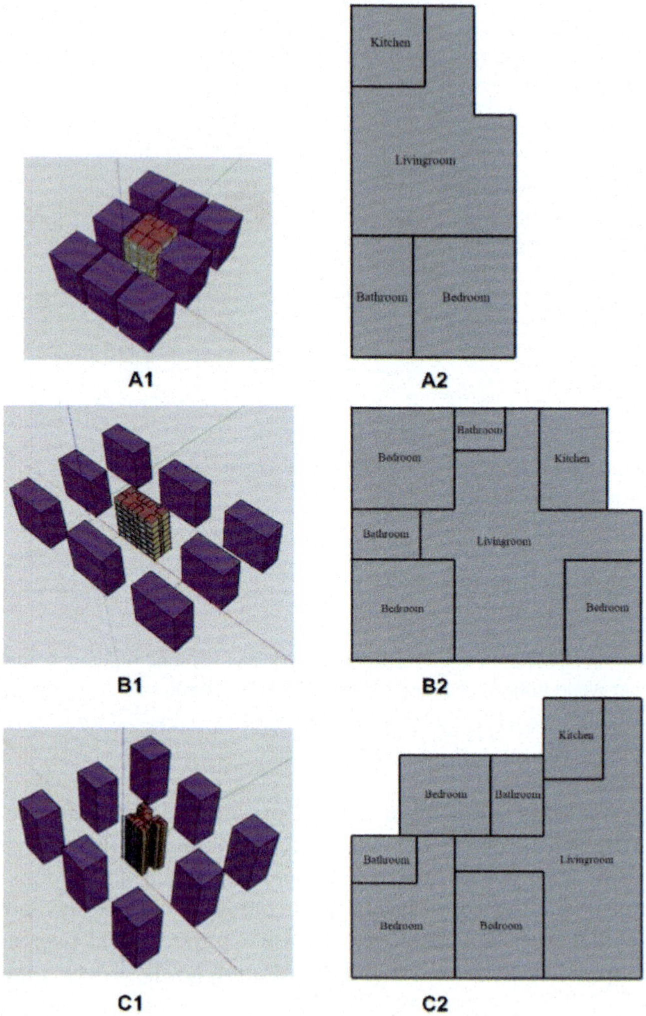

FIGURE 14.10 Three-dimensional models (left) and corresponding two-dimensional floor plans (right) of representative residential building types in Guangzhou: commodity housing (A1 -A2), old housing (B1 -B2), and urban village housing (C1 -C2) (Chen et al., 2024).

the highest resilience, even these buildings require further improvements to withstand prolonged power outages during extreme heat events.

The findings underscore the importance of designing more resilient buildings that effectively protect occupants from overheating risks, particularly in regions prone to extreme temperatures and frequent power outages. The study highlights the necessity of updating building codes to incorporate not only energy efficiency measures but also the capacity to maintain safe indoor conditions during crisis scenarios. Additionally, the research points to the need for urban adaptation plans, including the identification of high-risk areas and the implementation of mitigation strategies such as green infrastructure, natural ventilation enhancements, and shading systems.

The study's conclusions are also relevant to the energy sector, where emergency response strategies should prioritize power supply to the most vulnerable areas affected by overheating during extended blackouts.

14.9 NEXUS OF THERMAL RESILIENCE AND ENERGY EFFICIENCY IN BUILDINGS: A CASE STUDY OF A NURSING HOME

Sun et al. (2020) conducted a study on the relationship between energy efficiency and thermal resilience in buildings, using a nursing home in Florida as a case study. The facility experienced an HVAC system failure during Hurricane Irma in 2017, which resulted in the deaths of 12 residents. The study aimed to determine how energy efficiency improvements influence a building's ability to withstand extreme heat and to identify the most effective combinations of passive and active strategies to mitigate overheating risks.

The research followed a structured methodology that included data collection, model calibration, and the evaluation of passive and active thermal resilience strategies. This workflow is summarized in Figure 14.11. A 3D model of the case study building was developed using DesignBuilder (Figure 14.12).

The case study focused on a real nursing home that lost air conditioning for over 60 hours during the hurricane, causing indoor temperatures to reach life-threatening levels. The collected data included architectural drawings, records of previous renovations, and information about the HVAC system failure.

The analyzed building was a two-story facility covering 5,406 m^2, equipped with a central air conditioning system that shut down at 3:50 PM on September 10, 2017.

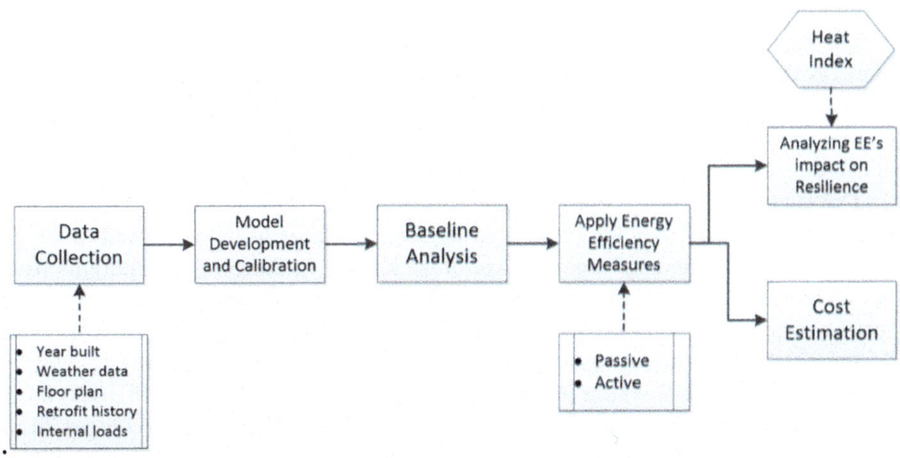

FIGURE 14.11 Workflow for assessing thermal resilience, including building data collection, model calibration, application of passive and active strategies, and cost estimation (Sun et al., 2020).

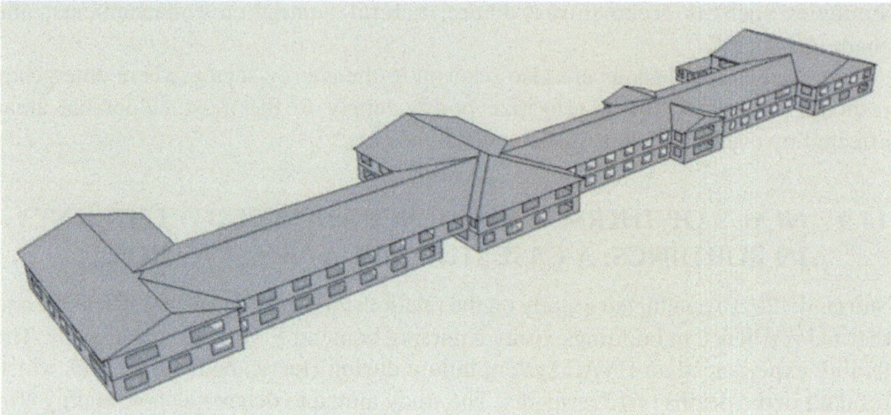

FIGURE 14.12 3D model of the case study nursing home in Florida developed in DesignBuilder for EnergyPlus-based thermal simulation and resilience analysis (Sun et al., 2020).

The outage lasted until 6:00 AM on September 13, during which indoor temperatures on the second floor reached 37.2°C, with the highest recorded heat index peaking at 49°C.

The model was calibrated using available data, including official government reports. A verification process was conducted by comparing simulated temperature trends with post-hurricane records. The results confirmed that the simulated peak indoor temperature of 37.3°C closely matched the recorded values, validating the model's reliability. This level of accuracy in temperature prediction demonstrates that the model effectively represents real-world thermal conditions within the building, making it a useful tool for assessing thermal resilience during extreme weather events.

The simulation results showed that once the HVAC system failed, indoor temperatures rose rapidly, particularly on the second floor. Within 24 hours, some rooms exceeded 35°C, and after 48 hours, temperatures in patient rooms peaked at 49°C. This posed a severe health risk, with 32.3% of the total time spent in conditions classified as dangerous or extremely dangerous, with heat index values surpassing 39.4°C.

The study examined the effectiveness of several energy efficiency and thermal resilience measures. Among passive strategies, natural ventilation proved to be the most effective, reducing the percentage of time spent in dangerous conditions from 32.3% to 1.2%. Reducing internal heat loads by limiting lighting and appliance use further lowered the overheating exposure to 6.5%. High-reflectivity window films, designed to reduce solar heat gain, had limited impact due to existing regulatory constraints in Florida, where the required solar heat gain coefficient (SHGC) is set at 0.25. As a result, the overall temperature reduction achieved with window films was only 25.2%.

Active cooling solutions, particularly thermal energy storage, demonstrated the highest effectiveness. A cold storage system with a 24-hour capacity successfully maintained temperatures below the dangerous threshold throughout the power outage.

In contrast, a smaller storage system with an 8-hour capacity provided only temporary relief, as temperatures quickly rose once the stored cooling energy was exhausted.

To further examine how local climatic conditions influence the effectiveness of resilience strategies, the study expanded its simulations to two additional cities, San Francisco and Chicago. In San Francisco, where relative humidity is lower, the heat index never reached critical levels, even though indoor air temperatures in some rooms exceeded 43°C. In Chicago, a five-day heat wave in 2012 resulted in extreme conditions, with the heat index reaching 60°C. In this scenario, none of the passive measures alone were sufficient to maintain safe conditions throughout the blackout. Even the most optimized combination of passive strategies reduced the time spent in dangerous conditions by only 73%, underscoring the necessity of active cooling solutions, such as emergency HVAC backup systems.

The study also assessed the financial costs of implementing thermal resilience measures. Passive strategies such as natural ventilation and internal heat load reduction were found to be the most cost-effective, requiring little to no investment. Meanwhile, cold storage systems were significantly more expensive, with costs ranging between 50 and 150 USD/m^2 depending on system capacity.

The findings of Sun et al. (2020) provide critical insights into the interaction between energy efficiency and thermal resilience in buildings. The study emphasizes that passive strategies such as ventilation and reducing internal heat sources can significantly mitigate overheating risks, but they are often insufficient in extreme conditions. In cases of prolonged power outages, active cooling solutions, including cold storage and backup HVAC systems, are necessary to prevent life-threatening indoor conditions.

The results highlight the need for incorporating thermal resilience requirements into energy efficiency standards, particularly for buildings housing vulnerable populations such as the elderly. The study concludes that future building regulations should not only focus on reducing energy consumption but also ensure that buildings can maintain safe indoor conditions during prolonged power disruptions.

14.10 MULTI-SCENARIO EXTREME WEATHER SIMULATOR APPLICATION TO HEAT WAVES: KO'OLAULOA COMMUNITY RESILIENCE HUB

Villa et al. (2024) conducted a study on the thermal resilience of buildings during extreme heat waves, utilizing the Multi-Scenario Extreme Weather Simulator (MEWS) to model future extreme weather conditions based on historical data and climate projections. The study focused on the newly designed Ko'olauloa Community Resilience Hub (KCRH) in Hawaii, a facility intended to operate both under normal conditions (NormOps) and in resilience mode (ResOps) during emergencies, such as power outages and extreme weather events. The goal was to assess the impact of increasing heat wave frequency and intensity on the building's energy demand and its ability to maintain thermal comfort under future climate scenarios.

The research was based on building energy modeling (BEM) using EnergyPlus and OpenStudio, enabling the simulation of indoor conditions across various climate scenarios. The MEWS framework (Multiple Extreme Weather Scenarios) utilized over 50 years of climate data from the National Oceanic and Atmospheric Administration (NOAA), along with climate projections from the Sixth Coupled Model Intercomparison Project (CMIP6). The analysis incorporated Shared Socioeconomic Pathways (SSP) scenarios, specifically SSP2-4.5, SSP3-7.0, and SSP5-8.5, to assess future climate risks and overheating trends.

The analytical process, illustrated in Figure 14.13, consisted of two main components. On the left, the historical climate data analysis was conducted once to calibrate

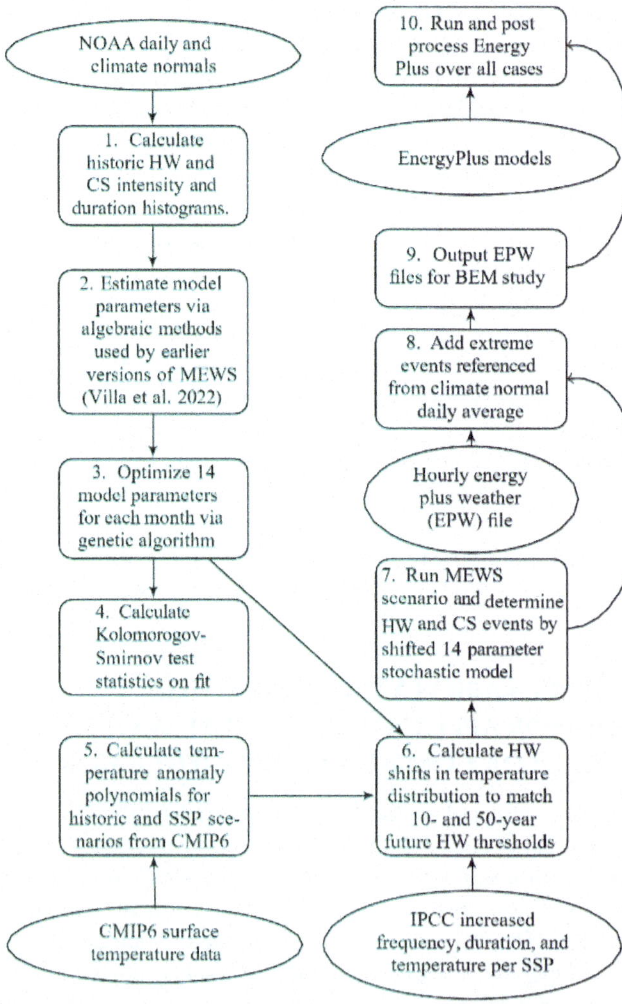

FIGURE 14.13 Overview of MEWS and BEM study process. Ovals portray analysis inputs and boxes indicate steps. The left-hand-side processis only executed once for historical fits, while the right-hand-side process is executed over SSP scenario, future year, HW FDI confidence interval, BEM model, and realization number (Villa et al., 2024).

the statistical model. On the right, iterative steps were performed for different SSP scenarios, forecast years (2020, 2040, 2060, and 2080), and confidence intervals for extreme weather conditions. The use of optimization algorithms enabled fine-tuning the model parameters to best represent the probability distributions of historical heat waves and their projected intensifications.

The use of optimization algorithms enabled fine-tuning the model parameters to best represent the probability distributions of historical heat waves and their projected intensifications.

MEWS generated 7,200 future weather files across different years and climate scenarios, which were used for 14,400 EnergyPlus simulations. This approach allowed for assessing future energy consumption and evaluating changes in the building's thermal resilience under both NormOps and ResOps modes.

The Ko'olauloa Community Resilience Hub (KCRH) was designed as a dual-purpose facility, operating as a community center under normal conditions and serving as an emergency shelter during extreme weather events. Located in Hau'ula, on the northeastern coast of Oahu, the building is particularly vulnerable to hurricanes, heat waves, and other climate-related hazards. Due to frequent flooding along Kamehameha Highway (Highway 83) during severe storms, KCRH plays a critical role in providing support and essential services to the local community during emergencies. To ensure operational continuity in the event of power outages, the facility incorporates a microgrid system capable of sustaining essential energy demands.

The building's energy model comprises 31 thermal zones, enabling detailed analysis of temperature distribution across its internal spaces. During ResOps mode, air conditioning is strategically limited to designated areas, optimizing energy use while maintaining thermal resilience under constrained conditions.

KCRH was modeled in detail using EnergyPlus and OpenStudio. As shown in Figure 14.14, the facility consists of three functional levels: a shaded parking and service level, a main entry floor housing public functions, and an upper level containing classrooms and staff workspaces. The zoning approach supports high-resolution thermal modeling under both standard and emergency operation scenarios.

Simulations indicated that under NormOps, the KCRH generally maintains stable indoor conditions, with temperatures rarely exceeding 32.2°C. This confirms that the building's design meets thermal comfort expectations and effectively mitigates overheating risks. However, in ResOps mode, where energy use and cooling are restricted, the number of hours exceeding this critical threshold increased significantly. In 2020, the percentage of time above 32.2°C in ResOps mode was 5%, rising to 29% under the most extreme 2080 climate scenarios. Moreover, in common areas without active cooling, the number of hours in the "Extreme Caution" range (32.2°C–39.4°C) increased from 300 to 1,800 hours per year in the SSP5-8.5 scenario.

Additionally, the analysis found that the average operative temperature in non-air-conditioned spaces under ResOps mode increased from 27.5°C in 2020 to 30.6°C in 2080 under SSP5-8.5. Compared to NormOps, where average temperatures were 1.2°C–2.3°C lower, this suggests a pressing need for additional passive cooling strategies to mitigate overheating effects.

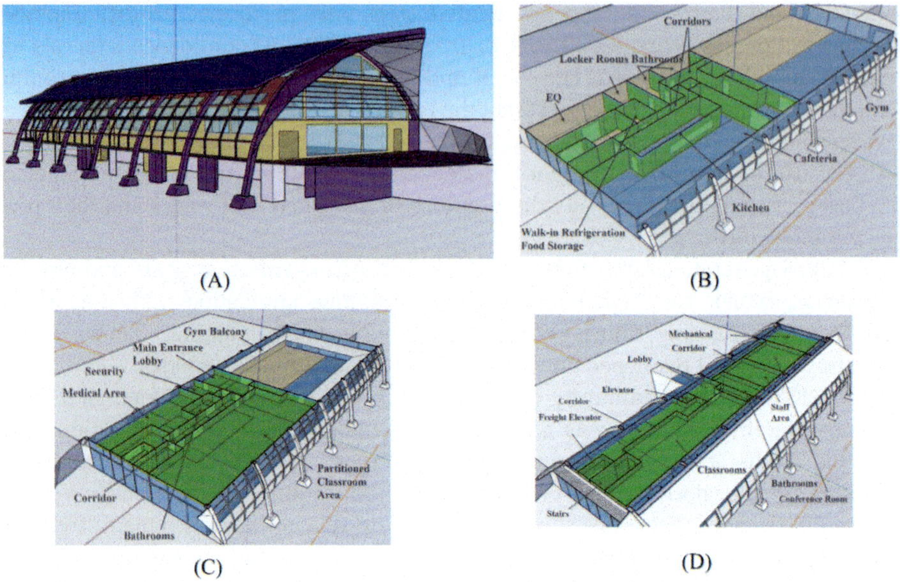

FIGURE 14.14 EnergyPlus model of the Ko'olauloa Community Resilience Hub (KCRH) in Hau'ula, HI, showing isometric view (A) and floor layouts: Service (B), entry (C), and upper level (D) (Villa et al., 2024).

The results indicate that as climate change progresses, the KCRH will face increasing challenges in maintaining thermal comfort. Climate projections suggest that as heat waves become more frequent and intense, the number of hours exceeding 32.2°C will rise dramatically. In the SSP5-8.5 scenario, the duration of these overheating conditions in ResOps mode increased sixfold compared to 2020.

Simulations also demonstrated a decline in the building's thermal resilience over time. Under SSP5-8.5, the average operative temperature in ResOps mode is projected to reach 29.8°C in 2060 and 30.6°C in 2080, whereas in NormOps, these values are expected to be 28.5°C in 2060 and 29.2°C in 2080. This growing dependence on active cooling highlights the increasing difficulty of maintaining comfort in a warming climate.

Energy consumption analysis revealed a 0.8% increase in annual cooling demand for NormOps between 2020 and 2080. In contrast, cooling energy use in ResOps mode declined by 26% due to restricted air conditioning. While this suggests improved energy efficiency, it also leads to increased indoor overheating.

Further projections indicate that during the hottest periods, the number of hours in the "Extreme Caution" zone will increase from 900 to 3,200 hours per year, while hours in the "Danger" zone (above 39.4°C) will rise from 50 to 275 annually. This escalation highlights the growing health risks associated with climate change and underscores the need to ensure priority access to air-conditioned spaces for vulnerable populations.

The study's findings emphasize that while KCRH is capable of maintaining acceptable indoor conditions under NormOps, its ability to sustain comfort under

ResOps will significantly decline in the future. The increasing number of hours above 32.2°C, rising exposure to "Extreme Caution" and "Danger" conditions, and higher operative temperatures indicate the necessity of implementing additional adaptation measures.

Recommendations include enhancing natural ventilation, optimizing shading systems, ensuring priority access to cooled spaces for at-risk populations, and increasing reliance on microgrids and renewable energy sources to reduce dependency on external power supplies. Future research should expand the analysis to different building typologies and climatic regions to tailor resilience strategies to diverse environmental conditions.

14.11 OVERHEATING ANALYSIS OF OPTIMIZED NEARLY ZERO-ENERGY DWELLING DURING CURRENT AND FUTURE HEAT WAVES COINCIDED WITH COOLING SYSTEM OUTAGE

Rahif et al. (2023) conducted a study on the impact of extreme heat waves on optimized nearly zero-energy buildings (nZEBs), focusing on simultaneous cooling system failures. The research analyzed the performance of benchmark terraced housing in Brussels, assessing the effectiveness of passive strategies in maintaining thermal comfort. The primary objective was to determine how optimized building designs can reduce energy consumption while improving indoor thermal conditions and whether these measures are sufficient to mitigate the effects of future heat waves and cooling system failures.

The study followed a two-stage approach, as outlined in Figure 14.15.

In the first stage, building parameters and passive strategies were optimized to minimize HVAC system energy consumption and enhance occupant comfort. This optimization utilized a genetic algorithm, specifically the Non-dominated Sorting Genetic Algorithm 2 (NSGA-II), to identify the most effective trade-offs among various design alternatives. A total of thirteen passive strategies were considered, including high airtightness, improved insulation, increased thermal mass, green roofs, and operable shading devices. The optimization process aimed to reduce HVAC energy consumption by 32% and improve thermal comfort by 46%.

The second stage focused on simulating the building's resilience to overheating during extreme heat waves. Simulations were conducted for three climate periods—historical (2001–2020), mid-century (2041–2060), and late-century (2081–2100)—analyzing the most severe heat events. Three optimized building variants were tested to assess their ability to maintain comfortable indoor conditions without mechanical cooling. The findings provided insights into the extent to which buildings could function passively and how climate change will affect future thermal resilience.

The reference building, depicted in Figure 14.16, represents a typical terraced house in Brussels, which was modernized in 2010 to meet nearly Zero-Energy Building (nZEB) standards. The renovations included facade insulation and the installation of photovoltaic panels, which generate an annual energy output of 3,000 kilowatt-hours.

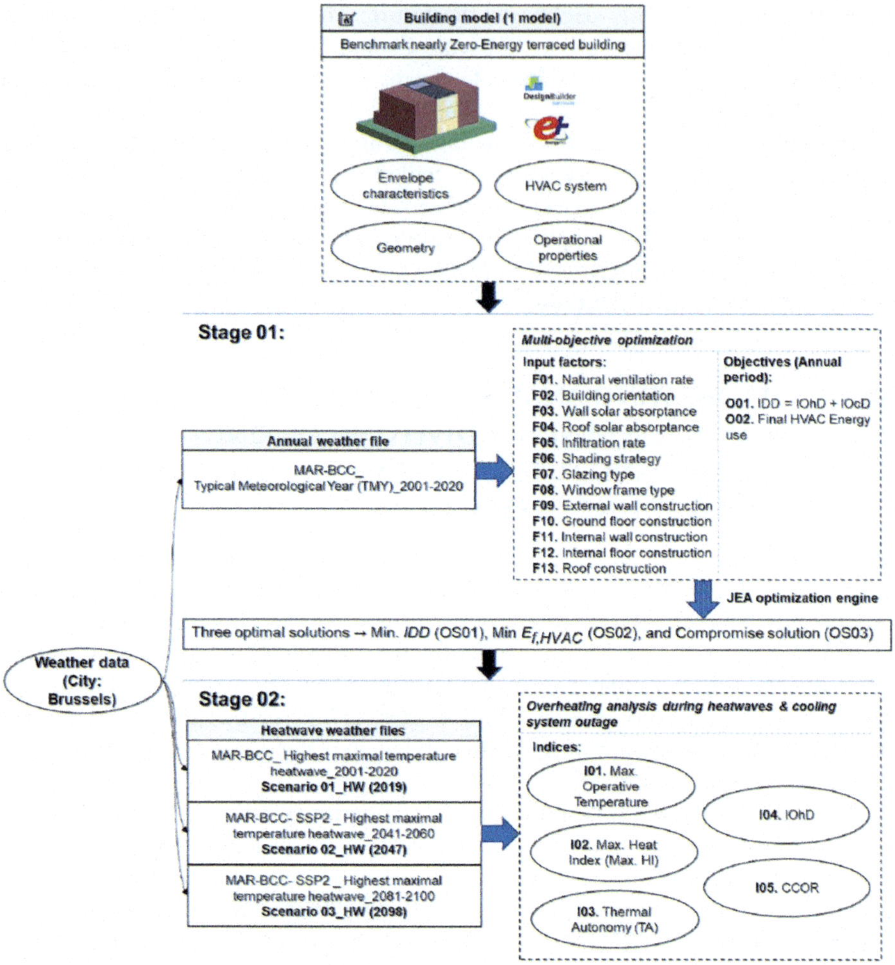

FIGURE 14.15 Study conceptual framework for evaluating thermal resilience under heat waves and cooling outages in an optimized nearly zero-energy terraced dwelling (Rahif et al., 2023).

The functional layout consists of four primary thermal zones: a living area (combined living room and kitchen), an office space, bedrooms, and transient-use areas such as hallways, bathrooms, and utility rooms. The building model was developed using DesignBuilder, which operates on the EnergyPlus simulation engine, enabling a detailed analysis of energy balances and indoor temperature dynamics under various climate conditions. The simulations incorporated realistic occupancy schedules, lighting patterns, and internal heat gains, following ISO 18253–2 standards and national energy statistics.

The study's findings indicated that none of the three optimized building variants could entirely prevent overheating under extreme heat wave scenarios and cooling system failures. During extreme conditions, indoor operative temperatures

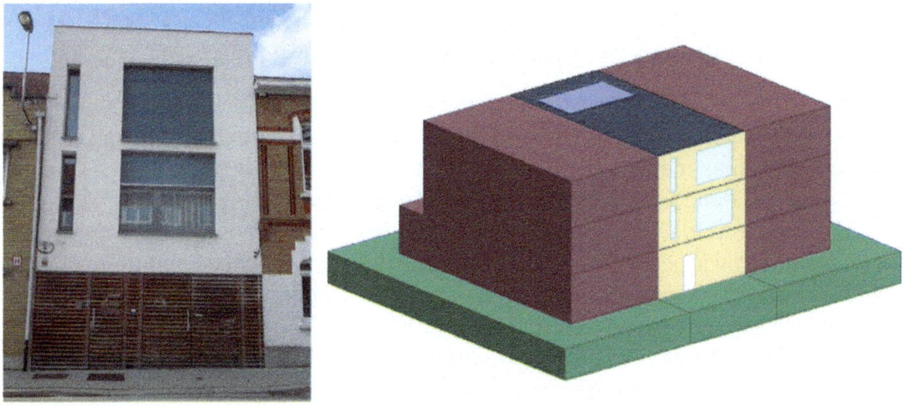

FIGURE 14.16 The benchmark nearly zero-energy terraced dwelling in Brussels: (Left) renovated south-facing façade; (Right) 3D model created in DesignBuilder for simulation (Rahif et al., 2023).

exceeded 29°C, posing a significant health risk to occupants. Furthermore, climate projections suggest a worsening trend, with maximum heat index values increasing by 0.28°C–0.49°C and peak operative temperatures rising by 1.34°C–2.33°C. Additionally, the thermal autonomy index, which measures a building's ability to maintain comfort without active cooling, declined by 17%–28%, highlighting the growing reliance on mechanical cooling systems.

The study also revealed key differences between the analyzed building variants. The OS01 variant, which provided the highest thermal comfort levels, also exhibited the highest energy consumption. In contrast, the OS02 variant, designed for minimal energy use, had the lowest capacity for maintaining thermal comfort. The OS03 variant balanced energy efficiency and occupant comfort, representing a compromise between the two extremes. These results indicate that even optimized nZEBs do not fully protect against extreme climate conditions, particularly in the face of rising global temperatures.

Based on the research findings, key passive strategies identified for improving heat wave resilience include natural ventilation, green roofs, and effective solar shading systems. While nZEBs meet current energy efficiency regulations, their design does not fully account for future climate conditions, necessitating the introduction of dynamic adaptation strategies such as night ventilation management, intelligent shading control, and increased thermal storage capacity.

The study also underscores the necessity of integrating resilience strategies into nZEB design, considering both climate variability and potential cooling system failures. Future developments may require the adoption of hybrid cooling solutions and enhanced adaptive capacity in building materials and systems.

In conclusion, the findings suggest that in future climates, dynamic temperature management strategies will be essential for improving building adaptation to rising heat wave risks. These insights have significant implications for future research and policy-making, emphasizing the need for comprehensive resilience strategies in the design and regulation of nZEBs.

14.12 A SIMULATION FRAMEWORK FOR ASSESSING THERMALLY RESILIENT BUILDINGS AND COMMUNITIES

Krelling et al. (2023) proposed a comprehensive simulation-based approach for assessing the thermal resilience of buildings and urban communities during extreme heat waves. Their methodology allows for the evaluation of buildings both at the individual level and within broader urban contexts, facilitating the identification of populations particularly vulnerable to overheating. The proposed framework incorporates three key aspects of thermal resilience: thermal resistance, failure resilience, and the capacity to restore thermal comfort. The simulation logic and key performance indicators are illustrated in Figure 14.17.

A set of performance indicators was employed to evaluate the thermal efficiency of buildings in relation to each stage of resilience. One of the key parameters was

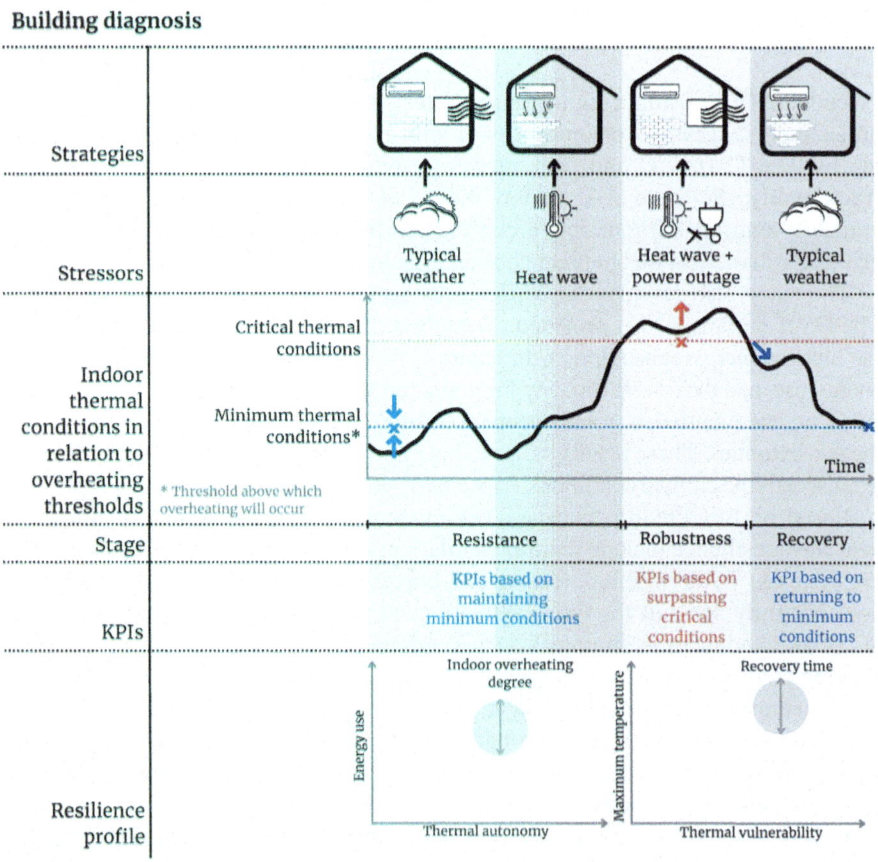

FIGURE 14.17 Thermal resilience simulation framework for single buildings, including key performance indicators across resistance, robustness, and recovery stages (Krelling et al., 2023).

thermal autonomy, representing the percentage of hours during which indoor conditions remained within an acceptable range without mechanical cooling.

To assess urban-scale resilience, the simulation framework was expanded to include groups of buildings and their aggregated performance under different stressors and climatic conditions.

A set of performance indicators was employed to evaluate the thermal efficiency of buildings in relation to each stage of resilience. One of the key parameters was thermal autonomy, which represents the percentage of hours during which indoor conditions remained within an acceptable temperature range without the need for cooling systems. The comfort range was defined as 18–26°C in Curitiba and Florianópolis and 18°C–28°C in São Luís. If these thresholds were exceeded, the building was considered to have failed to meet minimum thermal comfort requirements. Additional parameters included the degree of overheating, defined as the average deviation of operative temperature beyond the upper comfort limit, and thermal sensitivity, expressed as the percentage of hours during which indoor operative temperatures exceeded 30°C in Curitiba and Florianópolis and 32°C in São Luís. The thermal recovery time, indicating how long it took for indoor conditions to return to acceptable levels after exceeding the comfort threshold, was also assessed.

This extension of the model is shown in Figure 14.18, which reflects the broader resilience profile of building ensembles in varying city contexts.

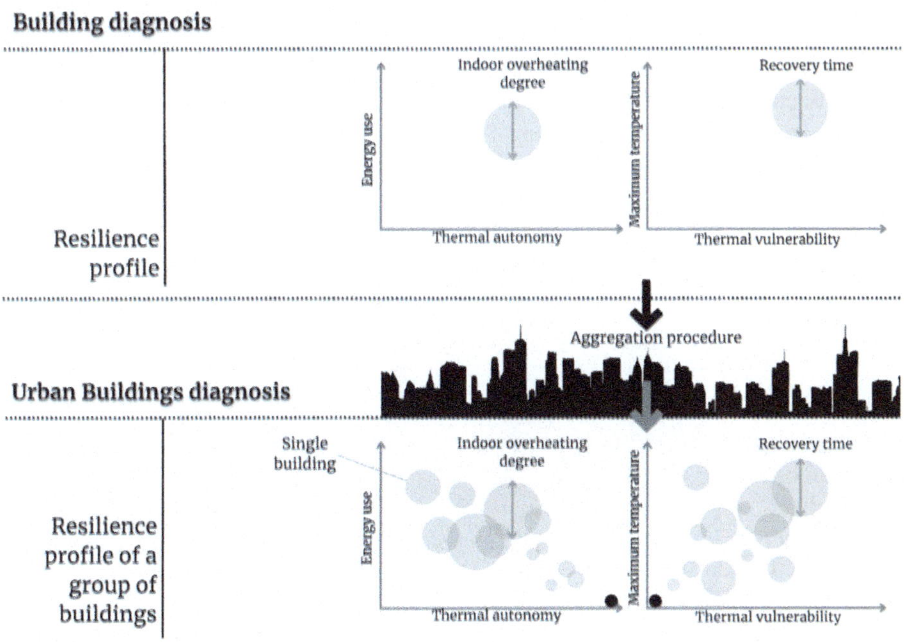

FIGURE 14.18 Aggregated thermal resilience framework for groups of buildings, used to assess urban-scale vulnerabilities and performance during extreme heat events (Krelling et al., 2023).

The study examined three Brazilian cities-Curitiba, Florianópolis, and São Luís-located in the south and northeast of the country, allowing for an evaluation of thermal resilience under different climate conditions. According to the Köppen-Geiger climate classification, Curitiba has a humid subtropical climate, Florianópolis a marine subtropical climate, and São Luís a tropical monsoon climate.

To establish a baseline for thermal resilience under standard climatic conditions, the study used year-long numerical simulations based on typical meteorological years. This procedure enabled comparisons between current conditions and extreme weather scenarios, such as heat waves and future climate projections.

The study focused on a sample of single-family homes, representing a typical low-standard residential design in Brazil, which closely reflects the existing housing stock. A significant portion, 86% of the sample, consisted of detached houses, making it highly representative of the predominant residential typology in the region.

The residential case study was based on a standardized low-income housing typology in Brazil. Its internal layout and spatial distribution of thermal zones are depicted in Figure 14.19, used as the geometric reference across all envelope variants simulated in the study.

A considerable 33% of Brazil's total housing stock does not meet building standards, with most relying on natural ventilation due to the limited availability of air conditioning. Passive cooling strategies play a crucial role in maintaining indoor comfort. The study assessed the thermal characteristics of buildings based on the variability of external envelope configurations. A set of 448 unique cases was generated to capture the impact of different material configurations by systematically comparing 14 types of external walls, 2 types of internal walls, and 16 different roof types. Each case retained identical geometry and occupancy profiles, ensuring standardized conditions and eliminating confounding variables in the evaluation of how external envelope characteristics influence thermal resistance.

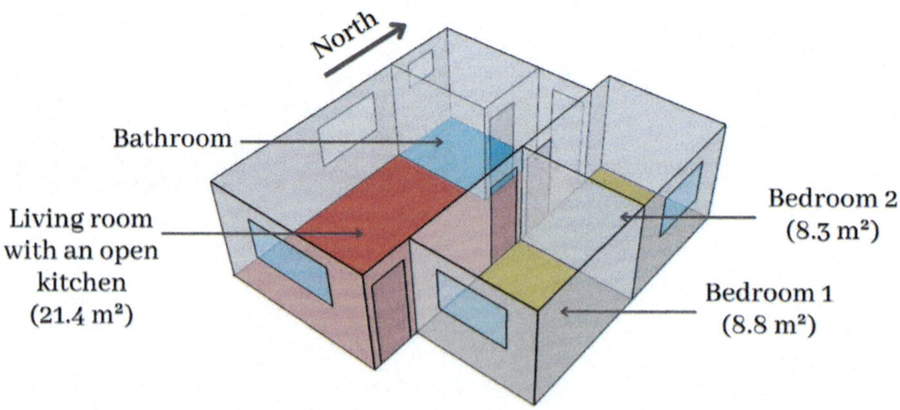

FIGURE 14.19 Representation of the reference case study dwelling used in simulations: A low-income house layout with labeled thermal zones and internal areas (Krelling et al., 2023).

The simulation results revealed significant variations in thermal resilience depending on the climate. In Curitiba, where conditions are mildest, most buildings maintained comfortable indoor temperatures year-round without requiring air conditioning. The thermal autonomy index approached 100%, indicating that buildings in this region had a high capacity to maintain stable temperatures without relying on mechanical cooling. Overheating risks were minimal, and occurrences of critical temperature exceedances were rare and short-lived.

In Florianópolis, which experiences a moderate climate, buildings demonstrated an average thermal autonomy of 69%, meaning that they could passively maintain comfortable conditions for nearly 70% of the time without cooling. The average overheating level was relatively low at 0.44°C, but in some cases, maximum temperatures exceeded 40°C, and thermal recovery times reached up to 200 hours.

In São Luís, characterized by an exceptionally hot climate, thermal resilience was significantly lower. Most buildings were entirely dependent on mechanical cooling systems, and the thermal autonomy index frequently dropped to 0%. The average overheating level was 1.2°C, with the most vulnerable buildings exceeding 2.5°C. On the hottest days, critical thermal conditions persisted for more than half of the day, and maximum indoor temperatures surpassed 42°C. Furthermore, thermal recovery times ranged from several hours to several weeks, depending on the building's thermal mass. High thermal inertia contributed to overheating, making it more difficult to dissipate accumulated heat.

The findings highlight the importance of adapting design strategies to specific climate conditions. In Curitiba, buildings with high thermal mass and effective insulation performed well, maintaining comfortable temperatures throughout most of the year without active cooling. In contrast, in São Luís, high thermal inertia contributed to overheating, suggesting that alternative strategies, such as increased natural ventilation and shading systems, would be more effective in mitigating heat accumulation.

14.13 NEW METRICS FOR THERMAL RESILIENCE OF PASSIVE BUILDINGS DURING HEAT EVENTS

Flores-Larsen et al. (2023) analyzed the thermal resilience of an office building located in Victorica, Argentina, during five major heat waves recorded in the summer of 2017–2018. The study combined in situ monitoring with dynamic thermal simulations, allowing for a thorough evaluation of passive strategies applied in the building's design. The geographical location and external view of the case study building are shown in Figure 14.20, while the internal layout and thermal zoning used in the simulations are presented in Figure 14.21.

During this period, maximum outdoor temperatures reached 42.3°C, and heat wave durations ranged from three to eight days, with the longest event having a global severity index of 13.5, reflecting both intensity and duration. The analysis combined real-world measurements and thermal simulations, enabling a comprehensive assessment of the effectiveness of passive design strategies.

FIGURE 14.20 Location and exterior view of the case study building in Victorica, Argentina. Left: Geographic location within the La Pampa region. Right: Street view of the INTA office (Flores-Larsen et al., 2023).

(Source: Google. Photo: Agencia de Extensión Rural Victorica, La Pampa.)

The studied building was designed with bioclimatic passive strategies, including massive, heavyweight walls, cross-ventilation, high thermal insulation, and solar pergolas. These features significantly reduced overheating effects. During the analyzed heat waves, indoor temperatures never exceeded 29°C, even when outdoor temperatures surpassed 42°C. The least severe heat wave, lasting three days with a peak temperature of 37.9°C, caused only minor increases in indoor temperatures. During the most intense heat wave, lasting eight days with a peak temperature of 42.3°C, indoor temperatures remained close to the upper thermal comfort threshold of 26.7°C.

One of the challenges observed was heat accumulation in the massive structural elements, which prolonged the time needed for the building to return to comfortable thermal conditions. After heat waves ended, elevated indoor temperatures persisted for one to three days, mainly due to heat retention in the heavyweight walls and insufficient nighttime ventilation.

The study highlighted the importance of accounting for the thermal history of a building. The introduction of the Accumulation Factor (γ_{IOD}) allowed the assessment of how previously stored heat influenced current indoor conditions. The analysis showed that effective management of nighttime ventilation could reduce cooling times and accelerate the return to thermal comfort.

Another key aspect was the Solar Radiation Factor (β_{IOD}), which helped quantify the influence of solar radiation on building overheating. The results indicated that on days with high solar exposure, indoor temperatures increased more rapidly, even when outdoor temperatures were moderate. For example, on days with solar radiation above 23.4 MJ/m^2, indoor temperatures approached the 26.7°C threshold, despite outdoor temperatures not reaching extreme levels.

The study concluded that the office building in Victorica demonstrated strong thermal resilience, effectively minimizing the impact of heat waves on indoor

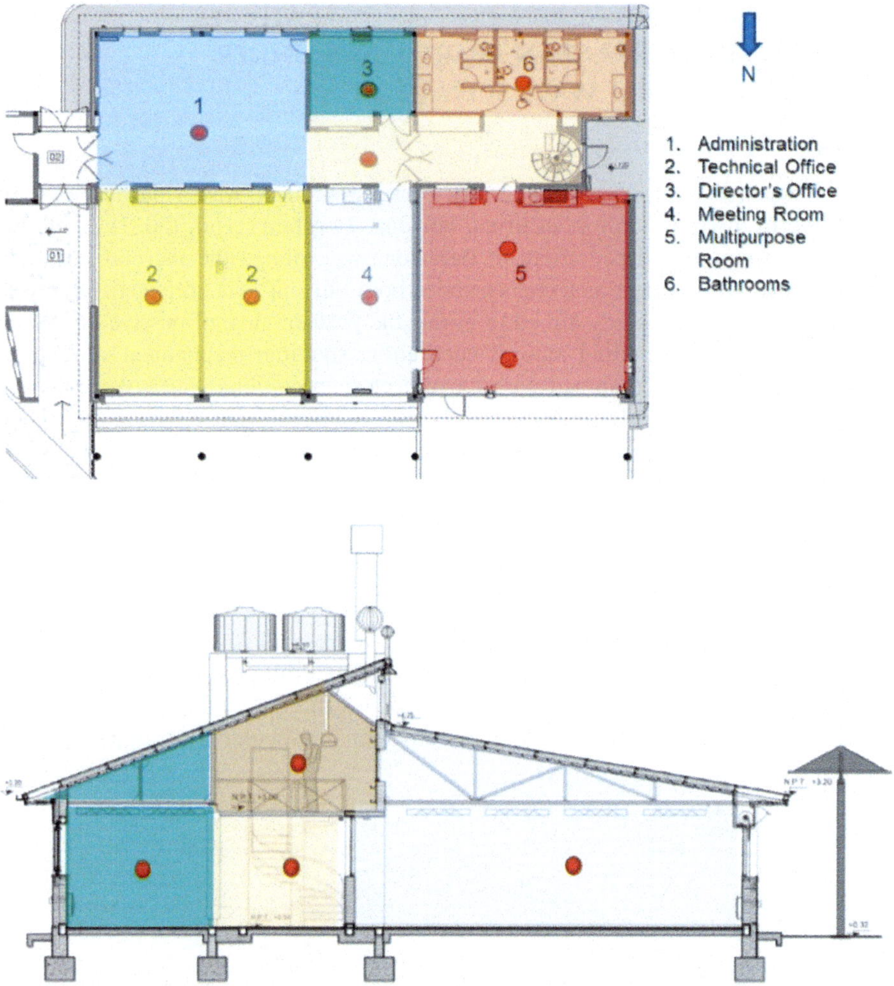

FIGURE 14.21 Top: Floor plan of the office building, with temperature and humidity sensor locations (red dots) and color-coded thermal zones used in simulations. Bottom: Building section highlighting the sloped roof and ventilated technical plenum (Flores-Larsen et al., 2023).

temperatures. However, the analysis revealed that heat accumulation in massive structural elements and insufficient nighttime ventilation reduced the effectiveness of passive strategies. The results suggest that improving nighttime ventilation and solar shading management is essential, particularly on days with high solar radiation. The introduction of the γ_{IOD} and β_{IOD} metrics provided a more detailed understanding of overheating dynamics and helped identify areas for improvement. The findings emphasize the importance of considering thermal history and solar radiation in evaluating building resilience, which can enhance future building design and retrofitting strategies in regions prone to extreme heat waves.

14.14 PASSIVE AND ACTIVE COOLING STRATEGIES IN MEDITERRANEAN APARTMENTS UNDER FUTURE HEAT WAVE SCENARIOS

14.14.1 Introduction

The increasing frequency and intensity of summer heat waves pose a growing challenge to thermal comfort in urban housing. Borghero et al. (2023) conducted a detailed simulation-based study in Barcelona to evaluate the thermal resilience of typical Mediterranean apartments under both current and projected future climatic conditions. The study aimed to assess the performance of passive and active cooling strategies, including natural ventilation, shading, mechanical ventilation, and air conditioning, combined with retrofitting interventions. Particular emphasis was placed on behavioral influence, energy use, and performance during future heat waves. The analysis evaluates both thermal comfort and energy demand under multiple climate scenarios (2018, 2022, 2030, and 2050), using a real-world apartment layout as the basis for simulation, as illustrated in Figure 14.22.

14.14.2 Methodology

The study involved 40 dynamic simulations of a representative apartment located on an intermediate floor of a multifamily residential building built between 1951 and 1980 in the Barcelona region. The layout includes three bedrooms, a living room, kitchen, and bathrooms, occupied by a three-person household. The original envelope

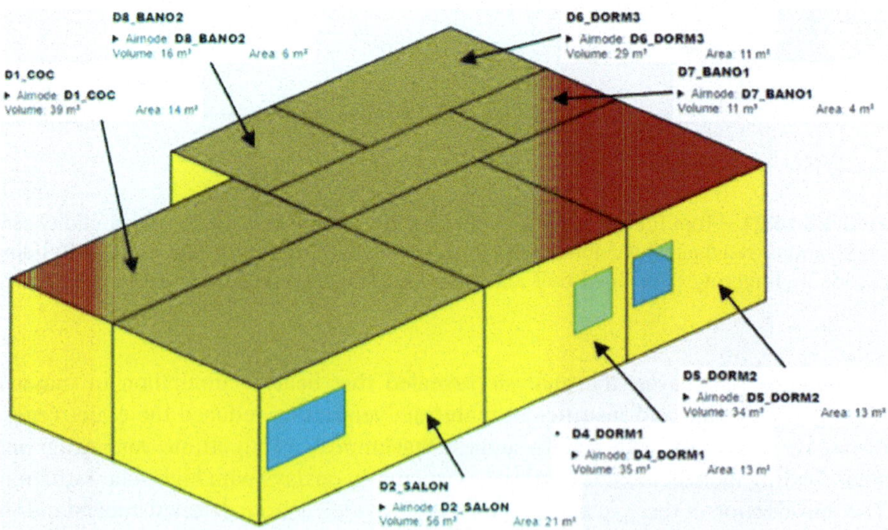

FIGURE 14.22 Floor plan of the reference apartment in Barcelona used for simulations, showing living areas and bedrooms served by the cooling strategies tested (Borghero et al., 2023).

is characterized by poor thermal insulation, which was further tested in a retrofitted variant featuring external insulation and upgraded double-glazed windows.

Five cooling strategies were simulated using TRNSYS 18:

1. Natural Ventilation (Fixed and Aware)—relying on cross-ventilation and solar shading with fixed or user-controlled window opening patterns.
2. Mechanical Ventilation (Fans)—employing ceiling and desk fans in key rooms, activated when indoor operative temperatures exceeded 28 °C.
3. Air Conditioning (Fixed and Aware)—a ducted system serving the living room and bedrooms, with the aware version incorporating adaptive use of ventilation in uncooled spaces.

Each strategy was tested in its base (original envelope) and retrofitted form. Weather data included observed summers from 2018 and 2022, as well as two projected scenarios (2030 and 2050) under the RCP4.5 climate trajectory. Heat wave periods were identified based on both local municipality thresholds (≥33.1°C for 3+ days) and the Excess Heat Factor (EHF). Performance was assessed using a comprehensive set of thermal resilience indicators, including the adaptive comfort model, Heat Index (HI), Humidex, Indoor Overheating Degree (IOD), Ambient Warmness Degree (AWD), and the Overheating Escalation Factor (α_{IOD}).

14.14.3 Results

The adaptive comfort model showed that passive strategies, especially natural ventilation with user awareness, maintained comfort for a substantial share of time during regular summers. In 2022, the aware natural ventilation scenario combined with retrofit kept indoor conditions within CAT I for 92% of the heat wave period. However, in the projected summers of 2030 and 2050, natural ventilation alone became increasingly ineffective, offering acceptable comfort only 13–35% of the time even when retrofit was applied.

Mechanical ventilation with fans extended comfort ranges modestly, particularly in intermediate conditions, but could not fully offset overheating during more intense future events. Air conditioning was the only strategy that consistently ensured high comfort levels across all years, though it incurred significant energy consumption, exceeding 13 kWh/day during peak periods. Aware user behavior reduced AC-related energy use by up to 15%, and envelope retrofit brought additional savings of 7–13%, especially under 2030 conditions.

Humidity-based indices such as the Heat Index and Humidex further confirmed the limitations of passive measures in high-moisture coastal conditions. In 2050 heat waves, only air-conditioned rooms remained consistently in the "Caution" or "Normal" HI bands. Non-conditioned spaces, such as kitchens, experienced discomfort even with active shading and ventilation.

Energy comfort trade-offs were mapped, revealing a clear separation between passive strategies (low energy, variable comfort) and active systems (high energy, high comfort). Retrofit and behavior modifications improved both axes, but could not close the performance gap under severe climatic stress.

14.14.4 Conclusion

This simulation-based assessment confirms that passive cooling strategies such as natural ventilation and shading remain viable options for reducing overheating in Mediterranean apartments under current summer conditions, particularly when supported by user awareness and retrofit measures. However, their effectiveness deteriorates significantly in projected 2030 and 2050 heat waves. Only mechanical air conditioning consistently delivered high thermal comfort under future scenarios, albeit at higher energy cost.

The study underscores the critical role of building envelope improvements and user behavior in enhancing resilience, but also highlights the structural limitations of passive approaches in the face of escalating climate extremes. It further supports a mixed-strategy design approach, combining passive techniques with efficient, user-aware active systems, and stresses the urgency of adaptive retrofit programs targeting vulnerable building typologies in southern Europe.

14.15 Benefits of Cool Skin and Ventilated Cavity Skin for Energy Efficiency and Heat Resilience in High-Rise Apartments

The increasing frequency and intensity of heat waves, combined with growing cooling demand, highlight the importance of façade systems that improve both energy efficiency and thermal resilience. Park et al. (2024) evaluated two façade configurations for these purposes: Cool Skin (CS), characterized by high reflectivity and low emissivity, and Ventilated Cavity Skin (VCS), which introduces a ventilated gap between layers of cladding. Their study, conducted on a typical South Korean high-rise residential building, assessed performance under both typical summer conditions and extreme heat waves with cooling system outages, providing insights into how advanced façade designs can simultaneously save energy and mitigate heat and grid stress.

Figure 14.23 shows the reference building model used in the simulations, including a 3D view and elevations for south, east, north, and west orientations. This visual reference helps to link the performance results to geometry and glazing exposure.

14.15.1 Methodology

The reference model represents a typical high-rise apartment block in Seoul. Geometry, window-to-wall ratios, and corridor-core layout were defined in OpenStudio and EnergyPlus using local climate data and prescribed heat wave episodes. Four envelope configurations were compared:

1. Reference wall
2. Cool Skin only
3. Ventilated Cavity Skin only
4. Combined CS + VCS

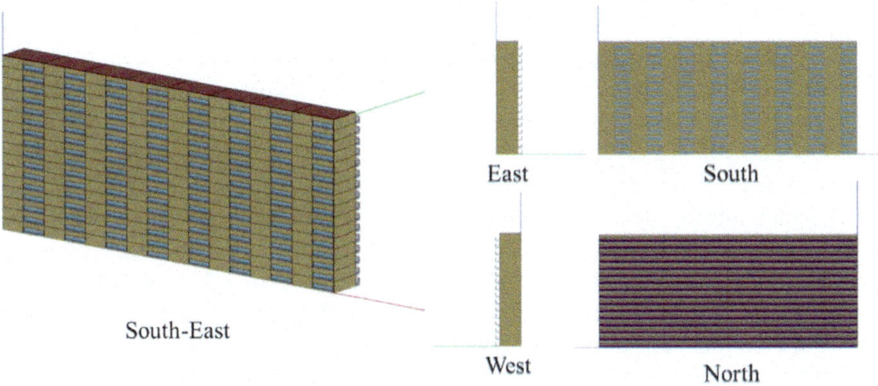

FIGURE 14.23 South Korean high-rise apartment reference model with 3D and four elevations by South, East, North, and West in OpenStudio. Sixteen floors and 128 units are used for simulations (Park et al., 2024).

The VCS configuration includes an additional cladding layer acting as a baffle and a ventilated cavity mounted over the existing wall. This arrangement promotes airflow in the cavity and reduces heat transfer to the interior, as illustrated in Figure 14.24.

Performance evaluation covered annual cooling energy use, peak cooling load reduction during the most demanding hours, and comfort metrics including PMV, SET, and Heat Index. For resilience analysis, the simulations examined indoor temperature rise during cooling system outages, hours exceeding discomfort thresholds, and reduction in exposure to high heat risk categories. Results were interpreted in the context of the geometry and orientation shown in Figure 14.24.

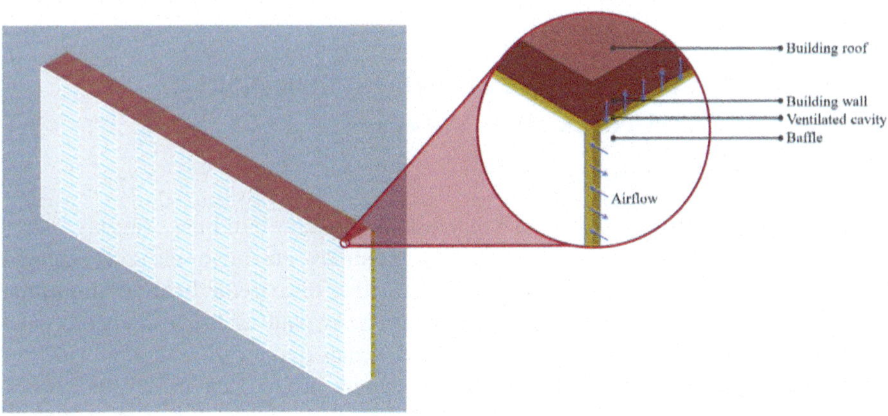

FIGURE 14.24 Ventilated cavity skin components and layout applied to the building envelope. The system adds a baffle and a ventilated cavity over the general wall to promote airflow and cut heat gains(Park et al., 2024).

14.15.2 Results

Under typical summer conditions, both façade systems reduced cooling demand compared to the reference case, with the combined configuration achieving the largest savings. Annual cooling energy use decreased by approximately 11% for CS, 7% for VCS, and 18% for CS + VCS. During peak afternoon hours, cooling load reductions reached 22% for CS + VCS, compared to 13% for CS and 9% for VCS.

Thermal comfort also improved. During the cooling season, CS + VCS maintained PMV within ±0.5 during occupancy hours, whereas the reference case was at ±0.8. During the hottest hours, SET decreased by up to 1.5°C, and Heat Index values were 2–3°C lower during heat waves, more frequently placing conditions in the Caution rather than Dangerous range.

Resilience testing under three-day heat waves with cooling outages showed a clear delay in indoor temperature rise. In the reference case, operative temperature exceeded 30°C after approximately 8 hours. With CS + VCS, the threshold was reached after about 14 hours. The number of discomfort hours with SET > 28°C dropped by about 38% compared to the reference. During outages, HI values in the combined configuration were 1.8°C–2.5°C lower, reducing exposure to the highest heat risk categories.

14.15.3 Conclusion

High-reflectivity, low-emissivity claddings and ventilated cavity façades deliver measurable energy and resilience benefits for high-rise residential buildings in hot summer climates. The best results were achieved with CS + VCS, combining solar gain reduction with active heat removal in the ventilated cavity. Figure 14.23 anchors the results in the geometric and orientation parameters of the 16-story reference building, while Figure 14.24 illustrates the VCS components and layout responsible for the enhanced performance. Applying such façade systems as passive support for cooling can reduce energy use, peak loads, and overheating risk during outages.

14.16 THERMAL RESILIENCE OF PUBLIC BUILDING ATRIUMS UNDER HEAT WAVES

14.16.1 Introduction

Atriums are a common feature of contemporary public buildings, serving as circulation hubs and social spaces. Their large volumes, glazing areas, and exposure to solar radiation make them particularly sensitive to extreme heat. With climate change intensifying heat waves in Chinese cities, evaluating the resilience of such spaces has become urgent. Zhang et al. (2025) conducted a full-scale field study on the thermal resilience of a university activity center in Shanghai during the record-breaking summer of 2024. Unlike simulation-only frameworks, this work combined continuous monitoring of air temperature and humidity with infrared thermography, providing rare empirical evidence of atrium performance under both closed and open ventilation states.

14.16.2 Building Description and Measurement Setup

The case study focused on a southwest-facing atrium within a newly built activity center in Jiading District, Shanghai (total floor area: 21,750 m^2). The atrium spans two floors, each 5.5 m high. The second floor cantilevers outward by 4.7 m, shading the ground level. A large skylight with oversized structural beams provides daylight while reducing direct solar penetration. Operable south-facing doors and top-hung windows enable cross-ventilation. Figure 14.25 illustrates the floor layouts, section, and exterior view of the building, as well as the positions of thermal sensors and infrared imaging points, which together framed the measurement campaign.

Measurements were conducted over 12 consecutive days, covering a severe heat wave when outdoor maxima peaked at 42.4°C. Two scenarios were monitored: (1) closed phase (all doors and windows shut), and (2) open phase (daytime cross-ventilation through fully opened doors and windows). Eight sensors recorded air temperature and humidity at one-minute intervals, supplemented by thermal images

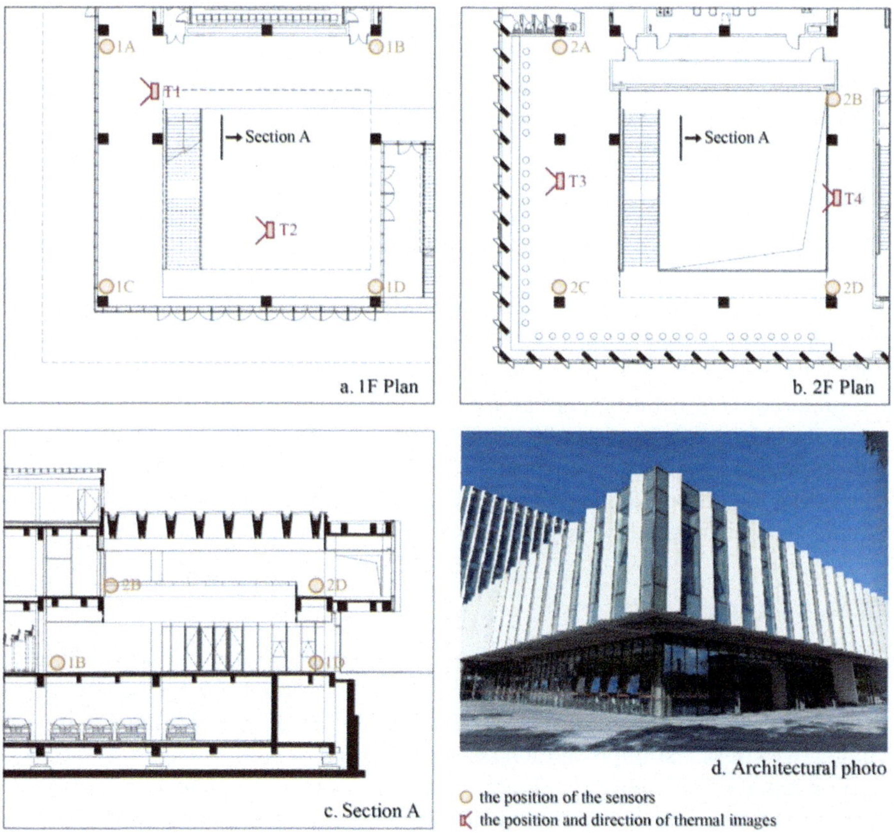

FIGURE 14.25 Plans, section, and exterior of the Shanghai atrium case study, showing sensor and thermal image positions used in the resilience assessment (Zhang et al., 2025).

taken hourly from four viewpoints (T1–T4). HVAC systems were kept off to capture the passive response of the atrium.

14.16.3 Results

In the closed phase, the atrium interior remained on average 7°C cooler than outdoors, with a maximum differential of 11°C. Peak indoor temperatures lagged outdoor maxima by up to four hours. Vertical stratification was evident: second-floor air temperatures were consistently 2°C–4°C higher than on the first floor. The Indoor Overheating Degree (IOD) values indicated that although peaks were delayed, discomfort accumulated progressively, with an average daily IOD nearly twice as high as in the ventilated case. The Average Weighted Discomfort (AWD) also reflected this tendency, as prolonged afternoon heat led to persistent discomfort at the upper level.

In the open phase, natural ventilation reduced cumulative heat loads and prevented long-term accumulation. IOD values fell sharply, indicating shorter and less intense exceedances of comfort thresholds. AWD also decreased, reflecting improved night-time recovery and reduced persistence of high indoor temperatures. However, greater fluctuations and spatial unevenness were observed: on some afternoons, second-floor peaks briefly surpassed 35°C, while the ground floor remained significantly cooler. Thermal images confirmed these patterns, showing dynamic but uneven cooling across vertical zones.

14.16.4 Discussion

The comparison of IOD and AWD across the two phases highlights a key resilience trade-off. The closed phase offered thermal stability and delayed peaks, yet led to heat accumulation and higher cumulative discomfort. The open phase, by contrast, produced greater short-term variability but prevented accumulation, lowering both IOD and AWD. This outcome is consistent with prior findings that emphasize the importance of natural ventilation in atria as a means of dissipating stored heat (Daniel et al., 2020; Li et al., 2024). Statistical analyses confirmed that resilience indices, such as α_{IOD}, improved from 0.23 in the closed state to 0.05 with ventilation, supporting the interpretation that cross-ventilation enhances long-term survivability despite local heterogeneity.

14.16.5 Conclusion

The Shanghai atrium study demonstrates that thermal resilience during heat waves depends on balancing stability and dissipation. The closed configuration buffered external peaks but trapped heat over time, while the open configuration reduced cumulative stress and improved IOD and AWD performance, even if vertical stratification remained a challenge. Figure 17.26 anchors these findings in the building's geometry and opening configuration, underlining the importance of cantilevers, skylights, and operable façades in moderating solar exposure and facilitating ventilation.

The evidence confirms that atrium spaces can maintain habitable conditions during extreme heat if cross-ventilation is effectively managed. Future resilience standards should account for both cumulative discomfort indices and spatial heterogeneity, integrating shading geometry, controllable skylights, and ventilation openings as essential features of climate-adaptive public architecture.

14.17 WINTER THERMAL BUILDING RESILIENCE OF SINGLE-FAMILY HOMES IN NOWY KISIELIN

14.17.1 Research Justification

Thermal resilience has emerged as a critical dimension of building performance in the context of intensifying climate variability and risks to energy infrastructure. While building regulations have traditionally emphasized energy efficiency during steady-state operation, numerous studies have shown that optimal performance under standard conditions does not guarantee adequate thermal safety during system outages or deliberate shutdowns (D'Agostino et al., 2023; Homaei & Hamdy, 2021). This limitation is particularly concerning in heating-dominated climates, where prolonged exposure to cold may pose serious health and safety risks.

Cold waves, polar vortex events, and unplanned heating outages have revealed the vulnerability of lightweight or highly insulated dwellings that lack sufficient passive buffering. Simulation-based approaches have attempted to capture resilience through metrics such as Resistance Time (RT), Degree of Disruption (DoD), and Hours above Safety Threshold (HST), but these tools are rarely validated against full-scale empirical data. Moreover, most frameworks do not adequately account for the combined effects of thermal inertia, preconditioning, and floor–soil interaction, particularly in buildings where horizontal slab insulation is omitted.

The potential role of thermal coupling with the ground deserves particular attention. Preliminary modelling studies suggest that preheated subsoil can act as a thermal reservoir, delaying indoor temperature decline during outages. Yet the interaction between envelope mass, ground coupling, and typology remains underexplored and under-verified. Existing indicators, such as DoD or the Weighted Unmet Thermal Performance (WUMTP), were developed primarily for short simulation windows and may not reflect the dynamics of multi-day outages. Their sensitivity to window length and tendency toward saturation limit comparability across scenarios. This underlines the need for physically grounded, time-dependent metrics that can be applied across typologies and climates.

Bridging this gap is essential if performance standards are to evolve beyond efficiency-based metrics. As climate conditions become increasingly volatile, buildings must be evaluated not only for average operation but also for their ability to maintain habitable conditions during crises. Resilience must be treated as a fundamental performance dimension, directly related to shelter, continuity, and life safety.

14.17.2 Objectives and Scope of the Study

The aim of this study was to empirically assess the passive thermal resilience of two full-scale single-family houses with contrasting construction logic during simulated winter heating outages. Both buildings, located in Nowy Kisielin in western Poland, have been described in detail in Section 13.2.7. They share identical floor plans, orientation, fenestration, and ventilation strategy, which ensures comparability. Building B1 is a medium-weight masonry house without insulation under the slab, enabling conductive ground coupling. Building B2 is a lightweight timber-frame structure with full slab insulation, thermally separated from the ground.

Two real-weather scenarios were selected to represent typical and extreme winter conditions. The Typical December Year (TDY) scenario recorded average outdoor temperatures of +0.7°C with solar gains of approximately 37 W/m^2. The Extreme Cold Year (ECY) scenario represented harsher conditions, with a mean of –4.5°C, nighttime lows reaching –10.1°C, and minimal solar input. In both cases, the buildings were operated with heating until steady equilibrium was reached, after which systems were deliberately switched off for 5–7 days.

The study employed both established and novel resilience indicators, including Resistance Time (RT), Intensity of Failure (IoF), Cooling Speed (CS), Recovery Speed (RS), Degree of Disruption (DoD), and Weighted Unmet Thermal Performance (WUMTP), as well as time-resolved descriptors such as $T(t_x)$ and T_x. The focus was strictly on passive performance: there were no occupants, no internal heat gains, and no operational interventions.

14.17.3 Methodology

14.17.3.1 Experimental Setup and Building Characteristics

Both experimental houses are located at the University of Zielona Góra's Science and Technology Park. Each has a usable area of 123 m^2, single-story layout with unheated attic, identical room configurations, and the same window-to-wall ratio. Mechanical ventilation was balanced at 0.6 ACH with 75% heat recovery efficiency. Heating was provided by distributed electric heaters, which were switched off at the start of the test.

Building B1 has masonry walls of cellular concrete blocks with external insulation of 20 cm mineral wool, internal partitions of silicate blocks, and a floor slab directly coupled to the ground following the 2019 modification. This design increased its Thermal Mass Parameter (TMP) to 467 kJ/m^{2}K. Building B2 is a lightweight timber-frame structure with mineral wool infill and full slab insulation, with a TMP of 192 kJ/m^{2}K.

14.17.3.2 Measurement Protocol

Two controlled outages were simulated. In December 2024 (TDY), both buildings were stabilized at 21°C for several days before heating was switched off for a five-day period. In February 2025 (ECY), the buildings had been heated continuously for a month before the shutdown, ensuring preconditioning of both the envelope and, in B1, the underlying soil. Indoor temperatures were monitored at five points per house and weighted by room volume. Outdoor air temperature, solar radiation, and humidity were logged continuously.

14.17.3.3 Indicators and Calculation

Indicators included:

- **RT (Resistance Time)**: duration until indoor air fell below 15°C.

- **IoF (Intensity of Failure)**: difference between initial and minimum indoor temperature.
- **CS (Cooling Speed)**: average rate of temperature drop.
- **RS (Recovery Speed)**: average rate of re-heating once systems were restored.
- **DoD (Degree of Disruption)**: integrated discomfort below threshold.
- **WUMTP (Weighted Unmet Thermal Performance)**: composite penalty index.
- **$T(t_x)$ and T_x**: time-based descriptors of temperature decline.

14.17.4 Results and Discussion

The winter outage experiments revealed a consistent performance gap between the two test houses. In the typical December scenario (TDY), B1 cooled slowly and remained above 15°C for 28.5 hours, while B2 crossed this limit after only 10.3 hours. During the February cold spell (ECY), B1 sustained habitable conditions for 11.0 hours, whereas B2 fell below 15°C after 8.0 hours and dropped to 6.5°C. When measured over the full outages, the difference was even more pronounced: in TDY, B1 stayed above 15°C for the entire 120-hour period (minimum 16.2°C), while B2 reached 12.1°C after 57 hours. In ECY, B1 remained above 15°C for 104 hours (minimum 14.1°C), whereas B2 crossed the threshold after 36 hours and cooled to 10.4°C.

Quantitative indicators reinforced these patterns. The Intensity of Failure (IoF) was 3.2 K for B1 and 6.9 K for B2 in TDY, widening to 6.1 K and 10.2 K in ECY. Cooling Speed in B2 was more than twice that of B1, confirming its lower thermal inertia. In February, preheated subsoil beneath B1 delayed early losses, acting as a thermal reservoir. After heating was restored, B2 warmed faster, but only because of its deeper temperature deficit; B1 required less energy for recovery.

Aggregate indices showed the same advantage. In TDY, Degree of Disruption (DoD) was negligible in B1 but exceeded 30 K·h in B2; in ECY, B2 rose above 75 K·h while B1 remained below 10 K·h. WUMTP penalized B2 more than three times higher. However, these measures proved sensitive to outage duration, whereas simple time-based descriptors offered clearer comparisons: at 24 and 48 hours, B2 was consistently 2.5–3.5 K cooler than B1.

These results correspond with earlier research. Rehman et al. (2024) observed similar patterns in simulations, though their fixed 31-hour window limited comparability. Homaei and Hamdy (2021) noted that DoD can distort long outages, and Wijesuriya et al. (2024) proposed HST as an alternative, though its binary nature reduces nuance. Dynamic time-based metrics therefore appear more informative. The importance of ground coupling is also evident: preheating warmed the soil under B1, which acted as a buffer, similar to effects reported in traditional masonry dwellings in Portugal (., 2015) and Italy (Cardinale et al., 2013). In contrast, small single-room structures often overheated or undercooled quickly, as observed by Roque et al. (2022).

The broader resilience literature confirms these findings. Laouadi et al. (2018) highlighted that indoor environments may amplify or moderate outdoor extremes more than climate itself. Rui et al. (2022) warned against relying on typical meteorological year data, and Liyanage et al. (2024) showed that Canadian homes can drop below 5°C indoors within four days of outage. By comparison, B1 maintained above 13°C for more than 60 hours, underscoring the resilience benefits of envelope mass and soil interaction.

14.17.5 Discussion in the Context of Previous Research

The findings of the Nowy Kisielin experiments reinforce and extend conclusions drawn from recent simulation-based studies. Rehman et al. (2024) demonstrated through modeling that resistance time and cooling speed are critical indicators of winter survivability, but their reliance on fixed 31-hour windows limits comparability with longer-duration events. In the present study, the medium-weight masonry house with ground coupling (B1) sustained habitable conditions for more than 100 hours in extreme cold, well beyond the time frames typically analyzed in such simulations. This highlights the importance of empirical data that capture multi-day dynamics rather than short-term performance snapshots. Homaei and Hamdy (2021) also pointed to the limitations of penalty-based indices such as the Degree of Disruption when applied over long intervals, a concern that was confirmed here, since DoD values escalated disproportionately with the length of the outage.

Wijesuriya et al. (2024) proposed the Hours of Safety Threshold as a simpler indicator for resilience evaluation. While this metric offers an intuitive classification, the binary safe or unsafe framework lacks nuance, particularly when buildings exhibit very different cooling speeds and recovery patterns. The present experiments therefore support the introduction of dynamic time-based descriptors such as T_x and $T(t_x)$, which proved more effective in capturing critical differences between the two houses. Rui et al. (2022) stressed that typical meteorological year datasets tend to underrepresent rare extremes and should not be the sole basis for resilience assessment. By integrating real weather events with full-scale monitoring, the Nowy Kisielin campaign provides rare empirical confirmation of this concern and demonstrates that lightweight construction can fail rapidly under natural cold stress conditions.

The role of ground coupling, often neglected in resilience literature, emerged as decisive in these experiments. The preheating of soil beneath B1 during standard operation created a thermal reservoir that slowed the rate of indoor cooling once heating was cut, especially during the February cold spell. This behavior corresponds with observations from vernacular dwellings in Portugal and southern Italy, where masonry envelopes with direct ground contact were found to maintain habitable temperatures even when outdoor conditions were severe (Cardinale et al., 2013; Fernandes et al., 2015). In contrast, smaller lightweight structures without sufficient mass or scale, such as those examined by Roque et al. (2022), tended to lose heat rapidly and provided little buffering capacity. These comparisons underline that the interaction between mass, scale, and soil contact is a critical determinant of passive winter resilience.

Laouadi et al. (2018) argued that indoor envelope dynamics can amplify or moderate outdoor extremes more strongly than the external climate itself. The present study confirms this principle, showing that two geometrically identical houses subjected to the same boundary conditions diverged sharply in their thermal resilience due to differences in construction logic. Liyanage et al. (2024) further reported that Canadian code-compliant houses can drop below 5°C within four days of heating loss. By contrast, the medium-weight masonry house in Nowy Kisielin remained above 13°C for more than 60 hours, a level of protection that would be underestimated in purely simulation-based analyses. Together, these results emphasize that empirical, long-duration monitoring is essential for developing resilience frameworks that are grounded in real building behavior.

14.17.6 Summary and Implications

The experiments carried out in Nowy Kisielin confirm that medium-weight masonry envelopes with controlled ground coupling can provide robust winter thermal resilience during prolonged heating outages. Building B1 maintained indoor temperatures above 15°C for 104 to 120 hours, depending on external severity, while the lightweight timber-frame Building B2 crossed this threshold within 36 to 57 hours. Resistance time, cooling speed, and intensity of failure all pointed to the decisive role of thermal inertia and subsoil interaction. Importantly, the findings show that lightweight structures, even when compliant with modern efficiency standards, may not offer adequate protection under outage conditions.

From a methodological perspective, time-based indicators such as T_x and $T(t_x)$ offered clearer and more transferable insights than penalty-based indices like DoD or WUMTP, which proved sensitive to observation length. These simplified descriptors capture the temporal dimension of resilience and should be integrated into future assessment frameworks to allow cross-comparison across building types, climates, and outage durations.

For design practice, the results suggest that incorporating thermal mass and, where feasible, controlled ground coupling can extend the period of habitable indoor conditions by two to three days under severe cold. Such design strategies provide occupants and emergency services with a longer response window and reduce the depth of post-outage recovery. Retrofit strategies might also consider maintaining slab-on-grade contact while strengthening perimeter insulation, balancing steady-state energy efficiency with enhanced outage resilience. Preconditioning before expected cold events could further improve thermal survivability in ground-coupled structures.

From a regulatory perspective, the findings point to the necessity of extending building codes beyond efficiency metrics toward resilience-based criteria. Current standards that evaluate only steady-state performance risk overlooking critical vulnerabilities during disruptions. Incorporating resilience indicators into regulations and certification systems would ensure that buildings are not only energy efficient under normal operation but also safe shelters during extreme events.

In sum, the Nowy Kisielin experiments provide strong empirical confirmation that moderate increases in thermal mass combined with ground coupling can transform the winter survivability of residential buildings. This evidence contributes to the broader effort of reframing building performance assessment from a static efficiency paradigm toward a dynamic resilience perspective that is indispensable under changing climatic and energy conditions.

14.18 SUMMER THERMAL BUILDING RESILIENCE OF SINGLE-FAMILY HOMES IN NOWY KISIELIN

14.18.1 Introduction and Experimental Design

As global temperatures continue to rise, summer heat waves have become both more frequent and more intense, posing significant challenges for thermal safety in residential buildings. These risks are amplified when heat waves coincide with power outages, rendering mechanical cooling systems unavailable. Under such compound stress conditions, a building's ability to passively maintain safe indoor temperatures becomes essential for occupant health and survivability. This chapter presents an empirical assessment of summer thermal resilience in two full-scale single-family homes of contrasting construction logic, monitored during natural heat waves in 2018 and 2019.

The two test houses, designated B1 and B2, have been previously described in detail in Section 13.2.7. They are geometrically identical, each with a usable floor area of 123 m^2, identical fenestration, and the same orientation and ventilation regime. B1 is a medium-weight masonry house, modified in 2019 by removing the original 30 cm of under-slab insulation to enable ground coupling. B2 is a lightweight timber-frame structure with full slab insulation, separated thermally from the soil. Both houses employ balanced mechanical ventilation at 0.6 ACH with no active cooling systems. The interiors were unoccupied, unfurnished, and free of internal gains during testing, ensuring that observed thermal differences could be attributed solely to envelope mass and ground interaction.

The monitoring campaign spanned two consecutive summers. In 2018, both houses retained insulated slabs, and thermal behavior was observed under a prolonged heat wave between July 20 and August 1, when average daily outdoor temperatures reached 23.5°C with peaks exceeding 30°C. In 2019, following the slab modification in B1, the buildings were again monitored during three separate heat waves, including a particularly intense event in late August when average daily maximums reached 30.9°C. Despite slightly cooler average conditions in 2019 (21.0°C vs 21.9°C in 2018), the duration and recurrence of heat waves created a more severe test environment for resilience evaluation.

Indoor air temperature was monitored at five representative points per building at 1.2 m above floor level, using calibrated sensors with a five-minute logging interval. Weighted averages were calculated according to room volume to ensure comparability across buildings. Ground temperatures beneath the slab of B1 were also measured at multiple depths, while meteorological data, including air temperature, solar radiation, wind speed, and humidity, were collected by an on-site weather station.

To evaluate passive resilience, several established and emerging indicators were applied. From the framework of Hamdy et al. (2017), the Indoor Overheating Degree (IOD), Average Weighted Discomfort (AWD), and Overheating Exposure Frequency (OEF) were calculated. The Accumulation Factor (γ_{IOD}), introduced by Flores-Larsen et al. (2023), was used to represent the influence of prior thermal history on daily overheating patterns and Heat Index Hazard Hours (HIHH), were used to evaluate occupant safety under blackout conditions. Time-resolved descriptors such as $T(t_x)$ and T_x were introduced to better represent the dynamic progression of overheating, overcoming limitations of static penalty-based indices.

14.18.2 Results and Discussion

14.18.2.1 General Thermal Behavior

The results revealed a consistent and substantial performance gap between the heavyweight masonry building (B1) and the lightweight timber-frame structure (B2). During the 2018 heat wave, both houses initially exhibited rising indoor temperatures in line with outdoor loads. However, B1 showed a slower rate of heating, maintaining indoor values consistently below those of B2. By the end of the 12-day period, B1 recorded an average Indoor Overheating Degree (IOD) of 309.6°C·h, compared to 985.2°C·h for B2. In practice, this meant that B1 experienced short exceedances above 26°C but rarely surpassed 28°C, whereas B2 consistently remained above both thresholds for extended periods.

The divergence widened in 2019 following the removal of slab insulation in B1. Across three consecutive heat wave episodes, B1 demonstrated near-complete thermal compliance. Indoor temperatures never exceeded 28°C for more than one hour, and during the most severe event in late August, mean daily peaks in B1 remained below 27°C, compared to over 31°C in B2. As a result, B1 recorded zero overheating hours across the summer, while B2 accumulated more than 730°C·h of IOD and exceeded comfort thresholds for over 147 occupied hours.

14.18.2.2 Health-Oriented Indicators

The evaluation using Tao et al. (2025) metrics further reinforced the gap between the two structures. In 2018, B1 achieved a three-star resilience rating, with exceedances limited to short intervals, while B2 failed to reach minimum safety criteria, with more than 154 hours above the WBGT limit of 28°C. In 2019, B1 achieved the highest possible four-star rating, with zero Hours of Exceedance (HE) and zero High-Intensity Heat Hours (HIHH). In contrast, B2 showed 117 hours above the WBGT threshold, with a cumulative IOD of 582.8°C·h. The persistence of high indoor temperatures in B2 created sustained exposure risks, confirming that lightweight envelopes are particularly vulnerable under prolonged heat stress.

A decisive shift in performance trajectory was observed in 2019 when direct thermal interaction between the building and the ground was enabled in B1 through removal of the horizontal insulation beneath the slab. This intervention altered the building's thermal boundary conditions without modifying geometry, envelope mass, glazing ratio, or internal layout. Following this change, overheating indicators in B1 were reduced to zero or near-zero values during the heatwave. Both IOD and HIHH remained at zero, while the Overheating Escalation Factor decreased to 0.17. More than 96 percent of occupied hours stayed within the adaptive comfort range, and no critical thresholds were exceeded. In contrast, B2, unchanged in configuration, continued to accumulate substantial thermal stress, with HIHH exceeding 600°C·h

14.18.2.3 Comparative Literature Context

The Nowy Kisielin experiments provide strong empirical confirmation of the role of envelope mass and ground coupling in moderating overheating during heat waves. Previous studies relied mainly on simulations. Hamdy et al. (2017) highlighted the role of thermal inertia in extending resistance times, but their analyses, based on typical meteorological years and short windows, understated the cumulative effect of consecutive hot days. The present results extend this principle by showing that mass alone is insufficient when thermally isolated from the ground, while the combination of mass and soil interaction can ensure comfort even during record-breaking summers.

Dynamic and cumulative indicators add clarity. Flores-Larsen et al. (2023) showed that daily overheating in passive buildings cannot be explained by AWD alone, emphasizing the role of prior thermal history and solar radiation. Their analysis revealed effective nightly dissipation in the heavyweight house compared with progressive heat build-up in the lightweight one. Similarly, High-Intensity Heat Hours (HIHH), introduced by Tao et al. (2025), when applied to the Nowy Kisielin data, showed that only the lightweight building accumulated hazard exposure. This supports observations by Sun et al. (2021), who reported that uncooled dwellings frequently exceed health thresholds during heat waves.

The vulnerability of lightweight and superinsulated structures has been documented in Europe and Asia. Rahif et al. (2023) and Dartevelle et al. (2021) reported rapid accumulation of heat stress in lightweight dwellings without cooling, while Yu et al. (2025) showed that efficiency-oriented superinsulated homes became liabilities during prolonged outages. The Nowy Kisielin data confirm that this is a structural weakness of low-mass construction.

At the same time, earlier full-scale experiments by Kuczyński et al.2021) demonstrated that integrating mass, shading, and night ventilation can reduce peaks by more than 7°C and eliminate overheating. The novelty of the present work lies in showing that mass combined with ground coupling alone can maintain thermal neutrality through extended heat waves.

Finally, as Laouadi et al. (2018) argued, indoor resilience often depends more on envelope dynamics than on outdoor climate. The sharp divergence between the two identical houses in Nowy Kisielin, differing only in construction logic, provides empirical proof of this principle and strengthens the case for resilience-oriented building standards.

14.18.3 Summary and Implications

The Nowy Kisielin experiments confirm that medium-weight masonry envelopes with ground coupling deliver robust summer thermal resilience during prolonged heat waves. Building B1 consistently maintained indoor temperatures below 28°C, while the lightweight timber-frame Building B2 accumulated substantial overheating, with IOD values and health-based indices indicating critical risks. These outcomes highlight that envelope mass and soil interaction are decisive in preventing cumulative thermal stress.

Methodologically, the results validate the usefulness of dynamic and cumulative indicators such as IOD and HIHH, which capture persistence and health relevance more effectively than static exceedance counts. Time-sensitive descriptors proved especially valuable in differentiating the two buildings under real climate conditions, demonstrating that resilience assessment must go beyond fixed thresholds.

For design practice, the evidence suggests that relatively modest increases in envelope mass, when combined with controlled ground coupling, can eliminate the need for mechanical cooling even in record summers. Such strategies provide low-cost, durable protection while reducing reliance on air conditioning. For lightweight structures, supplementary measures such as shading, night ventilation, or PCMs remain necessary to achieve comparable resilience.

From a regulatory perspective, these findings strengthen the case for integrating resilience-based metrics into building codes and certification systems. Current standards centered on efficiency do not address the risks posed by compound heat and blackout events. Incorporating indicators such as IOD, HIHH, and time-to-threshold measures would ensure that buildings are not only efficient in average operation but also safe under climatic extremes.

In sum, the Nowy Kisielin study demonstrates that summer resilience is a measurable, design-controllable outcome. Mass and ground coupling, often undervalued in modern construction, can transform thermal performance, offering both immediate comfort and long-term protection as heat waves intensify under climate change.

14.19 NOVEL CONTRIBUTIONS FROM THE NOWY KISIELIN FULL-SCALE EXPERIMENTS

The full-scale monitoring campaign conducted on two geometrically identical single-family houses in Nowy Kisielin, western Poland, provides rare empirical evidence on thermal resilience under real weather conditions. The study examined two construction logics: a lightweight timber-framed house with full slab insulation, and a medium-weight masonry house where the horizontal under-slab insulation was removed, enabling direct thermal coupling with the soil.

The summer results are particularly striking. Increasing the envelope mass to a level typical of more than 90% of newly constructed detached houses in Poland, and removing the subfloor insulation, completely eliminated the need for mechanical cooling, even under the most extreme global warming scenarios. During multi-day heat waves, indoor air temperatures in the masonry, ground-coupled house remained consistently below 28°C, while the lightweight insulated house showed prolonged exceedance hours and high accumulation indices. For example, in the hottest events of 2019, the lightweight building accumulated more than 700°C·h of overheating, whereas the heavyweight counterpart maintained thermal neutrality throughout the period.

Winter performance confirmed the same pattern. In simulated power outages, the ground-coupled masonry house maintained habitable conditions above 15°C for 104–120 hours, compared to only 36–57 hours in the lightweight reference. Resistance time, cooling speed, and weighted unmet performance indicators all demonstrated the buffering effect of soil heat storage and envelope inertia. In practical terms, the masonry house offered occupants two to three additional days of survivable indoor conditions during extreme cold spells without heating.

These outcomes highlight that the interaction between envelope mass and ground coupling can function as a dual-purpose resilience strategy, protecting occupants during both hot and cold extremes. Unlike isolated passive measures, which often provide benefits only under one set of climatic conditions, this configuration performed effectively across seasons. Such findings provide a strong empirical foundation for reframing resilience as a year-round design priority rather than a seasonal adaptation.

At the same time, the approach has clear boundaries. It was tested only on single-story detached dwellings in a temperate climate, and its broader application requires careful assessment of annual energy balances and prospective life-cycle carbon performance. Ground coupling, while beneficial during outages, may increase steady-state heat losses if not paired with appropriate perimeter insulation. These trade-offs underline the need for performance standards that evaluate buildings not only in terms of efficiency but also through dynamic resilience indicators capable of capturing both operational and crisis conditions.

Nevertheless, the empirical data from Nowy Kisielin confirm that a design strategy often considered marginal, modest thermal mass with direct ground coupling, can provide robust year-round resilience. By eliminating mechanical cooling needs in summer and extending habitable conditions in winter outages, this configuration demonstrates a viable pathway toward climate-adaptive building envelopes, provided that durability, context, and whole-life performance are thoroughly evaluated. More broadly, the study contributes to the international literature by showing that resilience-oriented design does not always require complex or high-cost technologies, but can be achieved through calibrated adjustments of basic construction logics.

REFERENCES

Amaripadath, D., Azar, E., Singh, M. K., & Attia, S. (2024). Heat exposure mitigation in renovated nearly zero-energy dwellings during concurrent heat waves and power outages. *Journal of Building Engineering*, *91*, 109655. https://doi.org/10.1016/j.jobe.2024.109655

Attia, S., Benzidane, C., Rahif, R., Amaripadath, D., Hamdy, M., Holzer, P., Koch, A., Maas, A., Moosberger, S., Petersen, S., Mavrogianni, A., Hidalgo-Betanzos, J. M., Almeida, M., Akander, J., Khosravi Bakhtiari, H., Kinnane, O., Kosonen, R., & Carlucci, S. (2023). Overheating calculation methods, criteria, and indicators in European regulation for residential buildings. *Energy and Buildings*, *292*, 113170. https://doi.org/10.1016/j.enbuild.2023.113170

Borghero, L., Clèries, E., Péan, T., Ortiz, J., & Salom, J. (2023). Comparing cooling strategies to assess thermal comfort resilience of residential buildings in Barcelona for present and future heatwaves. *Building and Environment*, *231*, 110043. https://doi.org/10.1016/j.buildenv.2023.110043

Cardinale, N., Rospi, G., & Stefanizzi, P. (2013). Energy and microclimatic performance of Mediterranean vernacular buildings: The Sassi district of Matera and the Trulli district of Alberobello. *Building and Environment*, *59*, 590–598. https://doi.org/10.1016/j.buildenv.2012.10.006.

Chen, S., Zhao, H., Luo, K., Li, Q., Nie, Z., & Zhao, L. (2024). Health risk assessment of residential overheating under the heat waves in Guangzhou. *Building and Environment*, *266*, 112089. https://doi.org/10.1016/j.buildenv.2024.112089

D'Agostino, D., Becchio, C., Crespi, G., & Corgnati, S. P. (2023). Assessment of passive and active buildings resilience to gas supply disruption in winter across European climates. *Sustainable Cities and Society*, *92*, 104461. https://doi.org/10.1016/j.scs.2023.104461

Dartevelle, M., Carlucci, S., Gaur, A. S., & Hamdy, M. (2021). Investigating overheating risk in lightweight buildings: A parametric simulation approach. *Building and Environment*, *198*, 107898. https://doi.org/10.1016/j.buildenv.2021.107898

Fernandes, J., Pimenta, C., Mateus, R., Silva, S. M., & Bragança, L. (2015). Contribution of Portuguese vernacular building strategies to indoor thermal comfort and occupants' perception. *Buildings*, *5*, 1242–1264. https://doi.org/10.3390/buildings5041242.

Flores-Larsen, S., Filippín, C., & Bre, F. (2023). New metrics for thermal resilience of passive buildings during heat events. *Building and Environment*, *230*, 109990. https://doi.org/10.1016/j.buildenv.2023.109990

Hamdy, M., Carlucci, S., Hoes, P.-J., & Hensen, J. L. M. (2017). The impact of climate change on the overheating risk in dwellings—a Dutch case study. *Building and Environment*, *122*, 307–323. https://doi.org/10.1016/j.buildenv.2017.06.031

Homaei, S., & Hamdy, M. (2021). Thermal resilient buildings: How to be quantified? A novel benchmarking framework and labelling metric. *Building and Environment*, *201*, 108022. https://doi.org/10.1016/j.buildenv.2021.108022

Krelling, A. F., Lamberts, R., Malik, J., & Hong, T. (2023). A simulation framework for assessing thermally resilient buildings and communities. *Building and Environment*, *245*, 110887. https://doi.org/10.1016/j.buildenv.2023.110887

Kuczyński, T., & Staszczuk, A. (2020). Experimental study of the influence of thermal mass on thermal comfort and cooling energy demand in residential buildings. *Energy*, *195*, 116984. https://doi.org/10.1016/j.energy.2020.116984

Kuczyński,T., Staszczuk, A., Gortych, M., & Stryjski, R. (2021). Effect of thermal mass, night ventilation and window shading on summer thermal comfort of buildings in temperate climate. *Building and Environment*, *204*, 108126. https://doi.org/10.1016/j.buildenv.2021.108126

Laouadi, A., Gaur, A., & Bartko, M. (2018). *Climate resilience of buildings: Overheating in buildings—literature review* (NRCC Report No. 56164). National Research Council Canada. https://doi.org/10.4224/40001234

Li, Y., Tao, X., Zhang, Y., & Li, W. (2024). Combining use of natural ventilation, external shading, cool roof, and thermal mass to improve indoor thermal environment: Field measurements and simulation study. *Journal of Building Engineering*, *86*, 108904. https://doi.org/10.1016/j.jobe.2024.108904

Liyanage, D. R., Hewage, K., Ghobadi, M., & Sadiq, R. (2024). Thermal resiliency of single-family housing stock under extreme hot and cold conditions. *Energy and Buildings*, *323*, 114809. https://doi.org/10.1016/j.enbuild.2024.114809

Park, J., Lee, K. H., Lee, S. H., & Hong, T. (2024). Benefits assessment of cool skin and ventilated cavity skin: Saving energy and mitigating heat and grid stress. *Building and Environment*, *247*, 111027. https://doi.org/10.1016/j.buildenv.2023.111027

Peri, G., Cirrincione, L., Mazzeo, D., Matera, N., & Scaccianoce, G. (2024). Building resilience to a warming world: A contribution toward a definition of "Integrated Climate Resilience" specific for buildings—literature review and proposals. *Energy and Buildings*, *315*, 114319. https://doi.org/10.1016/j.enbuild.2024.114319

Rahif, R., Kazemi, M., & Attia, S. (2023). Overheating analysis of optimized nearly zero-energy dwelling during current and future heatwaves coincided with cooling system outage. *Energy and Buildings*, *287*, 112998. https://doi.org/10.1016/j.enbuild.2023.112998

Rehman, H. U., & Hasan, A. (2023). Energy flexibility and towards resilience in new and old residential houses in cold climates: A techno-economic analysis. *Energies*, *16*(14), 5506. https://doi.org/10.3390/en16145506

Rehman, H. U., Nik, V. M., Ramesh, R., & Ala-Juusela, M. (2024). Quantifying and rating the energy resilience performance of buildings integrated with renewables in the Nordics under typical and extreme climatic conditions. *Buildings*, *14*(9), 2821. https://doi.org/10.3390/buildings14092821

Roque, E., Vicente, R., Almeida, R. M. S. F., & Ferreira, V. M. (2022). The impact of thermal inertia on the indoor thermal environment of light steel framing constructions. *Energies*, *15*, 3061. https://doi.org/10.3390/en15093061

Rui, B., Chang, W.-S., Yu, Y., Xu, Y., & Guo, H. (2022). Overheating of residential buildings in the severe cold and cold regions of China: The gap between building policy and performance. *Building and Environment, 225*, 109601. https://doi.org/10.1016/j.buildenv.2022.109601

Sengupta, A., Al Assaad, D., Kazanci, O. B., Shinoda, J., Breesch, H., & Steeman, M. (2024). Building and system design's impact on thermal resilience to overheating during heatwaves: an uncertainty and sensitivity analysis. *Building and Environment, 265*, 112031. https://doi.org/10.1016/j.buildenv.2024.112031

Siu, C. Y., O'Brien, W., Touchie, M., Armstrong, M., Laouadi, A., Gaur, A., Jandaghian, Z., & Macdonald, I. (2023). Evaluating thermal resilience of building designs using building performance simulation—a review of existing practices. *Building and Environment, 234*, 110124. https://doi.org/10.1016/j.buildenv.2023.110124

Sun, K., Specian, M., & Hong, T. (2020). Nexus of thermal resilience and energy efficiency in buildings: A case study of a nursing home. *Building and Environment, 177*, 106842. https://doi.org/10.1016/j.buildenv.2020.106842

Sun, K., Zhang, W., Zeng, Z., Hong, T., & Jain, R. K. (2021). Passive cooling designs to improve heat resilience of homes in underserved and vulnerable communities. *Energy and Buildings, 252*, 111383. https://doi.org/10.1016/j.enbuild.2021.111383

Tao, M., Gou, Z., & Ma, N. (2025). Assessment of thermal comfort and thermal resilience in dwellings during heat waves: A case study of a near-zero energy house. *Journal of Building Engineering, 109*, 113052. https://doi.org/10.1016/j.jobe.2025.113052

Villa, D. L., Lee, S. H., Bianchi, C., Carvallo, J. P., Azaroff, I., & Mammoli, A. (2024). Multi-scenario extreme weather simulator application to heat waves: Ko'olauloa community resilience hub. *Science and Technology for the Built Environment, 30*(4), 375–393. https://doi.org/10.1080/23744731.2023.2279467

Wijesuriya, S., Kishore, R. A., Bianchi, M. V. A., & Booten, C. (2024). Enhancing thermal resilience of US residential homes in hot humid climates during extreme temperature events. *Cell Reports Physical Science, 5*(6), 101986. https://doi.org/10.1016/j.xcrp.2024.101986

Yu, Z., Zhou, J., & Wang, J. (2025). Thermal resilience of lightweight buildings under future heatwaves in Northern China. *Energy and Buildings, 340*, 115990. https://doi.org/10.1016/j.enbuild.2025.115990

Zhang, G., Li, L., Yu, Y., Liu, J., & Zhang, Q. (2025). Thermal resilience of public building atriums under different states during heatwaves. *Buildings, 15*(4), 598. https://doi.org/10.3390/buildings15040598

15 Year-Round Thermal Efficiency

15.1 MOTIVATION FOR THE WINTER MONITORING

Previous studies conducted by the authors demonstrated the significant potential of two passive design strategies: increasing wall thermal mass and enhancing floor–ground thermal coupling through the removal of horizontal subfloor insulation (Gortych & Kuczyński, 2025; Gortych et. al, 2026). These strategies were evaluated during earlier monitoring campaigns in summer and transitional seasons, where they led to substantial improvements in indoor thermal stability and passive buffering capacity. Both approaches contributed to lower peak indoor temperatures, extended comfort periods without mechanical cooling, and improved resilience during warm conditions, including periods of elevated external heat or limited access to active systems.

While these results indicated clear advantages under warm-season conditions, it remained essential to examine whether the same design modifications would impose penalties during winter operation. In particular, concerns arose regarding the potential for increased conductive heat losses through uninsulated ground-contact floors. Similarly, the contribution of high wall thermal mass to reducing heating energy demand in buildings with continuous thermostat regulation had not been empirically assessed under real-world conditions (D'Agostino et al., 2024; Yannas & Rodríguez-Álvarez, 2020).

To investigate these issues, a dedicated winter monitoring phase was implemented to assess the influence of both strategies on heating performance. Two full-scale residential test buildings with comparable geometry, internal layout, and thermal insulation levels in roofs and walls, but differing in wall construction type and floor insulation, were monitored under controlled indoor setpoints. This experimental design allowed for an isolated evaluation of wall heat capacity and ground coupling under temperate winter conditions.

Building on these considerations, the following sections present the results of the winter monitoring campaign, focusing on the role of material thermal inertia and floor–soil interaction in shaping seasonal heating energy demand.

15.2 THERMAL MASS AND GROUND COUPLING: EMPIRICAL EVIDENCE FROM COMPARATIVE MONITORING

The influence of thermal mass and ground coupling on heating performance has been the subject of ongoing theoretical and simulation-based inquiry (Goia et al., 2015; Ibrahim & Pelsmakers, 2018; Liu et al., 2021), yet empirical evidence under real-world conditions remains relatively limited. In this context, the comparative monitoring of two test buildings, one with heavyweight masonry construction (B1) and another with lightweight timber-frame walls (B2), offers a nuanced perspective on how material thermal inertia and floor–soil interaction affect seasonal energy demand in temperate climates.

The initial phase of the study focused on the role of wall thermal mass during a controlled 70-day winter period. Both buildings were subject to identical boundary conditions, with stable heating setpoints and high levels of insulation in roofs and floors. Despite these similarities, Building B1 consumed 3.6% less heating energy than B2, a modest yet consistent advantage attributable to the higher heat capacity of its masonry walls. This reduction, though relatively small, suggests that thermal mass in walls can contribute to smoother daily energy profiles by dampening short-term fluctuations in internal temperature, even under continuous heating operation. However, the benefits of thermal mass in such fully regulated settings appear constrained, aligning with broader research indicating that its full potential is more readily observed under dynamic thermal regimes, such as intermittent heating or passive solar gain cycles (Costanzo et al., 2018; Yannas & Rodríguez-Álvarez, 2020).

In contrast, the second phase of the monitoring campaign revealed a much more pronounced effect when the thermal configuration of the floor system was altered. By removing the horizontal insulation beneath the slab in Building B1, the floor became directly coupled to the ground, enabling bidirectional heat flow between the building and the subsoil. Over a full heating season, this design change led to a 12.2% increase in heating energy demand in B1 compared to B2. Interestingly, this penalty did not emerge immediately. During the early winter months, both buildings exhibited similar cumulative energy use. The divergence became apparent only in the later part of the season, reflecting a delayed but accelerating performance gap.

This delay can be interpreted as evidence of seasonal ground buffering. The soil beneath the uninsulated slab in B1, having accumulated thermal energy during the summer and autumn, initially acted as a passive heat source, reducing early-season heat losses (Goldberg & Mosiman, 2015). As the season progressed and ground temperatures declined, the direction of heat flow reversed. The absence of floor insulation then allowed continuous conductive losses into the cooler soil, sharply increasing energy demand in the final months of winter. This dynamic behavior underscores the temporal asymmetry of ground–building heat exchange and highlights the limitations of static performance indicators, which may fail to capture such delayed effects (ISO 13370; Cox-Smith, 2016).

To further elucidate these phenomena, a year-long ground temperature monitoring campaign was conducted, providing high-resolution data on subsurface thermal behavior. Measurements taken at a depth of 1.5 meters under both buildings revealed that the soil beneath B1 remained 4 to 5°C warmer than that under B2 throughout the winter, confirming the significant influence of floor–ground thermal coupling. The warmer ground beneath B1 functioned as a temporary energy reservoir, delaying performance degradation but eventually becoming a thermal sink. These observations challenge the validity of widely used modeling approaches, such as those prescribed in ISO 13370, which rely on simplified assumptions of one-dimensional, steady-state heat flow and often neglect the mitigating role of vertical insulation layers (Cox-Smith, 2016; Liu et al., 2021).

Indeed, the discrepancy between predicted and measured performance was substantial. Standardized models estimated a heating penalty of nearly 24% for B1, while empirical monitoring showed only 12.2%. This difference likely stems from the presence of vertical perimeter insulation around B1's foundation, which reduced lateral soil-side losses and slowed ground cooling. Such detailing is often excluded or oversimplified in simulation frameworks, yet it played a critical role in delaying energy penalties over the course of the heating season (Goldberg & Mosiman, 2015; Liu et al., 2021).

These findings illustrate the importance of time-resolved, empirical assessment in evaluating the thermal performance of buildings with unconventional envelopes. Both the modest benefit of wall mass and the more pronounced effects of ground coupling exhibited strong seasonal dependencies, revealing mechanisms that are not adequately captured by conventional modeling techniques. Moreover, the results highlight the need to view soil not as a static thermal boundary, but as a dynamic element whose energy interactions evolve over time and influence both heating performance and potential summer resilience (De Masi et al., 2021; Roque et al., 2021).

15.3 RETHINKING THERMAL REGULATION IN BUILDING DESIGN

Empirical findings from the monitored buildings point to critical limitations in existing energy performance standards that rely primarily on static envelope metrics such as thermal transmittance. While U-values remain useful in characterizing steady-state heat transfer, they do not capture the dynamic, time-dependent behavior of thermal systems, particularly those involving high heat capacity or ground interaction. The observed performance of Building B1 illustrates this disconnect. Although it lacked horizontal subfloor insulation, its heating demand remained comparable to that of the fully insulated B2 during the early winter months. This was due to delayed conductive losses and elevated soil temperatures resulting from seasonal heat retention. Only later in the season did energy demand in B1 increase significantly, revealing a time-lagged penalty associated with direct ground coupling (Gortych & Kuczyński, 2025).

Such temporal effects were not anticipated by standardized simulation approaches, which assume homogeneity, steady conditions, and one-directional heat flow. Notably, ISO 13370 predicted a substantially higher seasonal penalty for the uninsulated slab than was actually observed. The discrepancy is largely explained by vertical perimeter insulation, which significantly reduced lateral losses and extended the effective buffering period of the ground. These findings show that simulation models and prescriptive norms often fail to capture critical aspects of envelope–ground interaction and transient thermal storage, thereby undervaluing hybrid or non-standard solutions (Cox-Smith, 2016; Liu et al., 2021).

Rather than reinforcing strict envelope requirements that exclude such strategies, regulations should move toward performance-based criteria that reflect actual energy use, thermal comfort, and resilience over time. This approach is consistent with recent Canadian proposals to embed explicit indoor temperature limits into building codes, for example, 26°C during heat events, and to prioritize passive measures before permitting active cooling systems (Siu et al., 2025). Such occupant-centered thresholds provide measurable, outcome-focused benchmarks and address the needs of vulnerable populations, moving beyond purely prescriptive component values. A parallel policy assessment for the United Kingdom reaches similar conclusions, highlighting gaps in current codes regarding overheating, the dominance of winter efficiency in policy, and the need to integrate passive-first design and indoor temperature thresholds to avoid rapid lock-in to mechanical cooling and rising peak demand (Hoggett et al., 2024).

Recent studies have also questioned the effectiveness of prescriptive regulations. Arriazu-Ramos et al. (2023) showed that newer Spanish apartment buildings constructed under updated energy codes did not perform better in terms of summer thermal comfort than pre-regulation dwellings. In fact, some certified dwellings exhibited the highest levels of indoor overheating during warm periods. Their follow-up urban-scale analysis (Arriazu-Ramos et al., 2024) found that standard insulation retrofits failed to reduce thermal stress in over 85,000 residential units in Pamplona, especially in top-floor dwellings without shading or cross-ventilation. Similarly, Palma et al. (2024) concluded that insulation-centric retrofitting, if not accompanied by passive cooling strategies, can increase summer overheating risk, particularly in regions undergoing climate reclassification or experiencing energy poverty.

The passive strategies investigated in the present study, namely thermal mass activation and ground coupling, are aligned with emerging regulatory priorities that emphasize thermal autonomy, climate resilience, and environmental equity. These solutions may not comply with conventional prescriptive models, yet they offer measurable benefits in terms of both operational performance and thermal buffering (Norouzi et al., 2023).

Several researchers, including those involved in the Canadian code revision process, advocate for integrating dynamic design principles into regulatory frameworks, where both winter and summer resilience are assessed using field-validated, time-resolved metrics. De Masi et al. (2021) highlight that envelope solutions optimized for winter loads may become problematic in climates with high solar exposure or limited night cooling potential. Yannas and Rodríguez-Álvarez (2020) also argue for incorporating passive cooling performance into the definition of envelope quality, particularly in regions where overheating has emerged as a dominant thermal risk.

Ultimately, the empirical findings from this study support a regulatory shift from component-level prescriptions to whole-building performance evaluation. Passive buffering, peak load reduction, and the ability to maintain comfort during power outages or extreme temperatures should be recognized as core attributes of high-performance design. A performance-oriented regulatory framework would better reflect the complex, dynamic realities of thermal behavior and offer designers greater flexibility to pursue solutions adapted to evolving climatic conditions.

Another argument for including non-standard passive solutions in performance criteria is their compatibility with the requirements of the energy transition. Buildings with high thermal capacity and controlled ground coupling can act as thermal buffers, extending the time between active heating or cooling cycles. These properties make it easier to align the building's demand profile with the variable availability of solar energy, which is absent at night, and wind power, which is inherently intermittent and often subject to prolonged lulls (Hoggett et al., 2024). In regulatory terms, this means that energy flexibility can be improved without the need for costly expansion of electrical or thermal storage systems, as also noted in recent UK-based assessments of building–grid integration (Hoggett et al., 2024).

15.4 REGULATORY DIVERGENCE AND CLIMATIC REALIGNMENT ACROSS EUROPE

While the technical limitations of prescriptive performance standards are increasingly evident, there is also a mismatch between climate trends and regulatory structures at the national level. Climatic convergence across European regions, characterized by milder winters in the north and intensified summer heat in the south, has not been accompanied by a corresponding convergence in regulatory criteria. Instead, insulation requirements, construction practices, and compliance thresholds continue to reflect historical heating needs rather than evolving environmental risks (D'Agostino et al., 2024; Yannas & Rodríguez-Álvarez, 2020).

A clear example lies in the treatment of ground-contact floors. In Poland, uninsulated slabs are prohibited in new residential construction. Yet in Portugal, where winter temperatures are comparable but summer conditions are considerably harsher, such configurations remain permitted and commonplace. As highlighted by Roque et al. (2021), these practices persist despite growing evidence that Portuguese buildings face significant summer thermal stress. Meteorological data confirm this convergence. During the 2019–2020 heating season, the average outdoor temperature in Zielona Góra (Poland) was 4.9°C, compared to 5.0°C in Tabuaço (Portugal), while the latter experienced substantially higher summer temperatures (Fernandes et al., 2015). This regulatory inconsistency reflects a broader lag between building standards and contemporary climatic realities.

Canadian regulatory debates provide a relevant parallel. Siu et al. (2025) note that in Canada, traditionally a heating-dominated climate, code discussions now explicitly address overheating risk, set measurable indoor safety thresholds, and encourage regionally tailored compliance paths. This includes integrating passive-first compliance hierarchies and outcome-based metrics into national codes, offering flexibility for designs that achieve verified year-round comfort and resilience even if they deviate from standard prescriptive assemblies. A similar diagnosis has been made in the United Kingdom, where Hoggett et al. (2024) identify a policy gap between decarbonization goals and the lack of overheating protections in building regulations, recommending mandatory summer comfort standards, passive-first design requirements, and the incorporation of performance-based overheating metrics to avoid rapid lock-in to mechanical cooling and escalating peak electricity demand.

As summer overheating emerges as a dominant thermal risk, regulatory divergence becomes increasingly difficult to justify, especially when rigid prescriptions prevent the use of passive strategies that could improve year-round resilience. In regions where summers are lengthening and power grid reliability is under pressure, conventional solutions that focus only on winter performance may prove inadequate or even counterproductive (Palma et al., 2024). De Masi et al. (2021) argue that well-insulated envelopes, while effective in minimizing winter losses, may underperform in hot climates unless complemented by passive cooling systems and adaptive design measures.

To address these gaps, regulatory systems must better reflect regional climate projections and operational realities. This includes allowing greater design flexibility in response to local thermal profiles and emerging risks such as blackout-induced discomfort. Harmonized yet climate-sensitive metrics, such as seasonal energy balance, summer overheating thresholds, and carbon intensity per square meter, should supplement or, where justified, replace fixed envelope criteria. The Canadian and UK examples show that integrating such metrics into codes is both feasible and beneficial for aligning building performance with long-term occupant safety and with the stability objectives of a power system that is increasingly supplied by variable renewables (Hoggett et al., 2024; Siu et al., 2025).

Recognizing the role of transitional configurations is particularly important for rural areas and retrofitted buildings, where full compliance with standardized insulation requirements may be technically or economically impractical. As shown in this study, hybrid strategies, such as ground coupling combined with vertical perimeter insulation, can delay performance degradation and enhance thermal autonomy without the need for active systems (Gortych & Kuczyński, 2025).

In summary, aligning building codes with evolving climate patterns requires a shift from uniform prescriptions to adaptive frameworks. Performance-based regulations that recognize thermal inertia, soil-buffering effects, and envelope responsiveness will be better suited to the demands of resilient design. As climate variability accelerates, the ability of buildings to passively moderate indoor temperatures under both heating and cooling conditions will become essential, not only for efficiency but also for occupant safety and long-term environmental integrity.

15.5 PASSIVE RESILIENCE AS A DESIGN PRIORITY IN A WARMING CLIMATE

The empirical results from the present study add to a growing body of evidence that passive design features such as thermal mass, ground coupling, effective solar control, and adaptable ventilation can make a decisive difference in maintaining safe indoor conditions during both seasonal extremes and energy supply interruptions. Building B1, with its medium-weight envelope and direct thermal contact with the ground, consistently demonstrated a capacity to buffer indoor temperatures well beyond that of the lightweight, fully insulated B2. This effect was evident not only in the winter monitoring phase, where ground coupling delayed early-season losses, but also in previous summer resilience studies by the authors (Gortych & Kuczyński, 2025; Gortych et. al, 2025), where the same configuration prevented overheating altogether during multi-day heat waves without mechanical cooling.

These outcomes align with findings from other research that challenges the notion of thermal efficiency as the sole determinant of building quality. Yu et al. (2025) showed that in cold-climate Chinese cities, insulation-centric designs often increase overheating risk, especially in lightweight constructions with low thermal mass. Similar patterns have been reported in European contexts, where insulation-heavy retrofits have sometimes exacerbated summer discomfort (Arriazu-Ramos et al., 2023, 2024; Palma et al., 2024). In parallel, institutional assessments (Sarabia-Escriva et al., 2024) reveal that official energy certification methods can underestimate future cooling loads, particularly in regions undergoing rapid climatic warming.

An important implication of these findings is that resilience, defined as the ability of a building to passively sustain habitable conditions during adverse events, should be considered a primary performance criterion alongside operational efficiency. This view is consistent with recent proposals to integrate explicit indoor temperature thresholds into building codes and to prioritize passive solutions as the first line of defense before resorting to active systems (Siu et al., 2025). Comparable recommendations have been advanced in the United Kingdom, where policy analysis highlights the absence of robust overheating protections in regulations and argues for passive-first design, summer comfort standards, and equity-focused safeguards to protect vulnerable occupants (Hoggett et al., 2024).

The Canadian code revision process offers a useful example, showing that performance-based metrics can be embedded into national regulations in ways that accommodate regional variation while maintaining clear safety outcomes. The present study provides the type of field-validated, time-resolved evidence needed to support similar shifts elsewhere. By capturing both the seasonal dynamics of ground–envelope interaction and the long-duration benefits of thermal inertia, these results contribute directly to the evidence base for performance-oriented regulation in a warming climate. This perspective is consistent with the proposals of Norouzi et al. (2023), who advocate for the inclusion of unconventional measures such as seasonal thermal storage, adaptive envelopes, and thermally zoned interiors within regulatory frameworks, recognizing their potential to enhance passive survivability and long-term thermal autonomy. In parallel, UK analyses caution that without such performance-based overheating criteria, housing policy risks a rapid lock-in to mechanical cooling with rising peak electricity demand and weakened decarbonization outcomes (Hoggett et al., 2024).

Furthermore, the combination of high thermal capacity and controlled ground coupling demonstrated in B1 has direct relevance for the integration of variable renewable energy sources into building operation. Such configurations enable the building to act as a thermal energy store, smoothing demand by absorbing surplus heat from solar generation during the day and releasing it at night, or by retaining warmth during periods of high wind power output and sustaining comfort through windless intervals (Hoggett et al., 2024). This synergy between passive thermal resilience and renewable energy integration reduces reliance on short-term electrical storage, enhances grid stability, and strengthens the role of buildings as active participants in the decarbonized energy system. In the UK policy context, this aligns with calls to couple passive-first building design with a power system increasingly supplied by variable renewables in order to manage peaks and support system adequacy while protecting occupant health (Hoggett et al., 2024).

15.6 SUMMARY AND PRACTICAL IMPLICATIONS

The results presented in this chapter, based on the authors' full-scale experimental research, provide rare empirical evidence on the year-round implications of integrating high wall thermal mass and direct floor–ground coupling into residential building design. In contrast to many simulation-based studies, these findings are grounded in long-term monitored data under real climatic conditions, enabling a more accurate assessment of seasonal trade-offs between heating energy use, cooling resilience, and passive thermal safety.

The monitoring confirmed that while removing horizontal subfloor insulation in Building B1 increased heating energy demand by 12.2% over a full season, this penalty was lower than predicted by standardized modeling methods such as ISO 13370. The discrepancy is explained by vertical perimeter insulation and the seasonal heat storage capacity of the ground, which delayed and moderated conductive losses. At the same time, the heavyweight masonry envelope reduced heating demand by 3.6% compared to the lightweight B2, illustrating that wall thermal mass can provide a measurable operational benefit even in continuously heated buildings.

When placed in the broader context of the authors' previous studies (Gortych & Kuczyński, 2025; Gortych et. al, 2025), the significance of these results becomes clear. The same ground-coupled slab that imposed a moderate winter penalty was shown in Sections 14.16 and 14.17 to deliver substantial resilience gains during both summer heat waves and winter heating outages. In practice, this means that the loss in seasonal efficiency is offset by an increase in passive survivability, extending habitable conditions far beyond those achievable in the lightweight, fully insulated counterpart. This integrated perspective challenges the conventional regulatory focus on heating efficiency as the primary performance target. A similar conclusion emerges from recent UK policy analysis, which warns that neglecting summer protection in codes risks a rapid lock-in to mechanical cooling with rising peak demand and weakened decarbonization outcomes (Hoggett et al., 2024).

The findings strongly support the growing body of literature advocating a shift from prescriptive, component-based regulation to performance-based frameworks. As highlighted by Siu et al. (2025) in the Canadian context, codes can be adapted to include explicit indoor temperature limits for both summer and winter conditions, prioritization of passive measures before active systems, and regionally flexible compliance paths. Such approaches directly address occupant safety, particularly for vulnerable populations, while allowing for innovative envelope configurations that deviate from standard prescriptions but deliver verifiable year-round comfort and resilience. UK recommendations align with this direction, calling for passive-first design and the incorporation of overheating metrics into performance requirements to safeguard summer comfort and system adequacy (Hoggett et al., 2024).

This aligns with recent evidence from multiple regions. Arriazu-Ramos et al. (2023, 2024) demonstrated that insulation-focused retrofits in Spain often failed to reduce overheating, particularly in top-floor dwellings without adequate shading or cross-ventilation. Palma et al. (2024) warned that winter-optimized envelopes can underperform in hot climates unless complemented by passive cooling systems. De Masi et al. (2021) and Yannas and Rodríguez-Álvarez (2020) similarly called for redefining envelope quality to include passive cooling performance. The present study adds to this discourse by showing, in measurable operational terms, how soil–structure interaction and thermal inertia influence not only energy use but also the duration of safe indoor conditions during extreme weather events.

From a practical standpoint, the authors' results suggest that:

- Moderate heating penalties may be acceptable in exchange for significant gains in passive thermal resilience, particularly in climates facing both winter cold spells and increasingly frequent summer heat waves.
- Hybrid ground-coupled configurations with vertical perimeter insulation can provide seasonal heat buffering, reducing predicted energy penalties and enhancing summer performance without mechanical cooling.
- Buildings with high thermal capacity and controlled ground coupling can function as effective thermal energy stores, supporting integration with variable renewable sources. This capability can reduce peak-hour energy demand, lower the need for electrical storage, stabilize grid operation, and maintain high levels of indoor comfort.
- Thermal mass remains a valuable design asset, not only for passive cooling but also for smoothing winter heating loads, especially under dynamic operation or intermittent supply scenarios.
- Performance-based regulatory metrics should be adopted, combining year-round energy demand, seasonal load distribution, and resilience indicators such as resistance time, overheating degree hours, and hours above safety thresholds.
- Policy frameworks should embed indoor summer temperature limits and passive-first compliance hierarchies to protect vulnerable occupants and to avoid a systemic lock-in to active cooling, in line with emerging Canadian and UK recommendations (Hoggett et al., 2024; Siu et al., 2025).

In conclusion, the authors' experimental evidence demonstrates that unconventional envelope strategies, when properly detailed, can deliver a balanced combination of winter efficiency, summer comfort, and passive survivability. In the context of a warming climate and increasing energy system stress, such approaches offer a viable path toward climate-adaptive building design. By recognizing thermal mass and controlled ground coupling as legitimate tools within performance-based frameworks, policymakers can enable innovative solutions that protect occupants, reduce dependence on mechanical systems, and support long-term environmental goals.

15.7 FROM OPERATIONAL METRICS TO WHOLE-LIFECYCLE ASSESSMENT

While the comparative thermal and energy performance analysis presented in this chapter offers important insights into the seasonal behavior of envelope strategies, energy efficiency alone no longer suffices as the dominant metric of building sustainability. As climate mitigation efforts expand beyond operational performance, there is a growing imperative to assess the full environmental impact of building design choices, including those that deviate from conventional norms, such as uninsulated ground slabs or thermally massive walls.

The two test buildings evaluated in this study, one lightweight with a fully insulated floor (B2) and the other medium-weight with a ground-coupled, uninsulated slab (B1), exhibited distinct seasonal energy profiles. However, these thermal differences must now be examined in the broader context of lifecycle environmental performance, particularly in terms of cumulative energy demand and long-term carbon emissions.

To that end, the next chapter presents a comparative Life Cycle Assessment of both buildings, integrating embodied energy and embodied carbon associated with construction materials and components, as well as operational energy use for heating and cooling under projected climate conditions. By accounting for both construction-phase impacts and long-term performance, this lifecycle approach provides a more comprehensive framework for evaluating the sustainability of passive design strategies.

Ultimately, understanding the environmental trade-offs between lightweight insulated systems and mass-based ground-coupled envelopes is essential for informing both design practice and regulatory evolution in the era of climate-responsive architecture.

REFERENCES

Arriazu-Ramos, A., Bes-Rastrollo, M., Sánchez-Ostiz Gutiérrez, A., & Monge-Barrio, A. (2023). Building parameters that influence overheating of apartment buildings in a temperate climate in Southern Europe. *Building and Environment*, *228*, 109899. https://doi.org/10.1016/j.buildenv.2022.109899

Arriazu-Ramos, A., Pons Izquierdo, J. J., Ramos Ruiz, G., Sánchez-Ostiz, A., & Monge-Barrio, A. (2024). Facing climate change in a temperate European city: Urban-scale diagnosis of indoor overheating and adaptation strategies for residential buildings. *Buildings*, *14*(5), 1423. https://doi.org/10.3390/buildings14051423

Costanzo, V., Fabbri, K., & Piraccini, S. (2018). Stressing the passive behavior of a Passivhaus: An evidence-based scenario analysis for a Mediterranean case study. *Building and Environment*, *142*, 265–277. https://doi.org/10.1016/j.buildenv.2018.06.035

Cox-Smith, I. 2016. Perimeter Insulation of Concrete Slab Foundations. BRANZ.

D'Agostino, D., Congedo, P. M., Albanese, P. M., Rubino, A., & Baglivo, C. (2024). Impact of climate change on the energy performance of building envelopes and implications on energy regulations across Europe. *Energy*, *288*, 129886. https://doi.org/10.1016/j.energy.2023.129886

De Masi, R. F., Gigante, A., & Vanoli, G. P. (2021). Are nZEB design solutions environmental sustainable? Sensitive analysis for building envelope configurations and photovoltaic integration in different climates. *Journal of Building Engineering*, *39*, 102292. https://doi.org/10.1016/j.jobe.2021.102292

Fernandes, J., Pimenta, C., Mateus, R., Silva, S. M., & Bragança, L. (2015). Contribution of Portuguese vernacular building strategies to indoor thermal comfort and occupants' perception. *Buildings*, *5*(4), 1242–1264. https://doi.org/10.3390/buildings5041242

Goia, F., Time, B., & Gustavsen, A. (2015). Impact of opaque building envelope configuration on the heating and cooling energy need of a single family house in cold climates. *Energy Procedia*, *78*, 2626–2631. https://doi.org/10.1016/j.egypro.2015.11.284

Goldberg, L. F., & Mosiman, G. E. 2015. High-Performance Slab-on-Grade Foundation Insulation Retrofits. NorthernSTAR.

Gortych, M., Staszczuk, A., Kuczyński, T. (2025). Empirical assessment of passive thermal resilience in buildings with varying heat storage capacity during heatwaves and power outages. *Energies*, *18*(18), 4871. https://doi.org/10.3390/en18184871

Gortych, M., & Kuczyński, T. (2025). *Winter thermal resilience of lightweight and ground-coupled mediumweight buildings: An experimental study during heating outages.* Energies, 18(15), 4022. https://doi.org/10.3390/en18154022

Gortych, M., Staszczuk, A., Kuczyński, T. (2026). Impact of Floor - Ground Coupling and Thermal Mass on Seasonal Heating Energy Use in Single - Storey Residential Buildings in a Temperate Climate. *Civil and Environmental Engineering Reports,* 36(1), 1–20. https://doi.org/10.59440/ceer/216402

Hoggett, R., Lowes, R., Ugalde-Loo, C. E., & Khosravi, F. (2024). Hot and cold: Policy perspectives on overheating and cooling in United Kingdom homes. *Energy Research & Social Science, 118,* 103803. https://doi.org/10.1016/j.erss.2024.103803

Ibrahim, A., & Pelsmakers, S. (2018). Low-energy housing retrofit in North England: Overheating risks and possible mitigation strategies. *Building Services Engineering Research and Technology, 39*(2), 161–172.

Liu, Z., Alterman, D., Page, A., Moghtaderi, B., & Chen, D. (2021). An experimental study on the thermal effects of slab-edge insulation for slab-on-grade housing in a moderate Australian climate. *Energy and Buildings, 235,* 110675. https://doi.org/10.1016/j.enbuild.2020.110675

Norouzi, M., Haddad, A. N., Jiménez, L., Hoseinzadeh, S., & Boer, D. (2023). Carbon footprint of low-energy buildings in the United Kingdom: Effects of mitigating technological pathways and decarbonization strategies. *Science of The Total Environment, 882,* 163490. https://doi.org/10.1016/j.scitotenv.2023.163490

Palma, R. M., Castro Medina, D., Guerrero Delgado, M. C., Sánchez Ramos, J., Montero-Gutiérrez, P., & Álvarez Domínguez, S. (2024). Enhancing the building resilience in a changing climate through a passive cooling roof: A case study in Camas (Seville, Spain). *Energy and Buildings, 321,* 114680. https://doi.org/10.1016/j.enbuild.2024.114680

Roque, E., Vicente, R., & Almeida, R. M. S. F. (2021). Opportunities of light steel framing towards thermal comfort in Southern European climates: Long-term monitoring and comparison with the heavyweight construction. *Building and Environment, 200,* 107937. https://doi.org/10.1016/j.buildenv.2021.107937

Sarabia-Escriva, E.-J., Jiménez-Navarro, J.-P., Soto-Francés, V.-M., & Pinazo-Ojer, J.-M. (2024). Assessing the energy performance certification effectiveness for the Spanish building stock in response to recent climate change data. *Energy and Buildings, 323,* 114816. https://doi.org/10.1016/j.enbuild.2024.114816

Siu, C. Y., Brown, S., Brideau, S., Ji, L., O'Brien, W., Ferguson, A., Greco, E., Touchie, M., McLellan, C., Lepage, R., & Dmytrenko, N. (2025). Exploring the role of building codes in protecting occupants from overheating: A Canadian perspective. *Energy and Buildings, 338,* 115673. https://doi.org/10.1016/j.enbuild.2025.115673

Yannas, S., & Rodríguez-Álvarez, J. (2020). Domestic overheating in a temperate climate: Feedback from London residential schemes. *Sustainable Cities and Society, 59,* 102189. https://doi.org/10.1016/j.scs.2020.102189

Yu, J., Zhao, B., Xu, Y., Jin, M., Wang, Y., Chang, W.-S., & Guo, H. (2025). Overheating causes in cold regions' residences under evolving building regulations: Focusing on field investigation and simulation of envelope performance. *Building and Environment, 281,* 113180. https://doi.org/10.1016/j.buildenv.2025.113180

16 Carbon Footprint Analysis of Single-Family Homes in Zielona Góra

The Impact of Climate Scenarios and Construction Choices

16.1 BACKGROUND AND RATIONALE

Under growing pressure to achieve climate neutrality, the residential building sector remains a major source of greenhouse gas emissions. While assessments have traditionally focused on operational energy, the significance of embodied emissions has increased, especially in energy-efficient constructions. In low-energy buildings, operational emissions may account for only 50–60% of the life cycle footprint, and in timber buildings even less (Andersen et al., 2022; Dani et al., 2022; Younis & Dodoo, 2022).

Heating has been identified as the dominant source of emissions in cold and temperate climates (Liu et al., 2016; Tettey et al., 2019). However, rising temperatures, urbanization, and evolving usage patterns are increasing the need for cooling, even in regions traditionally dominated by heating. Over long time frames, this shift can significantly affect the life cycle CO_2 balance.

Despite these developments, many LCA studies still rely on static climate and energy assumptions. Yet recent research shows that climate change and electricity grid decarbonization substantially alter the ratio of operational to embodied emissions (Allan & Phillips, 2021; Hart et al., 2021). As a result, linear baselines may misrepresent the environmental implications of design choices. As highlighted in recent work, integrating dynamic climate and energy scenarios is essential for accurate long-term assessments (Bruhn et al., 2023; Fischer et al., 2023).

This calls for refined approaches to building design that account for long-term environmental outcomes. Features such as thermal mass, ground coupling, and passive buffering capacity are becoming increasingly important. However, these attributes remain underrepresented in current LCA tools and certification systems. Even advanced frameworks such as BREEAM or LEED treat life cycle metrics as optional and rarely capture passive resilience.

Cooling-related emissions are often underestimated. In well-insulated lightweight buildings, future cooling demand may offset heating savings, especially in buildings

DOI: 10.1201/9781003645573-16

with low thermal inertia (Andersen et al., 2022; Dani et al., 2022; Kang et al., 2024). Many LCA models either exclude cooling or assume constant demand. As Hart et al. (2021) and Fischer et al. (2023) point out, this can obscure the true carbon impact of envelope design.

To address these gaps, building LCA should adopt more dynamic, climate-informed approaches. Passive thermal features, such as mass and ground coupling, must be better integrated, particularly in the context of renewable energy systems, where thermal inertia supports demand flexibility and grid stability.

This study contributes to that shift by comparing the life cycle carbon impact of two full-scale residential buildings with similar size and use but contrasting in their thermal characteristics. The goal is to inform prospective LCA frameworks that reflect future climate and energy dynamics.

16.2 LITERATURE REVIEW AND CONCEPTUAL FOUNDATIONS

16.2.1 Comparative LCA of Lightweight and Heavyweight Building Structures

Numerous comparative life cycle assessment (LCA) studies have evaluated the environmental performance of lightweight vs heavyweight building structures. Despite differences in scope and methodology, these works highlight how material choices affect both embodied and operational carbon emissions. Table 16.1 summarizes key findings, emphasizing contrasts between cross-laminated timber (CLT), concrete, and steel systems.

Overall, CLT and other engineered timber systems generally yield lower embodied carbon than reinforced concrete or steel. Guardigli et al. (2011) reported a 25–30% reduction in life cycle emissions for a timber-framed office building. Chen et al. (2020) found a 20.6% reduction in cradle-to-grave emissions for CLT and nearly 70% when accounting for end-of-life recovery (Module D). Dzhurko et al. (2024) observed 30–40% lower emissions for dowel-laminated timber (DLT) compared to concrete and brick systems, with significantly higher carbon storage capacity. De Araujo et al. (2023), in a global review, confirmed that CLT offers one of the lowest embodied carbon intensities among wood-based construction systems.

Several studies employed cradle-to-gate boundaries (e.g., Pierobon et al., 2019; Robertson et al., 2012), focusing solely on production-stage emissions, while others used cradle-to-grave models spanning 50 to 100 years (e.g., Andersen et al., 2021; Dolezal et al., 2021; Peñaloza et al., 2016; Tettey et al., 2019). In Norway, Andersen et al. (2022) and Andersen et al. (2022) found that CLT buildings performed better than concrete in 11 of 18 environmental impact categories, even when accounting for biogenic carbon and shifts in forest stocks. Sporchia et al. (2025) compared conventional LCA with emergy-based evaluation and concluded that while standard LCA showed only a 3% advantage for CLT, emergy methods revealed a 42.5% reduction in ecological load.

Operational energy results were more mixed. Peñaloza et al. (2016) reported 42% lower total GWP for CLT buildings under dynamic modeling, though operational emissions were relatively minor. Tettey et al. (2019) found that CLT structures had

TABLE 16.1
Summary of Comparative LCA Studies

Study	Building Types Compared	LCA Scope	Unit Emissions (kg CO_2/m^2 Unless Noted)	Relative Reduction	Location
Guardigli et al. (2011)	Concrete vs timber	Cradle-to-grave	230 vs 310	Timber: –25–30%	Europe
Robertson et al. (2012)	Concrete vs laminated timber	Cradle-to-gate	Not specified	Timber: –14–18%	Canada
Dolezal et al. (2021)	Concrete vs timber	Cradle-to-grave	Timber: >100 lower	Timber: –18%	Austria
Andersen et al. (2022)	Concrete vs CLT	Cradle-to-grave	Not specified	CLT: better in 11/18 indicators	Norway
Pierobon et al. (2019)	Concrete vs hybrid CLT	Cradle-to-gate	Not specified	CLT: lower GWP	USA (Seattle)
Andersen et al. (2022)	Concrete vs CLT	Consequential LCA	Not specified	CLT: lower total GWP	Norway
Dodoo et al. (2021)	Variants of CLT	Cradle-to-grave	119–279	Design-dependent	Sweden
Tettey et al. (2019)	Concrete vs CLT vs wood modules	Full LCA	CLT: –25–35%	CLT lowest	Sweden
Peñaloza et al. (2016)	Concrete vs CLT	Static & dynamic	218–281	CLT: –42%	Sweden
Dzhurko et al. (2024)	DLT vs brick & concrete	A1–A3 only	DLT: 61–83 vs 126	DLT: –30–40%	Germany
De Araujo et al. (2023)	CLT vs others	Systematic review	0.31 per kg material	CLT lower LCA	Global
Chen et al. (2020)	CLT vs concrete	Cradle-to-grave	CLT: –20.6% to –69.5%	CLT better in A–C+D	USA
Sporchia et al. (2025)	CLT vs concrete	LCA + emergy	1209.5 vs 1173.2 t	–3% LCA, –42.5% emergy	Italy
Hemmati et al. (2024)	CLT vs steel	A1–A4	152 vs 238	–19%	USA
Sah et al. (2024)	PCSP vs others	Review	10 m^2K/W (PCSP best)	Thermal mass superior	Global

reduced heating loads due to better insulation but lacked thermal inertia compared to concrete. Dodoo et al. (2021) highlighted design dependency in operational outcomes across CLT variants, with emissions ranging from 119 to 279 kg CO_2/m^2.

Only a few studies explicitly examined thermal inertia or ground coupling. Hemmati et al. (2024) analyzed a CLT dormitory and a steel equivalent and found that concrete and transport dominated embodied emissions, with a moderate displacement factor of 0.28. Sah et al. (2024) emphasized that precast concrete sandwich panels (PCSP) provided superior thermal storage and buffering, especially under dynamic conditions.

Building typology and local context also influenced performance. Tettey et al. (2019) compared three frame types (concrete, CLT, and modular wood) in a Swedish apartment building and found that CLT consistently performed best across life cycle stages. Similarly, Dolezal et al. (2021) showed that timber construction in Austria yielded more than 100 kg CO_2/m^2 lower emissions than its concrete counterpart.

In sum, timber-based construction generally outperforms concrete and steel in embodied emissions, particularly in cradle-to-gate or cradle-to-grave assessments. Long-term performance, however, depends on operational conditions, thermal design, and modeling assumptions. Only a small number of studies consider passive design strategies such as ground coupling and structural heat capacity, which may prove decisive under future climate scenarios. This study contributes to that emerging area by incorporating these features into a dynamic, climate-sensitive LCA framework.

16.2.2 Climate-Induced Shifts in Heating and Cooling Demand

While the studies reviewed above largely assess current or static boundary conditions, it is equally important to recognize that the operational balance of heating and cooling will shift substantially under future climate scenarios. This dynamic affects not only the ratio between operational and embodied emissions but also the comparative advantage of lightweight vs heavyweight structures. In order to integrate this dimension, a growing body of research has examined projected changes in heating and cooling demand under different climate pathways (Table 16.2).

16.3 LIMITATIONS OF CURRENT STANDARDS AND ASSESSMENT FRAMEWORKS

Despite advances in LCA for building evaluation, most standards and tools still rely on simplified assumptions. They emphasize static performance metrics, such as U-values or annual energy use, without accounting for climate variability or long-term operational change.

Even sophisticated LCAs often apply fixed energy use profiles that ignore the effects of warming and grid decarbonization (Hart et al., 2021). As a result, buildings considered efficient in the short term may generate higher emissions over time, especially with rising cooling needs.

Embodied carbon remains underrepresented. In traditional buildings, operational emissions dominate, but in energy-efficient or timber buildings, the embodied share can reach 40–70% (Chastas et al., 2018; Dani et al., 2022). Nonetheless, most

TABLE 16.2
Reported Changes in Heating and Cooling Demand in LCA Studies under Future Climate Scenarios

Authors (Year)	Country/Location	Time Horizon	Change in Heating Demand	Change in Cooling Demand
Jalali et al. (2023)	New Zealand	1999–2090 (RCP 4.5, 8.5)	Decrease by 15–71%	Increase by 40–380%
Ciancio et al. (2020)	19 cities in Europe	2020, 2050, 2080	Decrease by up to 45%	Increase by up to 142%
D'Agostino et al. (2022)	8 locations in Europe	2018 vs 2060 (RCP 8.5)	Decrease by 36–57%	Increase by 90–281%
Yan and Ogata (2025)	Japan (9 cities)	2050 and 2080 (SSP1–2.6 to SSP5–8.5)	Decrease by 13–32%	Increase by 48–136%
Duan et al. (2025)	Australia (10 cities)	2020–2080 (SSP2–4.5, SSP5–8.5)	Decrease by 35–81%	Increase by 13–85%
Chetouni et al. (2025)	Morocco (6 climate zones)	2050 (RCP 4.5 and 8.5)	Decrease by 23.5–48.9%	Increase by 25.5–54.5%
Tomrukcu and Ashrafian (2024)	Turkey (Istanbul, Izmir)	2050 and 2080 (RCP 4.5 and 8.5)	Decrease by 20–45%	Increase by 125–200%
Milić and Rohdin (2024)	Sweden (Linköping)	1980–2050 (projection)	Decrease by 24–27%	Not applicable (no AC)
Arul Babu et al. (2024)	India (8 cities)	1990s–2090s (RCPs)	Decrease by 34–100%	Increase by 20–179%
Piggot-Navarrete et al. (2025)	Canada (Quebec City)	2020–2080 (RCP 8.5)	Decrease by 25%	Increase by 91–116%
Haji Ismail et al. (2021)	UK (Reading, prefab.)	2017–2080 (A1B, A1F1)	Decrease by 3.6–17%, up to −34% (HDD)	Not applicable (no cooling systems)
Shen et al. (2025)	China (Beijing, Shanghai, Shenzhen)	2010–2099 (RCP2.6, 4.5, 6.0, 8.5)	Decrease in HDD by 6–98%; in Shenzhen HL decreased by 13–59%	Increase by 5–28% in Shenzhen, 27–84% in Shanghai, 32–90% in Beijing

CL, Cooling load; HDD, heating degree days; HL, heating load.

frameworks emphasize Module B6, while undervaluing emissions from material production (A1–A3) and end-of-life stages (C1–C4).

There is also little integration between LCA findings and design standards. In frameworks such as LEED or BREEAM, LCA is often optional or has limited influence, even though studies confirm that material and design choices have lasting carbon implications (Fischer et al., 2023; Younis & Dodoo, 2022).

Moreover, few standards require future climate or energy scenarios. Constant emission factors remain widely used, even though evidence shows declining grid intensity and changing demand patterns. Bruhn et al. (2023) argue that LCA must evolve into a prospective framework incorporating emission variability and decarbonization pathways.

Finally, passive attributes such as thermal mass or ground coupling are rarely included in assessments, even though they can significantly reduce cooling loads and improve long-term environmental performance, especially under climate stress.

16.4 RESEARCH AIM AND STRUCTURE OF THE CHAPTER

This chapter evaluates the long-term carbon performance of two real-world single-family buildings with identical size and layout but contrasting thermal configurations. Building B1 is a heavyweight masonry structure coupled to the ground without floor insulation. Building B2 is a lightweight timber-frame structure with full slab insulation. Both are located in the same climate zone and operated under uniform conditions.

The analysis adopts a cradle-to-grave scope over 75 years (2025–2100), integrating three Shared Socioeconomic Pathway (SSP) scenarios. It includes embodied emissions (A1–A3), operational emissions from heating and cooling (B6), and end-of-life impacts (C1–C4). A particular focus is placed on future cooling demand and the role of mass and ground coupling in mitigating operational emissions.

Unlike many previous studies based on abstract models (e.g., Allan & Phillips, 2021; Dani et al., 2022; Liu et al., 2016), this work is grounded in empirical energy monitoring and detailed material data, extended using prospective LCA methods (Bruhn et al., 2023). This approach allows for a robust evaluation of both current and future environmental trade-offs.

Section 2 outlines the buildings' specifications, LCA methodology, climate scenarios, and data sources. Section 3 presents operational and life cycle emissions under each scenario. Section 4 discusses the results in relation to prior studies and policy implications. Section 5 concludes with key findings and recommendations for climate-resilient design.

16.5 METHODS AND ASSUMPTIONS

This chapter extends the analysis of Buildings B1 and B2, previously examined in terms of thermal resilience and year-round energy efficiency, by conducting a cradle-to-grave life cycle carbon assessment over a 75-year period (2025–2100). The goal is to evaluate long-term environmental trade-offs associated with thermal mass and ground coupling strategies.

The LCA includes embodied carbon from construction materials (A1–A3), operational emissions from space heating and cooling (B6), and end-of-life impacts (C1–C4), in accordance with EN 15978. Three climate and energy scenarios were adopted based on SSP trajectories: SSP1–2.6, SSP4–6.0, and SSP5–8.5. The assessment period was divided into three subperiods (2026–2050, 2051–2075, and 2076–2100), with each phase incorporating projected changes in heating and cooling demand and the carbon intensity of electricity.

Heating demand was derived from empirical monitoring during the 2020–2021 heating season and scaled for full-year performance. For B1 and B2, base annual demand was set at 8,019 kWh and 7,149 kWh, respectively. These values were adjusted for progressive climate warming using reduction rates of 10%, 25%, and 40% across the three timeframes. Cooling demand was modeled only for B2 based on recorded summer overheating, starting at 1,008 kWh per year and increasing by 10%, 50%, and 80% in subsequent periods. Heating and cooling systems were assumed to operate electrically with a seasonal COP of 4.5. The electricity emission factor declined from 400 g CO_2/kWh in the first period to 100 g CO_2/kWh in the third.

Embodied emissions were calculated from detailed material inventories using Ecoinvent 3.7 and ICE v3.0 emission factors. Total emissions from production (A1–A3) were 40,036 kg CO_2 for B1 and 42,543 kg CO_2 for B2. End-of-life emissions (C1–C4) amounted to 1,821 kg CO_2 (B1) and 1,852 kg CO_2 (B2), reflecting higher combustibility in B2's timber components. These values correspond to 4.28 and 4.49 kg CO_2/m^2/year, respectively, over the full life cycle.

Additional embodied emissions from HVAC system replacements were included separately. Each building used a reversible air-to-air heat pump with a service life of 20 years. Over 75 years, this resulted in three replacements, adding 2,334 kg CO_2 equally in both cases.

All emissions were aggregated into three categories: operational (B6), non-operational (A1–A3 and C1–C4), and auxiliary (HVAC replacements). Total values were normalized to floor area (kg CO_2/m^2/year) and compared across climate scenarios to assess the long-term carbon impact of thermal mass and subfloor configuration.

The modeling approach follows prospective LCA principles, including dynamic operational modeling, scenario integration, and time-segmented aggregation. It aims to provide a more realistic foundation for comparing the environmental performance of passive design strategies over extended building lifespans under future climate conditions.

16.6 RESULTS AND COMPARATIVE EMISSIONS

The life cycle carbon performance of the two buildings revealed significant differences driven by their thermal configurations and responsiveness to future climate and energy scenarios. Over the 75-year assessment period (2025–2100), Building B1, a medium-weight masonry structure with ground coupling and no slab insulation, achieved a lower cumulative carbon footprint than Building B2, a lightweight timber-frame building with a fully insulated ground floor and greater cooling dependency.

16.6.1 Operational Emissions

Across all three SSP climate scenarios, operational emissions (Module B6) represented the largest share of total life cycle emissions. Under the mid-range SSP4–6.0 trajectory, cumulative operational emissions were 99,284 kg CO_2 for B1 and 108,731 kg CO_2 for B2, about 9.5% higher. The difference was primarily due to the inclusion of cooling demand in B2, which increased progressively and eventually offset its initial heating advantage.

In relative terms, B1's heating demand decreased by 33% between the first and third periods (2026–2050 vs 2076–2100), reflecting the projected decline in winter severity. B2, however, required additional mechanical cooling, which was modeled to increase by 80% by the end of the century. As a result, its operational emissions remained elevated despite lower winter loads.

16.6.2 Embodied and End-of-Life Emissions

The embodied carbon emissions (A1–A3) were 40,036 kg CO_2 for B1 and 42,543 kg CO_2 for B2, while end-of-life emissions (C1–C4) added 1,821 kg and 1,852 kg CO_2, respectively. Combined, these material-related emissions totaled 41,857 kg CO_2 for B1 and 44,395 kg CO_2 for B2, reflecting a 6.1% difference in favor of the heavyweight structure.

The higher embodied emissions in B2 were primarily due to more complex timber-framed wall assemblies and thicker slab insulation. In contrast, B1's monolithic concrete elements created a compact and thermally massive envelope with slightly lower emissions per unit area.

16.6.3 Auxiliary HVAC Emissions

Each building used a reversible heat pump system, replaced every 20 years. Over 75 years, this resulted in three replacements, adding 2,334 kg CO_2 equally in both cases. This value was treated as fixed in the overall balance.

16.6.4 Total Life Cycle Emissions

When all emissions were aggregated, including operational, embodied, end-of-life, and auxiliary components, Building B1 demonstrated superior performance across all scenarios. Under SSP4–6.0, total cradle-to-grave emissions were approximately:

- B1: 143,475 kg CO_2
- B2: 155,460 kg CO_2

This represents a 7.7% reduction for B1. Similar patterns were observed under SSP1–2.6 and SSP5–8.5, with the relative advantage of B1 increasing under scenarios with stronger cooling demand and slower grid decarbonization.

16.6.5 Emissions Per Square Meter

Normalized to the floor area (123 m^2), B1's total emissions were 1,167 kg CO_2/m^2, while B2 reached 1,265 kg CO_2/m^2. On an annual basis, this corresponds to 15.6 kg CO_2/m^2/year for B1 and 16.9 kg CO_2/m^2/year for B2.

These results place both buildings within the typical range for energy-efficient construction, but confirm that the addition of thermal mass and ground coupling can lead to measurable carbon reductions over the building lifespan.

16.7 DISCUSSION AND INTERPRETATION

The comparative life cycle carbon assessment provides several insights into the long-term trade-offs between thermal mass with ground coupling (Building B1) and lightweight insulated construction (Building B2). While B1 incurred a modest increase in heating-related emissions due to direct ground contact, this disadvantage was offset by its superior ability to limit cooling demand under future climate conditions.

Across the 75-year horizon, total life cycle emissions, including embodied, operational, and auxiliary components, were consistently lower for B1. Although the differences between the two buildings were not dramatic, B2 produced slightly higher cumulative CO_2 emissions in all SSP scenarios. This shows that mass and ground coupling do not penalize long-term performance; instead, they shift the balance between heating and cooling depending on the season and the climate trajectory.

These results are particularly relevant given anticipated warming trends. Numerous studies (e.g., Ciancio et al., 2020; D'Agostino et al., 2022; Shen et al., 2025) confirm that cooling demand in temperate climates will increase substantially, in many cases surpassing heating demand. In this context, the passive buffering provided by B1's heavier envelope and ground coupling emerges as a clear advantage that static or short-term assessments fail to capture.

The embodied carbon difference between the two buildings was modest and does not fully explain the overall life cycle advantage of B1. The decisive factor was the distribution of operational energy over time. B1 benefited from thermal inertia and a reduced reliance on mechanical systems, which became more significant as the climate warmed. These temporal effects align with arguments advanced by Bruhn et al. (2023) and Fischer et al. (2023), who call for prospective LCA models that explicitly integrate climate dynamics and passive performance.

Another important observation is the declining marginal value of floor insulation under long-term climate change. Insulation improves heating efficiency in the near term, but its cooling penalty grows with each decade. Thermal mass, by contrast, continues to provide resilience, especially in climates with mixed seasonal demands. This outcome is consistent with earlier findings from Chapters 13 and 14, where B1 consistently outperformed B2 in both summer and winter resilience tests.

The findings also support a re-examination of assumptions embedded in LCA tools and building codes. Current methodologies often favor insulation-centric design while neglecting the role of mass and soil–structure interaction. The results presented here suggest that broader definitions of environmental performance are required. Indicators should capture not only resistance metrics such as U-values, but also dynamic capacity and long-term adaptability.

Finally, the study demonstrates the importance of empirical calibration in prospective LCA. By incorporating measured operational data rather than relying only on archetypal models, the assessment is more robust and directly applicable to the Central European context. In this way, it helps bridge the gap between experimental research and policy-oriented frameworks.

16.8 SUMMARY

This chapter extended earlier work on thermal resilience and seasonal energy performance by embedding two experimental houses within a prospective life cycle carbon (LCA) framework. Three main results were confirmed. First, in winter outages, the ground-coupled masonry house maintained habitable conditions significantly longer than the lightweight insulated counterpart. Second, in summer heat waves, it avoided overheating without mechanical cooling by damping diurnal swings and releasing stored heat overnight. Third, in routine operation, the heating penalty from removing slab insulation was modest and partly offset by vertical perimeter insulation.

Across a 75-year horizon, the heavyweight ground-coupled house consistently achieved lower cradle-to-grave emissions. Under SSP4–6.0, its cumulative footprint was about 143,475 kg CO_2 compared with 155,460 kg CO_2 for the lightweight structure, a reduction of 7.7%. The difference was driven mainly by cooling demand: as summers warmed, B2 increasingly required mechanical cooling, while B1 maintained passive comfort.

Although the masonry structure relied on thicker external and internal walls, its embodied and end-of-life emissions were slightly lower than those of the lightweight house: about 41,857 kg CO_2 for B1 compared with 44,395 kg CO_2 for B2. This reflects the more complex wall assemblies of the timber-frame design and the conservative assumption applied here, which did not grant carbon-neutral credits for timber as many studies do. However, the decisive influence came from operational performance. Under SSP4–6.0, operational emissions reached about 99,284 kg CO_2 for B1 but 108,731 kg CO_2 for B2. Thus, cooling-related emissions in the lightweight insulated house exceeded its early heating advantage, producing a long-term penalty that dominated the life cycle balance.

The results highlight a simple mechanism. As winters become milder, the relative size of the heating penalty diminishes, while the avoidance of cooling demand compounds over decades. Ground coupling and envelope mass act as a slow thermal store, buffering fluctuations and aligning with renewable-based grids by reducing peak loads.

These outcomes underline the value of a prospective LCA lens. Static or single-year models cannot capture how buildings, energy systems, and climate co-evolve. By contrast, dynamic evaluation shows that a mass-based, ground-coupled envelope offers three benefits simultaneously: winter safety during outages, summer protection without active cooling, and lower life cycle emissions.

Policy discussions in Canada and the UK support this direction, warning that neglecting summer protection risks lock-in to mechanical cooling and higher peak electricity demand. The Nowy Kisielin findings provide empirical evidence that passive buffering strategies deserve regulatory credit.

The conclusions apply to two single-story houses in western Poland under specific assumptions. Broader transferability requires verification in other typologies, climates, and soil conditions, including geotechnical and hygrothermal analyses. Within these limits, however, the message is clear: a mass-based, ground-coupled envelope with perimeter insulation can extend passive survivability, suppress future cooling demand, and reduce life cycle carbon emissions.

16.8.1 Practical Implications

The analysis highlights several practical implications. First, resilience and carbon balance should be evaluated jointly. From the concept stage, both the duration of safe indoor conditions during outages and the total balance of operational and embodied emissions must be considered. In temperate and warming climates, buildings with higher envelope mass combined with controlled floor–ground heat exchange demonstrate lower life cycle emissions and greater resilience during both summer heat waves and winter heating outages.

Second, life cycle assessment should be prospective rather than static, based on evolving climate and energy transition scenarios. Only in this way can the future decline in heating demand and the concurrent rise in cooling needs be adequately reflected, together with the impact of grid decarbonization on total emissions. Closely related to this is the role of the building envelope as a thermal store. By combining wall mass, ground coupling, external shading, and night ventilation, buildings can buffer diurnal swings, reduce peak loads, and align more effectively with renewable-based energy systems.

Third, design detailing and durability are critical. Ground-coupled solutions require complementary vertical insulation and robust moisture protection, and their long-term reliability should be verified through hygrothermal analyses that evaluate heat and moisture transfer and the risk of condensation. The findings reported here apply specifically to single-story detached houses in a temperate climate; applying similar strategies to other building types or climates will require further empirical studies and calibrated numerical modelling.

Finally, regulatory frameworks and certification schemes should recognize the buffering effects of thermal mass and ground coupling. Performance-based indicators should be introduced to reward proven capacity to stabilize indoor conditions and to maintain safe environments during extreme events.

REFERENCES

Allan, K., & Phillips, A. R. (2021). Comparative cradle-to-grave life cycle assessment of low and mid-rise mass timber buildings with equivalent structural steel alternatives. *Sustainability, 13*(6), 3401. https://doi.org/10.3390/su13063401

Andersen, J. H., Rasmussen, N. L., & Ryberg, M. W. (2021). Comparative life cycle assessment of cross laminated timber building and concrete building with special focus on biogenic carbon. *Building and Environment, 204*, 108130. https://doi.org/10.1016/j.buildenv.2021.108130

Andersen, J. H., Rasmussen, N. L., & Ryberg, M. W. (2022). Comparative life cycle assessment of cross laminated timber building and concrete building with special focus on biogenic carbon. *Energy and Buildings*, *254*, 111604. https://doi.org/10.1016/j.enbuild.2021.111604

Arul Babu, A. D. C., Srivastava, R. S., & Rai, A. C. (2024). Impact of climate change on the heating and cooling load components of an archetypical residential room in major Indian cities. *Building and Environment*, *250*, 111181. https://doi.org/10.1016/j.buildenv.2024.111181

Bruhn, S., Sacchi, R., Cimpan, C., & Birkved, M. (2023). Ten questions concerning prospective LCA for decision support for the built environment. *Building and Environment*, *242*, 110535. https://doi.org/10.1016/j.buildenv.2023.110535

Chastas, P., Theodosiou, T., Kontoleon, K. J., & Bikas, D. (2018). Normalising and assessing carbon emissions in the building sector: A review on the embodied CO_2 emissions of residential buildings. *Building and Environment*, *130*, 212–226. https://doi.org/10.1016/j.buildenv.2017.12.032

Chen, Z., Gu, H., Bergman, R. D., & Liang, S. (2020). Comparative life-cycle assessment of a high-rise mass timber building with an equivalent reinforced concrete alternative using the Athena Impact Estimator for Buildings. *Sustainability*, *12*(11), 4708. https://doi.org/10.3390/su12114708

Chetouni, A., Idrissi Kaitouni, S., & Jamil, A. (2025). Climate change impacts on future thermal energy demands and indoor comfort of a modular residential building across different climate zones. *Journal of Building Engineering*, *102*, 111927. https://doi.org/10.1016/j.jobe.2025.111927

Ciancio, V., Salata, F., Falasca, S., Curci, G., Golasi, I., & de Wilde, P. (2020). Energy demands of buildings in the framework of climate change: An investigation across Europe. *Sustainable Cities and Society*, *60*, 102213. https://doi.org/10.1016/j.scs.2020.102213

D'Agostino, D., Parker, D., Epifani, I., Crawley, D., & Lawrie, L. (2022). How will future climate impact the design and performance of nearly zero energy buildings (NZEBs)? *Energy*, *240*, 122479. https://doi.org/10.1016/j.energy.2021.122479

Dani, A. A., Roy, K., Masood, R., Fang, Z., & Lim, J. B. P. (2022). A comparative study on the life cycle assessment of New Zealand residential buildings. *Buildings*, *12*(1), 50. https://doi.org/10.3390/buildings12010050

De Araujo, V., Aguiar, F., Jardim, P., Mascarenhas, F., Marini, L., Aquino, V., Santos, H., Panzera, T., Lahr, F., & Christoforo, A. (2023). Is Cross-Laminated Timber (CLT) a wood panel, a building, or a construction system? A systematic review on its functions, characteristics, performances, and applications. *Forests*, *14*(2), 264. https://doi.org/10.3390/f14020264

Dodoo, A., Nguyen, T., Dorn, M., Olsson, A., & Bader, T. K. (2021). Exploring the synergy between structural engineering design solutions and life cycle carbon footprint of cross-laminated timber in multi-storey buildings. *Wood Material Science & Engineering*. https://doi.org/10.1080/17480272.2021.1974937

Dolezal, F., Dornigg, I., Wurm, M., & Figl, H. (2021). Overview and main findings for the Austrian case study. *Sustainability*, *13*(13), 7584. https://doi.org/10.3390/su13147584

Duan, Z., Omrany, H., & Zuo, J. (2025). Impact of climate change on energy performance and energy conservation measures effectiveness in Australian office buildings. *Energy*, *319*, 134956. https://doi.org/10.1016/j.energy.2025.134956

Dzhurko, D., Haacke, B., & Haberbosch, A. et al (2024). Future buildings as carbon sinks: Comparative analysis of timber-based building typologies regarding their carbon emissions and storage. *Frontiers in Built Environment*, *10*, 1330105. https://doi.org/10.3389/fbuil.2024.1330105

Fischer, H., Aichholzer, M., & Korjenic, A. (2023). Ecological potential of building components in multi-storey residential construction: A comparative case study between an existing concrete and a timber building in Austria. *Sustainability*, *15*(8), 6349. https://doi.org/10.3390/su15086349

Guardigli, L., Monari, F., & Bragadin, M. A. (2011). Assessing environmental impact of green buildings through LCA methods: A comparison between reinforced concrete and wood structures in the European context. *Procedia Engineering, 21*, 1199–1206. https://doi.org/10.1016/j.proeng.2011.11.2131

Haji Ismail, F., Shahrestani, M., Vahdati, M., Boyd, P., & Donyavi, S. (2021). Climate change and the energy performance of buildings in the future—a case study for prefabricated buildings in the UK. *Journal of Building Engineering, 39*, 102285. https://doi.org/10.1016/j.jobe.2021.102285

Hart, J., D'Amico, B., & Pomponi, F. (2021). Whole-life embodied carbon in multistory buildings: Steel, concrete and timber structures. *Journal of Industrial Ecology, 25*(2), 403–418. https://doi.org/10.1111/jiec.13139

Hemmati, M., Messadi, T., Gu, H., Seddelmeyer, J., & Hemmati, M. (2024). Comparison of embodied carbon footprint of a mass timber building structure with a steel equivalent. *Buildings, 14*(5), 1276. https://doi.org/10.3390/buildings14051276

Jalali, Z., Shamseldin, A. Y., & Ghaffarianhoseini, A. (2023). Impact assessment of climate change on energy performance and thermal load of residential buildings in New Zealand. *Building and Environment, 243*, 110627. https://doi.org/10.1016/j.buildenv.2023.110627

Kang, Y., Shin, B., & Kim, S. (2024). Enhancing greenhouse gas emission reduction via innovative envelope strategies: A focus on cross-laminated timber buildings for climate adaptation. *Energy and Buildings, 317*, 114380. https://doi.org/10.1016/j.enbuild.2024.114380

Liu, Y., Guo, H., Sun, C., & Chang, W.-S. (2016). Assessing cross laminated timber (CLT) as an alternative material for mid-rise residential buildings in cold regions in China—a life-cycle assessment approach. *Sustainability, 8*(10), 1047. https://doi.org/10.3390/su8101047

Milić, V., & Rohdin, P. (2024). Exploring the effects of a warmer climate on power and energy demand in multi-family buildings in a Nordic climate. *Environmental Advances, 15*, 100502. https://doi.org/10.1016/j.envadv.2024.100502

Peñaloza, D., Erlandsson, M., & Falk, A. (2016). Exploring the climate impact effects of increased use of bio-based materials in buildings. *Construction and Building Materials, 125*, 219–226. https://doi.org/10.1016/j.conbuildmat.2016.08.041

Pierobon, F., Huang, M., Simonen, K., & Ganguly, I. (2019). Environmental benefits of using hybrid CLT structure in midrise non-residential construction: An LCA based comparative case study in the U.S. Pacific Northwest. *Journal of Building Engineering, 26*, 100862. https://doi.org/10.1016/j.jobe.2019.100862

Piggot-Navarrete, J., Blanchet, P., Cabral, M. R., & Cogulet, A. (2025). Impact of climate change on the energy demand of buildings utilizing wooden prefabricated envelopes in cold weather. *Energy and Buildings, 338*, 115714. https://doi.org/10.1016/j.enbuild.2025.115714

Riahi, K., R. Schaeffer, J. Arango, K. Calvin, C. Guivarch, T. Hasegawa, K. Jiang, E. Kriegler, R. Matthews, G.P. Peters, A. Rao, S. Robertson, A.M. Sebbit, J. Steinberger, M. Tavoni, D.P. van Vuuren, 2022: Mitigation pathways compatible with long-term goals. In P. R. Shukla, J. Skea, R. Slade, A. Al Khourdajie, R. van Diemen, D. McCollum, M. Pathak, S. Some, P. Vyas, R. Fradera, M. Belkacemi, A. Hasija, G. Lisboa, S. Luz, & J. Malley (Eds.), IPCC, 2022: Climate Change 2022: Mitigation of Climate Change. Contribution of Working Group III to the Sixth Assessment Report of the Intergovernmental Panel on Climate Change. Cambridge University Press. https://doi.org/10.1017/9781009157926.005

Robertson, A. B., Lam, F. C. F., & Cole, R. J. (2012). A comparative cradle-to-gate life cycle assessment of mid-rise office building construction alternatives: Laminated timber or reinforced concrete. *Buildings, 2*(3), 245–270. https://doi.org/10.3390/buildings2030245

Sah, T. P., Lacey, A. W., Hao, H., & Chen, W. (2024). Prefabricated concrete sandwich and other lightweight wall panels for sustainable building construction: State-of-the-art review. *Journal of Building Engineering, 89*, 109391. https://doi.org/10.1016/j.jobe.2024.109391

Shen, P., Ji, Y., Li, Y., Wang, M., Cui, X., & Tong, H. (2025). Combined impact of climate change and urban heat island on building energy use in three megacities in China. *Energy and Buildings*, *331*, 115386. https://doi.org/10.1016/j.enbuild.2025.115386

Sporchia, F., Bruno, M., Neri, E., Pulselli, F. M., Patrizi, N., & Bastianoni, S. (2025). Complementing emergy evaluation and life cycle assessment for enlightening the environmental benefits of using engineered timber in the building sector. *Science of the Total Environment*, *970*, 179030. https://doi.org/10.1016/j.scitotenv.2025.179030

Tettey, U. Y. A., Dodoo, A., & Gustavsson, L. (2019). Effect of different frame materials on the primary energy use of a multi-storey residential building in a life cycle perspective. *Energy and Buildings*, *185*, 259–271. https://doi.org/10.1016/j.enbuild.2018.12.017

Tomrukcu, G., & Ashrafian, T. (2024). Climate-resilient building energy efficiency retrofit: Evaluating climate change impacts on residential buildings. *Energy and Buildings*, *316*, 114315. https://doi.org/10.1016/j.enbuild.2024.114315

Yan, C., & Ogata, S. (2025). Changes in residential heating and cooling energy consumption in nine major cities of Japan's warm climate regions under future climate conditions. *Energy and Buildings*, *344*, 115990. https://doi.org/10.1016/j.enbuild.2025.115990

Younis, A., & Dodoo, A. (2022). Whole life cycle assessment of cross-laminated timber buildings: Insights, challenges, and recommendations. *Journal of Building Engineering*, *52*, 104482. https://doi.org/10.1016/j.jobe.2022.104482

17 Innovative Building Solutions for Future Climate Challenges

17.1 MULTILAYER PASSIVE COOLING STRATEGIES IN CONTEMPORARY ARCHITECTURE: GLOBAL CASE STUDIES

17.1.1 Introduction

Across the nine case studies examined in this chapter, the conceptual terrain introduced in Section 11.5—particularly the taxonomy of adaptive façades outlined by Tahmasbi et al. (2025)—reappears in multiple configurations and at a range of spatial scales. Although each project operates within distinct climatic, cultural and technological contexts, their environmental logics converge around a shared set of principles: modulation of solar radiation, filtration of daylight, harnessing of wind-driven and buoyancy-driven airflow, and the stabilizing influence of thermal mass. Whether delivered through high-tech kinetic elements, culturally derived porous screens, or vernacular mass-airflow interactions, these strategies position the façade not as a static barrier but as an ecological mediator.

The diversity of responses becomes clear when viewed through the typological lens provided in Figures 17.1–17.3. One Airport Square in Accra deploys a hybrid structural–environmental system in which an exoskeleton provides deep fixed shading, operable louvers tune daylight and airflow, and a ventilated envelope supports passive cooling under tropical conditions. The Gherkin extends this logic vertically: its aerodynamic form, spiraling ventilation channels, and atrial chimneys create a coupled wind–thermal system directly echoing the ventilated double-skin façade configurations discussed in Section 11.5. Commerzbank Tower builds on this lineage through an advanced DSF, daylight-driven atria, and Sky Gardens that function as seasonal thermal buffers, demonstrating how natural ventilation and daylighting can be integrated into a tall-building typology.

Doha Tower provides a culturally specific reinterpretation of the same façade principles. Its mashrabiya-inspired double skin delivers selective solar filtering, diffused daylight, and continuous cavity ventilation, translating a centuries-old shading device into a high-performance passive envelope tailored to Gulf conditions. Masdar City pushes façade intelligence to the scale of an urban district: photovoltaic canopies, porous mashrabiya cladding, narrow shaded streets, and a central wind tower collectively operate as a distributed environmental interface. Here, urban morphology becomes an extension of façade logic, forming a multi-scalar passive cooling system embedded in the street network, public realm, and building skins.

DOI: 10.1201/9781003645573-17

FIGURE 17.1 The Gherkin in London, showcasing its aerodynamic form and integration within the urban skyline (Foster and Partners, 2025).

The southern European cases introduce further variations. Torre Sevilla deploys orientation-tuned terracotta louvers as a continuous ventilated screen, providing fixed shading aligned with solar exposure and reinforcing convective air movement along the façade. Torre Iberdrola uses an actively ventilated double-skin system in which cavity air is pre-conditioned before entering the occupied zone, illustrating a hybrid DSF model adapted to temperate oceanic conditions and operating alongside landscape-based microclimate improvements.

The Artist Hotel in Bilbao demonstrates yet another synthesis. Its patterned stone façade acts as a climatic filter, transforming the envelope into a layered device combining shading, daylight diffusion and thermal buffering. Inside, a monumental stone column integrated with evaporative cooling and stack ventilation serves as a passive climatic engine. On the rooftop, stone-based mass moderates evening temperatures. The entire building functions as a multi-layer passive system in which mass, shading and ventilation reinforce one another—an approach closely aligned with vernacular courtyard and cave-dwelling logics described in Chapter 12.

FIGURE 17.2 Cutaway view of The Gherkin, revealing the vertical atriums and passive ventilation system (Foster and Partners, 2025).

In contrast, Eastgate Centre in Harare demonstrates that high performance does not require advanced materials or automated systems. Its textured, perforated façade, deep recesses, and biomimetic ventilation shafts—modeled on termite mound thermoregulation—use mass, porosity, and geometry to maintain stable indoor conditions with minimal mechanical assistance. This biomimetic strategy, which relies on the same aerodynamic and thermal principles observed in The Gherkin's airflow logic, reveals how ecological analogies can meaningfully expand the vocabulary of passive design.

When considered collectively, these case studies illustrate that high-performance façades operate simultaneously as technological interfaces and ecological mediators.

FIGURE 17.3 Technical cross-section of The Gherkin, illustrating the integration of its diagrid structure and passive climate control systems (Foster and Partners, 2025).

Some rely on automated shading, adaptive glazing, and BIPV integration; others on mass, excavation, porosity, and geometry derived from centuries of climatic adaptation. Yet, all share an underlying lineage with the vernacular strategies analysed in Chapter 12—courtyard cooling, wind capture, shaded streets, evaporative devices, and thermal mass—translated into contemporary materials and design methodologies.

The following sections examine each of the nine case studies in detail, demonstrating how they calibrate these shared principles to climate, orientation, program, and urban setting. Figures 17.1–17.3 provide a typological guide for interpreting the façade systems discussed here and for understanding their position within the broader continuum of passive cooling strategies presented across this book.

17.2 CASE STUDY: THE GHERKIN—BIOMIMETIC HIGH-RISE ARCHITECTURE AND NATURAL VENTILATION

17.2.1 Introduction

The Gherkin, officially known as 30 St Mary Axe, is one of the most recognizable skyscrapers in London's financial district and a landmark in sustainable high-rise architecture. Designed by Norman Foster and Foster + Partners and completed in 2003, the 180-meter-tall, 41-story tower integrates natural ventilation, aerodynamic form, and passive cooling strategies into its design, making it a benchmark for energy-efficient skyscrapers.

Unlike traditional high-rises, which rely heavily on mechanical HVAC systems, The Gherkin employs a biomimetic approach, drawing inspiration from natural airflow mechanisms to reduce energy consumption and carbon footprint. Its tapered cylindrical shape, diagrid structural system, and double-skin façade enhance thermal efficiency, structural stability, and daylight penetration, distinguishing it as a model for future high-rise developments.

Figure 17.1 presents an aerial view of The Gherkin, highlighting its curved, aerodynamic profile and its seamless integration into London's dense urban skyline. The building's shape and exoskeletal diagrid design play a crucial role in reducing wind loads and optimizing passive environmental strategies.

17.2.2 Atriums and Natural Ventilation

A defining feature of The Gherkin's environmental strategy is its natural ventilation system, which drastically reduces the need for mechanical air conditioning. Unlike conventional high-rises, which depend on sealed façades, The Gherkin incorporates six evenly distributed atriums that function as vertical air shafts, promoting cross-ventilation and passive cooling.

As shown in Figure 17.2, these atriums are strategically positioned to allow fresh air to move naturally through the building, creating a stack effect where warm air rises and is expelled, while cooler air is drawn in. This system reduces heat buildup, enhances air quality, and decreases energy consumption for cooling.

Figure 17.3 provides a technical cross-section of The Gherkin, illustrating how the atriums and passive ventilation shafts integrate into the tower's overall structural and environmental design.

The Building Management System (BMS) controls automated ventilation louvers embedded within the façade, which adjust dynamically in response to internal temperatures, CO_2 levels, and outdoor climate conditions. This enables the building to continuously optimize airflow and indoor comfort, reducing reliance on traditional air conditioning systems.

17.2.3 Double-Skin Façade and Thermal Regulation

The double-skin façade of The Gherkin is integral to its passive climate control strategy. Comprising two layers of high-performance glazing, separated by a buffer zone, this façade serves as an insulating barrier that helps regulate indoor temperatures while maximizing natural light penetration.

During winter, the outer glazing layer absorbs solar radiation, reducing heat loss and stabilizing internal conditions. In summer, the air trapped between the double-skin layers facilitates passive cooling, as warm air is channeled upward and expelled through the top of the building.

The façade also features adjustable sunshades, allowing for real-time optimization of solar exposure, which minimizes glare and overheating. The combination of high-performance glass and automated shading significantly reduces energy demand for heating and cooling.

Figure 17.4 captures an interior perspective of the double-skin façade, demonstrating how the structural exoskeleton and ventilation mechanisms interact to maintain a comfortable indoor environment while ensuring transparency and daylight access.

FIGURE 17.4 Interior perspective of The Gherkin's structural diagrid and double-skin façade (Foster and Partners, 2025).

17.2.4 Structural System and Open-Plan Office Layout

The Gherkin's diagrid exoskeleton is one of its most innovative engineering features, allowing for an open-plan office layout with minimal internal columns.

Unlike conventional high-rises, which rely on dense steel frames and vertical supports, The Gherkin's structural system distributes loads efficiently through a triangulated exoskeletal framework, reducing material usage while increasing spatial flexibility.

As illustrated in Figure 17.5, this diagrid system eliminates the need for internal support columns, enabling unobstructed office layouts with improved natural daylight penetration. This approach reduces material consumption by 50% compared to conventional steel-framed buildings, further enhancing the sustainability of the structure.

Figure 17.5 showcases a view from within The Gherkin's upper floors, highlighting how the diagrid structure integrates with panoramic city views, reinforcing the tower's connection to its urban surroundings.

17.2.5 The Glass Dome and Passive Ventilation

The Gherkin's glass dome, located at its peak, is a key component of the building's passive ventilation system. Rather than being completely enclosed, the dome functions as a natural exhaust system, allowing hot air to escape through a stack effect, improving air circulation across the entire tower.

Figure 17.6 presents the glass dome at the top of The Gherkin, illustrating its radial glazing pattern and integrated ventilation system. Unlike traditional skyscrapers,

FIGURE 17.5 View from inside The Gherkin's upper floors, highlighting open-plan office spaces and panoramic city views (Foster and Partners, 2025).

FIGURE 17.6 The glass dome at the top of The Gherkin, functioning as an essential component of the building's passive ventilation system (Foster and Partners, 2025).

which seal their upper floors, The Gherkin's ventilation design ensures that warm air is expelled efficiently, maintaining optimal thermal comfort within the tower.

17.2.6 Energy Performance and Sustainability Challenges

The synergy between atriums, automated ventilation, and the double-skin façade contributes to a 50% reduction in energy consumption compared to conventional high-rise office buildings. By drawing inspiration from natural airflow mechanisms, The Gherkin optimizes ventilation, daylight access, and passive thermal regulation, proving that high-rise buildings can be both energy-efficient and architecturally innovative.

However, despite its high-performance design, the natural ventilation system faced operational challenges. A malfunction in the automated ventilation system in 2005, combined with subsequent modifications to lease agreements, restricted the full realization of the passive cooling strategy. These setbacks highlight the complexities of implementing natural ventilation in densely occupied high-rise environments, particularly when building operations evolve over time.

17.2.7 Conclusion and Future Perspectives

The Gherkin remains one of the most influential examples of biomimetic architecture, demonstrating how natural airflow strategies can enhance energy efficiency and structural integrity in modern skyscrapers.

Through its aerodynamic form, integrated natural ventilation system, and high-performance double-skin façade, the tower has set new standards for sustainable high-rise design.

17.3 CASE STUDY: COMMERZBANK TOWER—SUSTAINABLE HIGH-RISE ARCHITECTURE

17.3.1 Introduction

In the face of increasing energy efficiency requirements and the need for sustainable development, high-rise buildings present one of the greatest challenges for ecological architecture. The Commerzbank Tower in Frankfurt, designed by Foster + Partners, is one of the first skyscrapers to fully implement passive strategies aimed at reducing energy consumption and enhancing occupant comfort.

Standing 259 meters tall with 56 floors, completed in 1997, this building integrates natural ventilation, daylight optimization, winter gardens (Sky Gardens), and a double-skin façade, which collectively contribute to reducing energy consumption by up to 60% compared to conventional high-rise buildings (Lotfabadi, 2015).

Figure 17.7 presents an exterior view of Commerzbank Tower against the Frankfurt skyline, highlighting its distinctive triangular form and glass façade, both of which play a crucial role in the building's energy efficiency.

17.3.2 Passive Energy Strategies in Commerzbank Tower

The building employs four primary passive strategies that work together to minimize energy consumption and enhance occupant comfort: direct solar gain and daylight optimization, natural ventilation and air circulation, Sky Gardens as thermal buffers and ecological microclimates, and a double-skin façade as a climate control system. These strategies were meticulously designed to reduce CO_2 emissions and alleviate the burden on HVAC systems while ensuring a high standard of thermal and visual comfort for occupants.

17.3.3 Direct Solar Gain and Daylight Optimization

A fundamental aspect of the Commerzbank Tower's passive design is the maximization of daylight access through strategic orientation and geometric design. The triangular floor plan and the southwest orientation at 161° ensure even daylight distribution throughout the office spaces. The extensive use of glass in the façade and

FIGURE 17.7 Commerzbank Tower in Frankfurt, exterior view showcasing its distinctive triangular form and extensive glass façade (Foster and Partners, 2025).

atrium directs natural light deep into the building, thereby minimizing reliance on artificial lighting.

The Sky Gardens serve as an additional light source, facilitating more efficient light distribution across office spaces while simultaneously contributing to internal thermal regulation. The external shading systems and adjustable blinds further assist in mitigating overheating in the summer while allowing for passive solar heating in the winter (Lotfabadi, 2015).

Figure 17.8 illustrates the building's sectional model, depicting the arrangement of the atriums and the integration of Sky Gardens to optimize daylight penetration. Figure 17.9 provides a technical sectional drawing of the building, emphasizing the spatial organization of office areas and the positioning of winter gardens to balance internal climatic conditions.

FIGURE 17.8 Cross-section of Commerzbank Tower illustrating the integration of atriums and Sky Gardens in its passive design strategy (Foster and Partners, 2025).

17.3.4 Natural Ventilation and Air Circulation

The Commerzbank Tower incorporates an innovative natural ventilation strategy, which reduces energy consumption for cooling and ventilation by up to 60% annually (Lotfabadi, 2015). The central atrium serves as a ventilation chimney, facilitating the natural movement of air throughout the structure. The incorporation of operable windows on multiple floors enables controlled airflow regulation, thereby minimizing the dependency on mechanical ventilation.

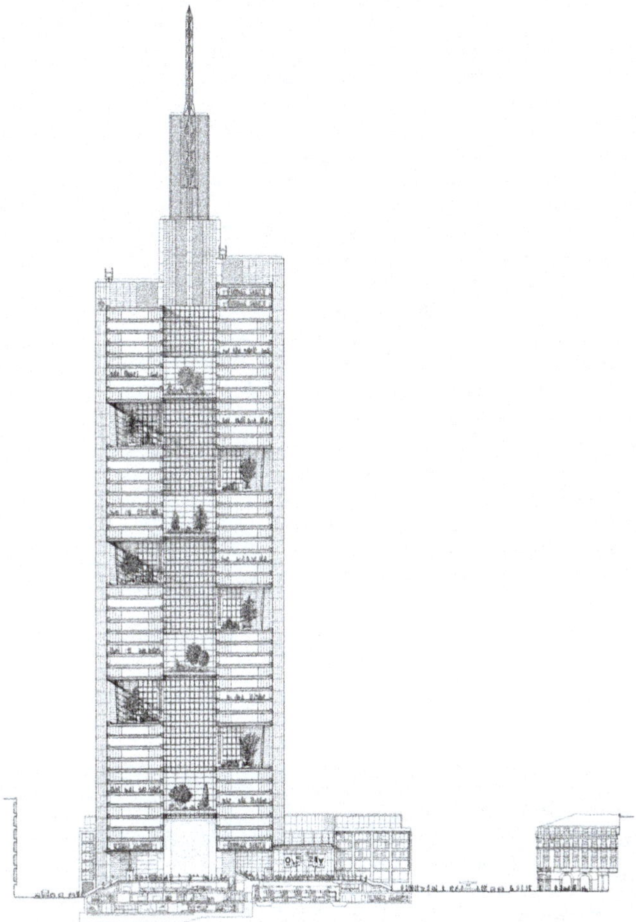

FIGURE 17.9 Technical cross-section of Commerzbank Tower emphasizing passive solar design elements, including natural ventilation and daylight distribution (Foster and Partners, 2025).

The difference in air pressure between the atrium and the surrounding spaces drives air circulation, with warm air rising and being expelled through the upper vents, while cooler air is drawn in from lower levels. This solar chimney effect, although partially implemented in Commerzbank Tower, could be further developed to enhance summer cooling and reduce the demand for mechanical air conditioning (Lotfabadi, 2015).

Figure 17.10 demonstrates the building's summer ventilation scheme, illustrating how the natural movement of air helps to regulate indoor temperatures through passive cooling mechanisms.

Figure 17.11 presents the winter ventilation strategy, showcasing how warm air retention and controlled air circulation contribute to maintaining optimal thermal conditions during colder months.

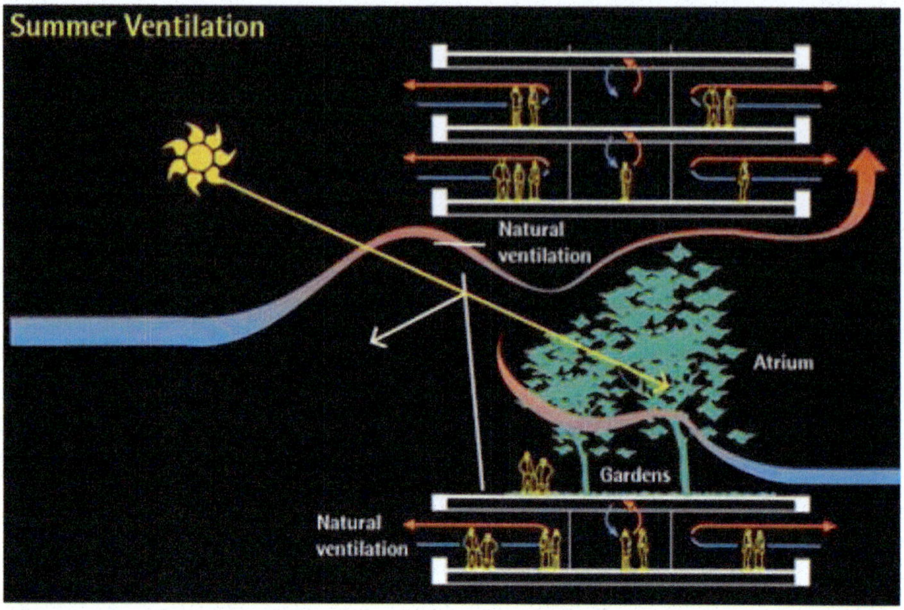

FIGURE 17.10 Summer ventilation strategy in Commerzbank Tower, illustrating natural airflow and cooling mechanisms. (Lotfabadi, 2015).

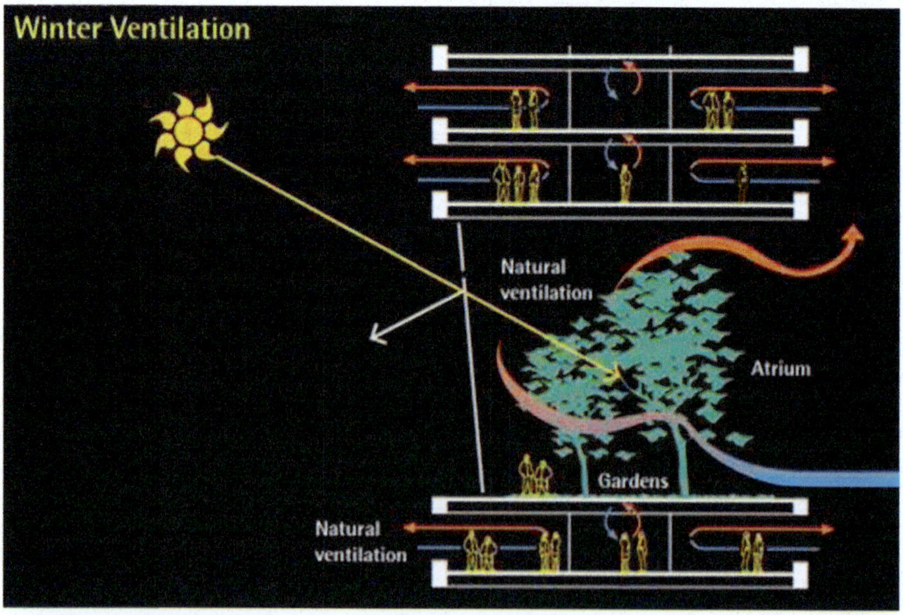

FIGURE 17.11 Winter ventilation strategy in Commerzbank Tower, depicting airflow patterns that optimize thermal efficiency. (Lotfabadi, 2015).

17.3.5 Sky Gardens as Thermal Buffers and Ecological Microclimates

The Sky Gardens, strategically positioned at varying levels throughout the building, function as natural thermal regulators and microclimates, enhancing both indoor air quality and thermal stability. These green spaces play a vital role in improving occupant well-being and reducing the building's environmental footprint.

The gardens contribute to thermal regulation by absorbing and redistributing heat as needed, ensuring a stable indoor temperature. They also improve air quality by absorbing CO_2 and increasing humidity levels, which creates a healthier indoor environment. Additionally, the presence of vegetation within the atrium supports natural air purification processes, further enhancing occupant comfort.

The Sky Gardens also optimize ventilation efficiency, as their strategic placement encourages cross-ventilation, thereby reducing the need for mechanical HVAC systems (Lotfabadi, 2015).

Figure 17.12 showcases photograph of the building's interior, highlighting the Sky Gardens and their contribution to environmental performance and occupant well-being.

17.3.6 Double-Skin Façade for Passive Climate Control

The Commerzbank Tower features an advanced three-layered glass façade, which enhances thermal insulation and facilitates natural ventilation. The inner glazing

FIGURE 17.12 Interior view of the Commerzbank Tower's Sky Gardens, illustrating how integrated greenery within the atrium supports passive climate regulation, natural shading, and improved indoor environmental quality (Foster and Partners, 2025).

layer includes operable windows, allowing occupants to adjust airflow manually. The outer glazing layer acts as a protective barrier, reducing thermal fluctuations, while the ventilated cavity between the two layers enables controlled air circulation, further improving energy efficiency.

This system effectively reduces heat loss in the winter while mitigating excessive solar gain in the summer, ensuring thermal comfort year-round. Compared to conventional high-rise buildings, the Commerzbank Tower achieves a 25–30% reduction in energy consumption related to heating and cooling (Lotfabadi, 2015).

17.3.7 Thermal Storage Mass and Energy Efficiency

While Commerzbank Tower does not extensively utilize thermal mass storage, the integration of denser materials such as stone or ceramics in the Sky Gardens' flooring could improve heat retention and release. The absence of significant heat-storing elements means that heat gains from solar radiation are not fully optimized.

Incorporating thermal mass into the flooring or atrium surfaces would allow heat to be absorbed during the day and released at night, contributing to more consistent indoor temperatures. Although this approach was not widely adopted in Commerzbank Tower, it remains an opportunity for further enhancing the passive performance of high-rise buildings (Lotfabadi, 2015).

17.3.8 Comparative Energy Efficiency Assessment

Lotfabadi (2015) highlights that conventional skyscrapers primarily rely on HVAC systems, leading to high energy consumption levels. In contrast, the passive strategies in Commerzbank Tower reduce the need for mechanical heating and cooling by up to 60%, translating into significant energy savings.

Post-occupancy studies indicate that the actual energy consumption of the building is 20% lower than originally projected, further confirming the effectiveness of its passive design strategies. This data underscores how thoughtfully implemented passive techniques can drastically reduce operational energy demands, positioning Commerzbank Tower as a model for future high-rise architecture.

17.3.9 Conclusion

The Commerzbank Tower exemplifies how passive design strategies can be effectively integrated into high-rise buildings to optimize energy efficiency and occupant comfort. By leveraging natural ventilation, daylighting optimization, and thermal buffering through Sky Gardens, the building achieves substantial energy reductions while maintaining high indoor environmental quality.

The findings presented by Lotfabadi (2015) demonstrate that passive solar design is not only a viable approach for high-rise buildings but also a critical factor in ensuring their long-term sustainability. The Commerzbank Tower stands as a benchmark for future skyscrapers, proving that energy efficiency and modern architecture can coexist harmoniously.

17.4 CASE STUDY: DOHA TOWER—INTEGRATION OF TRADITIONAL AND MODERN ARCHITECTURAL STRATEGIES FOR CLIMATE ADAPTATION

17.4.1 Introduction

Doha Tower, also known as Burj Doha, was completed in 2012, quickly establishing itself as an iconic architectural landmark in Qatar. Located in the West Bay business district of Doha, the tower was designed by French architect Jean Nouvel and financed by H. E. Sheikh Saud Bin Mohammed Bin Ali Al Thani's investment firm. The primary objective of the project was to develop a modern, energy-efficient office building that simultaneously honors the cultural heritage of the region. The construction, initiated in 2005 and spanning seven years, culminated in a structure that seamlessly integrates advanced engineering solutions with traditional design principles.

Figure 17.13 presents an exterior view of Doha Tower, offering a contrasting example of passive design strategies in high-rise architecture.

While Commerzbank Tower employs natural ventilation and Sky Gardens to enhance energy efficiency, Doha Tower utilizes an advanced double-skin façade combined with mashrabiya-inspired solar shading to mitigate intense solar radiation and minimize heat gain in Qatar's extreme climate. The juxtaposition of these two

FIGURE 17.13 General view of Doha Tower's double-skin facade, integrating passive solar shading and cultural symbolism into its design (Ateliers Jean Nouvel. Photo © CSCEC or All Rights Reserved).

buildings highlights divergent yet complementary approaches to sustainable skyscraper design, tailored to distinct environmental and cultural contexts.

The investors and designers of Doha Tower aimed to develop a visually distinctive building while incorporating technological innovations suited to the extreme climatic conditions of the Persian Gulf region. As the demand for energy-efficient buildings continues to rise in areas characterized by high temperatures and intense solar exposure, this project stands as an exemplar of architectural ingenuity. By integrating traditional Islamic architectural elements, such as mashrabiya, with cutting-edge engineering solutions, Doha Tower effectively merges aesthetics with functionality, demonstrating the potential for sustainable high-rise architecture in hot and arid climates.

17.4.2 Structural and Functional Characteristics

Doha Tower stands 238 meters tall and has 46 floors, with a cylindrical footprint of 45 meters in diameter. The building consists mainly of office spaces, while the top floor houses a penthouse with a private entrance. A key feature of the interior is its open-plan layout, free of internal columns. This was achieved through an innovative structural approach that places the reinforced concrete exoskeleton behind the facade, allowing for greater flexibility in interior arrangements. The building's core, which houses elevators and mechanical systems, is located in the center of the structure, further optimizing floor space.

17.4.3 Double-Skin Facade and Solar Shading

A defining feature of Doha Tower is its double-skin façade, which seamlessly integrates energy efficiency with a strong cultural identity. The inner layer, composed of a high-performance glass curtain wall, provides thermal insulation while allowing ample natural light penetration. This system is complemented by an external aluminum screen, inspired by traditional Islamic mashrabiya patterns. Beyond its aesthetic role, this intricate lattice structure serves as a critical element of passive cooling, mitigating the challenges posed by Qatar's intense solar radiation.

As illustrated in Figure 17.14, the aluminum lattice fully wraps the cylindrical tower, functioning as a dynamic shading system. The density of perforations varies across the façade in response to solar exposure, ensuring targeted protection throughout the day. The eastern and western facades, which experience the most intense sunlight, feature 60% shading density, effectively minimizing heat gain. The southern façade, exposed to moderate solar radiation, has a 40% coverage, while the northern side, which receives the least direct sunlight, incorporates 25% opacity to balance daylight access and thermal comfort.

The exterior shading screen, a signature element of Doha Tower, consists of four interlocking butterfly-shaped components, each adorned with intricate geometric perforations of varying scales. This layered shading system is instrumental in regulating daylight penetration and heat transfer, ensuring that the interior remains naturally illuminated without excessive overheating. The gradient perforation strategy

FIGURE 17.14 Close-up of the mashrabiya façade (Ateliers Jean Nouvel. Photo © CSCEC or All Rights Reserved).

in the mashrabiya screen further enhances passive cooling, dynamically adjusting opacity levels based on orientation to optimize daylight use while preventing overheating.

From a distance, the façade appears uniform and seamless, reinforcing the tower's iconic cylindrical profile. Upon closer inspection, the complex layering of shading elements becomes more apparent, creating a sophisticated interplay of light and shadow throughout the structure. By reducing solar heat gain and lowering cooling demands, this adaptive shading system exemplifies an effective passive strategy, meticulously tailored to the climatic conditions of Doha.

17.4.4 Climate Adaptation and Energy Efficiency

The double facade system provides multiple energy-efficient benefits. It reduces glare and regulates daylight penetration, minimizing the need for artificial lighting.

FIGURE 17.15 Interior view of the maintenance and ventilation corridor, demonstrating the chimney effect cooling strategy within the tower's double façade (Ateliers Jean Nouvel. Photo © CSCEC or All Rights Reserved).

It creates a thermal buffer zone, reducing heat transfer between the external climate and the building's interior. It also integrates reflective glass, which further optimizes thermal insulation and daylight diffusion.

One of Doha Tower's key passive cooling mechanisms is the natural ventilation system embedded in its double facade, which significantly reduces heat accumulation and improves overall energy efficiency. As shown in Figure 17.15, a 2-meter-wide service corridor located between the inner glass curtain wall and the outer mashrabiya screen plays a crucial role in both climate control and building maintenance.

The ventilated cavity within the double facade supports a passive chimney effect, where hot air rises naturally and exits through dedicated openings at the top of the tower, drawing in cooler air from the shaded lower sections of the facade. This continuous airflow cycle reduces reliance on mechanical cooling, optimizing indoor comfort while lowering energy demands. By preventing excessive heat buildup within the building envelope, this system significantly enhances the overall resilience of the structure to extreme climatic conditions.

Beyond its role in passive cooling, the service corridor also functions as a protective buffer, reducing dust accumulation on the glass curtain wall, a common issue in Qatar's arid climate. As shown in Figure 17.15, this accessible space allows workers to maintain both the inner and outer layers of the facade without disturbing building operations. The mashrabiya panels, constructed from aluminum

FIGURE 17.16 The perforated dome structure at the tower's crown, acting as a light diffuser and shading element while reinforcing the building's cultural identity. ((Ateliers Jean Nouvel. Photo © CSCEC or All Rights Reserved))

and stainless steel, ensure long-term durability, resisting thermal expansion, corrosion, and sand abrasion, making Doha Tower a benchmark in resilient high-rise architecture.

As illustrated in Figure 17.16, the natural ventilation process is finalized at the perforated dome, which serves as an exhaust point for the hot air collected within the facade system. Cool air enters the lower sections of the tower's double skin, replacing the rising warm air and creating a continuous natural ventilation cycle that enhances thermal comfort across all floors. This effect minimizes reliance on mechanical cooling, optimizing indoor conditions while significantly lowering energy consumption.

17.4.5 Cultural Identity and Architectural Recognition

The perforated dome at the top of the tower, shown in Figure 17.19, extends the mashrabiya system to the highest point of the structure, reinforcing Doha Tower's connection to traditional Middle Eastern architecture. The dome regulates sunlight at the upper levels, ensuring a diffused light effect that prevents overheating while maintaining visual transparency. At the base, a 25-meter-wide pergola provides a shaded pedestrian entrance, integrating passive cooling at the ground level. Additionally, 40% of the surrounding site is landscaped, contributing to localized cooling and improved microclimate regulation. In 2012, Doha Tower was awarded the CTBUH Best Tall Building Worldwide Award, recognizing its fusion of sustainable strategies and cultural identity.

17.4.6 Conclusion and Future Prospects

Doha Tower exemplifies climate-responsive high-rise architecture, integrating adaptive solar shading through the variable-opacity mashrabiya screen, passive ventilation via the chimney effect within the double facade, and diffused daylight management through the perforated dome and shading elements. Its innovative double facade system serves as a model for future high-rise buildings in extreme climates, proving that passive strategies can successfully merge with cultural heritage and modern engineering.

As urbanization in hot and arid climates continues, Doha Tower serves as a precedent for future skyscrapers, demonstrating that sustainability and cultural expression can be seamlessly integrated into modern architectural design.

17.5 CASE STUDY: MASDAR CITY—SUSTAINABLE URBAN DEVELOPMENT IN EXTREME CLIMATES

17.5.1 Introduction

Masdar City, designed by Foster and Partners, is one of the most advanced urban sustainability projects in the world. It seamlessly integrates traditional vernacular architecture with cutting-edge technology to create a carbon-neutral city that operates with minimal energy consumption. The city serves as a model for climate-adaptive urban design, addressing the unique challenges of extreme desert environments through a combination of passive cooling techniques, high-performance materials, and renewable energy systems.

Inspired by historical desert cities, such as Shibam in Yemen, Masdar City employs compact urban layouts, shaded streets, and natural ventilation to mitigate excessive heat gain and create a comfortable pedestrian experience. The city's urban master plan was developed using detailed solar and wind analysis, ensuring that streets and public spaces are optimally shaded and naturally ventilated. The city's northwest-southeast orientation minimizes urban heat accumulation while facilitating natural airflow, a strategy that significantly reduces outdoor temperatures compared to conventional cities in the region.

Figure 17.17 presents Masdar Institute at dusk, showcasing its low-rise, energy-efficient design that incorporates passive cooling strategies and traditional Arabic architectural elements. The integration of a wind tower, visible above the complex, highlights the city's commitment to natural ventilation as a primary cooling method.

17.5.2 Facade Design and Mashrabiya-Inspired Shading Systems

The residential and commercial buildings in Masdar City incorporate mashrabiya-inspired shading elements, blending aesthetic and functional benefits. These geometric perforated panels, commonly used in Middle Eastern architecture, play a critical

FIGURE 17.17 The Masdar Institute at dusk, featuring its distinct low-rise, energy-efficient design integrating passive cooling strategies (Foster and Partners, 2025).

role in solar shading, allowing filtered daylight to enter while blocking direct sun exposure. This design reduces heat gain, improves natural ventilation, and enhances indoor thermal comfort.

The integration of shading devices and passive cooling strategies is evident in Figure 17.18, which depicts a shaded courtyard at Masdar Institute. The large mashrabiya panel on the right allows airflow while diffusing sunlight, preventing excessive heat accumulation. The ventilated façade on the left features small, uniformly spaced rectangular openings, which enhance cross-ventilation while maintaining privacy. Above, a semi-transparent solar canopy further reduces heat gain, ensuring a cool and comfortable outdoor environment.

Figure 17.19 offers an upward perspective of the ventilated façade and solar shading system, highlighting the advanced airflow regulation mechanisms. The metallic panels, arranged at a 45-degree angle, optimize natural ventilation while minimizing solar heat exposure. The small rectangular openings covered with a translucent membrane prevent dust infiltration, which is particularly crucial in the desert climate. Above, integrated photovoltaic panels provide solar shading while simultaneously generating renewable energy, further reinforcing the sustainability model of Masdar City.

The design approach of Masdar City extends beyond individual buildings to urban-level thermal management. Figure 17.20 illustrates an internal courtyard where shading devices and vegetation combine to create a cooler microclimate. The incorporation of narrow shaded streets, colonnades, and tree-lined walkways follows the traditional principles of Middle Eastern urban planning, which prioritize passive cooling in pedestrian environments. The combined effect of shading, vegetation, and

FIGURE 17.18 A shaded courtyard within Masdar Institute, demonstrating passive cooling techniques through mashrabiya facades, ventilated surfaces, and solar shading (Foster and Partners, 2025).

strategic building orientation reduces outdoor temperatures, making public spaces significantly more comfortable than those in conventional city layouts.

17.5.3 Integrated Passive Cooling Strategies

Masdar City relies on a holistic approach to passive cooling, incorporating shading, evaporative cooling, and natural ventilation into its urban fabric.

In Figure 17.21, the courtyard at Masdar Institute exemplifies this strategy, demonstrating how water features, vegetation, and shading devices work together to

FIGURE 17.19 An upward perspective of the ventilated façade and solar shading system at Masdar Institute, illustrating how architectural elements contribute to passive cooling and daylight diffusion (Foster and Partners, 2025).

FIGURE 17.20 A courtyard at Masdar Institute integrating passive cooling through vegetation, water features, and an advanced shading structure with geometric paneling (Foster and Partners, 2025).

FIGURE 17.21 A courtyard at Masdar Institute demonstrating multiple passive cooling strategies, including vegetation, water features, solar shading, and natural ventilation elements (Foster and Partners, 2025).

enhance thermal comfort. The circular water feature in the foreground facilitates evaporative cooling, reducing surrounding temperatures. Lush green trees and plants provide natural shading, while the perforated mashrabiya panels on the left allow filtered light and airflow into the interior spaces. On the right, a façade with angled ventilation louvers optimizes cross-ventilation, reducing the need for mechanical air conditioning. The solar panels atop the structures further contribute to energy efficiency by providing shade and generating electricity.

A defining feature of Masdar City's passive cooling strategy is its wind tower, depicted in Figure 17.22. This 45-meter-high structure, inspired by traditional Middle Eastern wind catchers, directs cooler air from higher altitudes into public spaces, while expelling hot air from the lower pedestrian zones. This mechanism lowers ambient temperatures, improving thermal comfort without relying on conventional HVAC systems.

17.5.4 Renewable Energy and Photovoltaic Systems

Masdar City's commitment to sustainability extends beyond passive cooling techniques; renewable energy plays a central role in powering the city. The extensive use of photovoltaic (PV) panels ensures that Masdar operates with a net-zero carbon footprint. These panels are integrated into building rooftops, shading canopies, and external structures, maximizing their energy-harvesting potential while reducing direct solar exposure on pedestrian pathways.

FIGURE 17.22 A modern wind tower at Masdar Institute, optimizing passive ventilation and air circulation within the urban environment (Foster and Partners, 2025).

Thermal analysis has demonstrated that due to the combination of the wind tower, shading systems, and reflective surfaces, temperatures at street level in Masdar City are approximately 33°C, compared to 57°C in central Abu Dhabi under similar conditions. This remarkable difference underscores the effectiveness of passive strategies in mitigating urban heat stress and improving microclimatic conditions.

17.5.5 High-Performance Materials and Insulation Strategies

The materials selected for Masdar City were carefully engineered to withstand extreme desert conditions while ensuring high energy efficiency. Many research and commercial buildings feature ETFE (ethylene tetrafluoroethylene) cushions, which provide thermal insulation while maintaining controlled daylight penetration.

All structures incorporate high-performance insulated building envelopes, reducing air leakage and preventing the loss of cooled air from indoor spaces. Unlike conventional buildings in Abu Dhabi, where significant cooling energy is wasted due to poor insulation, the high-sealing techniques used in Masdar City significantly improve efficiency. The window-to-wall ratio has also been optimized, reducing glazing areas by approximately 30%, concentrating windows on lower floors where shading is most effective.

17.5.6 Impact on Urban Sustainability and Future Prospects

Masdar City stands as a benchmark for sustainable urban development, proving that modern cities can operate with minimal environmental impact. By combining passive cooling techniques, high-performance materials, and renewable energy systems, Masdar achieves substantial reductions in energy consumption and carbon emissions while maintaining a high standard of living for residents and workers.

Figure 17.23 presents a pedestrian-friendly narrow street within Masdar Institute, illustrating how shading, ventilation, and urban design combine to create a

FIGURE 17.23 A pedestrian-friendly narrow street within Masdar Institute, showcasing passive cooling through shading, ventilation, and urban design strategies (Foster and Partners, 2025).

comfortable and walkable environment. The perforated mashrabiya panels, curved balconies, and solar shading elements minimize heat absorption, ensuring thermal comfort in the public realm.

Masdar City provides a compelling case study for how future cities can be designed to harmonize sustainability, tradition, and technological advancement—serving as a model for climate-adaptive urban development worldwide.

17.6 CASE STUDY: ONE AIRPORT SQUARE—SUSTAINABLE DESIGN IN A TROPICAL CLIMATE

17.6.1 Introduction

One Airport Square, located in Accra, Ghana, is a landmark mixed-use development that integrates passive cooling strategies, natural ventilation, and sustainable material use to create a climate-responsive architectural solution. Designed by Mario Cucinella Architects (MCA) and completed in 2016, the building exemplifies energy-efficient office and commercial spaces in tropical climates.

Drawing inspiration from vernacular West African architecture, One Airport Square utilizes solar shading, an optimized building envelope, and elevated terraces to reduce energy consumption and maintain indoor comfort in Accra's hot and humid climate. The project demonstrates how modern architectural principles and traditional passive strategies can work in harmony to create a high-performance, low-energy building suited for tropical environments.

Figure 17.24 presents an exterior view of One Airport Square, highlighting its overhanging terraces and perforated exoskeleton, which function to reduce heat gain and optimize airflow within the building.

17.6.2 Climate-Responsive Design and Passive Cooling Strategies

The tropical climate of Accra presents significant challenges due to high temperatures and humidity, necessitating effective cooling strategies. One Airport Square was designed to minimize reliance on mechanical ventilation and air conditioning through passive cooling techniques that naturally regulate temperature and airflow.

The building's stepped façade and cantilevered terraces create a layered shading effect, preventing direct sunlight from penetrating interior spaces while promoting natural ventilation. The external envelope incorporates shading panels, operable louvers, and overhangs, which reduce solar radiation while maintaining daylight penetration. The open-air circulation spaces further enhance passive cooling by facilitating cross-ventilation and minimizing heat buildup.

Figure 17.25 provides a closer view of the building's façade, showcasing the combination of structural shading and adjustable louvers. The reinforced concrete exoskeleton serves as a passive shading device, reducing the impact of direct solar radiation. Meanwhile, the operable louvers allow occupants to dynamically regulate airflow and daylight levels, adapting to changing external conditions. This combination of terraces and louvers ensures that both solar heat and wind patterns are efficiently utilized to maintain indoor thermal comfort.

FIGURE 17.24 Exterior view of One Airport Square, showcasing its overhanging terraces and exoskeleton structure designed for passive cooling (Mario Cucinella Architects S.U.R.L., 2025).

FIGURE 17.25 Ground-level perspective of One Airport Square, highlighting the exoskeleton structure, shaded walkways, and passive cooling features (Mario Cucinella Architects S.U.R.L., 2025).

17.6.3 Ventilation and Natural Air Circulation

One Airport Square employs a sophisticated natural ventilation system, ensuring that interior spaces remain cool with minimal energy input. Unlike conventional buildings that rely on mechanically controlled HVAC systems, the structure is designed to optimize airflow through its geometry and materiality.

Figure 17.26 presents a diagrammatic representation of the building's natural air circulation system, illustrating how prevailing winds enter through the façade, move through the atrium, and exit through roof vents. This stack effect facilitates passive cooling, continuously drawing fresh air into the interior while expelling warm air. The open-air atrium at the center of the building functions as a natural chimney, ensuring consistent airflow across multiple floors.

The elevated ground level, supported by a raised concrete podium, allows air to circulate underneath the building, further enhancing ventilation efficiency and preventing heat accumulation at the base of the structure. This approach is a modern adaptation of traditional tropical architecture, which prioritizes elevated buildings and shaded spaces for improved airflow and thermal comfort.

17.6.4 Integration of Light, Ventilation, and Water Management

One Airport Square's sustainability approach extends beyond passive cooling to include natural daylighting and water conservation strategies. The building integrates

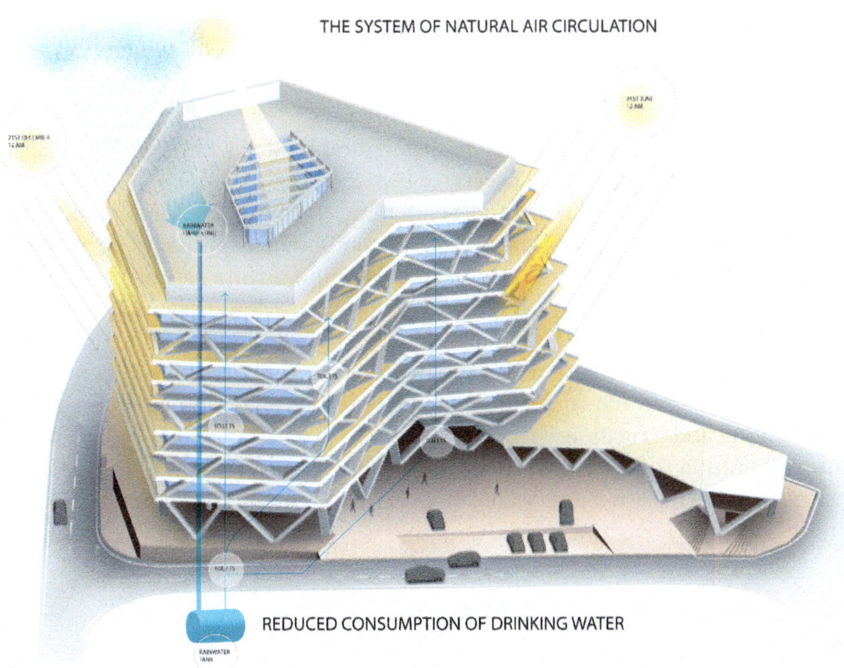

FIGURE 17.26 Diagram illustrating the natural air circulation and water conservation strategies at One Airport Square (Mario Cucinella Architects S.U.R.L., 2025).

FIGURE 17.27 Light and natural ventilation diagram for One Airport Square, illustrating airflow dynamics, stack effect, and daylighting strategies (Mario Cucinella Architects S.U.R.L., 2025).

light wells and stepped terraces to ensure that natural light reaches deep into interior spaces while preventing glare and excessive heat gain.

Figure 17.27 illustrates how natural ventilation, daylight penetration, and water conservation systems interact as an integrated environmental strategy. The diagram highlights the role of the central atrium in enhancing airflow, allowing warm air to rise and be expelled while fresh air enters through the lower levels. The stepped terraces diffuse natural daylight, reducing the need for artificial lighting while preventing overheating. The diagram also shows the rainwater harvesting system, where water is collected from the roof, filtered, and stored in underground reservoirs for reuse in cooling and irrigation.

The integration of multiple passive strategies ensures that One Airport Square operates with low energy consumption while maintaining optimal indoor conditions.

17.6.5 Structural System and Material Efficiency

The structural system of One Airport Square is designed for both durability and environmental performance. The reinforced concrete exoskeleton, inspired by traditional Ghanaian weaving techniques, provides both structural stability and solar

FIGURE 17.28 Architectural section of One Airport Square, illustrating its stepped terraces, ventilation shafts, and passive cooling framework.

protection. The triangular grid of the façade allows for natural ventilation, reducing the reliance on artificial cooling.

The building's materials were sourced locally, minimizing embodied energy and supporting regional economies. The high thermal mass of concrete helps regulate indoor temperatures by absorbing heat during the day and releasing it at night, thereby reducing peak cooling loads.

Figure 17.28 presents an architectural section of the building, illustrating how the stepped terraces, ventilation shafts, and exoskeleton framework contribute to the overall passive cooling strategy(Mario Cucinella Architects S.U.R.L., 2025).

17.6.6 Energy Performance and Operational Efficiency

One Airport Square significantly reduces operational energy demand through a combination of passive cooling, daylighting, and energy-efficient systems. The design achieves:

- 40% reduction in energy consumption compared to conventional office buildings in Accra.

- Minimized reliance on air conditioning, as the passive ventilation system keeps indoor temperatures within a comfortable range.
- Improved indoor air quality due to the natural air circulation strategy, reducing CO_2 buildup and humidity levels.

By prioritizing natural ventilation and daylighting, the building enhances occupant comfort while lowering overall energy costs, making it an economically and environmentally sustainable model for future urban developments.

17.6.7 Conclusion

One Airport Square is a milestone in sustainable architecture in West Africa, proving that high-rise commercial buildings can be designed to adapt to local climatic conditions using passive strategies.

- The project's success demonstrates that Passive cooling strategies can significantly reduce energy demand in tropical climates.
- Locally sourced materials contribute to lower embodied carbon and greater resilience.
- Water conservation strategies enhance resource efficiency and long-term sustainability.
- Adaptive structural and façade designs improve comfort while maintaining architectural identity.

17.7 CASE STUDY: TORRE SEVILLA, SEVILLE

Torre Sevilla, also known as Torre Pelli, is the tallest skyscraper in Seville and Andalusia, reaching 180.5 m in height. Designed by César Pelli and completed in 2015, it forms part of the Centro Comercial Torre Sevilla complex that includes offices, a hotel, retail areas, and landscaped public spaces. The project demonstrates how passive cooling strategies can be integrated into high-rise design in a hot Mediterranean climate.

17.7.1 Terracotta Louvers and Double-Skin Façade

The tower is wrapped by horizontal terracotta louvers, which form an external layer in front of the glazed façade. This creates a continuous double-skin system along the entire height of the building. The ventilated buffer reduces solar gains, stabilizes indoor thermal conditions, and filters daylight to improve visual comfort. As shown in Figures 17.29 and 17.30, the louver system is calibrated according to expected solar radiation. Greater density and depth are applied on façades exposed to intense solar gain, while less shading is provided where radiation is weaker. This selective design provides optimal thermal protection while maintaining daylight and views. The louvers also act as an architectural filter framing the panorama of Seville, giving the façade both climatic and aesthetic functions.

FIGURE 17.29 Torre Sevilla with terracotta louvers filtering sunlight.

17.7.2 Commercial Podium and Public Space

The podium accommodates retail functions and is connected to landscaped plazas. As shown in Figure 17.31, vegetation, seating areas, and water features create cooler microclimates by providing shade and enabling evaporation. This arrangement reduces the urban heat island effect and integrates the commercial part of the complex into its urban setting. The podium façades also employ terracotta louvers, ensuring consistency with the tower.

17.7.3 Green Roofs and Photovoltaics on the Podium

The podium roofs are covered with extensive gardens and photovoltaic panels. As shown in Figure 17.32, the green roofs act as thermal buffers, lowering surface temperatures, insulating interior spaces, and storing rainwater. The photovoltaic panels supply renewable energy to the complex. Together, they demonstrate how passive and active strategies can be combined to improve sustainability.

17.7.4 Membrane Canopy for Shading and Ventilation

The crown of the tower is protected by a tensile membrane canopy. As shown in Figure 17.33, this structure reduces direct solar radiation while allowing diffuse daylight. Its open geometry enables air circulation so that warm air rises and escapes, producing a chimney effect that reinforces natural ventilation. The canopy functions

FIGURE 17.30 Detail of façade with terracotta louvers shading glazing.

as both a shading device and a ventilating system, improving comfort in the observation terraces and reducing heat loads on the upper interiors.

17.7.5 Synthesis

Torre Sevilla demonstrates that a high-rise in a hot Mediterranean climate can operate as an integrated environmental system. The terracotta louvers form a continuous double-skin façade calibrated to solar orientation. The podium combines shaded plazas with vegetation and water. The podium roofs employ green surfaces and photovoltaic panels. The tower crown is protected by a tensile membrane canopy that shades and ventilates. Each of these strategies is tied to a specific part of the complex, yet together they form a coherent system of passive cooling and climate responsiveness. The design illustrates how contemporary engineering can reinterpret

FIGURE 17.31 Podium with shaded plazas and landscaped areas.

FIGURE 17.32 Podium roofs with green surfaces and photovoltaic panels.

FIGURE 17.33 Tensile membrane canopy shading and ventilating crown.

vernacular principles of shading, ventilation, and thermal buffering in the context of high-rise urban architecture.

17.8 CASE STUDY: TORRE IBERDROLA, BILBAO

17.8.1 General Context

Torre Iberdrola is the tallest skyscraper in the Basque Country, rising to 165 m. Designed by César Pelli and completed in 2011, it became a landmark of the Bilbao skyline and a central piece of the Abandoibarra regeneration project. The tower anchors the Elcano Diagonal axis and defines Plaza Euskadi together with two lower residential blocks. Its triangular prism volume with curved façades tapers toward the sky, evoking a transparent obelisk visible across the city (Figure 17.34).

The building was the first in Europe to achieve LEED Platinum Core & Shell and Interiors certification. Beyond architecture, it plays a role in the Biscay Startup Bay ecosystem, where public policies, corporations, and startups collaborate under an Open Innovation framework.

17.8.2 Entrance and Climatic Transition

The main entrance of the tower is framed by a large glazed canopy and a reflective metallic cladding at the base (Figure 17.35). The canopy shades visitors from direct sun and rain, while the curved reflective surface limits heat accumulation on the lower façades. This transition space mitigates external climatic impacts before entry into the lobby.

FIGURE 17.34 Torre Iberdrola in the Bilbao skyline.

FIGURE 17.35 Main entrance with canopy and reflective base.

17.8.3 Active Double-Skin Façade

The double-skin façade is the key passive feature of Torre Iberdrola. A ventilated cavity between two glass layers creates a thermal buffer that reduces solar gains in summer and heat loss in winter. Ventilation grilles at the base form part of the system (Figure 17.36). View of a curved glass façade with vertical profiles, clearly showing the structure of the double-layer façade. (Figure 17.37).

17.8.4 Lobby as Climatic and Social Buffer

The entrance hall is a monumental glazed lobby opening toward Plaza Euskadi. It functions as a climatic transition space, filtering daylight while reducing radiation through the double façade. Inside, vegetation such as olive trees and planted beds contributes to evaporative cooling, improved air quality, and a psychological sense of freshness. The lobby demonstrates how environmental performance and social representation can merge into a single climatic buffer.

17.8.5 Sustainable and Passive Strategies

Torre Iberdrola combines passive and active strategies into a coherent environmental system.

- The double-skin façade acts as a thermal buffer, reducing solar gain and heat loss.

FIGURE 17.36 Lower façade with ventilation grilles.

FIGURE 17.37 Detail of curved glass façade with vertical profiles, showing the visible double-skin system.

- The ventilation system preconditions air in the façade cavity, assisted by vertical shafts.
- The landscaped lobby provides cooling, shading, and air quality benefits.
- The park of Campa de los Ingleses surrounding the tower contributes to shading, evapotranspiration, and stormwater retention.
- The building relies on 100% renewable electricity and incorporates systems for gray water recovery and rainwater reuse.

Together, these measures deliver more than 20% energy savings compared to conventional towers, securing the LEED Platinum rating.

17.8.6 Integration with the Regional Innovation Ecosystem

The tower also serves as a symbolic and functional node in the Biscay Startup Bay entrepreneurial ecosystem, promoted by the Provincial Council of Bizkaia. This framework connects corporations such as Iberdrola with startups, research institutions, and accelerators under the logic of Open Innovation. Torre Iberdrola represents both sustainable architecture and the corporate face of a region investing in clean energy, innovation, and entrepreneurship.

17.8.7 Synthesis

Torre Iberdrola demonstrates how high-rise design can merge sustainable architecture, passive strategies, and policy frameworks. The double-skin façade forms an active climatic envelope. The entrance canopy and reflective base create a shaded transition. The lobby with vegetation operates as a climatic buffer. Surrounding parks improve urban microclimate. Combined with renewable energy supply and water recycling, these features made Torre Iberdrola the first building in Europe with LEED Platinum Core & Shell and Interiors certification. Beyond the technical measures, its integration into the Biscay Startup Bay ecosystem highlights the convergence of sustainable design and regional innovation strategies in Bilbao.

17.9 CASE STUDY: THE ARTIST HOTEL, BILBAO

The Artist Hotel, located directly opposite the Guggenheim Museum, is a striking example of how contemporary architecture can integrate mass, shading, and water into a coherent passive system. Designed as a luxury urban hotel, it combines sculptural expression with environmental logic, making its form an active climatic device rather than a neutral background.

17.9.1 Façade Facing the Guggenheim

The main façade of the hotel is oriented toward the Guggenheim Museum. Built of pale stone with patterned reliefs and perforations, it filters light and provides self-shading (Figure 17.38). The articulation of the surface creates alternating areas of sun and shadow, limiting overheating while ensuring diffuse illumination inside. The stone cladding has significant thermal inertia, absorbing heat slowly and releasing it gradually, which stabilizes indoor temperatures. This design recalls the logic of traditional mashrabiya screens, where geometry and material together create a filter for radiation, airflow, and light.

17.9.2 Monumental Stone Column in the Atrium

Inside the atrium stands a monumental cylindrical stone column, approximately 2–2.5 meters in diameter, rising continuously from the ground floor to the roof (Figure 17.39). At its base is a shallow pool that provides water for sprays, keeping

FIGURE 17.38 Façade of the Artist Hotel facing the Guggenheim.

the column surface moist. This irrigation triggers evaporative cooling, lowering the air temperature and increasing humidity. The mass of the column absorbs heat during the day, acting as a giant buffer. Together with the vertical geometry of the atrium, it creates a stack effect: warm air rises and escapes upwards, while cooler, denser air flows into the space at lower levels. The result is a hybrid device combining mass storage, evaporation, and natural ventilation.

17.9.3 Stone Surfaces on the Rooftop Terrace

On the rooftop terrace, which accommodates a restaurant and bar, stones of varied sizes cover the floor and define seating areas (Figure 17.40). They serve aesthetic and functional roles, acting as anchors for furniture and extending the interior's design language into the exterior. Yet their environmental role is also clear. The stones accumulate heat during the day, reducing surface temperature peaks on the terrace, and release it slowly at night, moderating the cooling rate of the space.

For diners, this buffering of temperature fluctuations creates a more comfortable microclimate, turning the terrace into a thermally resilient outdoor room. The same stone material is also used in cages that support tables, as shown in Figure 17.41

FIGURE 17.39 Monumental stone column in the atrium of the Artist Hotel.

reinforcing visual continuity across the building. While primarily aesthetic, the considerable mass of these elements contributes to buffering effects, both physically and in the psychological perception of solidity and coolness.

17.9.4 Synthesis

The Artist Hotel demonstrates how contemporary architecture can merge environmental performance with expressive form. The façade facing the Guggenheim functions as a climatic filter, moderating light and heat. The monumental stone column in the atrium combines mass, evaporation, and ventilation to regulate the building's central microclimate. The rooftop stones extend this logic outdoors, stabilizing comfort conditions for guests in an exposed space. Together, these three layers of design illustrate that intelligent façades, monumental interior devices, and landscape materials can be orchestrated into a single passive cooling system, transforming the entire hotel into a climate-responsive organism.

FIGURE 17.40 Rooftop terrace tables anchored in stone-filled cages.

17.10 CASE STUDY: EASTGATE CENTRE—BIOMIMETIC VENTILATION AND THERMAL REGULATION IN SUBTROPICAL CLIMATES

17.10.1 Introduction

The Eastgate Centre, located in Harare, Zimbabwe, is a pioneering example of biomimetic architecture, demonstrating how buildings can efficiently regulate their indoor climate without reliance on mechanical air conditioning. Designed by architect Mick Pearce in collaboration with Arup Engineers, the mixed-use complex—comprising office spaces and retail facilities—was completed in 1996 and remains a benchmark for energy-efficient design in hot climates.

Inspired by the self-cooling capabilities of termite mounds, Eastgate Centre employs passive ventilation and thermal mass storage to maintain comfortable indoor temperatures with significantly lower energy consumption compared to conventional buildings. By using natural air circulation and heat absorption rather than HVAC systems, the building reduces operational energy costs by 35% and cooling energy demands by up to 90% (Pearce, 2025).

Figure 17.42 presents an exterior view of Eastgate Centre, illustrating its distinctive ventilated façade, exposed concrete structure, and shading elements, all of which contribute to the building's passive climate control system.

FIGURE 17.41 Rooftop terrace perimeter with stone surface and planter boxes.

17.10.2 Biomimetic Inspiration and Climate Adaptation

Eastgate Centre is located in a subtropical highland climate, where daytime temperatures often exceed 30°C, but nights remain relatively cool. Conventional buildings in this climate struggle with heat buildup, requiring energy-intensive cooling systems.

Taking inspiration from African termite mounds, which regulate their internal climate using a network of ventilation channels, Eastgate Centre mimics this natural thermoregulation strategy. The building's design incorporates passive ventilation shafts and chimneys that create a stack effect to draw hot air out while pulling cool air in. The thick concrete walls with high thermal mass store heat during the day and release it at night, regulating temperature fluctuations. The strategically placed shading elements minimize direct solar radiation and prevent overheating.

Figure 17.43 illustrates a schematic cross-section of the ventilation system, showing how warm air rises and exits through rooftop exhaust vents while cooler air is drawn in at lower levels to maintain continuous air movement.

FIGURE 17.42 Exterior view of Eastgate Centre, illustrating its passive cooling design and biomimetic ventilation system (Pearce, 2025).

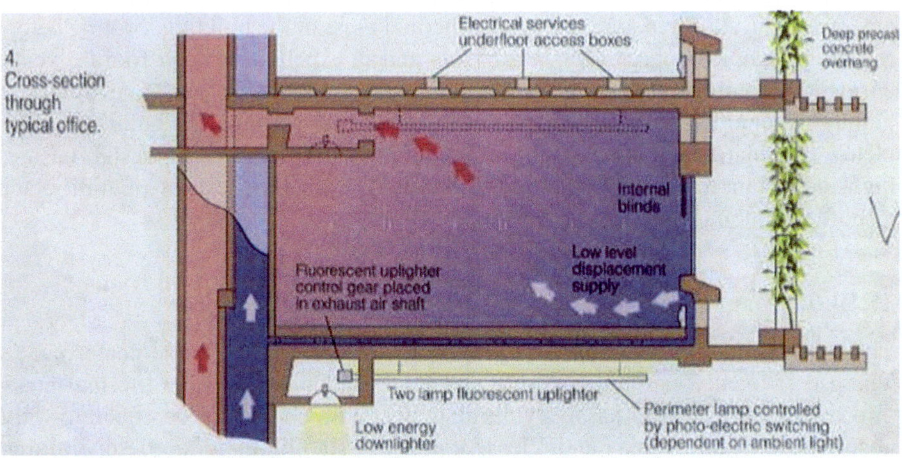

Section through typical office Eastgate, Harare

FIGURE 17.43 Cross-section of Eastgate Centre's ventilation system, illustrating its biomimetic airflow mechanism (Pearce, 2025).

17.10.3 The Role of External Façade in Passive Cooling

The façade of Eastgate Centre is not just an aesthetic feature but a critical component of its passive cooling system. According to Mick Pearce (2025), the design of a building's exterior significantly influences its thermal performance. The façade follows principles of heat absorption and radiation that are evident in nature. Smooth surfaces absorb more heat during the day and are less efficient at releasing heat at night, making them unsuitable for passive cooling in hot climates. In contrast, rough or irregular surfaces absorb less heat by day but are better at radiating heat at night, helping to maintain a stable indoor environment.

Eastgate Centre incorporates these principles by using a highly textured façade, featuring deep recesses and projections. This design reduces heat gain during the day while improving heat dissipation at night. The rough surface of the building helps break up solar radiation, reducing its intensity before it reaches the interior. The building's façade, inspired by natural adaptations seen in desert-dwelling organisms, ensures thermal efficiency without the need for mechanical cooling systems.

Figure 17.44 captures this principle in action, illustrating how the façade's geometry and material properties contribute to the building's self-regulating thermal system.

17.10.4 Natural Ventilation and Passive Cooling Strategies

The ventilation strategy of Eastgate Centre functions much like a living organism, constantly regulating its internal climate. The façade is perforated with ventilation openings, allowing for continuous air exchange between the interior and exterior environments.

The system operates by drawing in cool air at ground level, where temperatures are lower. The air circulates through the thermal mass of the building, cooling before reaching occupied spaces. Warm air rises and is expelled through rooftop vents, ensuring a continuous flow of fresh air. This process enables Eastgate Centre to maintain indoor temperatures 3–5°C cooler than conventional office buildings, completely eliminating the need for air conditioning. Additionally, the design ensures high indoor air quality, preventing the buildup of stale air, excess humidity, and airborne pollutants.

17.10.5 Structural and Material Considerations

The building's structural materials play a crucial role in its thermal performance. The use of locally sourced concrete and brickwork provides high thermal mass, allowing the building to passively regulate heat gain and loss. The concrete walls absorb heat during the day and release it at night, stabilizing indoor temperatures. High ceilings and open-plan spaces optimize air circulation, preventing heat buildup. The modular layout ensures consistent airflow throughout different zones of the building.

FIGURE 17.44 External façade of Eastgate Centre, illustrating its role in passive cooling through geometric surface variation.

By using local materials, Eastgate Centre minimized construction costs and environmental impact, making it an affordable yet highly sustainable solution. Figure 17.45 captures an interior view of Eastgate Centre, showcasing its structural columns, ventilated facades, and shaded circulation spaces, all contributing to passive cooling and climate-responsive design.

17.10.6 Green Façade

In addition to its textured concrete façade, Eastgate Centre incorporates green façade elements that further enhance its passive cooling performance. Vegetation is integrated directly into the building's balconies (Figure 17.46), creating vertical layers

FIGURE 17.45 Interior view of Eastgate Centre, showcasing its structural elements and passive cooling strategies (Pearce, 2025).

of shading and contributing to evaporative cooling. The climbing plants reduce solar exposure on the concrete surfaces, improve microclimatic comfort around the façade, and reinforce the building's biomimetic approach to climate regulation.

17.10.7 Measured Energy Efficiency and Cost Savings

According to Mick Pearce's (2025) presentation, Eastgate Centre has proven its long-term sustainability with quantifiable energy savings:

- Passive cooling has reduced energy demand for air conditioning by 90%.
- The building uses 35% less total energy compared to similar commercial buildings.

FIGURE 17.46 Vegetated façade of Eastgate Centre, illustrating the role of greenery in passive cooling (Pearce, 2025).

- Rent is 20% lower than comparable office spaces due to lower operational costs.
- The internal temperature remains between 23 and 25°C year-round, despite extreme external variations.
- Thermal imaging studies confirm that indoor temperature fluctuations are minimal, validating the effectiveness of passive cooling strategies.

These statistics confirm that biomimetic design is not just a theoretical concept but a functional, measurable solution to reducing energy consumption in extreme climates.

17.10.8 Lessons for Future Sustainable Architecture

Eastgate Centre remains a leading case study for biomimetic and passive design, proving that high-rise buildings can be both sustainable and cost-effective without relying on mechanical HVAC systems. Several broader and transferable design principles emerge clearly from the performance of this project, including: (i) biomimicry as a practical climate solution, drawing on natural systems to inform efficient and adaptive architecture; (ii) the replacement of mechanical HVAC systems by passive cooling strategies in hot climates, resulting in lower energy use and operational costs; (iii) the use of locally sourced materials to enhance sustainability and reduce construction costs and embodied carbon; and (iv) the careful balancing of thermal mass and ventilation to ensure stable indoor temperatures.

17.11 SUMMARY: INNOVATIVE BUILDING SOLUTIONS FOR FUTURE CLIMATE CHALLENGES

As climate change intensifies and cities continue to expand, architecture faces the urgent task of producing buildings that are not only energy-efficient but also resilient under increasingly extreme thermal conditions. The case studies presented in this chapter—spanning diverse climates, urban settings, and building typologies—demonstrate how contemporary design can integrate advanced engineering with time-tested strategies rooted in vernacular tradition. They show that the most effective climate-responsive architecture emerges not from isolated technological solutions, but from the intelligent combination of passive processes, material performance, and contextual adaptation.

Across all buildings examined, several consistent patterns emerge. High-performance façades act not merely as environmental separators, but as active regulators of heat, light, and airflow. Adaptive shading systems, multi-layer glazing, mashrabiya-inspired screens, and photovoltaic surfaces modulate solar gain, improve daylight quality, and support natural ventilation. When calibrated to orientation, sun paths, and prevailing wind directions, these envelopes deliver measurable reductions in energy use while enhancing thermal comfort.

Equally significant is the persistence of passive ventilation and shading strategies that originated centuries ago: stack-effect towers, ventilated courtyards, deep loggias, windcatchers, shaded canyons, and heavy thermal envelopes. These remain among the most reliable means of controlling heat loads and promoting indoor air movement—particularly when mechanical cooling is unavailable or undesirable. At the urban scale, projects such as Masdar City confirm that compact street networks, shaded pedestrian corridors, vegetation, and controlled orientation can reduce heat-island intensity and produce comfortable microclimates even in extreme desert environments.

Where performance was strongest, multiple systems—façade control, ventilation pathways, thermal inertia, evaporative elements, vegetated buffers, and district-level shading—were conceived as a single integrated environmental engine. In such cases, architectural form, structural logic, mechanical assistance, and material selection worked together rather than functioning as independent layers.

17.11.1 Linking Contemporary Strategies with Vernacular Principles

The case studies presented here confirm a strong conceptual continuity between the vernacular architectures examined in Chapters 10–12 and contemporary high-performance buildings. Although their material expression and technological complexity differ, the underlying climatic logic remains strikingly consistent. Several parallels illustrate this continuity:

- Wind-driven ventilation:
  Wind towers in Yazd find their contemporary analogues in Masdar City's 45-metre wind catcher and in the pressure-driven atrial chimneys of The Gherkin, both of which operate as calibrated devices for inducing buoyancy- and wind-assisted airflow.
- Courtyard-based microclimates:
  The cooling patios of Marrakesh, the shaded courtyards of Seville, and the sky-wells of Jiangnan reappear in the atrium systems of Commerzbank Tower and the internal cooling courts of Masdar Institute, where vertical voids moderate temperature and support natural ventilation.
- Shaded urban canyons:
  The self-shading alleys of Mardin or Zhongshan Old Town re-emerge in Masdar's pedestrian network, where narrow, shaded streets and colonnades reduce radiant load and channel breezes through the urban fabric.
- Thermal mass strategies:
  The heavy masonry of Andalusian or Canarian dwellings is conceptually mirrored in the high-mass cores and interior stone devices of The Artist Hotel and Eastgate Centre, where thermal inertia stabilises indoor temperatures across diurnal peaks.

These correspondences demonstrate that modern climate-responsive architecture often extends—not replaces—vernacular environmental intelligence.

17.11.2 Multilayer Integration

Across the case studies, effective thermal resilience was consistently achieved through multilayer environmental systems, in which multiple passive strategies reinforced each other. The most robust configurations combined:

- Outer shading layers:
  Louvers, mashrabiya screens, overhangs and recesses that reduce incident solar radiation and lower façade temperatures.
- Ventilated cavities:
  Double-skin façades, porous screens and ventilated service corridors that buffer heat and support stack- or wind-driven convection.
- Interior airflow structures:
  Atria, sky-wells, chimneys, and multi-storey ventilation shafts that promote vertical air movement and regulate pressure differentials.
- Thermal-mass components:
  Structural cores, stone surfaces, and high-mass elements that dampen diurnal temperature fluctuations and assist night-flush cooling.
- Urban microclimate modifiers:
  Shaded streets, vegetation, water features, and wind corridors that temper outdoor conditions and support building-scale ventilation.

Performance gains were strongest when these components operated as a single, coupled environmental engine rather than as isolated devices. The synergy among layers—airflow, mass, shading, and microclimate—proved more influential than any individual measure.

17.11.3 General Principles for Climate-Adaptive Design

When integrated across scales, the observations from these case studies converge into a set of broader principles:

- Contextual responsiveness.
  The most successful strategies were finely calibrated to local sun paths, wind regimes, humidity levels, material availability, and cultural patterns of use. Performance gains were maximised when architectural form was aligned with environmental drivers rather than imposed independently of them.
- Synergistic integration.
  Shading, ventilation, daylighting, thermal buffering, and evaporative processes performed best when designed as interdependent systems. Benefits accumulated through their combined operation, not through isolated devices.
- Scale coupling.
  Building-level performance is reinforced when supported by neighbourhood-scale measures—such as shaded street networks, vegetation corridors, and wind-aligned massing—that extend microclimatic resilience beyond individual structures.
- Dialogue between tradition and innovation.
  Vernacular strategies offer robust, field-tested principles, while contemporary engineering provides precision, optimisation, and advanced materials. The most effective designs synthesize both, producing environmentally responsive architectures that are culturally resonant and technologically adaptable.

These findings underscore that climate-adaptive design is not a matter of applying a fixed checklist of passive measures. Instead, it requires selecting, combining, and refining strategies according to context, while ensuring that systems can evolve over time as climatic, technological, and operational conditions change.

17.11.4 Toward a Broader Understanding of Passive Resilience

The case studies reviewed in this chapter demonstrate that resilience is not a static attribute but a dynamic capacity—the ability of buildings to maintain comfort and functionality across a wide range of conditions, including extremes that may intensify in the coming decades. This capacity is strengthened when designs incorporate redundancy, diversity of passive systems, and responsiveness to environmental change.

These insights align closely with the broader themes developed throughout this book—from long-term literature synthesis, field monitoring, and thermal-comfort modeling, to the life-cycle analyses of contrasting building types. Together, they reinforce a central message: passive resilience is no longer a desirable supplement but a foundational requirement of sustainable, future-ready architecture.

By bridging heritage, ecology, and technological innovation—and by responding dynamically to a rapidly changing climate—contemporary architecture can develop solutions that are both culturally grounded and scientifically advanced.

REFERENCES

Lotfabadi, P. (2015). Analyzing passive solar strategies in the case of high-rise buildings. *Renewable and Sustainable Energy Reviews*, *52*, 1340–1353. https://doi.org/10.1016/j.rser.2015.07.189.

Pearce, M. (2025). Personal communication (March 4, 2025).

Tahmasbi, F., Khdair, A. I., Aburumman, G. A., Tahmasebi, M., Thi, N. H., & Afrand, M. (2025). Energy-efficient building façades: A comprehensive review of innovative technologies and sustainable strategies. *Journal of Building Engineering*, *99*, 111643. https://doi.org/10.1016/j.jobe.2024.111643

Mario Cucinella Architects S.U.R.L. (2025). Personal communication (March 4, 2025).

Foster and Partners (2025). Personal communication (March 12, 2025).

18 Synthesis and Implications

18.1 CONTEXT AND PURPOSE OF THE SUMMARY

The analyses presented in this book have demonstrated that passive strategies are not merely supplementary measures but a foundation of building design in the twenty-first century. The review of literature, empirical research, prospective life cycle analyses, and case studies from different regions of the world confirm that their importance is growing in the face of intensifying extreme events. The effectiveness of these strategies, however, depends on climatic, social, and cultural contexts, and their full value emerges only when they are part of an integrated design approach. This conclusion aims to summarize the most important findings of the book, indicate their implications for practice and regulation, and highlight areas that require further research.

18.2 SYNTHESIS OF KEY FINDINGS

The collected results confirm that passive strategies can provide thermal comfort in everyday use as well as resilience during crisis conditions. This conclusion rests on several observations:

- Scientific evidence consistently confirms the effectiveness of measures such as shading, natural ventilation, thermal mass, evaporative cooling, and high-albedo surfaces in mitigating overheating and reducing dependence on mechanical cooling.
- Full-scale field studies have provided unique data showing that appropriately configured thermal mass and building envelopes improve both daily comfort and passive survivability during weather extremes.
- Life cycle analyses have demonstrated that short-term energy balances are insufficient; a full evaluation must include future climate changes and the transformation of energy systems.
- The comparison of vernacular and contemporary architecture has shown that the durability of passive solutions stems primarily from their adaptation to local environmental and cultural conditions.

Together, these findings confirm that buildings must be conceived as dynamic systems functioning across different time scales, from daily and seasonal rhythms to multi-decadal perspectives.

DOI: 10.1201/9781003645573-18

18.3 IMPLICATIONS FOR PRACTICE AND REGULATION

The analyses presented in this book indicate that the effectiveness of passive solutions depends on their full integration into the design process. In practice, this means that such strategies should inform basic decisions on orientation, form, and building structure, rather than being treated as secondary additions. It is also crucial to balance seasonal performance so that improved winter efficiency does not lead to vulnerability to overheating in the summer.

Traditional architecture developed a wide range of methods for passive microclimatic regulation. It employed massive envelopes to stabilize thermal conditions, diverse shading and light-filtering systems to protect against excessive solar exposure, and spatial configurations that facilitated natural ventilation and air exchange. Water and vegetation were also essential elements, providing evaporative cooling and reducing ambient temperatures. These groups of strategies, reinterpreted and enhanced by modern technologies, can serve as the foundation of climate-resilient, low-emission architecture that is harmoniously integrated into its environmental and cultural context.

Regulatory frameworks should adapt to this approach. In cold and temperate climates, in addition to traditional requirements concerning winter performance, it is necessary to include criteria relating to building performance during heat waves and energy outages. Contemporary standards are increasingly based on performance-oriented methods that assess the effectiveness of the entire building, rather than imposing prescriptive limits on individual components. Requirements should include such measures as the duration of safe indoor temperatures, the number of hours exceeding critical thresholds, thermal autonomy, and the ability to provide passive survivability. Their inclusion in assessment systems will ensure that design responds not only to typical seasonal loads but also to the growing variability of climate.

18.4 RESEARCH PRIORITIES

The collected analyses point to several areas where further research is needed. First and foremost, long-term monitoring is necessary across different climatic zones, encompassing a variety of building typologies and socio-economic contexts. It is also important to gain a deeper understanding of the interactions between passive and active solutions, including predictive control and user behavior, in order to develop optimized hybrid configurations.

Another priority is the development of dynamic modelling tools that more accurately represent the interactions between weather conditions, occupants, and building elements. At the same time, research is required into synergies at the urban scale, including the influence of greenery, ventilation corridors, and reflective surfaces on the urban microclimate.

Social and cultural aspects also cannot be overlooked, since they determine the acceptance and effectiveness of strategies inspired by tradition. These require in-depth studies that take into account local identity and user expectations.

Research perspectives should also expand through the development of prospective life-cycle analyses that consider both climate change and the transformation of energy systems. A key task remains the creation of standardized and empirically validated indicators of thermal resilience, applicable in design practice and regulatory frameworks.

18.5 CONCLUDING REMARKS

First, the results confirm that passive strategies provide an effective means of reducing the risk of overheating and improving thermal comfort. In cold and temperate climates, appropriate configurations make it possible in many cases to maintain acceptable indoor conditions without mechanical systems. In warmer regions, passive solutions serve a complementary function, reducing the operating time of air conditioning and lowering the environmental and energy burdens associated with active cooling.

Second, the life-cycle perspective demonstrates that the evaluation of building projects cannot be based solely on short-term energy indicators. Prospective analyses show that solutions that may entail minor losses in heating performance today can, over the long term, deliver significant environmental benefits by limiting the growth of cooling demand in a warming climate with decarbonizing energy systems.

Third, full-scale studies combined with life-cycle assessments showed that configurations based on increased thermal mass and controlled ground coupling significantly enhance the thermal resilience of buildings. Such solutions extend the period during which safe indoor conditions are maintained during winter outages and summer heat waves, while also leading to more favorable emission balances across the life cycle. Transferring this approach to other building typologies and soil conditions requires further empirical verification.

Fourth, the findings highlight the necessity of reformulating regulatory frameworks. Regulations focused solely on reducing heat losses in winter no longer adequately address the rising risk of overheating. Modern requirements should assess the ability of buildings to maintain thermal comfort and thermal resilience throughout the entire year, including during extreme events. Flexibility of context-specific solutions and their integration with urban-scale measures must be recognized.

Fifth, the analyses reveal the continuity between vernacular strategies and contemporary design practice. Historical methods of microclimatic regulation, encompassing spatial layouts, shading and ventilation systems, as well as the use of water and vegetation, confirm that adaptation to environmental conditions has always been a foundation of architecture. Contemporary reinterpretations of these strategies, supported by new materials and technologies, make it possible to create resilient, low-emission buildings that are consistent with the local cultural context.

19 Toward a Climate-Responsive Design Paradigm

19.1 INTRODUCTION

Chapter 18 synthesized the findings of this book and outlined their implications for policy, regulation, and future research. This final chapter takes a different perspective. It focuses on design practice by comparing vernacular and contemporary case studies, integrating the lessons of Chapters 12 and 17 with the empirical and environmental analyses of Chapters 13 to 16. Its aim is to provide a comparative design framework that architects, planners, and engineers can apply across diverse climatic regions.

19.1.1 Hot Arid Climates

In hyper-arid Adrar, Ksar Tamentit stabilized indoor conditions by combining compact urban grain, massive adobe walls, and scheduled nighttime ventilation. Courtyards and semi-covered passages extended these conditions into the shared fabric, creating cooler microclimatic corridors even under outdoor peaks above 45°C. A similar grammar is evident in Masdar City, where narrow street canyons, calibrated shading systems, and an urban wind tower extend climatic moderation beyond the building scale. Yazd illustrates the coupling of windcatchers, shaded courtyards, and evaporative devices, producing a settlement-wide engine of airflow and cooling. Contemporary architecture translates this logic into advanced façade systems, as in Doha Tower, where the double skin combines a variable-density mashrabiya screen with a ventilated buffer cavity that also supports stack effect. In each case, solar control is provided by layered screens rather than single planes, and ventilation is primarily driven by pressure and buoyancy. Where water is scarce, shading and mass form the primary condition, while evaporation is deployed as an auxiliary mechanism.

19.1.2 Mediterranean Climates

Mediterranean examples link thermal inertia with the regulating capacity of courtyard microclimates. Andalusian cave dwellings maintain stable indoor temperatures of 18 to 22°C throughout the year by relying on geological mass, with ventilation supported by chimneys and small façade openings. In Marrakesh, arcades, riads, and shaded passages transform the dwelling into a self-ventilating and light-filtering

DOI: 10.1201/9781003645573-19

system, integrating earthen mass with layered filters and voids. Contemporary projects apply these same principles with different materials and scales. Torre Sevilla deploys a continuous terracotta louver screen, calibrated by orientation, as a double climate filter, while its podium landscapes combine shading, vegetation, and water to cool external spaces. The Artist Hotel integrates mass, shading, and evaporation at building scale: stone façades filter radiation, a monumental stone column acts simultaneously as heat store and convective shaft, and rooftop stone fields buffer exterior comfort. Across these cases, mass is consistently paired with filters that modulate radiation and with vertical voids that sustain airflow. The terracotta arrays, patterned screens, and ventilated cavities of contemporary architecture perform the same role as the arcades and cloisters that stabilized courtyards in Córdoba and Seville.

19.1.2.1 Humid and Tropical Climates

In Jiangnan and Zhongshan, thermal resilience depends on coupled wind–thermal networks. Sky-wells and narrow alleys regulate pressure and temperature, while waterways reduce ambient temperature and guide breezes through dense settlement fabrics. Soil-adjacent walls and floors provide thermal sinks that stabilize lightweight timber structures. The First Granary of Xuan'en illustrates the integration of these devices: a double-skin system of masonry and timber, buffer aisles, raised floors, and a tall central volume supporting stack ventilation. Summer indoor temperatures remained below 25°C with limited daily variation, while winter interiors were warmer and drier than outdoors, all without active systems. In Accra, One Airport Square applies similar logics in a different tropical setting, combining terraces, operable louvers, shaded ground levels, and a central stack to sustain airflow and limit solar gain. The continuity between vernacular and contemporary examples is clear. Where mass is limited or counterproductive, the buffer is created by cavities, operable filters, and soil contact. Ventilation is sequenced in time and layered in section, while water functions either as a settlement-scale boundary condition or as façade-level filtration combined with stack-assisted air paths.

19.1.3 Comparative Lessons

The cross-climatic analysis reveals several consistent patterns. Vernacular settlements stabilized thermal conditions by coupling envelope mass with radiative filters, airflow devices, and landscape elements. Contemporary projects achieve comparable outcomes with calibrated louvers, double skins, and vegetated podiums, demonstrating that while the material palette has evolved, the logic of integration persists. Historical buildings relied on overlapping layers of defense—thick walls, courtyards, shaded lanes—that provided redundancy against climatic fluctuations. Modern examples reproduce this effect with ventilated cavities, buffer zones, and planted canopies. Operability remains fundamental. Shutters, vents, and nightly ventilation schedules structured performance in Ksar Tamentit and Xuan'en; today, adjustable louvers, blinds, and automated elements fulfil the same role, ensuring adaptability to diurnal and seasonal changes. Continuity is also cultural and material. Adobe, stone, and timber gave vernacular dwellings both resilience and identity.

Contemporary terracotta screens, patterned façades, and ventilated claddings extend this dual role of performance and cultural expression.

19.1.3.1 Resilience under Extremes

The empirical studies in Chapters 13 to 16 confirmed that the benefits of passive strategies extend beyond average conditions and become decisive during extreme events. Thermal mass, shading, and ventilation systems delayed overheating during severe heat waves, extending the period of habitable indoor conditions when outdoor temperatures exceeded 40°C. In cold episodes, ground coupling and envelope inertia maintained indoor stability without auxiliary systems, prolonging thermal autonomy during outages. These outcomes mirror the long-term endurance of vernacular forms such as adobe houses in Adrar or timber and soil-coupled dwellings in Jiangnan, which survived centuries of climatic variability. Modern projects adopting similar systemic principles demonstrated equivalent resilience in monitored and modelled performance, underscoring the continuity between historic adaptation and contemporary design.

19.1.3.2 Sustainability and Policy

The comparison also points to broader environmental and institutional considerations. Vernacular architecture relied on locally available stone, earth, or timber, minimizing embodied energy and aligning construction with ecological context. Contemporary projects that emulate the performance of adobe walls or courtyard systems through double skins and terracotta screens must be evaluated not only for thermal benefits but also for embodied emissions, recyclability, and life-cycle carbon footprint. Analyses in Chapters 15 and 16 demonstrated that pairing passive strategies with low-carbon materials can reduce both operational and embodied emissions, aligning building resilience with climate mitigation. At the settlement scale, vernacular systems functioned as collective arrangements in which narrow streets, courtyards, and shading devices worked in concert. Contemporary urban contexts require design guidance that promotes similar coordination, from blue–green ventilation corridors to distributed networks of sky-wells and planted courtyards.

19.1.4 Design Guidance

Taken together, the cases from Chapters 12–17 and the empirical findings of Chapters 13–16 suggest a coherent approach to design. The foundation is a climate-anchored base of mass, shading, and airflow. Orientation and geometry should be calibrated to local wind roses and solar paths. Stability is reinforced through redundancy in layering, with cavity envelopes, vertical voids, and shaded outdoor rooms. Operability must be maintained at multiple scales, from shutters and louvers to courtyards and urban canyons. Where water is available, it can be used for evaporative assistance; in humid climates, airflow and soil buffering are more effective than heavy mass. At the urban scale, blue–green corridors and interconnected voids can extend comfort beyond individual buildings, creating field conditions that regulate microclimate across districts.

The vernacular cases illustrate an environmental logic refined by centuries of exposure to local climates. Contemporary projects extend this logic with tools of precision, scalability, and integration. Empirical evidence confirms their capacity to provide resilience during extremes, while life-cycle analyses underline their role in reducing both operational and embodied emissions.

Together, these perspectives chart a trajectory from heritage to innovation that offers designers in all climates a framework for creating buildings that are environmentally responsive, socially adaptable, and resilient to future extremes. In this sense, passive strategies are not supplementary measures but one of the fundamental foundations of sustainable architecture in the twenty-first century.

Index

Note: **Bold** and *italic* page numbers refer to **tables** and figures, respectively.

J

K

L

M

N

R

S

T

U

V

For Product Safety Concerns and Information please contact our EU representative GPSR@taylorandfrancis.com Taylor & Francis Verlag GmbH, Kaufingerstraße 24, 80331 München, Germany

Printed and bound by CPI Group (UK) Ltd, Croydon, CR0 4YY
14/09/2026
02216242-0001